ACS SYMPOSIUM SERIES **406**

Supercritical Fluid Science and Technology

Keith P. Johnston, EDITOR
The University of Texas

Johannes M. L. Penninger, EDITOR
Akzo Salt and Basic Chemicals bv.

Developed from a symposium sponsored
by the American Institute of Chemical Engineers
at the American Institute
of Chemical Engineers Annual Meeting
Washington, DC,
November 27–December 2, 1988

American Chemical Society, Washington, DC 1989

Library of Congress Cataloging-in-Publication Data

Supercritical fluid science and technology
Keith P. Johnston, editor, Johannes M. L. Penninger, editor.

Developed from a symposium sponsored by the American Institute of Chemical Engineers at the American Institute of Chemical Engineers Annual Meeting, Washington, DC, November 27–December 2, 1988.

p. cm.—(ACS Symposium Series, 0097–6156; 406).
Includes index.

ISBN 0–8412–1678–9
1. Supercritical fluid extraction—Congresses.

I. Johnston, Keith P., 1955– . II. Penninger, Johannes M. L., 1942– . III. American Institute of Chemical Engineers. IV. American Institute of Chemical Engineers. Meeting (1988: Washington, D.C.). V. Series.

TP156.E8S84 1989
660'.2842—dc20 89–17521
CIP

PRINTED IN THE UNITED STATES OF AMERICA

ACS Symposium Series

M. Joan Comstock, *Series Editor*

Foreword

The ACS SYMPOSIUM SERIES was founded in 1974 to provide a medium for publishing symposia quickly in book form. The format of the Series parallels that of the continuing ADVANCES IN CHEMISTRY SERIES except that, in order to save time, the papers are not typeset but are reproduced as they are submitted by the authors in camera-ready form. Papers are reviewed under the supervision of the Editors with the assistance of the Series Advisory Board and are selected to maintain the integrity of the symposia; however, verbatim reproductions of previously published papers are not accepted. Both reviews and reports of research are acceptable, because symposia may embrace both types of presentation.

Contents

DESIGN OF COMMERCIAL PLANTS

Preface

IN THE EARLY 1980S, THE EMPHASIS OF RESEARCH AND DEVELOPMENT in supercritical fluid (SCF) science and technology was on extraction of commodity chemicals and synthetic fuels. Since about 1984 the field has undergone a kind of *perestroika,* in that attention has shifted toward more complex and valuable substances that undergo a much broader range of physical and chemical transformations. A great deal of innovation took place in studies of reactions, separations, and materials processing of polymers, foods, surfactants, pharmaceuticals, and hazardous wastes. As a result, SCF science has become interdisciplinary and is studied by chemical engineers, chemists, food scientists, materials scientists, and researchers in biotechnology and environmental control. In most cases, SCF technology is too expensive; to replace conventional separations for commodity chemicals; however, it presents a variety of specialized opportunities to produce new and improved products.

The purpose of the book is to provide a detailed treatment of SCF science and technology from a broad perspective, from the molecular level to the level of applied technology. The first section provides an understanding of the molecular structures of SCF solutions that has emerged recently from spectroscopic and computer simulation studies. It is followed by a section on phase behavior, particularly for complex molecules such as polymers and food components. The next section discusses a new topic, the formation of reverse micelles and gels in SCFs. This section also discusses interactions of SCFs with polymers, which is of interest for the impregnation and purification of polymers.

The use of SCFs as media for chemical reactions has increased during the past few years, as discussed in the next section. The large partial molar volumes of solutes near the critical point result in unusually large volumes of activation and large variations of certain reaction rate constants and selectivities with pressure. The following section on rate processes describes relatively novel crystallization processes that have commercial promise and transport properties in SCFs. The last two sections discuss a variety of food, pharmaceutical, and environmental applications and provide an in-depth treatment of the design of commercial plants.

We would like to thank the authors; Phil Davidson, John O'Connell, and Henry McGee for their strong commitment to making this symposium possible; Joe Dobbs, Phil Davidson, Carl Lira, Steve Paspek, Michael Klein, William Flarsheim, Pablo Debenedetti, Ted Randolph, Syed Rizvi, and Rodger Marentis for chairing sessions; and the more than 60 people who reviewed manuscripts.

KEITH P. JOHNSTON
Department of Chemical Engineering
University of Texas
Austin, TX 78712

JOHANNES M. L. PENNINGER
Akzo Salt and Basic Chemicals bv.
7550 GC Hengelo
Netherlands

May 30, 1989

Chapter 1

New Directions in Supercritical Fluid Science and Technology

Keith P. Johnston

Department of Chemical Engineering, The University of Texas, Austin, TX 78712

An overview of new research directions is presented. The domain of this field has grown significantly with advances in separations, reactions, and materials processing of complex substances such as polymers, surfactants, and biomolecules. The field has encompassed a large number of areas in engineering and the chemical, physical, and biological sciences, which will be discussed. In the U.S. new commercial processes include coffee decaffeination, hops extraction, catalyst regeneration, extraction of organic wastes from water, and supercritical fluid chromatography. These applications complement older technologies such as residuum oil supercritical extraction (ROSE), carbon dioxide enhanced oil recovery, and reaction processes for the production of polyethylene and primary alcohols in supercritical fluid ethylene. The interest in environmental applications is increasing rapidly. Given the experience gained in developing commercial plants in Europe, the U.S., and now Japan and Korea, it would be expected that the time lag between research and commercialization will diminish.

Not long ago, research in supercritical fluid (SCF) science and technology emphasized energy savings, synfuels processing, and the thermodynamics of phase equilibria for commodity chemicals. The field dealt primarily with extractions of relatively simple molecules that dissolve in fluids such as carbon dioxide. In the renaissance of the last five years, attention has shifted to significantly more complex molecules, undergoing much broader types of physical and chemical transformations. A key innovation has been to explore the effect of a SCF on other phases besides the SCF phase, which often contain higher value substances, such as polymers, biomolecules, or growing crystals. This has extended the scope of the field markedly, as only a limited number of substances are soluble in carbon dioxide.

0097–6156/89/0406–0001$06.00/0

A number of examples illustrate the great deal of innovation that has occurred in the last five years. These include challenging fractionations of foods (such as ω-3 fatty acids) and pharmaceuticals (for example sterols), crystallization (including rapid expansion from supercritical solutions(1,2), retrograde crystallization (3) and gas anti-solvent crystallization(4)), purification of polymers, impregnation of polymers, manipulation of polymer morphology and porosity, manipulation and control of chemical reactions (even in SCF water), extraction and oxidation of hazardous wastes, spectroscopic studies of solvent strength and structure, computer simulation, separations and reactions in reverse micelles, and electrochemical synthesis. This extensive research activity has been complemented by advances in plant design, which are discussed in detail in this symposium. Together, these developments in research and design should lead to some innovative applications, particularly in the food, polymer and pharmaceutical industries, and in environmental protection.

For most applications, it is not possible to treat supercritical fluid separations, for example extraction, as a well-defined unit operation as is the case for simpler processes such as distillation. Instead, research is often needed to characterize the important properties for each specific separation or reaction. However, the recent advances in the molecular understanding of SCF solutions provide some general themes that can be utilized in a semi-quantitative manner to evaluate the potential of both research and applications.

Characterization of supercritical fluids: solvent strength and selectivity

The solvent strength of a supercritical fluid (compressed gas) may be adjusted continuously from gas-like to liquid-like values, as described qualitatively by the solubility parameter. The solubility parameter, δ (square root of the cohesive energy density) (5), is shown for gaseous, liquid, and SCF carbon dioxide as a function of pressure in Figure 1. It is a thermodynamic property which can be calculated rigorously as

$$\delta = \left(\frac{u^{ig} - u}{v}\right)^{1/2} = \left(\frac{h^{ig} - RT - h + Pv}{v}\right)^{1/2} \tag{1}$$

although this point has been missed in many papers. For a given fluid, this plot resembles that of density versus pressure. The δ for gaseous carbon dioxide is essentially zero whereas the value for liquid carbon dioxide is comparable with that of a hydrocarbon. At -30 C, there is a large increase in δ upon condensation from vapor to liquid. Above the critical temperature, it is possible to tune the solubility parameter continuously over a wide range with a small isothermal pressure change or a small isobaric temperature change. This ability to tune the solvent strength of a

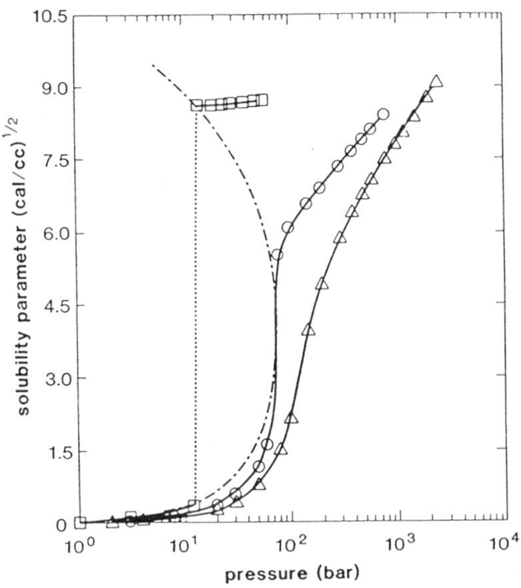

Figure 1. Solubility parameter of CO_2 (□: −30 °C, ○: 31 °C, Δ: 70 °C).

supercritical fluid is its unique feature, and it can be used to extract, then recover, products as explained in the book by McHugh and Krukonis which provides an excellent introduction to the field(6).

In order to understand solubilization of a solid in a SCF, consider the very similar tetracyclic sterols: cholesterol, stigmasterol, and ergosterol, each of which contains a single OH group (see Figure 2). It may seem surprising that the selectivity of carbon dioxide for cholesterol versus ergosterol (ratio of solubilities) is almost two orders of magnitude(7). To explain this result, it was necessary to measure their vapor pressures, which are as low as 10^{-10} bar (10^{-6} torr), using a gas saturation technique. The selectivity simply follows the ratio of vapor pressures, even though the vapor pressures are negligibly small.

To further understand the role of solute-solvent interactions on solubilities and selectivities, it is instructive to define an enhancement factor as the actual solubility, y_2, divided by the solubility in an ideal gas, so that $E = y_2 P/P_2^{sat}$. This factor is a normalized solubility, because it removes the effect of the vapor pressure, providing a means to focus on interactions in the SCF phase. In carbon dioxide at 35 C and 200 bar, E is about 7 for all three of the above sterols. In fact, enhancement factors do not vary much for many types of organic solids. As shown in Figure 3, E's fall within a range of only about 1.5 orders of magnitude for substances with a variety of polar functional groups, even though the actual solubilities (not shown) vary by many orders of magnitude. This means that solubilities, and also selectivities, in carbon dioxide are governed primarily by vapor pressures and only secondarily by solute-solvent interactions in the SCF phase. An exception is strong bases such as ammonia that can react with carbon dioxide.

SCF carbon dioxide is a lipophilic solvent since the solubility parameter and the dielectric constant are small compared with a number of polar hydrocarbon solvents. Co-solvents(also called entrainers, modifiers, moderators) such as ethanol have been added to fluids such as carbon dioxide to raise the solvent strength while maintaining it's adjustability. Most liquid cosolvents have solubility parameters which are larger than that of carbon dioxide, so that they may be used to increase yields, or to decrease pressure and solvent requirements. A summary of the large increases in solubility that may be obtained with a simple cosolvent is given at the top of Table I. Cosolvents, unlike carbon dioxide, can form electron donor-acceptor complexes (for example hydrogen bonds) with certain polar solutes to influence solubilities and selectivities beyond what would be expected based on volatilities alone. Several thermodynamic models have been developed to correlate and in some cases predict effects of cosolvent on solubilities(8,9). They are used extensively in SCF research and development.

Molecular structure of supercritical fluid solutions

The understanding of the molecular interactions and structure in SCF solutions has grown considerably in the last five years, and is reviewed in greater

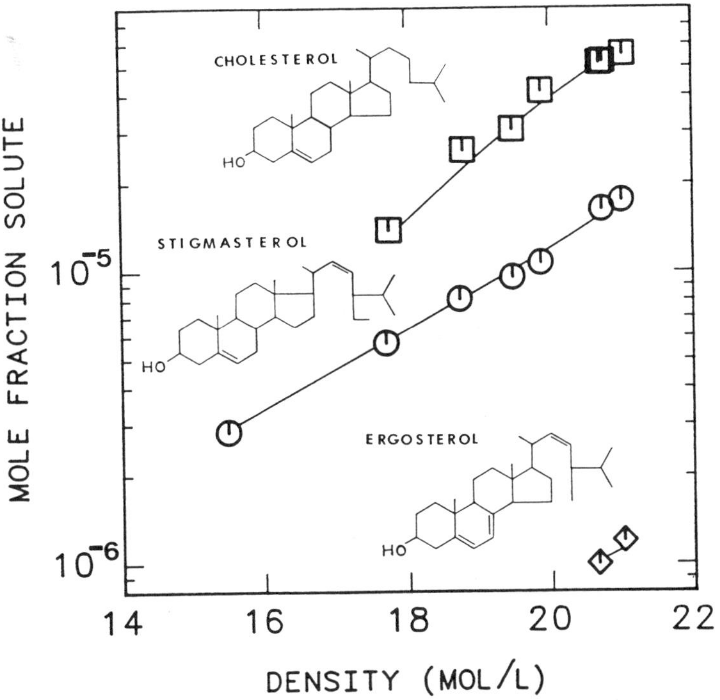

Figure 2. Solubility of sterols in pure CO_2 at 35 °C (data are correlated with the HSVDW [7]).

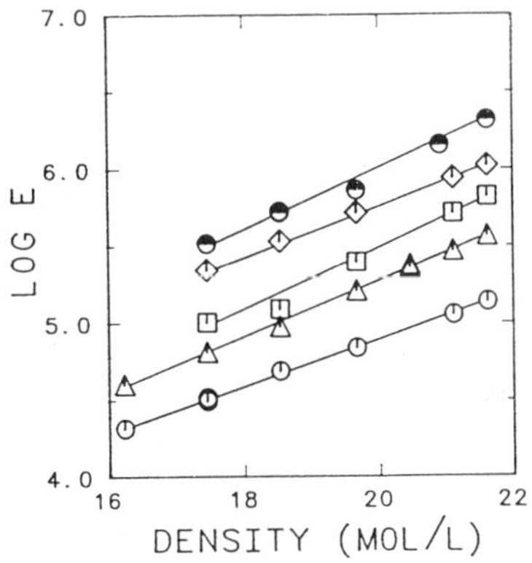

Figure 3. Enhancement factor for solids with a variety of polar functionalities in CO_2 at 35 °C (○: hexamethylbenzene, Δ: 2-naphthol, □: phthalic anhydride, ◇: anthracene, ●: acridine). (Reprinted from ref. 8. Copyright 1987 American Chemical Society.)

Table I. Effects of cosolvents including simple liquids, complexing agents and surfactants on solubilities in SCFs at 35 C

Solute	co-solvent	$\frac{y}{y_{binary}}$
acridine	3.5% CH_3OH	2.3
2-aminoben-zoic acid	3.5% CH_3OH	7.2
cholesterol	9% CH_3OH	100
hydroquinone	2% tributyl-phosphate	>300
hydroquinone	.65% AOT, $W_o = 10$* 6% octanol	>200
tryptophan	.53% AOT, $W_o = 10$* 5% octanol	>>100

SOURCE: Data are from references 7 and 8.
* W_o is the water-to-surfactant ratio.

detail in a following chapter(9). In the highly compressible region, a SCF solvent condenses about a solute (which is usually much larger and more polarizable) so that the solute's partial molar volume can be on the order of -10,000 cc/mole. Thermodynamically, this explains a key unique feature of SCF solutions, that is the large pressure effects on phase behavior and chemical reaction rates. For a solvent with a molar volume of 100 cc/mole, a $\bar{v}_2$ of -10,000 cc/mole indicates directly that an average of 100 solvent molecules condense about each solute. This solvent condensation or clustering may be described by a simple equation in terms of the isothermal compressibility of the solvent using Kirkwood-Buff solution theory(9). It is quite satisfying that the conclusions of a number of studies of clustering are consistent, given the variety in the methodologies: partial molar volume data, Kirkwood-Buff theory, UV-visible and fluorescence(spectral shift) data, integral equations, and computer simulation data. Another type of clustering in SCF solutions, solute-solute clustering, has been discovered recently using fluorescence studies(10) and integral equations(11).

Reactions in and with supercritical fluids

Many of the desirable properties of supercritical fluids which are used in separation processes can be exploited in systems undergoing chemical reaction. Less than 5% of the SCF technology literature deals with chemical reactions, although this relatively new area is growing considerably because of applications such as oxidation of hazardous wastes in supercritical water.

A number of interesting opportunities for performing reactions in supercritical fluid solvents will be discussed:

1. The solvent strength of a SCF may be manipulated using pressure and/or temperature to adjust reaction rates by changing rate and equilibrium constants, or concentrations of reactants and products. The latter is due to the large changes in concentrations that occur in the critical region.

A small change in the pressure of a SCF can produce a large change in the solvent strength (as measured by solubility parameter or solvatochromic polarity scales(9)), which can cause a large thermodynamic solvent effect on a rate or equilibrium constant. This phenomenon is unique to SCFs. An extremely pronounced pressure effect was discovered for the rate constant of the unimolecular decomposition of α-chlorobenzyl methyl ether in supercritical 1,1-difluoroethane(12)

$$C_6H_5CHCl\text{-}O\text{-}CH_3 \rightarrow C_6H_5CHO + CH_3Cl \qquad (2)$$

A change in pressure of only 15 bar increased the rate constant by an order of magnitude, because the density and thus the solvent strength increased significantly. This solvent effect can be explained in an alternative but more complex manner by using transition state theory. For a unimolecular reaction, $A = A^{\ddagger} \rightarrow$ products, the activation volume may be expressed as

$$\Delta v^{\ddagger} = -RT(\partial \ln k_x/\partial P)_T = \bar{v}_A{}^{\ddagger} - \bar{v}_A \qquad (3)$$

where k_x is the rate constant based on mole fraction units. In conventional liquid solvents activation volumes are up to ± 30 cc/mol. In a highly compressible SCF, $\bar{v}_i$s of solutes can reach thousands of cc/mol negative. As a result, activation volumes, which are differences in $\bar{v}_i$s, can also be pronounced. For the above ether decomposition reaction, activation volumes were observed as low as -6000 cc/mol, demonstrating the capability of SCFs to adjust rate constants.

Large pressure effects were also found for the Diels-Alder reaction of isoprene and maleic anhydride(13), the tautomeric equilibria of 2-hydroxypyridine and 2-pyridone at infinite dilution in propane and 1,1-difluoroethane(14), for the reversible redox reaction of the I_2/I^- couple in supercritical water(15), and for other examples in the chapter on reactions. Although these studies have made considerable progress in understanding these pressure effects in terms of solute-solvent interactions and solvent compressibility, further work is needed in this new area.

2. Selectivities to certain products may likewise be adjusted using pressure or temperature. In SCF water, reaction chemistry is governed by free-radical (homolytic) mechanisms for an ion product of water, K_w, less than 10^{-14}, and by ionic (heterolytic) mechanisms for larger values of K_w(16). This describes the observation that decomposition reactions shift from pyrolysis to hydrolysis as the density of water and thus K_w is increased(17). In the near-critical region, K_w can be manipulated by small changes in temperature and pressure to control the reaction selectivity.

In the photochemical dimerization of isophorone in SCF fluoroform and carbon dioxide, it is possible to study pressure effects on regioselectivity as well as stereoselectivity by observing the concentrations of the three primary dimers(18). Here regioselectivity refers to head-to-head versus head-to-tail oxygen configurations for the dimers, while stereoselectivity refers to syn versus anti configuration. The dipole moments of the regio-isomers are very different, while those of the stereoisomers are about the same. As a result, the regioselectivity is influenced by both solvent polarity (solute-solvent attractive interactions), and solvent reorganization (solute-solvent repulsive interactions), while the stereoselectivity is influenced only by solvent reorganization. Studies of additional reactions would help elucidate the emerging understanding of pressure effects on selectivity.

3. Reactions may be integrated with SCF separation processes to achieve a large degree of control for producing a highly purified product. Reaction products could be recovered by volatilization into, or precipitation from, a SCF phase. The classic example is the high pressure production of polyethylene in the reacting solvent SCF ethylene(6). The molecular weight distribution may be controlled by choosing the temperature and pressure for precipitating the polymer from the SCF phase.

4. The use of a supercritical fluid may provide a means to perform a single phase homogeneous reaction, instead of a multiphase reaction(19). For example, organics may be oxidized with oxygen in SCF water, in the one phase region with minimal mass transfer limitations. This improves both energy and destruction efficiencies.

5. Transport properties are improved in that diffusivities are higher and viscosities are lower than in liquids.

6. SCFs offer several benefits in heterogeneous catalysis. They can act as a vehicle to improve transport in catalysts by reducing capillary condensation and pore plugging(20). SCFs have been used to accelerate gas-solid reactions by

removing reactants or products from a solid phase(21). In the chlorination of alumina in SCF CCl_4, the reaction accelerates as density increases due to dissolution of the product $AlCl_3$ from the reactive surface.

SCFs offer a nonaqueous environment which can be desirable for enzymatic catalysis of lipophilic substrates. The lipophilic substance cholesterol is 2 to 3 orders of magnitude more soluble in CO_2-cosolvent blends than in water(7). In CO_2 based blends, it may be oxidized to cholest-4en-3one, a precursor for pharmaceutical production using an immobilized enzyme(22). The enzyme polyphenol oxidase has been found to be catalytically active in supercritical CO_2 and fluoroform (23). The purpose of using a SCF is that it is miscible with one of the reactants-oxygen. Lipase may be used to catalyze the hydrolysis and interesterification of triglycerides in supercritical CO_2 without severe loss of activity(24). These reactions could be integrated with SCF separations for product recovery.

7. SCFs may be used to perturb reactions gently in order to study reaction mechanisms and solvent effects, simply by a modest change in the pressure. Here the SCF would be used as a tool to obtain information about a reaction, whereas, it is used to improve the performance of a reaction in each of the above cases. The perturbation could be due to changes in solvent polarity or viscosity(25). The advantage of using a SCF is that the the reaction may be perturbed over a continuum with a single solvent instead of with a series of liquid solvents with differing molecular functionality.

Complex molecules in supercritical fluid science and technology: polymers, surfactants and biomolecules

SCF technology has spread quickly from molecules such as naphthalene to more complex substances such as polymers, biomolecules, and surfactants. Supercritical fluids can be used to reduce the lower critical solution temperature of polymer solutions in order to remove polymers from liquid solvents(6,26). The technology has been extended to induce crystallization of other substances besides polymers from liquids, and has been named gas recrystallization(4). In other important applications, SCF carbon dioxide has been used to accomplish challenging fractionations of poly(ethylene glycols) selectively based on molecular weight as discussed in this symposium, and of other polymers(6).

A new use for supercritical fluids is to swell glassy polymers so that they may be impregnated rapidly with additives such as plasticizers, pharmaceuticals, pigments, and other additives(27). The inverse of this concept is also important, that is SCFs may be used to purify polymers by extracting residual solvent and monomers(6). Both the impregnation and purification processes take advantage of enhanced solute diffusion by up to 9 orders of magnitude, which results from plasticizing a glassy polymer with a SCF(27). This increase is due to the increased freedom of motion of the swollen polymer chains. The swelling of polymers using pure SCFs has been measured only recently(28, 29,30); this information is useful for the extension to multicomponent systems, which contain solutes in addition to the polymer and SCF solvent. A final application is the manipulation of polymer morphology or porosity by tailoring the depressurization of a SCF.

An important factor that dictates the feasibility of both polymer impregnation and purification processes is the equilibrium distribution coefficient of the solute between the polymer phase and the SCF phase. The distribution coefficient of toluene between cross-linked silicone rubber and carbon dioxide has been explored quantitatively as a function of temperature over a wide range in pressure up to 250 bar(31). It is adjustable over a continuum in the highly compressible region of carbon dioxide as shown in Figure 4. The behavior was predicted quantitatively using only information from binary systems. This type of thermodynamic understanding of phase behavior in ternary systems containing polymers and SCFs is rare, but very beneficial for guiding practical applications.

Surfactants are another example of complex molecules that have interesting properties in systems containing SCFs. Carbon dioxide, even when doped with a co-solvent such as ethanol, is incapable of dissolving a measurable amount of hydrophilic substances such as proteins. The same is true for other SCFs with critical temperatures under 100 C. However, SCFs can solubilize hydrophilic molecules by forming complexes, aggregates, reverse micelles, or microemulsions. An example is given in Table I where an amino acid, which is very insoluble in ethane, can be solubilized with an anionic surfactant, sodium di-2-ethylhexyl sulfosuccinate (AOT). In other breakthroughs described in this symposium, viscous gels have been formed in SCF solutions, and emulsion polymerization has been carried out in reverse micelles. The recent advances involving polymers, surfactants, and biomolecules open up large new domains for research in SCF science and technology.

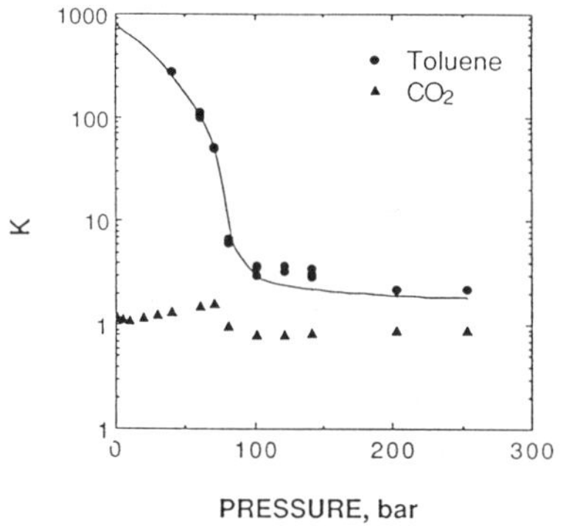

Figure 4. Large change in the distribution coefficient of toluene between silicone rubber and CO_2 based on volume fraction in the critical region (—: predicted value).

Acknowledgment

Acknowledgement is made to the donors of the Petroleum Research Fund, administered by the American Chemical Society, the Camille and Henry Dreyfus Foundation for a Teacher-Scholar Grant, and the Separations Research Program at the University of Texas.

Literature Cited

1. Krukonis, V. 1984, AIChE Annual Meeting, San Francisco
2. Mohamed R.S.; Halverson, D.S.; Debenedetti, P.G. this symposium.
3. Chimowitz, E.H.; Pennisi, K.J. AIChE J., 1986, 32, 1665.
4. Gallagher, P.M.; Coffey. M.P.; Krukonis, V.J.; Klasutis, N. this symposium.
5. Barton, A. Handbook of Solubility Parameters and Other Cohesive Parameters, CRC: Boca Raton, FL, 1983.
6. McHugh, M. A.; Krukonis, V. J. Supercritical Fluid Extraction; Principles and Practice, Butterworths; Boston, Mass, 1986.
7. Wong, J. M.; Johnston, K. P. Biotech. Progress 1986, 2, 29.
8. Dobbs, J.M.; Johnston, K.P. Ind. Engr. Chem. Res. 1987, 26, 1476.
9. Johnston, K.P.; Kim, S.; Combes, J., this symposium.
10. Brennecke, J.F.; Eckert, C.A.; this symposium.
11. Cochran, H.D.; Lee, L. L.; this symposium.
12. Johnston, K. P.; Haynes, C. AIChE J., 1987, 33, 2017.
13. Paulaitis, M.E.; Alexander, G. C. Pure & Appl. Chem. 1987, 59, 61.
14. Peck, D.G.; Mehta A. J.; Johnston K. P. J. Phys. Chem. 1989, in press.
15. Flarsheim, W.M.; Bard, A.J.; Johnston, K.P. J. Phys. Chem. 1989, in press.
16. Antal M. J.; Brittain, A.; DeAlmeida C.; Ramayya, S.; Roy J.C. ACS Symp. Series 329, 1986, 77.
17. Townsend, S.H.; Abraham, M.A.; Huppert, G.L.; Klein, M.T.; Paspek, S.C. Ind. Eng. Chem. Res. 1988, 27, 143.
18. Hrnjez, B.J.; Mehta, A.J; Fox, M.A.; Johnston, K.P. J. Am. Chem. Soc. 1989, 111 2662.
19. Subramaniam B.; McHugh M. A. Ind. Eng. Chem. Process Des. Dev. 1986, 25, 1.
20. Barton, P.; Kasun, T.J. In Supercritical Fluid Technology, Penninger, J. et al. Ed.; Elsevier, Amsterdam, 1985, 435.
21. Herrick, D.E.; Holder, G.D.; Shah, Y.T. AIChE. J. 1988, 34, 669.
22. Randolph, T.W.; Blanch, H.W.; Prausnitz, J.M. AIChE J. 1988, 34, 1354.
23. Hammond, D.A.; Karel, M.; Klibanov, A.M. Applied Biochemistry and Biotechnology 1985, 11, 393
24. Nakamura, K.; Chi, Y.M.; Yamada, Y.; Yano, T. Chem. Eng. Comm. 1986, 45, 207.

25. Aida T.; Squires T. G. ACS Symp. Series 329; 1987, 58.
26. McHugh, M.A.; Guckes, T.L. Macromolecules 1985, 18, 674.
27. Berens, A. R.; Huvard, G. S.; Korsmeyer, R.W. this symposium.
28. Fleming, G. K.; Koros W.J. Macromolecules 1986, 19, 2285.
29. Liau, I. S.; McHugh, M.A. Process Technology Proceedings 3 (Supercritical Fluid Technology), J. M. L. Penninger, et al.,Ed.; Elsevier, Amsterdam, 1985, 415.
30. Wissinger, R. G.; Paulaitis, M.E. J. Polym. Sci., Part B: Polym. Phys. 1987, 25, 2497.
31. Shim, J.J.; Johnston, K.P. AIChE J. 1989, in press.

RECEIVED May 1, 1989

MOLECULAR INTERACTIONS AND STRUCTURE

Chapter 2

Fluorescence Spectroscopy Studies of Intermolecular Interactions in Supercritical Fluids

Joan F. Brennecke and Charles A. Eckert

Department of Chemical Engineering, University of Illinois at Urbana–Champaign, Urbana, IL 61801

Much of the unusual behavior in supercritical fluids (SCF) may be related to the formation of loose aggregates or clusters. The existence of these clusters would explain supercritical phenomena including enhanced solubility, synergistic effects with mixed solutes, and even cosolvent or entrainer effects. The strong solute/solvent interactions are particularly pronounced in the vicinity of the critical point, where the fluid is most compressible and very large negative solute partial molar volumes have been measured. We present the results of fluorescence spectroscopy studies of dilute aromatic solutes in near-room temperature SCF solvents. Information on the strength of solute-solvent interactions is extracted from intensity changes as a function of proximity to the critical point, solvent polarity and solute functionality. In addition, the spectroscopic studies reveal the importance of solute/solute interactions in supercritical fluid solutions, even at very low concentrations. This suggests that our concept of infinite dilution for highly asymmetric solutions may need to be reconsidered. The results of these spectroscopic studies are likely to be useful for the development of a more realistic equation of state to describe supercritical fluid solutions that can be used to tailor solvents for optimal extraction selectivity.

Supercritical fluids (SCF) possess a variety of characteristics that make them particularly attractive for separation processes. These include both high diffusivity, low viscosity and high solubilities of heavy organic solutes. A wide range of applications have been proposed and are compiled in several review articles and monographs (1-4).

The design of separation processes would be facilitated greatly by accurate mathematical models of the thermodynamics. Unfortunately, most standard equations of state do not represent adequately the phase behavior in the near supercritical region and in general were not developed with highly asymmetric solu-

0097–6156/89/0406–0014$06.00/0

tions in mind. These equations include simple cubics like the Peng-Robinson (5), perturbation equations like the Augmented van der Waals, Carnahan Starling van der Waals and their modifications (6,7) and lattice gas models (8-11). Therefore, a better understanding of the true intermolecular interactions is necessary to develop realistic models.

A strong indication of the unusual behavior in supercritical fluid solutions was measurement of the partial molar volume at infinite dilution of several solutes in SCF ethylene and CO_2 at pressures of 50 to 250 bar and temperatures of 12 C to 45 C (12,13). At high reduced pressures, where the solvents are incompressible, the infinite dilution partial molar volumes were slightly positive. Very sharp negative dips in v_2^∞ were observed for solutes in the compressible region of the solvent, which is near the solvent critical point. These negative values were extremely large in magnitude (-1000 to -16,000 cc/gmol) and were largest for the isotherms closest to the critical temperatures. See for example Figure 1.

The extremely large negative infinite dilution partial molar volumes suggest the "condensation" of many solvent molecules when a molecule of solute is added to solution. This can be envisioned as the collapse of the solvent shell about the solute or the formation of solute/solvent clusters in solution. One way of modeling this phenomenon is the use of density dependent local composition (DDLC) mixing rules (14-16). Another method of looking at such a process might be a chemical theory description. Chemical theory hypothesizes the formation of complexes in solution,

$$B + n\,A \xrightarrow{K} BA_n$$

where B is the solute, A is the solvent, and n is the number of solvent molecules around a solute. An equilibrium constant, K is associated with the reaction and is given by

$$K = \frac{z_{ABn}\,\alpha_{ABn}}{z_B\,\alpha_B\,(z_A\,\alpha_A)^n}$$

where the z's are the "true" mole fractions in solution and the α's are the activity coefficients of the true species. A zeroth-order approximation of this theory which assumes no physical interactions was used to qualitatively represent the form of the infinite dilution partial molar volume curve (13).

A number of experimental studies using spectroscopy have addressed the formation of clusters in SCF solutions. Kim and Johnston first suggested the link between the local density and the isothermal compressibility (17), showing that the local density is highest in the region of highest compressibility near the critical point. This idea was used to interpret and obtain local densities from absorption data of phenol blue in SCF's (18). Yonker and coworkers (19) have studied the wavelength of maximum absorption of a chromophore in supercritical fluids and used that information to determine Kamlet-Taft π^* values as a function of solvent density in the SC region. Kajimoto and coworkers (20) have used both absorption and fluorescence to look at the complicated system of (N,N-dimethylamino)benzonitrile, which forms a charge transfer (CT) complex in addition to the normal fluorescence. The Stokes shift and ratio of the CT to

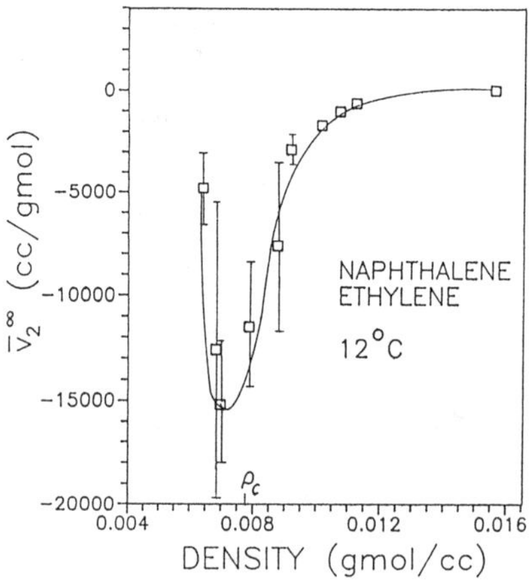

Figure 1. Infinite dilution partial molar volume of naphthalene in SCF ethylene at 285 K. (Reprinted from ref. 12. 1986 American Chemical Society.)

normal fluorescence were used to develop a simple aggregation model which uses a Langmuir-type adsorption description of the clustering phenomenon.

Additional work has been done using UV absorption measurements to quantify the clustering around solutes in solute/fluid/entrainer systems (21,16). The shift in the wavelength of maximum absorption is used to determine the local composition about the solute. This vicinity is shown to be enriched with the entrainer, especially in the highly compressible region nearest the critical point.

There have been a number of modeling efforts that employ the concept of clustering in supercritical fluid solutions. Debenedetti (22) has used a fluctuation analysis to estimate what might be described as a cluster size or aggregation number from the solute infinite dilution partial molar volumes. These calculations indicate the possible formation of very large clusters in the region of highest solvent compressibility, which is near the critical point. Recently, Lee and coworkers have calculated pair correlation functions of solutes in supercritical fluid solutions (23). Their results are also consistent with the cluster theory.

Donohue and coworkers (24) recognized a different type of clustering: the formation of complexes between solutes in SCF's and "entrainers" (1-5%) added to the SCF. Entrainers such as methanol or acetone have been used to enhance solubility by specific interactions with the solute (25,26). In their Associated Perturbed Anisotropic Chain Theory both association between alcohol molecules and solvation of the solute by the alcohol are taken into account. However, no complex formation is assumed to occur between the solute and the SCF solvent.

Solute/solute interactions are also important in SCF solutions, as observed by Kurnik and Reid and Kwiatkowski and coworkers (27,28), in the synergistic effects on the solubilities of mixed solutes. When a physical mixture of naphthalene and benzoic acid is extracted with SCF CO_2 the solubilities of both components is greater than the solubility of either pure component. This phenomenon occurs at concentrations as low as 10^{-2} to 10^{-3} mol fraction. Frequently, equation of state models assume "infinite dilution" of the solutes due to the low solubilities. Clearly, recognition and a better understanding of these solute/solute interactions are important for the prediction of SC phase equilibria.

In this paper we present new results of fluorescence spectroscopy studies of dilute organics in pure supercritical fluids. We compare those results to make observations about the strength of solute/solvent interactions in solution, especially near the critical point, as well as the importance of solute/solute interactions even at extremely low concentrations.

EXPERIMENTAL

A dilute mixture of a solute in supercritical fluid is introduced into a high pressure optical cell, equipped for 90 degree detection with a 1.3 cm pathlength. The windows are 6 mm thick fused quartz discs. The pressure in the optical cell is measured with a Texas Instruments model 140 pressure gauge, which has an accuracy of $\pm$0.2 bar. The temperature in the optical cell is controlled with a custom built precision temperature controller that is good to $\pm$0.02 C. The temperature is measured by recording the resistance of an Omega type 44032 thermistor. The heating and cooling elements are Melcor model CP1.4-71-06L Peltier coolers. These serve as heaters by simply switching the leads to the modules. Four modules are fastened to the bottom of the cell.

The spectrometer assembly incorporates a 1000 Watt xenon arc lamp with Kratos power supply, lenses, and monochromators. Detection is with a Hamamatsu 1P-28 photomultiplier tube, powered by a Keithley model 247 high voltage supply. A stepper motor on the emission monochromator is computer controlled and coordinated with the computer data acquisition of the signal from a Keithley model 414a picoammeter.

The schematic of the high pressure assembly used to introduce the sample into the optical cell is shown in Figure 2. The gas is compressed from the cylinder into the 2-liter mixing or solvent vessel with a Haskel model AG-152 air-driven compressor. Both vessels are connected to a vacuum pump, used to evacuate the vessel of air before it is filled with supercritical fluid because oxygen will effectively quench the fluorescence of most aromatic fluorophores. The solute is introduced into the mixing vessel by either weighing out the amount or, when very low concentrations are investigated, coating an inert substrate. Depending on the vapor pressure of the solute, the sample is either placed in the mixing vessel or in a special solute chamber located in the inlet line to the vessel. In this second case, the solute dissolves as the vessel is pressurized. The pressure in the mixing vessel is measured with a 1 kbar Heise gauge, which is accurate to ± 1 bar. The temperature is recorded with a thermistor as described earlier for the optical cell. The solvent vessel is necessary to fill the optical cell with pure solvent before the mixture is introduced; otherwise, the solute would fall out of solution when introduced into the cell. The cell is simply flushed with mixture until all the solvent is displaced and a spectrum can be recorded.

The carbon dioxide used was Linde anaerobic grade, which has a minimum purity of 99.99% and a maximum O_2 concentration of 10 ppm. The CP grade ethylene, which has a minimum purity of 99.5%, was obtained from Scott Specialty Gases, who analyzed the gases to insure that O_2 concentrations were below 5 ppm. The fluoroform has a 98% purity and was graciously supplied by E. I. DuPont de Nemours & Co., Inc. It was purified to remove oxygen by passage through a series of two Alltech Oxy-traps. Pyrene, naphthalene, and dibenzofuran were obtained from Aldrich with 99+% purity. Carbazole was obtained from Aldrich with 99% purity and was recrystallized before use. Unless specified otherwise, all chemicals were used without further purification.

RESULTS AND DISCUSSION

The systems studied were pyrene in SCF CO_2, ethylene and CHF_3 and naphthalene, dibenzofuran and carbazole in SCF CO_2. The spectra yield information on both solute/solvent and solute/solute interactions. A typical spectrum of a dilute pyrene sample is shown in the lower half of Figure 7. Pyrene was chosen for the initial experiments because it has been extremely well investigated in organic liquids and the solvent effect on the spectra is well documented (29).

Two aspects of the fluorescence spectra are important in determining the strength of solute/solvent interactions. The first is particular to some nonfunctional polycyclic aromatics. The intensity of the first peak is an excellent measure of the

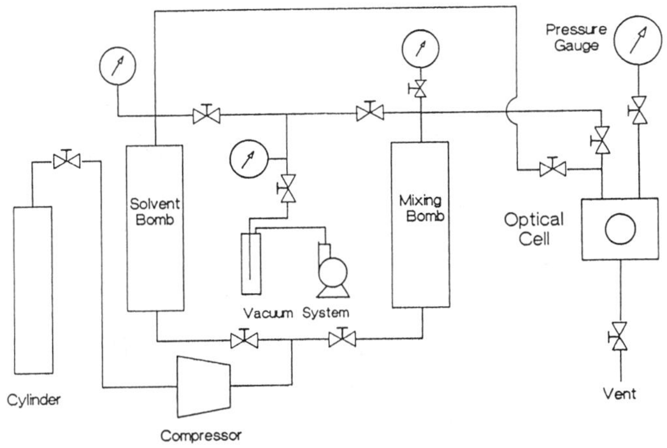

Figure 2. Schematic of high pressure assembly.

strength of the solute/solvent interactions. The transition corresponding to that peak is not allowed by symmetry arguments. However, if there is strong interaction between the solute and the solvent, the symmetry is disrupted and that transition takes place with greater intensity (30). Therefore, a higher value indicates strong solute/solvent interactions. The height of this peak is recorded as a ratio to a distinct stable peak in the spectrum. On this basis, for pyrene we chose the ratio of the height of the first to the third peak (I_1/I_3) and the first to the fourth peak for naphthalene (I_1/I_4). Another measure of the strength of solute/solvent interactions is the overall intensity of the spectrum. When a molecule is promoted to the excited state it can undergo several processes. One of these is fluorescence. Another is the loss of energy by radiationless transfer to the surrounding molecules. The closer the solvent is packed around the solute, the more energy it will lose in radiationless transfer by thermal dissipation to the surrounding molecules. Therefore, the overall intensity of the fluorescence is a good indicator of the strength of molecular interactions. In all cases we shall compare peak intensities at their maxima so any small wavelength shifts in the spectra will not bias the results.

The ratio of the first to third peaks (I_1/I_3) of pyrene in SCF CO_2, C_2H_4, and CHF_3 are shown in Figure 3 as a function of reduced density. As expected, the values are much higher for the polar CHF_3. Along a constant mol fraction isotherm the I_1/I_3 is lower near the critical point and increases as the pressure is raised up to a hundred bar above the critical point for all three fluids. However, this is a result of the changing density, which is lower near the critical point. A superior way to compare the data is at constant density, near and far from the critical point, as shown in Figure 4 for two isotherms of naphthalene in CO_2 at the same mol fraction. The I_1/I_4 for just 4 C above the critical point of carbon dioxide (T_c=31 C) is much greater than 14 C above the critical point, indicating much stronger solute/solvent interactions near the critical point. Identical results are observed for pyrene in all three supercritical fluids studied.

The overall intensity of the fluorescence is significantly lower near the critical point. This is demonstrated for pyrene in SCF ethylene in Figure 5, where the intensity is plotted as a function of solvent density at two temperatures, one close to and one 26 C above the critical temperature.
This indicates increasing solute/solvent interactions as the critical point is approached, due to increased radiationless energy loss. This drop in intensity is also observed for dibenzofuran and carbazole in SCF CO_2. One would expect the curves to converge at higher pressures where liquid densities are approached, as seen in Figure 4. However, the intensities maintain a significant differential. We are examining the data carefully in an attempt to explain the observations in terms of excimer formation and solute quenching. In addition, the overall fluorescence intensity is significantly less in SCF's than in liquid cyclohexane, where clustering is not expected to occur, at corresponding solute concentrations. In Figure 6 a SCF isotherm at constant mol fraction (but changing solution density as the critical point is approached) is compared to the intensity that might be expected from liquid cyclohexane samples. The intensities in SCF's are much lower due to more radiationless loss, especially near the critical point.. This lends further support to the theory that the solvents cluster strongly around the

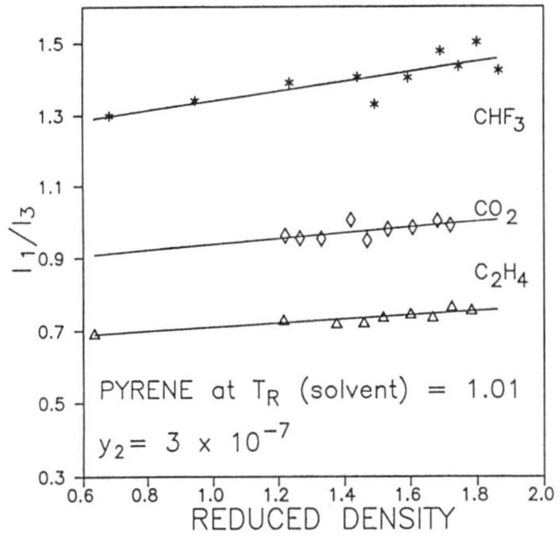

Figure 3. Comparison of I_1/I_3 for pyrene in three different supercritical fluids.

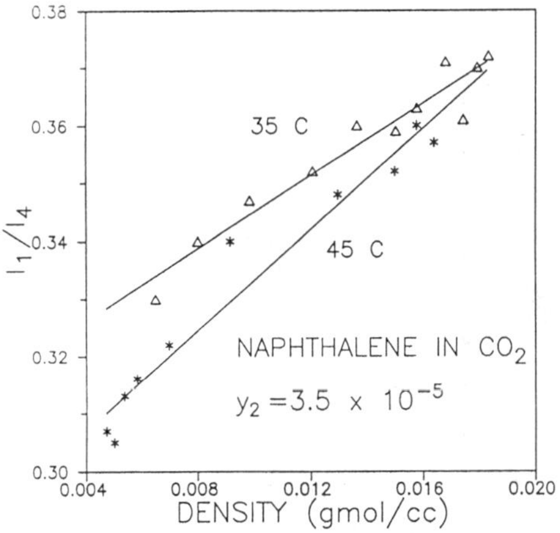

Figure 4. Comparison of I_1/I_4 of naphthalene near and far from the critical point.

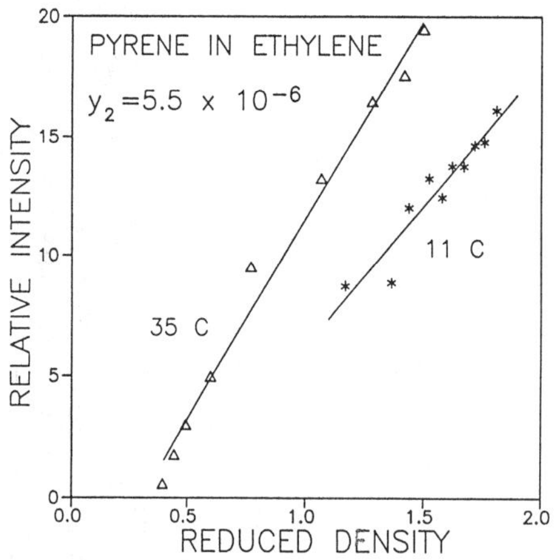

Figure 5. Decrease in overall intensity near the critical point of pyrene in ethylene.

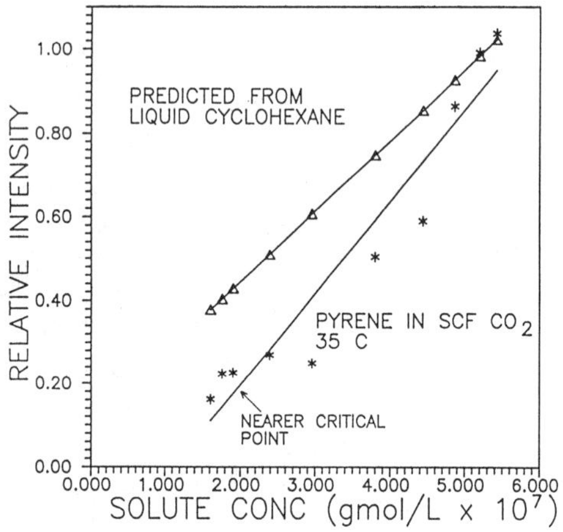

Figure 6. Lower overall intensity of pyrene in supercritical CO_2 compared with liquid cyclohexane.

solute in SCF solutions. Another possible explanation for intensity changes is that the fluorescence rate or absorption extinction coefficient changes with the conditions in the supercritical fluid. However, these quantities are normally independent of temperature so one could not explain the large observed intensity changes on that basis. We are currently conducting absorption measurements to verify the assumption of temperature independence.

Finally, it has been shown that there exists an anomaly between the observed values of the pyrene I_1/I_3 in SCF's and what would be expected from corresponding liquid organics (31). Values of I_1/I_3 have been correlated for liquid organics with the Kamlet-Taft π^* scale by the following relationship (32):

$$I_1/I_3 = 0.64 + 1.109\ \pi^*$$

The Kamlet-Taft π^* scale is based on a linear free energy relationship and measures the strength of the polarity and polarizability of the solvent. π^* values in SC fluids have been measured by Yonker and coworkers (19,21) with values for CO_2 ranging from -0.5 to -0.1. This would predict I_1/I_3 values of 0.4 to 0.5, which are much lower than those observed. Therefore, we observe I_1/I_3 values indicative of much stronger solute/solvent interactions than those predicted, suggesting a serious difference between the interactions in SCF's and those observed in normal liquid organics.

Solute/solute interactions are revealed by the formation of excimers. Excimers are excited state dimers that result in a broad structureless band at significantly longer wavelengths than the normal fluorescence. While they are not dimers in the sense of a ground state complex, their existence does indicate that there is sufficient interaction in the approximately 10^{-6} second lifetime of the excited state (33) to form the excited state complex. We have observed the formation of pyrene excimers even at extremely low concentrations in supercritical fluids. Figure 7 shows the spectra of pyrene in SCF CO_2 at two concentrations. At a mol fraction of just 5.5×10^{-6} significant excimer formation takes place. These solute/solute interactions are important because a large fraction of the solute molecules fluoresce as excimers. In fact, reasonable assumptions about the relative ability of monomers and excimers to fluoresce leads to an estimate of about 50% of the solute molecules fluorescing as excimers in just 8×10^{-6} mol fraction. In addition, this excimer formation is somewhat greater than one observes in liquids. For example, the 5.5×10^{-6} mol fraction sample in Figure 7 corresponds to a 5×10^{-5} Molar solution. To obtain a similar level of excimer formation in liquid cyclohexane requires a 2×10^{-3} Molar solution (34). This strongly suggests that the common practice of neglecting solute/solute interactions in models when the solute is below 1-2 mol % is unsubstantiated, and that our concept of infinite dilution in SCF's may need to be reevaluated. In addition, the excimer formation of pyrene in fluoroform is significantly less than in carbon dioxide or ethylene. This is shown in Figure 8 at similar solution densities for fluoroform and ethylene. Clearly, the stronger solute/solvent interactions shown in Figure 3 preclude the possibility of strong solute/solute interactions that would result in the excimer formation.

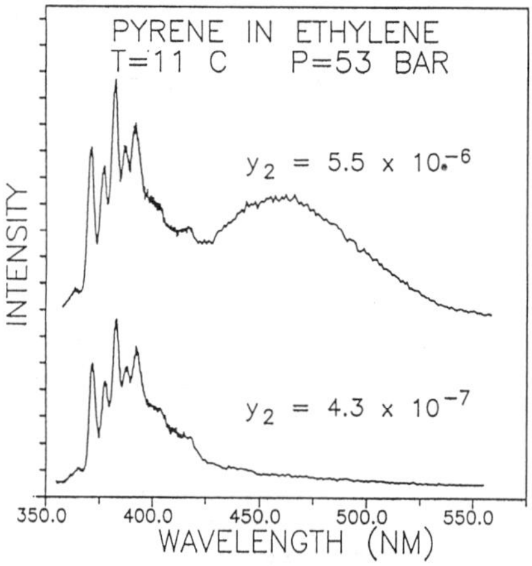

Figure 7. Excimer formation in dilute supercritical fluid solutions.

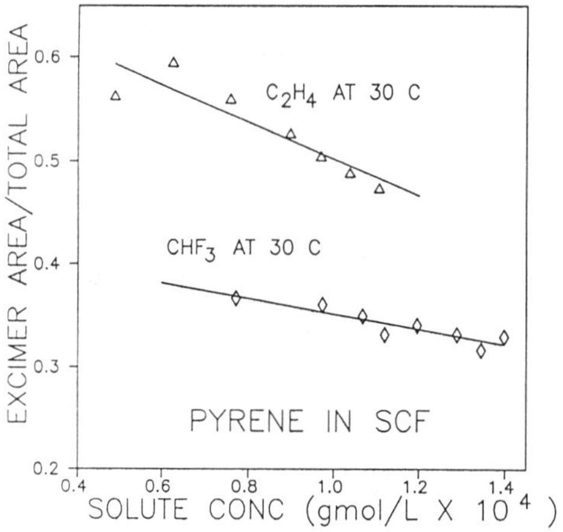

Figure 8. Excimer formation of pyrene in SCF C_2H_4 and CHF_3 at similar solution densities.

CONCLUSIONS

In this paper we have presented spectroscopic evidence of unusual intermolecular interactions in supercritical fluids. The strength of the solute/solvent interactions is much stronger near the critical point of the solvent, indicating an aggregation or clustering of the solvent around the solute. This is deduced from relative intensity ratios in the spectra and overall fluorescence intensities. In addition, the formation of excimers in dilute solutions indicates the importance of solute/solute interactions, even at concentrations as low as 5 x 10^{-6} mol fraction.

These studies are important for two reasons. First, they provide better understanding of the fundamentally different interactions that are occurring in the SCF solution. This understanding can be used to guide the development of equation of state models to include clustering or aggregation and reflect the true interactions in solution. Second, they provide a technique to estimate the parameters necessary for the models so that no new adjustable parameters will be required. Using a physical/chemical theory model these experiments are a tool to determine the complex equilibrium constants and the dependence of those constants on the temperature, solute functionality, solvent polarity and proximity to the critical point. In this way, an equation of state model that describes the true interactions in the SCF solutions can be developed based on the qualitative observations and quantitative measurements from these spectroscopic studies. Finally, the experiments can be extended easily to more complex solutes, higher temperature solvents and entrainer systems. These represent systems that will be important in the design of practical extraction and separation processes.

ACKNOWLEDGMENTS

This project has been financed in part with federal funds as part of the program of the Advanced Environmental Control Technology Research Center - University of Illinois at Urbana-Champaign, which is supported under cooperative agreement CR 806819 with the Environmental Protection Agency. The contents do not necessarily reflect the views and policies of the Environmental Protection Agency nor does the mention of trade names or commercial products constitute endorsement or recommendation for use. Additional funding was provided by the Department of Energy and the National Science Foundation. The authors would like to acknowledge the very helpful advice of Dr. Curt Frank, Department of Chemical Engineering, Stanford University.

LITERATURE CITED

1. Paulaitis, M. E.; Krukonis, V. J.; Kurnik, R. T.; Reid, R. C. Rev. Chem. Eng. 1982, 1(2), 179-250.
2. Paulaitis, M. E.; Penninger, J. M. L.; Gray, R. D.; Davidson, P., Eds., Chemical Engineering at Supercritical Fluid Conditions, Ann Arbor Science: Ann Arbor, MI, 1983.
3. McHugh, M. A.; Krukonis, V. J., Supercritical Fluid Extraction: Principles and Practice, Butterworths: Boston, 1986.
4. Eckert, C. A.; Van Alsten, J. G.; Stoicos, T. Environ. Sci. Tech. 1986, 20(4), 319-325.
5. Peng, D.-Y.; Robinson, D. B. Ind. Eng. Chem. Fundam. 1976, 15(1), 59- 64.

6. Johnston, K. P.; Eckert, C. A. AIChE J. 1981, 27(5), 773-779.
7. Johnston, K. P.; Ziger, D. H.; Eckert, C. A. Ind. Eng. Chem. Fundam. 1982, 21, 191-197.
8. Vezzetti, D. J. J. Chem. Phys. 1982, 77(3), 1512-1516.
9. Vezzetti, D. J. J. Chem. Phys. 1984, 80(2), 868-871.
10. Koningsveld, R.; Kleintjens, L. A.; Diepen, G. A. M. Ber. Bunsenges. Phys. Chem. 1984, 88, 848-855.
11. Gilbert, S. W.; Eckert, C. A. Fluid Phase Equilibria 1986, 30(41), 41-47.
12. Eckert, C. A.; Ziger, D. H.; Johnston, K. P.; Kim, S. J. Phys. Chem. 1986, 90(12), 2738-2746.
13. Eckert, C. A.; Ziger, D. H.; Johnston, K. P.; Ellison, T. K. Fluid Phase Equilibria 1983, 14, 167-175.
14. Vidal, J. Ber. Bunsenges. Phys. Chem. 1984, 88, 784-791.
15. Johnston, K. P.; Kim, S.; Wong, J. M. Fluid Phase Equilibria 1987, 38, 39-62.
16. Kim, S.; Johnston, K. P. AIChE J. 1987, 33(10), 1603-1611.
17. Kim. S.; Johnston, K. P. In Supercritical Fluids; Squires, T. G.; Paulaitis, M. E., Eds.; ACS Symposium Series No. 329; American Chemical Society: Washington, DC, 1987; pp 42-55.
18. Kim, S.; Johnston, K. P. Ind. Eng. Chem. Res. 1987, 26, 1206-1213.
19. Yonker, D. R.; Frye, S. L.; Kalkwarf, D. R.; Smith, R. D. J. Phys. Chem. 1986, 90, 3022-3026.
20. Kajimoto, O.; Futakami, M.; Kobayashi, T.; Yamasaki, K. J. Phys. Chem. 1988, 92, 1347-1352.
21. Yonker, C. R; Smith, R. D. J. Phys. Chem. 1988, 92, 2374-2378.
22. Debenedetti, P. G. Chem. Eng. Sci. 1987, 42, 2203-2212.
23. Cochran, H. D.; Pfund, D. M.; Lee, L. L. Proc. Intern. Sym. Supercritical Fluids, Nice, France, Oct. 1988, p 245.
24. Walsh, J. M.; Ikonomou, G. D.; Donohue, M. D. Fluid Phase Equilibria 1987, 33, 295-314.
25. Van Alsten, J. G. Ph.D. Thesis, University of Illinois, Urbana, 1986.
26. Schmitt, W. J.; Reid, R. C. Fluid Phase Equilibria 1986, 32, 77-89.
27. Kurnik, R. T.; Reid, R. C. Fluid Phase Equilibria 1982, 8, 93-105.
28. Kwiatkowski, J.; Lisicki, Z.; Majewski, W. Ber. Bunsenges. Phys. Chem. 1984, 88, 865-869.
29. Dong, D. C.; Winnik, M. A. Photochem. Photobiol. 1982, 35, 17-21.
30. Nakajima, A. Bull. Chem. Soc. Japan 1971, 44, 3272-3277.
31. Brennecke, J. F.; Eckert, C. A. Proc. Intern. Sym. Supercritical Fluids, Nice, France, Oct. 1988, p 263.
32. Dong, D. C.; Winnik, M. A. Can. J. Chem. 1984, 62, 2560-2565.
33. Turro, N. J., Modern Molecular Photochemistry; Benjamin/-Cummings Publishing Co., Inc.: Menlo Park, CA, 1978.
34. Birks, J. B., Photophysics of Aromatic Molecules; Wiley-Interscience: New York, 1970, p. 302.

RECEIVED May 1, 1989

Chapter 3

Solvation Structure in Supercritical Fluid Mixtures Based on Molecular Distribution Functions

Henry D. Cochran[1] and Lloyd L. Lee[2]

[1]Chemical Technology Division, Oak Ridge National Laboratory, Oak Ridge, TN 37831

[2]Department of Chemical Engineering, University of Oklahoma, Norman, OK 73019

On a molecular level, solubility enhancement and large, negative partial molar volumes near critical points are characterized by buildup of longer-range local solvent density around solute molecules. The magnitude of these clusters has been inferred from partial molar volume data. We have studied such clusters and related solution properties using molecular distribution functions calculated by an accurate integral equation theory. We observe the clustering as long-range correlations in the fluctuations of solvent molecules about a solute molecule. Partial molar volume is predicted in reasonable agreement with experiment, and solute-solute aggregation is also predicted as suggested by recent excimer fluorescence results. We conclude that integral equation theories are useful in revealing solvation structures of supercritical solutions.

The remarkable properties of supercritical fluids have stimulated considerable research interest because of practical applications as solvents for separations processes or reaction media and also because of the considerable challenges supercritical solutions present for modeling and theory. These challenges include the following: 1) fluid properties change drastically near and become nonclassical very near critical points (CPs), 2) conditions of interest and importance span a wide density range, 3) solutions are typically very dilute, 4) solvent and solute are frequently very different in molecular size and interaction energy, and 5) molecular correlations in fluids become long-ranged and fluctuations become large near a CP.

In current studies of supercritical mixtures, there is an outstanding hypothesis in the search for a molecular-scale mechanism that underlies all the unusual behavior associated with nearness to the CP. This hypothesis is the clustering of solvent molecules around the solute molecules. A surprising result from a recent fluorescence spectroscopy study shows the possibility of formation of solute-solute aggregates that occur near the CP. We propose in this study to clarify the understanding of solute-solvent clustering and solute-solute aggregation near CPs in supercritical solutions.

The fluid microstructure is of central interest in the theoretical study of supercritical solutions for several reasons. Bulk fluid properties can be obtained from knowledge of the fluid structure and the intermolecular potential through distribution function theory. In addition, it has recently become clear, both from experimental evidence and theoretical analyses (1–5) that solvation structure changes rapidly near a CP.

Eckert et al. (1) interpreted partial molar volume data for supercritical solutions as indicating the collapse of ca. 100 solvent molecules about a solute molecule. Kim and Johnston (2) interpreted the solvent shift in the UV absorption of phenol blue dissolved in supercritical ethylene to suggest a local solvent density surrounding a solute molecule more than 50% greater

0097–6156/89/0406–0027$06.00/0

than the bulk solvent density. Debenedetti (4) has used Kirkwood-Buff (6) fluctuation theory to infer from experimental partial volume data (1) that a solute molecule is solvated to form clusters of ca. 100 solvent molecules for naphthalene dissolved in supercritical CO_2 at CP, for example.

Several recent studies provided additional experimental evidence for the clustering phenomenon. Kajimoto et al. (3) interpreted absorption and fluorescence spectra of a polar solute in a polar, supercritical solvent in terms of a simple aggregation model to describe the effects of solvent-solute clustering. Brennecke and Eckert (7 and ACS Symp. Ser. in press) measured the fluorescence spectra of pyrene dissolved in supercritical carbon dioxide and ethylene. Strong enhancement of the symmetry-forbidden first singlet-singlet transition close to the critical point is taken as indicative of very strong solvent-solute interaction (probably only nearest neighbor interactions). Furthermore, the presence of a peak ascribed to the excimer dimer even at bulk pyrene mole fraction 10^{-5} suggests very strong solute-solute interactions in the supercritical solution, as well. An investigation of the disproportionation of toluene over ZSM-5 to form xylene (Collins, N. A. ; Debenedetti, P. G. ; Sundaresan, S. AIChE J. in press) tested the occurrence of solvation. Since toluene acts as both reactant and solvent, near the CP toluene clustered around the para-isomer, thus reducing the chances of secondary isomerization into other isomers.

In previous work (5,8,9) we have shown that models based on distribution function theory yield a priori predictions of the qualitative behavior of supercritical solutions and yield empirical fits to data which show excellent accuracy for solubility and partial molar volume.

An understanding of clustering at a molecular level will have to come from molecular distribution functions, especially the solvent-solute correlation function. These correlation functions could be obtained from neutron scattering experiments, computer simulation, or integral equation theories. No scattering experiments of supercritical solutions have been carried out for this purpose. Computer simulation can shed interesting light and is discussed elsewhere (See Petsche, I. B.; Debenedetti, P. G. J. Chem. Phys. submitted). For infinitely dilute solute species, the structure (particularly the long-range structure near a CP) is more easily calculated by integral equations.

We propose the study of Lennard-Jones (LJ) mixtures that simulate the carbon dioxide-naphthalene system. The LJ fluid is used only as a model, as real CO_2 and $C_{10}H_8$ are far from LJ particles. The rationale is that supercritical solubility enhancement is common to all fluids exhibiting critical behavior, irrespective of their specific intermolecular forces. Study of simpler models will bring out the salient features without the complications of details. The accurate HMSA integral equation (10) is employed to calculate the pair correlation functions at various conditions characteristic of supercritical solutions. In closely related work reported elsewhere (Pfund, D. M. ; Lee, L. L. ; Cochran, H. D. Int. J. Thermophys. in press and Fluid Phase Equilib. in preparation) we have explored methods of determining chemical potentials in solutions from molecular distribution functions.

Distribution Function Theory

Fluid microstructure may be characterized in terms of molecular distribution functions. The local number of molecules of type α at a distance between r and $r+dr$ from a molecule of type β is $\rho_\alpha 4\pi r^2 g_{\alpha\beta}(r)\,dr$ where $g_{\alpha\beta}(r)$ is the spatial pair correlation function. In principle, $g_{\alpha\beta}(r)$ may be determined experimentally by scattering experiments; however, results to date are limited to either pure fluids of small molecules or binary mixtures of monatomic species, and no mixture studies have been conducted near a CP. The molecular distribution functions may also be obtained, for molecules interacting by idealized potentials, from molecular simulations and from integral equation theories.

Statistical mechanics gives relationships between the distribution functions and the bulk properties of fluids. The total internal energy of a fluid is given by the energy equation, the pressure is given by the virial equation, and the isothermal compressibility is given by the compressibility equation, see e. g., Ref. 11. Through the Kirkwood-Buff formulas (6), the compressibility, the partial molar volumes, and the derivatives of the chemical potential with respect to number density are obtained; for simplicity of notation these are presented for binary mixtures.

$$\rho kT\chi_T = \Omega/\Phi,$$

$$\rho\bar{v}_\beta = [1 + \rho_\alpha (G_{\alpha\alpha} - G_{\alpha\beta})]/\Phi,$$

and

$$kT\left(\frac{\partial \rho_\alpha}{\partial \mu_\beta}\right)_{T,\mu_\alpha} = \delta_{\alpha\beta}\rho_\alpha + \rho_\alpha\rho_\beta G_{\alpha\beta},$$

where

$$\Omega = 1 + \rho_\alpha G_{\alpha\alpha} + \rho_\beta G_{\beta\beta} + \rho_\alpha\rho_\beta \left(G_{\alpha\alpha}G_{\beta\beta} - G_{\alpha\beta}^2\right),$$

$$\Phi = [\rho_\alpha + \rho_\beta + \rho_\alpha\rho_\beta (G_{\alpha\alpha} + G_{\beta\beta} - 2G_{\alpha\beta})]/\rho,$$

and $G_{\alpha\beta}$ is the Kirkwood fluctuation integral,

$$G_{\alpha\beta} \equiv \int_0^\infty dr\,[g_{\alpha\beta}(r) - 1]\,4\pi r^2.$$

We have previously shown (8) that the solubility of a pure, incompressible solid, B, in a supercritical solvent, A, is given by

$$\frac{d\rho_B}{\rho_B} = [G_{AB} + v_B^s + \rho_B v_B^s (G_{BB} - G_{AB})]\frac{dP}{kT}$$

where v_B^s is the solid molecular volume. Useful equations for dilute, supercritical solutions result from appropriate approximations in the above expression; see Ref. 8.

Kim and Johnston (2) and Debenedetti (4) have defined the size of the cluster of solvent A molecules about a solute B molecule in dilute solution by the following:

$$\xi_{AB} \equiv \rho_A G_{AB}^\infty$$

where the superscript ∞ signifies the value of the quantity at infinite dilution; ξ_{AB} expresses the number of solvent A molecules surrounding solute B molecule in excess of the bulk average.

Calculations

In this work we have calculated pair correlation functions and derived properties for systems with interaction potentials representative of supercritical solutions by solution of the Ornstein-Zernike (OZ) equation using the HMSA closure (10) and the efficient algorithm of Labik et al. (12). Independent and corroborative calculations using the RHNC closure (13) for the same mixtures are omitted for brevity. Although quantitative values of distribution functions from the two theories differed slightly, the general features are the same; the RHNC theory gave somewhat higher peaks, but they occur at the same intermolecular separation.

We used the following Lennard-Jones parameters for CO_2–$C_{10}H_8$:

	ϵ/k (K)	σ (Å)
CO_2–CO_2	225.3	3.794
$C_{10}H_8$–$C_{10}H_8$	554.4	6.199
CO_2–$C_{10}H_8$	353.4	4.997

The HMSA closure (10) has been used to solve the OZ equation; it is an interpolation between the soft mean spherical approximation at small r and the HNC closure at large r.

The interpolating parameter has been chosen to achieve consistency between the isothermal compressibility from the compressibility equation and that from the virial equation. The calculations in this work would have been impractical without the use of an efficient and robust algorithm for solution of the OZ equation. We employed the approach proposed by Labik, et al. (12). For these results we used 2048 grid points in real space, $0.005\,\sigma_{AA}$ step size, 64 Newton-Raphson terms within 10^{-5} tolerance on Newton-Raphson and 10^{-5} tolerance on direct iterations. Because of concern about the long range of correlation functions near the CP, we also explored calculations with 512 and 1024 grid points and with step size 0.010, 0.015, and $0.020\,\sigma_{AA}$. Calculations with range (number of grid points times step size) $10.24\,\sigma_{AA}$ or greater and step size $0.010\,\sigma_{AA}$ or smaller gave essentially the same results.

The calculations reported in this work were for very dilute, $x_B = 10^{-9}$, solutions. In other work (5) we found the results to be unaffected for solute mole fractions from 10^{-9} to 10^{-4}. Dilute solutions are characteristic of many experimentally-studied supercritical solutions. The low solute concentration does, however, tend to reduce the accuracy of the solute-solute correlation functions, $g_{BB}(r)$; so, we regard them as of qualitative value only.

This calculational method has been found to yield internal energy, pressure, and distribution functions in excellent agreement with simulation results for pure and mixed LJ fluids as well as fluids obeying other simple force laws (10). However, it should be noted that tests of the method have not included the highly compressible, near-critical states such as those studied in the present calculations where both simulations and integral equation calculations would encounter difficulties. Furthermore, we know of no direct way to test the accuracy of the long range structure calculated by the integral method at present.

Results and Discussion

Figure 1 shows the predicted spatial pair correlation functions, $g_{\alpha\beta}$, vs. reduced separation distance, $r^* \equiv r/\sigma_{AA}$, for a typical, dilute supercritical solution using the LJ parameters given above and at $T^* \equiv kT/\epsilon_{AA} = 1.37$ ($T = 308.4$K), $\rho^* \equiv \rho\sigma_{AA}^3 = 0.40$ ($\rho = 0.00732$Å^{-3}), and $x_B = 10^{-9}$; the calculated pressure for this state is 8.7 MPa. The clustering of solvent molecules about a dissolved solute molecule, as remarked earlier, is seen in the solvent-solute pair correlation function with first, second, and even third solvation shells. The long-rangedness of $g_{AB}(r)$ is clearly in evidence. Similar behavior is observed for the solute-solute pair correlation function also.

Figures 2a and 2b show how the predicted solvent-solute pair correlation function, g_{AB}, varies with density. At the highest density ($\rho^* = 0.6$) the structure is liquid-like with first, second, third, ... maxima/minima oscillating about 1.0. The size of the solvent-solute cluster for this state was calculated to be about -1 solvent molecule; the presence of one solute molecule at this state excludes about one solvent molecule.

At lower density ($\rho^* = 0.5$) the peak maxima are lower and the peak minima are shallower. Again the long range maxima and minima oscillate about 1.0. The size of the solvent-solute cluster for this state was calculated to be about +2 solvent molecules; one solute molecule is surrounded by about two excess solvent molecules compared with the bulk average concentration.

At density ($\rho^* = 0.4$) near the CP (the critical point of the pure LJ solvent is at about $T^* = 1.35$ and $\rho^* = 0.35$) the oscillations subside into a persistent long-range correlation, and the pair correlation function exhibits a long-range tail that is greater than 1.0. See particularly Figure 2b which clearly shows this long-range structure. The solvent-solute cluster for this state was calculated to be +14.65 solvent molecules. At states farther from the CP the solute molecule excluded solvent molecules, but near the CP there is a large buildup of longer-range local solvent density around a solute molecule.

In Figure 3 we have plotted the number of excess solvent molecules within a sphere of radius $L(\equiv r/\sigma_{AA})$ around a solute molecule, $\rho\, G_{AB}(L)$, vs. L for the same states as Figure 2.

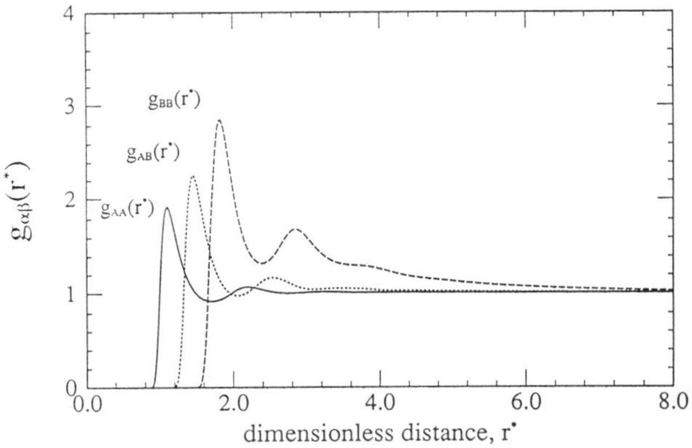

Figure 1: Pair correlation functions at $T^* = 1.37$ and $\rho^* = 0.40$. For the LJ fluid the critical point occurs at about $T^* = 1.35$ and $\rho^* = 0.35$.

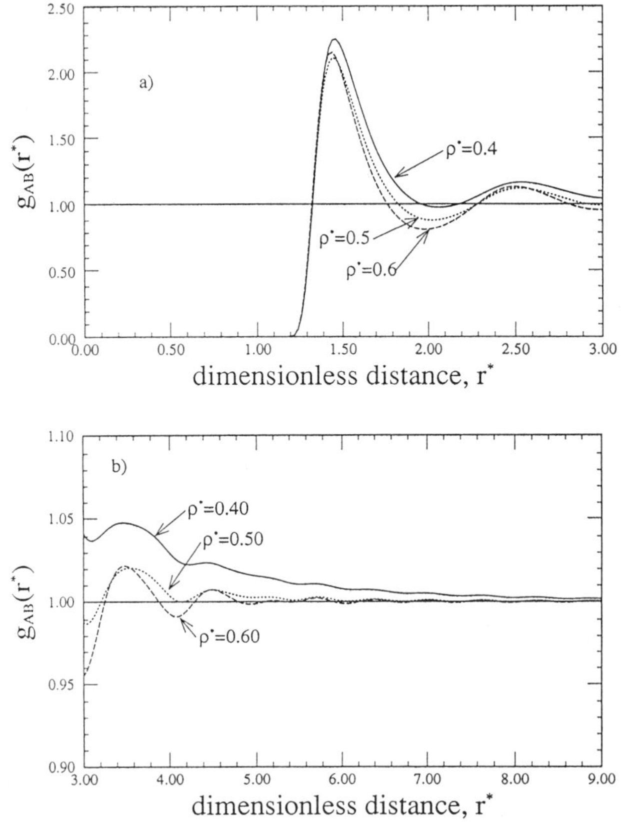

Figure 2: Solute-solvent pair correlation function at $T^* = 1.37$ and $\rho^* = 0.40, 0.50$, and 0.60. a) Short range structure; b) long range structure (note change of axes).

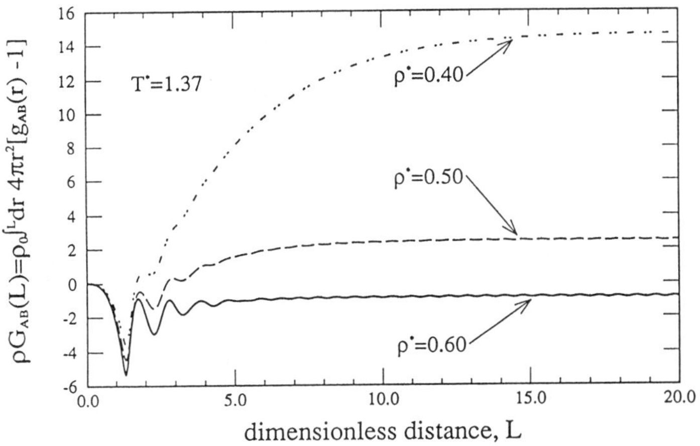

Figure 3: Number of excess solvent molecules within a sphere of radius L around a solute molecule. Same conditions as Figure 2.

$$\rho_A G_{AB}(L) = \rho_A \int_0^L dr\, 4\pi r^2 [g_{AB}(r) - 1]$$

This graph shows contributions to the fluctuation integral from the successive neighborhoods.

It is evident that for $\rho^* = 0.40$, the state close to the critical point, the solvent-solute cluster has grown to almost 15 excess solvent molecules. The contribution from the first solvation shell can be estimated; about 4 excess solvent molecules contribute to the first solvation sphere. It is also evident that the cluster consists mainly of a diffuse, longer-range buildup of local solvent density about the solute which extends 10–14 solvent diameters from the solute. Whether this cluster maintains its identity for a significant period of time or is merely a statistical phenomenon cannot be answered from these equilibrium calculations. Molecular dynamics calculations suggest that the clusters maintain their structural integrity for significant periods although the molecules constituting the cluster change identity more rapidly (See Petsche, I. B.; Debenedetti, P. G. J. Chem. Phys. submitted). Nevertheless, regardless of their lifetime, these clusters are characteristic of solvation structure near the CP and are responsible for the striking behavior of thermodynamic properties near the CP.

Figure 4 shows how the size of the solvent-solute cluster varies with density at $T^* =$ 1.37, 1.41, and 1.46 (35, 45, and 55 °C). As density increases from zero toward the CP the size of the cluster grows rapidly. Because the pair correlation functions become longer and longer in range as the CP is approached, there is a limit beyond which calculations by this technique fail to converge. As a practical matter, the range of the present calculations ($\leq 40.96\,\sigma_{AA}$) limits convergence at $T^* = 1.37$ to densities below 0.15 and densities above 0.4. The rapid decline in the cluster size as density is increased away from the CP is also evident. At temperatures higher above the CP the cluster is smaller and the structure is shorter in range.

These calculations support the notion of a large solvent-solute cluster near the CP as suggested from spectroscopic results (2,3,7). The estimation (1,4) that the cluster approaches 100 solvent molecules based on the partial molar volume data of Eckert et al. (1) for naphthalene in supercritical carbon dioxide at this temperature also appears entirely consistent with the predictions from the integral equation theory.

Figure 5 shows the partial molar volume of naphthalene in supercritical carbon dioxide predicted by theory compared with the data of Eckert et al. (1). Neither naphthalene nor carbon dioxide would be expected to be described quantitatively by the LJ potential. Furthermore, no binary interaction parameter was used with the Lorentz-Berthelot estimates for the CO_2–$C_{10}H_8$ interaction. Thus, the degree of agreement between effectively a priori prediction and experiment is judged to be very satisfactory.

Figures 6a and 6b show the solute-solute pair correlation function at the same conditions as the solvent-solute functions in Figures 2a and 2b. The aggregation of the solute molecules is quantitatively stronger than the solvation structure shown in Figure 2. A quantitative interpretation in terms of the local density of solute molecules surrounding a given solute molecule has not been given for the band attributed to a solute-solute excimer dimer in the fluorescence spectra of Brennecke and Eckert (7); nevertheless, their qualitative interpretation suggesting a significant solute-solute aggregation near the CP appears to be supported by these results.

Conclusions

The solvation structures which we have determined for the supercritical solutions of LJ molecules appear to be in agreement with the recent theoretical and experimental suggestions of solvent-solute clustering and solute-solute aggregation near the CP. Quantitative testing of theoretical models for solvation structure of supercritical solutions (such as that presented here) may become possible if recent efforts at simulation of supercritical solutions (elsewhere in this volume) prove successful. Clearly, the LJ model used in our theoretical studies to date does not provide an adequate representation of real molecular interactions. However, the method we have demonstrated is potentially capable of application with much more accurate (and complicated) potential functions, and meaningful quantitative interpretation of experiments may then become possible.

In summary, then, we conclude that integral equation theories are useful in revealing

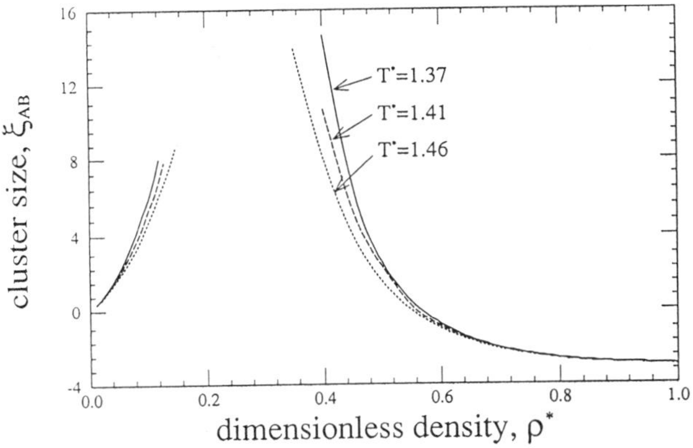

Figure 4: Size of solvent-solute cluster vs. density at $T^* = 1.37$, 1.41, and 1.46.

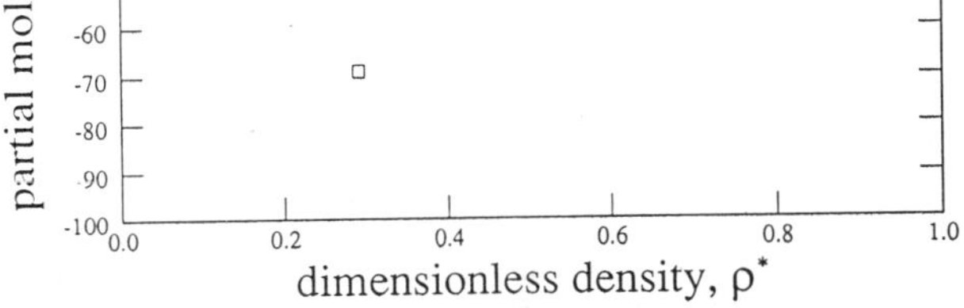

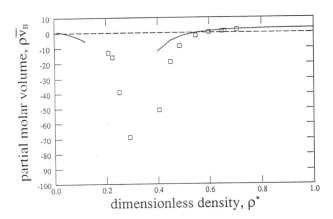

Figure 5: Calculated and experimental (1) solute partial molar volume vs. density at $T^* = 1.37$.

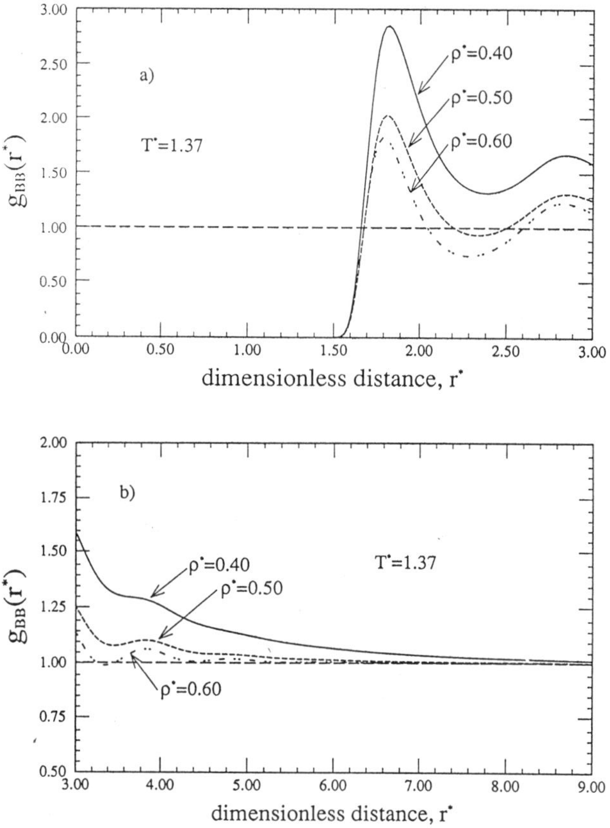

Figure 6: Solute-solute pair correlation function. Same conditions as Figure 2. a) Short range structure; b) long range structure (note change of axes). The curve at $\rho^* = 0.4$ shows a double peak with shallow minimum indicating solute-solute aggregation.

the solvation structures in supercritical solutions. Molecular distribution functions for states representative of typical supercritical solutions exhibit a longer range buildup of local solvent density around solute molecules near the CP. This local structure is consistent with the clustering of solvent molecules about solute molecules which has been hypothesized to occur near a CP. The size of the predicted solvent-solute cluster has been found to be in reasonable agreement with partial molar volume data.

The solute-solute distribution functions determined in this work suggest a structure that is consistent with the aggregation of solute molecules suggested by recent excimer fluorescence spectra. Again, the integral equation results cannot reveal whether the solute-solute aggregates persist for significant time.

Acknowledgment

The authors wish to acknowledge their indebtedness to D. M. Pfund for computer programming. Profs. Eckert and Debenedetti kindly shared their results in advance of publication. This work has been supported in part by the Chemical Sciences Division, Office of Basic Energy Science, U. S. Department of Energy, under Contract No. DE-AC05-840R21400 with Martin Marietta Energy Systems, Inc. and subcontract with the University of Oklahoma.

Legend of Symbols

A	=	component A, solvent, CO_2 [-]
B	=	component B, solute, $C_{10}H_8$ [-]
G	=	Kirkwood fluctuation integral [Å^3]
$g()$	=	pair correlation function [-]
k	=	Boltzmann constant [PaÅ^3/°K]
L	=	dimensionless radius of solvation sphere [-]
P	=	pressure [Pa]
r	=	separation distance [Å]
s	=	solid phase [-]
T	=	temperature [°K]
V	=	total volume [Å^3]
v	=	volume per molecule [Å^3]
x	=	mole fraction [-]
α	=	arbitrary species index
β	=	arbitrary species index
δ	=	Kronecker delta [-]
ϵ	=	LJ energy parameter [PaÅ^3]
μ	=	chemical potential [PaÅ^3]
ξ	=	cluster size [-]
ρ	=	number density [Å^{-3}]
σ	=	LJ size parameter [Å]
Φ	=	function in Kirkwood-Buff formulas [-]
χ_T	=	isothermal compressibility [Pa^{-1}]
Ω	=	function in Kirkwood-Buff formulas [-]
$^\infty$	=	value of quantity at infinite solute dilution
$^-$	=	partial molar quantity
*	=	reduced quantity

Literature Cited

1. Eckert, C. A. ; Ziger, D. H. ; Johnston, K. P. ; Ellison, T. K. Fluid Phase Equilib. 1983, 14, 167 and J. Phys. Chem. 1986, 90, 2738.
2. Kim, S. ; Johnston, K. P. Ind. Eng. Chem. Res. 1987, 26, 1206 and AIChE J. 1987, 33, 1603.
3. Kajimoto, O. ; Futakami, M. ; Kobayashi, T. ; Yamasaki, K. J. Phys. Chem. 1988, 92, 1347.
4. Debenedetti, P. G. Chem. Eng. Sci. 1987, 42, 2203 and Debenedetti, P. G. ; Kumar, S. AIChE J. , 1988, 34, 645.
5. Cochran, H. D. ; Pfund, D. M. ; Lee, L. L. Proc. Int. Symp. on Supercritical Fluids 1988, M. Perrut, ed. , Tom. 1, 245 and Sep. Sci. and Tech. , 1988, 23, 2031.
6. Kirkwood, J. G. ; F. P. Buff, J. Chem. Phys. 1951, 19, 774.
7. Brennecke, J. F. ; Eckert, C. A. Proc. Int. Symp. on Supercritical Fluids 1988, M. Perrut, ed. , Tom. 1, 263.
8. Cochran, H. D. ; Lee, L. L. ; Pfund, D. M. , Fluid Phase Equilib. 1987, 34, 219 and ibid. 1988, 39, 161.
9. Cochran, H. D. ; Lee, L. L. , AIChE J. 1987, 33, 1391 and ibid. 1988, 34, 170.
10. Zerah, G. ; Hansen, J. -P. , J. Chem. Phys. 1986, 84, 2336.
11. Lee, L. L. Molecular Thermodynamics of Nonideal Fluids; Butterworths, Stoneham, MA, 1988, Chapter V.
12. Labik, S. ; Malijevsky, A. ; Vonka, P. , Mol. Phys. 1985, 56, 709.
13. Lado, F. Phys. Rev. A 1973, 87, 2548.

RECEIVED May 1, 1989

Chapter 4

Gibbs-Ensemble Monte Carlo Simulations of Phase Equilibria in Supercritical Fluid Mixtures

A. Z. Panagiotopoulos

School of Chemical Engineering, Cornell University, Ithaca, NY 14853–5201

A novel technique for molecular-level computer simulations, the Gibbs-ensemble Monte Carlo methodology, is applied to the calculation of phase equilibria in supercritical fluid systems. The Gibbs method is based on performing a simulation in two regions in a way that ensures that the criteria for equilibrium between coexisting phases are satisfied in a statistical sense. Lennard-Jones intermolecular potentials are used in the calculations, with parameters obtained from pure component thermodynamic properties to represent carbon dioxide, acetone and water. Calculated binary phase diagrams are in good agreement with experimental data. Ternary calculations are only in qualitative agreement with experiment. Additional results obtained by varying the intermolecular potential parameters for the unlike pair interactions illustrate the effect of intermolecular forces on phase behavior.

Thermodynamic modelling of phase equilibria is an important part in the successful development and operation of physical separation processes. The prediction of phase equilibria for supercritical fluid extraction applications presents special challenges, not shared by more conventional separation systems. The challenges arise as a result of operation at high, and highly variable, pressures and the inherently asymmetric character of the mixtures involved that contain components of very different size or volatility. A substantial effort in recent years has been devoted to the development of improved equations-of-state and mixing rules that can describe the often complex phase equilibria encountered in supercritical extraction operations. While much progress has been made, existing models allow interpolations and moderate extrapolations of experimental data and cannot be used in the absence of experimental information. An alternative to macroscopic, phenomenological models is provided by molecular simulation methods which provide a direct link between the intermolecular forces and macroscopic phase behavior.

0097–6156/89/0406–0039$06.00/0

Computer simulation techniques have been used since their inception for the calculation of basic thermodynamic and structural properties of liquids (1). While it is relatively easy to calculate simple configurational properties such as the energy and pressure of a fluid, quantities that are related to the entropy or free energy have been considered very difficult to obtain from simulation. In recent years, improved techniques for the estimation of free energies have appeared (a recent review is given in 2). The Widom test particle method (3) is the most widely used method, and recent developments (Deitrick *et al.*, *J.Chem. Phys.* in press) have pointed to possible significant improvements in the accuracy and speed of the technique. Several successful predictions of phase equilibria of mixtures have been reported for atomic (4) and molecular fluids (5) using this method. Molecular simulation results for the solubility of solids in supercritical fluids have appeared in (6). The computer time requirements for these calculations are high because of the need to perform a large number of simulations at different densities and compositions, a number which increases rapidly with the number of components in a mixture. The semigrand ensemble approach developed by Kofke and Glandt (7) eliminates this last problem but still requires a series of simulations for the determination of a single phase coexistence point. The recently proposed Gibbs-ensemble Monte Carlo simulation method (8), is a significant improvement relative to previously described methods, as it always requires only a single simulation per coexistence point.

The Gibbs Method

Basic Concepts. The methodology for the determination of phase equilibria in mixtures using the Gibbs method has been presented in detail elsewhere (8,9,10). The essence of the technique is to perform a simulation in two distinct regions (e.g. a liquid and a gas region), each in periodic boundary conditions with images of itself. Three types of perturbation are performed in a way that ensures that the conditions for phase equilibrium between the phases are satisfied in a statistical sense: (a) displacements within each region, considered separately, to satisfy internal equilibration (b) volume rearrangements, to satisfy equality of pressures and (c) particle transfers between the two regions, to satisfy the condition of equality of chemical potentials. The corresponding criteria for the acceptance of each move are:

(a) particle displacements

$$P_{move} = \min\left[1 , \exp(-\beta\Delta E) \right] \tag{1}$$

where ΔE is the configurational energy change resulting from the trial displacement.

(b) volume change steps

In the constant-*NVT* ensemble the volume changes in the two regions are equal and opposite, so that total system volume is conserved:

$$P_{vol} = \min\left(1, \exp\left(-\beta\left[\Delta E^{I} + \Delta E^{II} - N^{I}kT\ell n\frac{V^{I}+\Delta V}{V^{I}} - N^{II}kT\ell n\frac{V^{II}-\Delta V}{V^{II}}\right]\right)\right) \quad (2)$$

In the constant-*NPT* ensemble the two regions undergo independent volume changes:

$$P_{vol} = \min\left(1, \exp\left(-\beta\left[\Delta E^{I} + \Delta E^{II} - N^{I}kT\ell n\frac{V^{I}+\Delta V^{I}}{V^{I}} - N^{II}kT\ell n\frac{V^{II}+\Delta V^{II}}{V^{II}} + P(\Delta V^{I}+\Delta V^{II})\right]\right)\right) \quad (3)$$

(c) Particle transfer steps

$$P_{transfer} = \min\left(1, \exp\left(-\beta\left[\Delta E^{I} + \Delta E^{II} + kT\ell n\frac{V^{II}(N_i^{I}+1)}{V^{I}N_i^{II}}\right]\right)\right) \quad (4)$$

The Gibbs technique has been used to predict vapor-liquid, liquid-liquid and osmotic equilibria for binary Lennard-Jones mixtures (9), phase transitions for fluids in pores (11), and phase equilibria for quadrupolar systems (Stapleton *et al.*, *Mol. Simulation*, in press).

Extension for Mixtures with Large Differences in Molecular Size. For pure fluids and simple mixtures, the methodology described in the previous subsection works well, even for densities close to triple-point liquid densities. However, for mixtures with components that have large differences in molecular size, the method fails at high densities, because most attempts to transfer a molecule of the larger species into a densely packed liquid are unsuccessful. In a recent paper (10), a modification was proposed to handle highly asymmetric mixtures. Only the particle transfer step is affected. We designate the component with the smallest size in a binary mixture as component 2. Component 1 has a significantly larger size, and therefore it is difficult to obtain successful transfers of this component between regions. In the modified method, direct transfers between the two regions (accepted using the criterion given in equation 4) are only attempted for component 2. For component 1, the transfer step involves the following: In one of the two regions, a randomly selected particle of species 2 becomes a particle of species 1. At the same time, the inverse procedure is applied to the other region: a randomly selected particle of species 1 becomes a particle of species 2. The move is accepted with a probability:

$$P_{exchange} = \min\left(1, \exp\left(-\beta\left[\Delta E^{I} + \Delta E^{II} + kT\ell n\frac{V^{II}(N^{I}{}_{1}+1)}{V^{I}N_1^{II}} + kT\ell n\frac{V^{I}(N_2^{II}+1)}{V^{II}N_2^{I}}\right]\right)\right) \quad (5)$$

The following conditions between the chemical potentials of the two components in the two regions are then satisfied in a statistical sense:

$$\mu_2^{I} = \mu_2^{II} \qquad \text{(from transfer step, equation 4)}$$
$$\mu_2^{I} - \mu_1^{I} = \mu_2^{II} - \mu_1^{II} \qquad \text{(from exchange step, equation 5)} \tag{6}$$

which are sufficient to ensure the equality of chemical potentials of both components in the two phases. The method can be easily generalized for systems with more than two components.

The advantage of the proposed modification is that is it much more efficient to attempt to increase the size of an existing molecule, than to attempt to place a molecule at a completely random position. In this respect the proposed modification is similar to the semigrand ensemble Monte Carlo simulation technique (7).

Model Potentials

Molecular simulation methods start from a description of the intermolecular forces of a system. For all but the simplest molecules, quantitative information on intermolecular potentials is not available. For this reason one has to resort to approximate, analytically convenient intermolecular potential functions and obtain the parameters by fitting experimental results. Although the need for fitting seems at least partly to negate some of the advantages of molecular simulation techniques over phenomenological approaches, the hope is that the fitted intermolecular potential parameters would be transferrable from system to system, and be applicable for a wide range of process conditions.

Our goal in this work is to explore the effects of intermolecular forces on the phase behavior of supercritical systems. For this purpose, it is preferable to use simple intermolecular potential functions, with as few parameters as possible. The Lennard-Jones (6,12) intermolecular potential function,

$$U_{ij}(r) = 4\epsilon_{ij} \left[\left(\frac{\sigma_{ij}}{r}\right)^{12} - \left(\frac{\sigma_{ij}}{r}\right)^{6} \right] \tag{7}$$

was used in this work to describe interactions among all components.

Three molecular species were selected for study. These are:

(a) Carbon dioxide, a commonly used supercritical solvent.

(b) Acetone, a relatively volatile (liquid) solute significantly larger in terms of molecular size from CO_2.

(c) Water, a species often present in mixtures to be separated or as an entrainer.

Because of current limitations of the Gibbs simulation methodology, no equilibria involving solid phases could be studied. Equilibria involving solids can be studied by indirect chemical potential calculation techniques (6) by restricting attention to the fluid phase and assuming a value for the vapor pressure of the solid.

The pure component intermolecular potential parameters used in this study are shown in Table I. They were obtained as follows: for carbon dioxide, we fitted the experimental critical temperature and pressure (12) , using data from (8) for the critical constants of the Lennard-Jones (LJ) system ($T_c^* = 1.31$, $P_c^* = 0.13$). For acetone, a single point for the vapor pressure and liquid density at T = 333 K was fitted. For water, we used vapor pressure data at room temperature and critical pressure data. Specifically for water, the LJ potential does not give a good representation of the strong directional forces in the real system, and thus only a limited temperature range could be successfully fitted. With this exception (the predicted saturated liquid volume for water is higher than the experimental value by 50%), the parameters in Table I give a reasonably good representation of the thermodynamic properties of the pure components around room temperature.

Table I. Pure component intermolecular potential parameters

Component	ϵ/k_B (K)	σ (nm)
Carbon Dioxide	232.2	0.384
Acetone	443	0.467
Water	540	0.358

The basic molecular characteristics of these species are reflected by the potential parameters given in Table I. Thus, the energy parameter ϵ (proportional to the component critical temperature) increases from carbon dioxide to water, while the parameter σ (molecular diameter) is the highest for acetone.

In addition to pure component parameters, mixture calculations require the estimation of the unlike-pair interaction parameters. These were obtained in this study using the Lorenz-Berthelot rules:

$$\epsilon_{ij} = \delta_{ij}(\epsilon_{ii}\ \epsilon_{jj})^{1/2} \tag{8}$$

$$\sigma_{ij} = \frac{\sigma_{ii} + \sigma_{jj}}{2} \tag{9}$$

A parameter, δ_{ij}, appears in the combining rule for the energy parameter. It describes the deviations of the unlike-pair interactions from the geometric-mean rule. Values less than 1 indicate that the unlike pair interactions are less favorable compared to the like-pair interactions.

Simulations of Binary Systems

Acetone / Carbon Dioxide. A molecular simulation study for this binary system has appeared in (4). In that study, good qualitative, but not quantitative, agreement was obtained between simulation and experimental results. The differences were ascribed to the inadequacy of the LJ potential to represent the properties of carbon dioxide and acetone. However, in that work, the pure component potential parameters for acetone were not optimized. The simulations in this work were performed using the potential parameters in Table I and the Lorenz-Berthelot combining rules with $\delta_{ij} = 1$.

The results from the Gibbs-ensemble simulations and a comparison with experimental data (13) and a commonly used cubic equation of state are presented in Figure 1. All results are at T = 313 K. The constant-*NPT* form of the Gibbs technique with the particle interchange algorithm (Equation 5) was used in all the simulations. Total system size was 400 particles, a number previously shown to give results independent of system size for similar calculations (9). Runs of 10^6-2×10^6 MC steps (trial moves) were performed per coexistence point, with 4×10^5-7×10^5 discarded for equilibration. As in previous studies, a cycle of particle displacements was followed by one or two volume change attempts and a series of between 100 and 700 transfer and species identity interchange trials. In the constant-*NPT* methodology the coexistence pressure is specified in advance, and thus there is no uncertainty associated with it. The estimated errors for the compositions were obtained by dividing the production period of each run into 10 blocks, and calculating the standard deviation of the block-average compositions. Results from the Monte Carlo simulations are in agreement with experimental data within one or two standard deviations of these block averages. The estimated uncertainties of the compositions given in Figure 1 are typical of what can be expected from the Gibbs simulation method. The accuracy of the simulation calculations using the Gibbs method is comparable to that of experimental measurements of medium to low accuracy (± 0.02 in mole fraction of a major component). It should be noted that in order to reduce the estimated uncertainties by a factor of 2, a simulation run approximately 4 times longer would be required.

Effect of Unlike-Pair Interactions on Phase Behavior. No adjustment of the unlike-pair interaction parameter was necessary for this system to obtain agreement between experimental data and simulation results (this is, however, also true of the cubic equation-of-state that reproduces the properties of this system with an interaction parameter $k_{ij}=0$). An interesting question that is ideally suited for study by simulation is the relationship between observed macroscopic phase equilibrium behavior and the intermolecular interactions in a model system. Acetone and carbon dioxide are mutually miscible above a pressure of approximately 80 bar at this temperature. Many systems of interest for supercritical extraction processes are immiscible up to much higher pressures. In order to investigate the transition to an immiscible system as a function of the strength of the intermolecular forces, we performed a series of calculations with lower strengths of the unlike-pair interactions. Values of $\delta_{ij} = 0.90$, 0.80, 0.70 were investigated.

Results from these calculations with varying strength of the unlike-pair interactions are presented in Fig. 2. Clearly, if one artificially varies the strength of the unlike-pair interactions, the systems that result are hypothetical, and no longer represent real mixtures of acetone and CO_2. For simplicity, we will still refer to the components of these hypothetical mixtures as acetone and CO_2. The results are expressed as mole fractions of CO_2 in the acetone-rich (liquid) phase and mole fractions of acetone in the CO_2-rich (fluid) phase, as functions of pressure. Miscible behavior was observed for $\delta_{ij} = 1$ (the base case) and 0.90, while immiscibility that persists for pressures significantly higher than the critical pressure of pure CO2

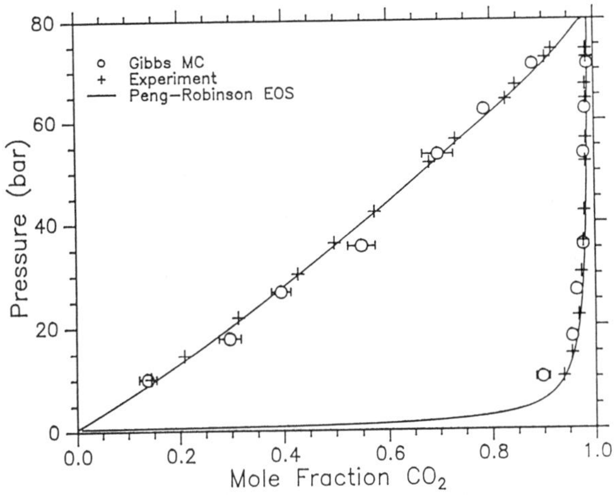

Figure 1. Phase equilibria for the system acetone / carbon dioxide at T = 313 K. Gibbs-ensemble Monte Carlo results are with $\delta_{ij} = 1$. Experimental results are from (13).

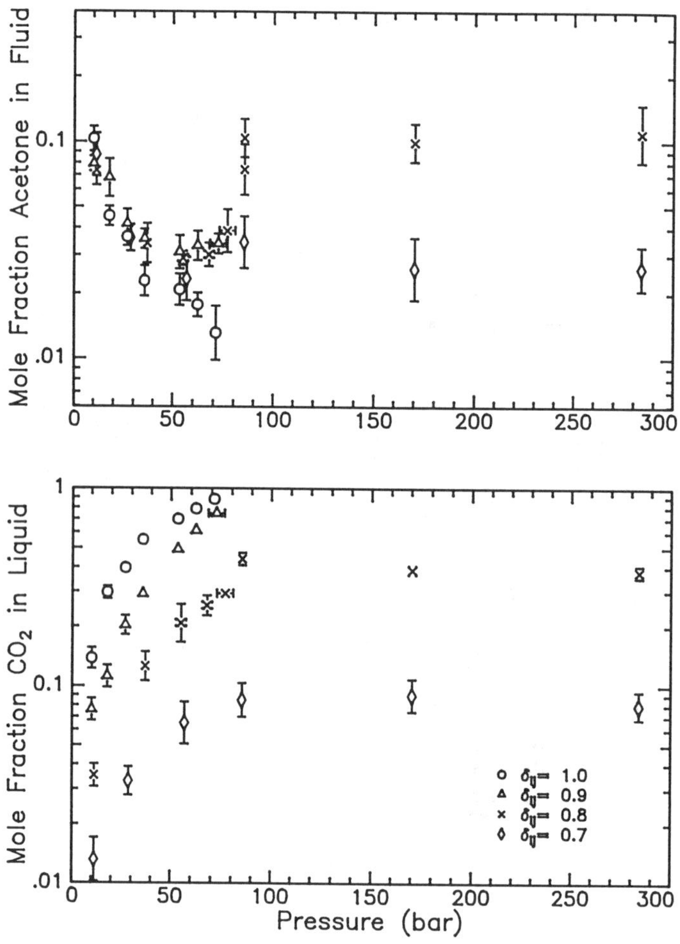

Figure 2. Gibbs-ensemble Monte Carlo results for a system with pure component parameters corresponding to acetone and carbon dioxide at T = 313 K, with varying δ_{ij}.

was observed for δ_{ij} = 0.80 and 0.70. The mole fraction of CO_2 in the acetone-rich liquid phase is a monotonic function of the unlike-pair interaction strength at all pressures. A rather small reduction of a few percent in the magnitude of unlike-pair interaction forces has a very strong effect on the concentration of CO_2 in the liquid phase. The solubility of CO_2 in the liquid phase is constant to within the simulation uncertainty for pressures higher than approximately 100 bar.

The mole fraction of acetone in the liquid phase is not a strong function of intermolecular interactions for pressures less than approximately 80 bar. For the immiscible systems, the shape of the mole fraction versus pressure curve is characteristic of solubility curves of solids in supercritical solvents (14), with a minimum around the pure solvent critical point. The effect of changing the intermolecular interactions is in the expected direction: the solubility of acetone in the fluid phase is lower (by a factor of 5) for the system with δ_{ij} = 0.70 relative to the one with δ_{ij} = 0.80. Again, a few percent change in the magnitude of the unlike-pair interactions has a greatly amplified effect on the solubility.

Carbon Dioxide / Water. Because of the use of Lennard-Jones potentials, which clearly do not describe several important characteristics of the water molecule, it was a matter of concern whether a good representation of the binary phase behavior for the system CO_2/water could be achieved. A high value of ϵ is required to obtain an approximation of the strong attractive forces present in liquid water. Thus, a temperature of 333 K corresponds to a reduced temperature of 0.62 for "LJ-water". This is below the triple-point temperature of the Lennard-Jones system (estimated at a reduced temperature of 0.68, 15). Difficulties with the Gibbs simulation method are expected at these low temperatures where the density of one of the phases becomes so high that no successful transfers are possible. For these reasons, simulations of binary phase equilibria for this system were performed at higher temperatures. The unlike-pair interaction parameter, δ_{ij}, was fitted to experimental data. The simulation results for the optimum value δ_{ij} = 0.81, shown in Table II, are in good agreement with experimental data, within their relatively large uncertainties. This is somewhat surprising, given the clear inadequacy of the LJ potential for describing the properties of liquid water, but may be related to the decreasing importance of hydrogen-

Table II. Results for the carbon dioxide / water binary system

T	P		CO_2 mole fraction		
(K)	(bar)	x_{MC} †	y_{MC} †	x_{EXP} ‡	y_{EXP} ‡
423	500	0.044±0.025	0.920±0.028	0.034	0.937
473	200	0.024±0.008	0.831±0.023	0.024	0.845
473	500	0.045±0.006	0.831±0.017	0.045	0.860
473	1000	0.066±0.022	0.826±0.016	0.061	0.855
473	1500	0.075±0.008	0.860±0.013	0.073	0.850

† With unlike pair interaction parameter δ_{ij} = 0.81
‡ From (16)

bonding at these temperatures. This inadequacy may result in δ_{ij} being a stong function of temperature if determined over a wide temperature range.

Similar simulation methodologies as for the carbon dioxide / acetone system were followed. A significantly higher number of attempted transfers/interchanges (1000-2000 per cycle) and a larger number of Monte Carlo steps ($3\text{-}4 \times 10^6$) were required for this system because of the higher densities of the coexisting phases.

Simulations of Ternary Systems

Simulations of ternary systems were performed using the pure component parameters in Table I and the cross parameters for the systems acetone/ CO_2 and water/CO_2 determined previously (δ_{ij} = 1 and 0.81 respectively). Because of expected difficulties similar to the ones mentioned for the water/CO_2 system, no attempt was made to simulate the system acetone/water near room temperature. Thus, we set the acetone/water interaction parameters to the values from the Lorenz-Berthelot rules with $\delta_{ij}=1$. Direct simulations of ternary phase equilibria have not been previously reported to the best of our knowledge.

A single temperature and pressure point (T = 333 K , P = 150 bar) for which experimental data have been reported (17) was selected for the calculations. The simulations proceeded in a similar manner as for the binary systems. Because of the low reduced temperature and high coexistence density of the water-rich phases, the simulations were quite long, requiring $3\text{-}5 \times 10^6$ Monte Carlo steps. A total number of 500 particles was used in the two regions. The number of attempted transfers was 2,000 per MC cycle of 500 attempted displacements.

The results of the calculations are compared with the experimental data in Figure 3a. The calculated phase envelope is qualitatively similar to the experimentally determined one for this system reported in (17). The main quantitative differences are in the slope of the tie lines, which run almost parallel to the CO_2/water side from the simulations, whereas they slope towards the water corner in the experimental results. This implies that the selectivity of the supercritical fluid for acetone is lower for the Monte Carlo results. To investigate the sensitivity of the calculated phase diagram to the acetone/water interaction parameter, we also performed a series of calculations with δ_{ij} for the acetone / water pair equal to 0.90. The results are shown in Figure 3b. The unfavorable acetone/water intermolecular interactions produce a significant change in the slope of the tie lines, with the selectivity of the supercritical fluid for acetone being greatly enhanced. The points at high water concentrations for both series of calculations are of doubtful accuracy, since the acceptance ratio of the transfer steps into the dense water phase was very low (only a few tens of successful transfers were obtained; for the δ_{ij} = 0.9 series, the concentrations of acetone and CO_2 in the water-rich phase were still decreasing at the end of the simulation runs). We include these points in Figure 3 because the overall shape of the calculated phase diagram is not significantly affected by these uncertainties.

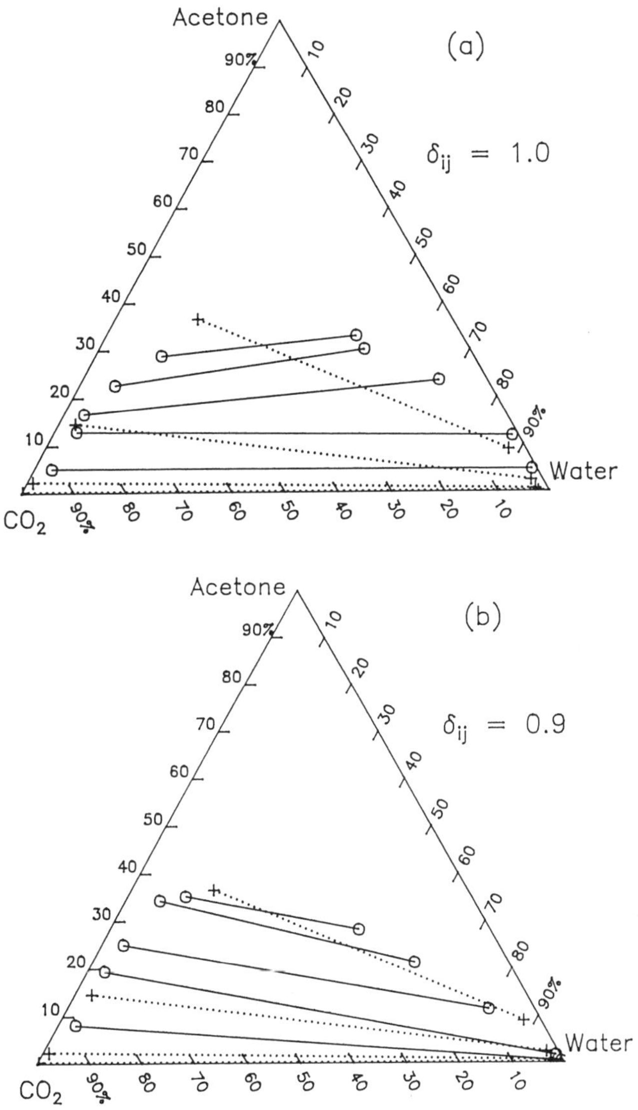

Figure 3. Gibbs-ensemble Monte Carlo (o——o) and experimental (+····+) results for the ternary system acetone / carbon dioxide / water at T = 333 K and P = 150 bar. Parameter δ_{ij} for the acetone-water pair is (a) δ_{ij} = 1.0 (b) δ_{ij} = 0.90. Experimetal results are from (17).

Discussion

The calculations reported in this paper and a related series of publications (8,9,10) indicate that it is now quite feasible to obtain reasonably accurate results for phase equilibria in simple fluid mixtures directly from molecular simulation. What is the possible value of such results? Clearly, because of the lack of accurate intermolecular potentials optimized for phase equilibrium calculations for most systems of practical interest, the immediate application of molecular simulation techniques as a replacement of the established modelling methods is not possible (or even desirable). For obtaining accurate results, the intermolecular potential parameters must be fitted to experimental results, in much the same way as parameters for equation-of-state or activity coefficient models. This conclusion is supported by other molecular-simulation based predictions of phase equilibria in similar systems (6). However, there is an important difference between the potential parameters in molecular simulation methods and fitted parameters of thermodynamic models. Molecular simulation calculations, such as the ones reported here, involve no approximations beyond those inherent in the potential models. The calculated behavior of a system with assumed intermolecular potentials is exact for any conditions of pressure, temperature or composition. Thus, if a good potential model for a component can be developed, it can be reliably used for predictions in the absence of experimental information.

Despite the important advances in recent years in methods for direct prediction of phase equilibria, molecular simulations still require significant amounts of computing power. Table III gives the computing time requirements for some of the calculations reported in this paper. The Cray X/MP is a state-of the-art supercomputer and the Sun 3/160 is an engineering workstation. While the computer time requirements of the Gibbs method are a small fraction of the corresponding requirements of previously available free-energy calculation techniques, these requirements are still far from small.

Table III. Computer time requirements

7 points on binary CO_2/acetone isotherm	1 hr on CRAY X/MP 5 days on Sun 3/160 FPA
5 points on ternary diagram at const. T and P	2 hrs on CRAY X/MP 10 days on Sun 3/160 FPA

In our opinion, the most significant potential contributions of molecular simulation techniques for predictions of phase equilibria are in obtaining exact results for model systems that can be used for (a) developing and testing of improved theoretical models (such as statistical-mechanics based equations of state and mixing rules) to a much higher level of rigor than that provided by experimental results and (b) investigating the qualitative effect of fundamental molecular parameters (such as size, energies of attraction) on the observed macroscopic behavior.

Acknowledgements

This material is based upon work supported by the U.S. National Science Foundation (grant CBT-8708734). Supercomputer time was provided by the Pittsburgh Supercomputing Center. Additional computational recourses were provided by the Cornell Theory Center.

Literature Cited

1. Allen, M.P., and Tildesley, D.J., *Computer Simulation of Liquids*, Clarendon Press, Oxford, 1987.

2. Frenkel, D., "Free-energy computations and first-order phase transitions", in G. Cicotti and W.G. Hoover, "*Molecular Dynamics Simulations of Statistical Mechanical Systems*", Proceeding of the International School of Physics "Enrico Fermi", course 97, Elsevier, 1986.

3. Widom, B., *J. Chem. Phys.* 1963, 39, 2808-2812.

4. Panagiotopoulos, A.Z.; Suter, U.W.; Reid, R.C., *Ind. Eng. Chem. Fundam.*,1986, 25, 525-35.

5. Fincham, D.; Quirke, N.; Tildesley, D.J., *J. Chem. Phys.*, 1986, 84, 4535-46.

6. Shing, K.S.; Chung, S.T., *J. Phys. Chem.*, 1987, 91, 1674-81.

7. Kofke, D.A.; Glandt, E.D., *J. Chem. Phys.*, 1987, 87, 4881-90.

8. Panagiotopoulos, A.Z., *Molec. Phys.*, 1987, 61, 813-26.

9. Panagiotopoulos, A.Z.; Quirke, N.;Stapleton, M.;Tildesley, D., *Molec. Phys.*, 1988, 63, 527-45.

10. Panagiotopoulos, A.Z., *Int. J. Thermophysics*, 1988 (in press).

11. Panagiotopoulos, A.Z., *Molec. Phys.*, 1987, 62, 701-19.

12. Reid, R.C.; Prausnitz, J.M.; Poling, B.E., *The Properties of Gases and Liquids*, 4th ed., McGraw-Hill, NY 1987.

13. Katayama, T.; Oghaki, K.; Maekawa, G.; Goto, M.; Nagano, T., *J. Chem. Eng. Jpn.* 1975, 8, 89-92.

14. Paulaitis, M.E.; Krukonis, V.J.; Kurnik R.T.; Reid, R.C. *Rev. Chem. Eng.*, 1983, *1*, 181-250.

15. Hansen, J.-P.; Verlet, L., *Phys. Rev.*, 1969, 184, 151-61.

16. Tödheide, K.; Franck, E.U., *Zeit. Phys. Chem. Neue Folge*, 1963, 37, 387-401.

17. Panagiotopoulos, A.Z; Reid, R.C., *Supercritical Fluids*; Squires, T.G. and Paulaitis, M.E., Eds; ACS Symposium Series No. 329; American Chemical Society: Washington, DC, 1987; pp 115-129.

RECEIVED May 2, 1989

Chapter 5

Spectroscopic Determination of Solvent Strength and Structure in Supercritical Fluid Mixtures: A Review

Keith P. Johnston[1], Sunwook Kim[2], and Jimmy Combes[1]

[1]Department of Chemical Engineering, The University of Texas, Austin, TX 78712
[2]Department of Chemical Engineering, Ulsan University, Korea

UV-visible, fluorescence, and IR spectroscopy have been used to characterize the solvent strength of pure and mixed supercritical fluid solvents, and to study solute-solvent interactions. The use of spectroscopic probes for the determination of clustering of pure and binary supercritical fluids about solutes is discussed. Spectroscopic studies of solvent strength and solute-solvent interactions are valuable for the development of molecular thermodynamic theory, engineering models, and for the molecular design of separation and reaction processes.

In reaction and separation processes, "solvent strength" is an important criterion in choosing an appropriate solvent. The term solvent strength is rather general and has been characterized by the dipole moment, dielectric constant, refractive index, solubility parameter, and polarizability per volume. Solvatochromic scales, which are based on shifts in the wavelength of maximum absorption for various indicators(1), are one of the most widely used and successful measures of solvent strength . This article reviews solvatochromic and other spectral data, which have become available only recently for supercritical fluids (SCFs). Other recent reviews discuss solution theory for the prediction of phase behavior(2,3). Our first objective is to provide an understanding of the solvent strength of a wide variety of SCFs using spectral data from several laboratories. The second objective is to provide a comprehensive analysis of spectroscopic studies of the degree of clustering in both pure and mixed SCF solvents.

The first understanding of the solvent power of a pure SCF came from solubility data, primarily for solids. However, at a given density, solubility differences among various solids are governed primarily by vapor pressure and

0097–6156/89/0406–0052$06.00/0

only secondarily by solute-solvent interactions in the SCF phase(4). This concept is best illustrated with the enhancement factor, E, which is the solubility, y_2, normalized by the solubility in an ideal gas, that is $E = y_2P/P_2^{sat}$. Here the vapor pressure effect is removed which provides a means to focus on solute-solvent interactions. For a relatively nonpolar SCF such as CO_2, solubilities often decrease by orders of magnitude as polar functional groups are added to a solute molecule. However, the enhancement factors for a series of solids of varying polarity fall typically within a range of only 1.5 orders of magnitude(4). Therefore, the enhancement factor and not the solubility should be used to understand solute-solvent interactions (solvent strength). Unfortunately, accurate enhancement factors are known for only a limited number of solids, since vapor pressures are often inaccurate or unavailable in the range of 10^{-6} to 10^{-2} Pa.

To understand solvent strength more effectively, an alternative approach is needed which focuses on the interactions in the SCF phase, that is a spectroscopic approach. In liquid solvents, spectroscopic solvatochromic scales are used to correlate and in some cases to predict solubility phenomena, solvent effects on reaction rate and equilibrium constants, free energies and enthalpies of acid-base complexes, retention indices in chromatography, absorption behavior in IR, NMR, ESR, and UV-visible spectroscopy, and finally, physiological and toxicological quantitative structure-activity relationships(1,5). Because of the widespread applicability of solvatochromic scales, they will play an important role in guiding the design and development of SCF technology, especially since relatively little is known about SCF solvents. It will be shown that solvatochromic parameters are influenced by the local solvent environment near the solute and thus describe solvent strength more effectively than bulk properties such as the solubility parameter or dielectric constant.

A unique feature of SCFs is that solvent molecules condense about solutes in regions where the compressibility is large. This physical condensation is often called clustering. Compressibility and clustering, in turn, influence macroscopic properties. For example, the partial molar volume of a solute such as naphthalene reaches thousands of mL/mol negative in a highly compressible SCF solvent(6,7). The small compressible fluid molecules condense about the highly polarizable solute, such that the local density is much larger than the bulk solvent density. This condensation or contraction is the cause of the highly negative partial molar volumes, which become more pronounced as the isothermal compressibility increases. These clusters are bound by van der Waals forces and are thus very different from other types of aggregates such as clathrates which are bound by specific chemical forces. Their size can reach values on the order of 100 molecules, which means that they extend over many coordination shells. While the partial molar volume, a macroscopic property, provides evidence of clustering, more detailed information has been obtained recently using spectroscopic techniques which probe solute-solvent interactions directly.

Solvent strength of supercritical fluids based on solvatochromic scales

The purpose of these scales is to provide a guide for choosing SCF solvents and co-solvents to achieve a desired solvent strength, for example in a separation or reaction process. An example of a solvatochromic scale is presented in Figure 1 for the UV-vis absorption of phenol blue in ethylene as a function of density at two temperatures (8). The scale is defined as the transition energy, $E_T = hc/\lambda_{max}$, where λ_{max} is the wavelength of maximum absorption. This probe is a red-shift probe in that the E_T shifts to lower energy as the polarity of the solvent increases. In ethylene E_T decreases with solvent density, indicating an increase in the solvent strength. Since ethylene has a dipole moment of zero, the solvent strength may be expressed as the number of polarizable molecules per unit volume. Near the critical density (7.75 gmol/L) small changes in pressure affect large changes in density and solvent strength. This adjustability of a SCF is one of its most important advantages in separation processes.

At a given density, the data indicate that the solvent strength is essentially independent of temperature, consistent with the fact that the polarizability is also independent of temperature. The "thermochromic effect" is minimal at constant density. However, it is significant at constant pressure in regions where density is a strong function of temperature.

Various properties that are often used to characterize solvent strength are listed for SCF C_2H_4, CF_3Cl, and liquid n-hexane in Table I. Because each property is tabulated for a constant value of E_T (= 52.0 kcal/mole), it should not vary among the solvents if it is to be a good indicator of solvent strength. Notice that it takes enormous pressures for the SCFs to achieve the solvent strength of hexane. Solvent density (either mass or molar density) is a mediocre indicator of solvent strength for various fluids, as is evident based on the large difference between CF_3Cl and n-hexane. However, for a given SCF, density is a much more meaningful indicator than pressure and has been used to correlate solubilities(9,10). The solubility parameter and the polarizability/volume (α/v) are good indicators of solvent strength in that they are relatively constant at constant E_T. The important conclusion is that the solvent strength is a function of both density and polarizability, and can be described by the product of the two, that is α/v. The use of polarizability or density alone is insufficient, although this concept has not been well-recognized in the literature.

With the use of solvatochromic probes, other non-specific forces (dispersion, dipole-induced dipole, and dipole-dipole) and specific acid-base forces have been explored in SCF solvents. In an effort to compare liquid and supercritical carbon dioxide, Hyatt(11) measured UV-visible spectra of several solvatochromic probes. There was little difference between the E_T in the liquid and SCF states; however, the data can not be interpreted fully since the density and the pressure were not given at the supercritical condition. The results indicated that the

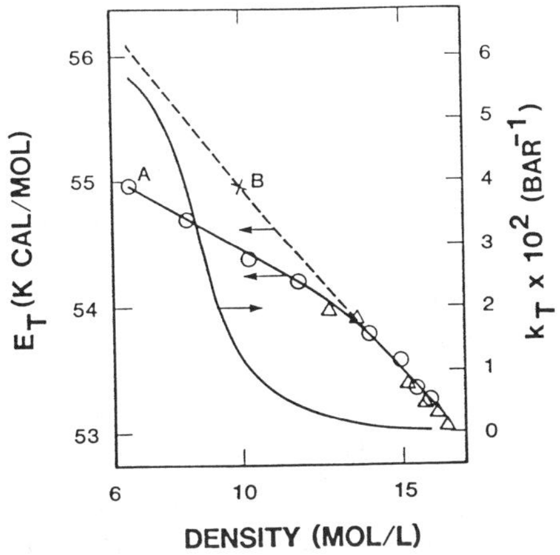

Figure 1. Transition energy (E_T) of phenol blue in ethylene and isothermal compressibility of ethylene versus density (O: 25 °C, Δ: 10 °C, - - - : calculated E_T at 25 °C using McRae model). (Reprinted from ref. 8. Copyright 1987 American Chemical Society.)

Table I. Properties of SCF solvents versus n-hexane at a constant value of solvent strength as defined by E_T for phenol blue[(1)]

Solvent	T	P	ρ	δ[(2)]	α/v[(3)]
	(° C)	(bar)	(g/cc)	$(cal/cc)^{1/2}$	
C_2H_4	25	1700	0.57	7.0	0.052
CF_3Cl	40	1300	1.95	6.9	0.051
$n\text{-}C_6H_{14}$	25	1	0.66	7.3	0.057

(1) E_T = 52.0 kcal/mol
(2) Hildebrand solubility parameter
(3) polarizability per volume

polarizability/volume of CO_2 is even less than that of fluorocarbons. In order to investigate the acidity of CO_2, shifts in the IR C=O stretching frequency were determined for acetone and cyclohexanone. The acidity of CO_2 was insignificant as it gave the same results as nonpolar non-acidic liquid solvents. Based on shifts in the N-H stretching frequency for pyrrole, it was found that CO_2 exhibited a modest hydrogen bond basicity on the order of an ether or acetate.

The association of decanol has been investigated in CO_2 using near-IR spectroscopy(12). The concentrations of the monomeric and associated decanol were determined independently as a function of pressure and temperature.

Sigman et al.(13) determined π^* (dipolarity-polarizability) and β (basicity) parameters for supercritical carbon dioxide as a function of density. As the density of SCF CO_2 is decreased, the π^* values decrease smoothly towards the value obtained in a vacuum. The π^* parameters were the average values of ten different probes in carbon dioxide, while the π^* values of Hyatt(11) and Yonker and co-workers(14,15,16,17) were obtained using a single probe. The trends in the π^* parameters were similar for the various studies, which suggests that the use of a single probe is reasonable in view of the complexities involved in performing experiments at elevated pressures.

Spectroscopic solvatochromic studies have been performed as a function of temperature and pressure for a variety of pure supercritical fluid solvents in addition to CO_2. Kim and Johnston (6,18) measured shifts in the absorption wavelength of phenol blue in SCF CO_2, C_2H_4, CHF_3 and CF_3Cl (see Figure 2). At a given reduced density and temperature, the solvent strengths (E_Ts) are similar for CO_2 and ethylene, but are much higher for SCF fluoroform due to the acidic hydrogen. The solvent strength is a little weaker for CF_3Cl than for CO_2 and C_2H_4 because it has a lower molar critical density. Phenol blue, which is quite sensitive to hydrogen bond (proton) donors, is useful for measuring acidities. The insignificant acidity of dry SCF CO_2 was consistent with the above IR results for the ketones. This explains the fact that the acidity of CO_2 has a limited effect on the solubility of the weakly basic solid acridine, relative to ethane at the same density. However, the acidity of CO_2 does become important in the presence of strong bases such as ammonia and amines, as solid carbamates are formed. The solvent strength of CHF_3 (at a density of 9 mol/L) is comparable to n-hexane and may be adjusted to equal that of CCl_4 by raising density to 17 mol/L. The E_T value of fluoroform approaches that of CCl_4 because of its acidity, even though it has a smaller polarizability/volume . It appears that the polarity component of solvent strength for CHF_3 is considerably larger than for other fluids with critical temperatures below 50 C, which is interesting for practical applications

Smith et al.(14) studied other polar fluids, including ammonia (T_c = 133 C), which has a much higher critical temperature than the above fluids. Ammonia's π^* value varies from that of n-hexane to tetrahydrofuran. At a given value of reduced temperature and pressure, it is a much more potent solvent than CO_2 because of its greater polarity and the additional thermal energy.

Acid-base interactions have been explored as a function of density in SCFs using solvatochromic probes. The basicity of SCF CO_2 was found to be relatively constant with respect to density(13), although the polarizability/volume is a linear function of density. The acidity of SCF CHF_3 was also observed to be density insensitive, over a range from 4 to 18 mol/L(8). Although it appears that hydrogen-bond interactions become fairly saturated at a relatively low density, further experimental and theoretical work is required to understand this behavior more quantitatively.

The final application of solvatochromic solvent strength scales is the correlation of reaction rate and equilibrium constants in SCF solvents. Solvatochromic scales are often quantitative indicators of the solvent effect on rate constants for a variety of reaction mechanisms(1). In a SCF, this solvent effect may be achieved conveniently with a single solvent using pressure. Based on solvatochromic data, it was predicted that an activation volume can reach thousands of mL/mol in a SCF(8). This prediction was confirmed for various types of reactions(19-21). For example, the solvatochromic parameter E_T for phenol blue

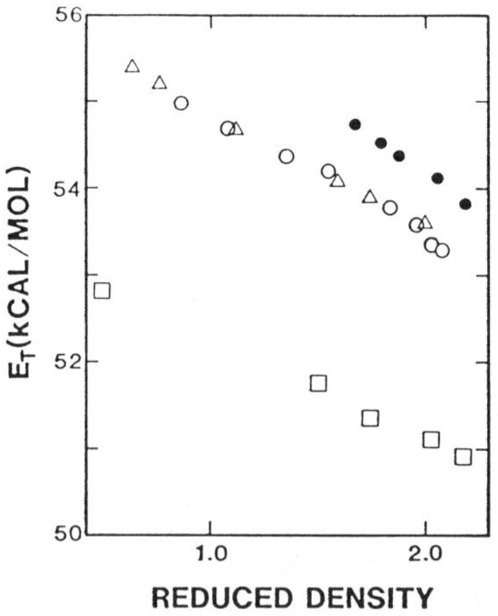

Figure 2. E_T of phenol blue as an indicator of solvent strength for various supercritical fluids (○: C_2H_4 – 25 °C, Δ: CO_2 – 45 °C, □: CHF_3 – 40 °C, •: CF_3Cl – 30 °C). (Reprinted from ref. 18. Copyright 1986 American Chemical Society.)

has been used to correlate linearly the reaction rate constant for the unimolecular decomposition of α-chlorobenzyl-methyl ether in SCF 1,1-difluoroethane(19), the selectivity for the parallel Diels-Alder additions of methyl-acrylate and cyclopentadiene in SCF CO_2(20), and the equilibrium constant for the tautomerization of 2-hydroxypyridine and 2-pyridone in SCF 1,1-difluoroethane(21). Since few reactions have been studied at SCF conditions, these scales provide a useful basis to guide future studies.

Solute-solvent interactions and clustering in the microscopic solute environment

Pure solvents Both UV-vis and fluorescent probes, which exhibit solvatochromic shifts, have been used to study clustering. Solvatochromic shifts are caused by the same types of solute-solvent intermolecular forces (i.e. dispersion, induction, and dipole-dipole forces) that influence solubilities, interacting over the same range. Consequently, the values of clustering determined spectroscopically are appropriate for considering the effect of clustering on solubilities.

The degree of clustering may be determined by a direct comparison of the experimental value of the solvatochromic parameter, for example, E_T, with the value which is calculated for a homogeneous polarizable dielectric. This calculation has been done using the McRae-Bayliss model(22) which is based on the continuum reaction field theory of Onsager.

The determination of the solvent local density is illustrated in Figure 1 for the probe phenol blue in ethylene. At higher densities, where the fluid is relatively incompressible, the data and model agree. At lower densities, the solvatochromic shift is larger (E_T is smaller) than the calculated value because of clustering of the highly compressible solvent about the solute. At a given value of solvent strength (E_T), the local density is given by point B for the bulk density at point A. This general technique for the determination of clustering, as a deviation from the calculated linear behavior, will be applied below for additional UV-visible and fluorescence solvatochromic studies.

The spectroscopic data may be interpreted using an equation that was derived form Kirkwood-Buff solution theory. The number of solvent molecules in excess of the bulk value, n_2^e, may be defined in terms of a local volume about the solute V_{12} as

$$n_2^e = (\rho_{12}^l - \rho) \cdot V_{12} \tag{1}$$

where ρ_{12}^l is the local solvent density about the solute. The fundamental equation which describes clustering is(8,18,23)

$$(\rho_{12}^l - \rho) \cdot V_{12} = \rho\,(kT \cdot k_T - \bar{v}_2^{\infty}) \tag{2}$$

or equivalently (24)

$$n_2^e = \rho\,(kT \cdot k_T - \bar{v}_2^{\infty}) \tag{3}$$

where k_T is the isothermal compressibility, and $\bar{v}_2^{\infty}$ is the partial molar volume of the solute at infinite dilution. Since $\bar{v}_2^{\infty}$ scales as k_T, clustering also scales as k_T.

This clustering theory may be tested using solvatochromic data. By assuming a linear relation between v_2^{∞} and k_T, in accordance with experimental results(6,25), an important result is obtained

$$(\rho_{12}^l - \rho)/\rho = a'k_T + b' \tag{4}$$

where a' and b' are functions of temperature. This result describes the local densities for phenol blue in ethylene as shown in Figure 3, and for other supercritical fluids which do not form hydrogen bonds(8,19). At the lowest pressure, that is $P_r = 1.33$, the compressibility is largest and the local solvent density exceeds the bulk density by 60%. The cluster sizes reach 100 solvent molecules based on partial molar volume data for naphthalene in carbon dioxide(24). As pressure increases, this density differential or clustering decays rapidly, as it scales directly with k_T. At high densities, the fluid is like a liquid in that the free volume, compressibility, and thus clustering are all much smaller.

The distinction between n_2^e and the local density, ρ_{12}^l, can be confusing, yet it is very important. To illustrate this, the local density, ρ^l, of ethylene ($T_c = 9$ C) about phenol blue is plotted versus pressure at 25 C in Figure 4. Consider an increase in pressure from 50 bar ($P_r \sim 1$) to 200 bar. Notice that ρ^l and thus the number of solute-solvent interactions increase monotonically primarily because the bulk density increases. However, the clustering, which is proportional to $\rho^l - \rho$, diminishes and eventually disappears. Notice that n_2^e is not just a cluster size, but a cluster size in excess of the bulk density. Again, the number of solute-solvent interactions increases with bulk density even though the degree of clustering decreases. This is a subtle, yet very important observation. For example, consider the magnitude of the attractive part of the chemical potential of a solute. It increases approximately in proportion to the local density (for a van der Waals fluid, μ_i^{att} is proportional to ρ), even in regions where clustering decreases. Therefore μ_i^{att} is related more closely to local density than to n_2^e. This distinction between n_2^e and ρ^l is important in evaluating the effect of clustering on thermodynamic and kinetic properties.

Solvatochromic π^* parameters have been obtained for SCF CO_2, SF_6, C_2H_6, N_2O and NH_3 as a function of density, and have been analyzed using the Onsager reaction field function(15), i.e. $L(n2)=(n2-1)/(2n2+1)$ as shown in Figure 5. In order to explain the plot of π^* versus the reaction field, two straight lines were

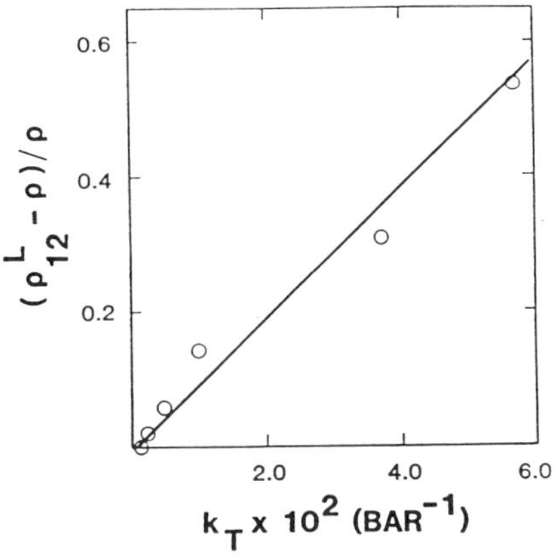

Figure 3. Clustering of ethylene about phenol blue as a linear function of the isothermal compressibility at 25 °C. (Reprinted from ref. 8. Copyright 1987 American Chemical Society.)

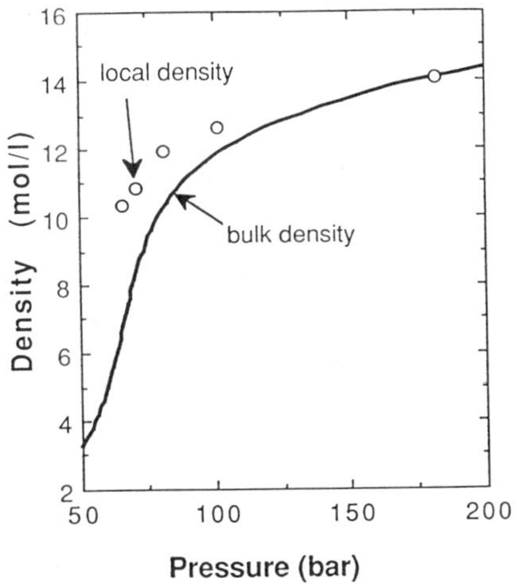

Figure 4. Comparison of local density and bulk density versus pressure based on the solvatochromism of phenol blue in ethylene at 25 °C (data from ref. 8.).

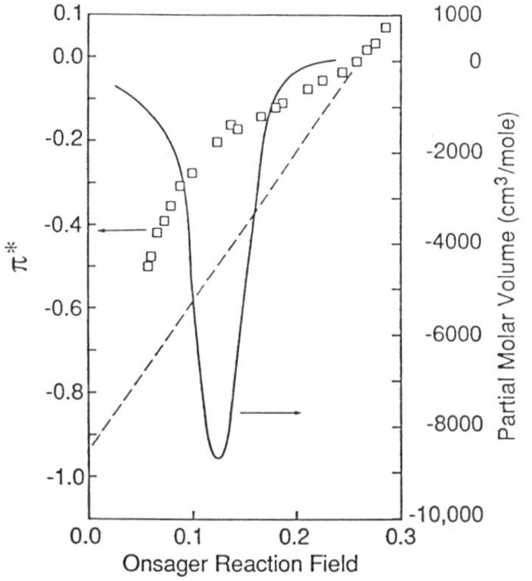

Figure 5. Clustering of CO_2 about a solute-consistency between spectroscopic data and partial molar volume data (□: π^* data for 2-nitroanisole at 35 °C from ref. 15, - - -: Onsager-McRae reaction field theory, —: $\bar{v}_2$ data for naphthalene at 35 °C from ref. 6).

constructed (not shown in the figure) with a discontinuous intersection. The intersection occurred at a density below the minimum in $\bar{v}_i$ for naphthalene in CO_2. A theoretical basis has not been provided for this type of discontinuity in the one-phase SCF region.

Sigman et al. (13) considered their π^* data to be a continuous function of density. The data in Figure 5 are re-evaluated by assuming continuous behavior. A linear asymptote is shown for the incompressible region at the highest densities, as expected for Onsager reaction field theory. Again, the degree of clustering is given by the difference between the data and the asymptotic line for the homogeneous fluid. Rigorously, the comparison would be made with density as the abscissa, but the results would be very similar. It is a maximum at the point where k_T is a maximum and decreases with density (as does k_T) in both directions. In Figure 1, the spectral E_T data were obtained only at pressures above the maximum compressibility, so the decay in clustering was determined only in one direction. This interpretation of the π^* data is consistent with the Kirkwood-Buff theory in eqs. 2-4, the $\bar{v}_i$ data for naphthalene in CO_2, the phenol blue data in Figure 1, and the fluorescence data which will be discussed next.

Clustering has also been observed in fluorescence spectroscopy based on spectral shifts for the charge transfer emission of (N,N,-dimethylamino)benzonitrile (DMABN) in SCF CHF_3(26). The dipole moment of DMABN is enormous (7.6D) and is much larger than for the solvatochromic probes described above. A Langmuir adsorption model was used to describe the clustering of solvent as a function of density. In Figure, 6, the data are interpreted with the same local density approach that was described in Figure 1. The results show that the degree of clustering, that is the difference between the experimental and calculated values of the spectral shift, is extremely pronounced. In the critical region, the clustering increases then decreases rapidly as a function of the solvent compressibility, in accordance with the model, eqs 2-4. The maximum degree of clustering, $(\rho^l - \rho)/\rho$, is 1.5 which is more than twice as large as that for phenol blue. This is due to the much higher polarity of DMABN, which in turn causes larger solute-solvent intermolecular forces.

The relationship between clustering and solute-solvent interaction forces may be explained thermodynamically . We begin with the triple product relationship

$$\bar{v}_2^{\infty} = v\, k_T \cdot n\, (\partial P/\partial n_2)_{T,V,n_1} \quad (5)$$

Substitution of eq 5 into eq 3 yields

$$n_2^e = \rho\, k_T\, [kT - (n/\rho)\, (\partial P/\partial n_2)_{T,V,n_1}] \quad (6)$$

Substitution of the attractive part of the van der Waals equation of state yields

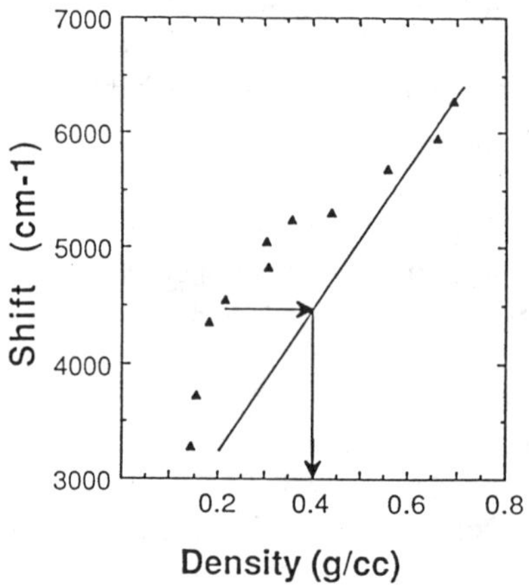

Figure 6. Clustering of CHF_3 about (dimethylamino)benzonitrile at 50 °C based on solvatochromic shifts in fluorescence (▲: data from ref. 26, —: Onsager reaction field theory).

$$n_2^e \sim \rho\, k_T\, (kT + 2\, a_{12}\rho) \quad (7)$$

where a_{12} is the solute-solvent interaction constant. An important concept is that the degree of clustering is related quantitatively to the solute-solvent interaction strength multiplied by the isothermal compressibility. This is the reason that clustering is greater about the highly polar solute DMABN than phenol blue at a given value of k_T.

Recently, fluorescence spectra have been obtained for naphthalene and pyrene in supercritical fluid carbon dioxide and ethylene, respectively (27,28). The technique is exciting in that it provides the first information concerning solute-solute clustering at extremely low concentrations. Solute-solvent clustering was also characterized. Another benefit of this approach is that it is possible to study a wide variety of solutes, for example those for which solubilities and partial molar volumes are known. Solute-solute clustering, as well as solute-solvent clustering, has been observed recently using computer simulation and integral equation calculations of radial distribution functions(29,30,31).

The fluorescence data, specifically ratios of emission intensities, I_1/I_3, were used to show that solute-solvent clustering increases as the temperature approaches the critical temperature (I_1/I_3 is a measure of solvent strength). In liquid organic solvents, it is known that the ratio I_1/I_3 may be correlated linearly with another solvent strength parameter, π^*, by the relationship

$$I_1/I_3 = 0.64 + 1.109\, \pi^* \quad (8)$$

It was found that the calculated values of I_1/I_3 from this equation and the π^* data of Yonker and co-workers were lower than the experimental values(27), which indicates clustering. The analysis is complicated by the fact that the π^* values for SCF CO_2 are also influenced by clustering, as was explained above in the discussion of Figure 5. It appears that there is a greater degree of clustering about pyrene than 2-nitroanisole (a π^* indicator), which is consistent with the larger polarizability for pyrene.

Mixed solvents The addition of a small amount of a co-solvent to a supercritical fluid can increase solubilities of certain substances from several percent to several orders of magnitude (32,33,34,35). Spectroscopic data, which have been obtained recently, indicate that preferential solvation by a co-solvent contributes to the large increases. The co-solvents acetone, methanol, ethanol, and n-octane, were investigated by Kim and Johnston(36) using the solute phenol blue as a solvatochromic indicator. In Figure 7, it is apparent that the red shift (solvent strength) exceeds the value which is obtained from linear behavior, i.e. the concavity is positive. This means that the local concentration of co-solvent near the solute

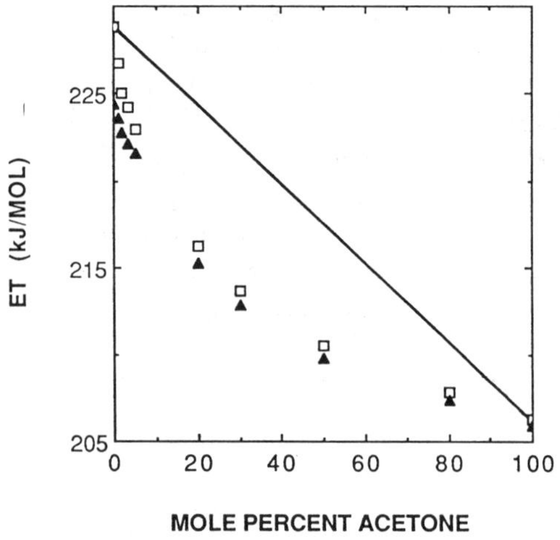

Figure 7. Transition energy of phenol blue in binary mixtures of acetone and CO_2 at 35 °C (□: 80 bar, ▲: 200 bar [ref. 36]).

exceeds the bulk value. The data were used to estimate local compositions of CO_2 and a co-solvent about an infinitely dilute solute using a procedure which will be described briefly.

The residual transition energy of a solute in a pure solvent i, ΔE_{Ti}, is defined by

$$\Delta E_{Ti} = E_{Ti} - E_T^{ideal\ gas} \tag{9}$$

at constant temperature and pressure. It is assumed that each solute-solvent pair interaction causes a fixed solvatochromic shift equal to $\Delta E_{Ti}/z_{i2}$ where z_{i2} is the coordination number for a pure solvent i about the solute, component 2. The interactions between the primary solvent(1) and co-solvent(3) are neglected since this is a second order effect compared with the solvent-solute and co-solvent-solute interactions. For a given component of the solvent, i, the residual transition energy is a product of two factors: the transition energy for a solute-solvent pair interaction, $\Delta E_{Ti}/z_{i2}$, and the number of i molecules about the solute, $x_{i2}z_m$, where x_{i2} is the local mole fraction of i about the solute in the first coordination shell and z_m is the coordination number for the solvent mixture about the solute. The sum of the contributions for each component of the solvent gives the residual E_T in the mixture, that is

$$\Delta E_{Tm} = x_{12}\, z_m\, \Delta E_{T1} / z_{12} + x_{32}\, z_m\, \Delta E_{T3} / z_{32} \tag{10}$$

Experimental solvatochromic data, ΔE_{Tm}, ΔE_{T1} and ΔE_{T3}, and theoretical expressions for z_m/z_{12} and z_m/z_{32} may be used to calculate the local compositions (local mole fractions) x_{12} and $x_{32} = 1 - x_{12}$.

For the cosolvents, acetone, ethanol, methanol, and octane, the estimated local compositions exceed the bulk values since the co-solvent interacts more strongly with the solute than does SCF CO_2(36). This preferential solvation by the co-solvent is presented in Figure 8 using the normalized function $X_Q = (x_{32}/x_{12})(x_1/x_3)$, which is unity when the local and bulk concentrations become the same. For this density range, the compressibility is largest at the lowest density and decreases with density. The amount of local ordering increases with an increase in solvent compressibility, which is consistent with computer simulation results for square-well molecules(37). For the phenol blue-acetone-CO_2 system, these local compositions agreed with those calculated using a density dependent local composition model(36,38), along with binary solute-CO_2 and solute-cosolvent interaction parameters, which were obtained independently from solubility data. This suggests that the inclusion of density dependent local composition mixing rules can improve models of phase behavior in highly compressible polar supercritical fluid solutions.

A simplified version of eq 10 was used to estimate the local compositions of SCF CO_2 and 2-propanol about the solvatochromic probe 2-nitroanisole (39)

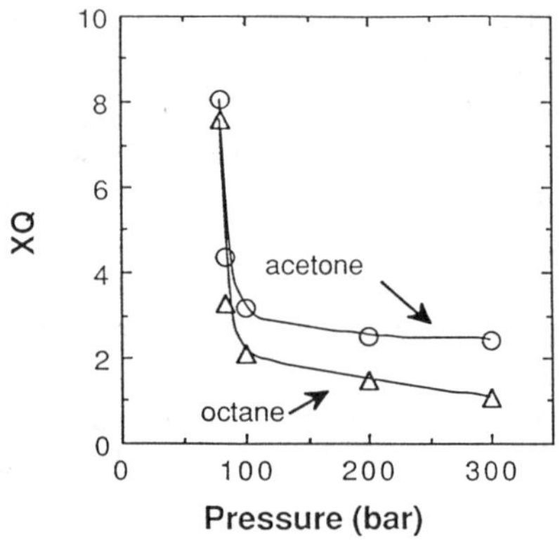

Figure 8. Preferential solvation of phenol blue by a cosolvent versus CO_2. The ordinate $X_Q = (x_{32}/x_{12})(x_1/x_3)$ is normalized, as it goes to unity when local compositions equal bulk values. ($x_3 = 0.01$.)

$$\pi_m^* = x_{12}\,\pi_1^* + x_{32}\,\pi_3^* \tag{11}$$

where π_m^* is the Kamlet-Taft solvatochromic parameter for the mixture, and π_1^* and π_3^* are for the pure liquids. The parameters π^* and ΔE_T are similar kinds of parameters in that both represent a difference from a reference value. Eq. 11 is based on the approximation that the coordination numbers are the same for the SCF mixture and the pure liquid solvents, i.e. z_m/z_{12} and z_m/z_{32} are unity. However, the coordination number in a SCF mixture varies with density, especially in the near critical region. Nevertheless, eq 11 simplifies data analysis and produces trends that are consistent with eq 10 and computer simulation data.

Conclusions

The wide variety of spectroscopic studies in the last five years provides us with a much sounder understanding of solvent strength and solute-solvent interactions in the SCF state. This has improved the ability to design solvent and cosolvent systems at the molecular level to achieve desired objectives in separation and reaction process.

Solute-solvent clustering, as determined both from $\bar{v}_2$, and solvatochromic shifts for UV-visible and fluorescent probes, has been described by a unified theoretical result, eq 2. The degree of clustering is related quantitatively to the solute-solvent interaction strength multiplied by the isothermal compressibility. Large pressure effects on chemical potentials, solubilities, rate constants, and equilibrium constants are all a manifestation of and can be related to the isothermal compressibility.

Acknowledgment

Acknowledgement is made to the National Science Foundation under grant no. CBT 8513784, the Camille and Henry Dreyfus Foundation for a Teacher-Scholar Grant, and the Separations Research Program at the University of Texas.

Literature Cited

1. Reichardt, C. Solvent Effects in Organic Chemistry; Verlag Chemie, Weinheim, 1988.
2. Johnston, K.P.; Peck, D.G.; Kim, S. Ind. Engr. Chem. Res. 1989, in press.
3. Lira, C.T., Am. Chem. Soc. Symp. Ser. 1988, no. 366, 1.
4. Dobbs, J.M.; Johnston, K.P. Ind. Engr. Chem. Res. 1987, 26, 1476.
5. Taft, R.W.; Abraham, M.H.; Doherty, R.M.; Kamlet, M. J. Nature London 1985, 313(31), 384.
6. Eckert, C. A.; Ziger, D. H.; Johnston, K. P.; Kim, S. J. Phys. Chem. 1986, 90, 2738.

7. Eckert C.A.; Ziger D.H.; Johnston, K.P.; Ellison, T.K. Fluid Phase Equilibria 1983, 14, 167.
8. Kim, S.; Johnston, K. P. Ind. Eng. Chem. Res. 1987, 26, 1206.
9. Schmitt, W. J., The Solubility of Monofunctional Organic Compounds in Chemically diverse Supercritical Fluids, Ph.D. Diss., M.I.T. (1984).
10. Kumar, S.K.; Johnston, K.P. J. Supercritical Fluids, 1988, 1, 15.
11. Hyatt, J.A. J. Org. Chem. 1984, 49, 5097.
12. Schneider, G.M.; Ellert, J., Haarhaus, U.; Holscher, I.F.; Katzenskiohling G.; Kopner, A.; Kulka, J.; Nickel, D.; Rubesamen, J.; Wilsch, A. Pure and Applied Chem. 1987, 59, 1115.
13. Sigman, M.E.; Lindley, S.M.; Leffler, J.E. J. Am. Chem. Soc. 1985, 107, 1471.
14. Smith, R.D.; Frye S.L.; Yonker C.R.; Gale, R.W. J. Phys. Chem. 1987, 91, 3059.
15. Yonker, C.R.; Smith, R.D. J. Phys. Chem. 1988, 92, 235.
16. Yonker, C.R.; Frye, S.L; Kalkwarf, D.R.; Smith, R.D. J. Phys. Chem. 1986, 90, 3022.
17. Frye, S.L.; Yonker, C.R.; Kalkwarf, D.R.; Smith, R.D. ACS Symp Ser. 1986, No. 329, 29.
18. Kim, S.; Johnston, K.P. ACS Symp. Ser. 1986, No. 329, 42.
19. Johnston, K.P.; Haynes, C. AIChE J. 1987, 33, 2017.
20. Kim S.; Johnston K. P. Chem. Eng. Comm.,1987, 63, 49.
21. Peck, D.G.; Mehta A. J.; Johnston K. P. J. Phys. Chem. 1989, in press.
22. McRae, E.G. J. Phys. Chem. 1957, 61, 562.
23. Kim, S. Molecular Thermodynamics at Supercritical Fluid Conditions: Solvent Effects on Reaction Kinetics and Separation Processes, Ph. D. Diss., Univ. Texas (1986).
24. Debenedetti, P. G. Chem. Eng. Sci. 1987, 42, 2203.
25 van Wasen, U.; Schneider, G.M. J. Phys. Chem. 1980, 84, 229.
26. Kajimoto O.; Futakami M.; Kobayashi T.; Yamasaki K. J. Phys. Chem. 1988, 92, 1347.
27. Brennecke, J.F.; Eckert, C.A. Proc. Internt'l. Symp. Supercritical Fluids, Soc. Francaise de Chemie 1988, p. 263.
28. Brennecke, J.F.; Eckert, C.A., this symposium.
29. Cochran, H.D.; Lee, L.L., this symposium.
30. Shing, K. S.; Chung, S. T. J. Phys. Chem. 1987, 91, 1674.
31. Debenedetti, P.G.; Mohamed, R.S. Proc. Internt'l. Symp. Supercritical Fluids, Soc. Francaise de Chemie 1988, 335.
32. van Alsten, J. G. Ph.D. Diss., Univ. Illinois (1986).
33. Dobbs, J. M.; J. M. Wong; R. J. Lahiere; Johnston, K.P. Ind. Eng. Chem. Res. 1987, 26, 56.
34. Schmitt, W.J.; Reid, R.C. Fluid Phase Equilibria 1986, 32, 77.
35. Walsh, J. M.; Ikonomou, G.D.; Donohue, M.D. Fluid Phase Equilib. 1987, 33, 295.
36. Kim, S.; Johnston, K.P. AIChE J. 1987, 33, 1603.
37. Lee, K. H.; S. I. Sandler; Patel N.C. Fluid Phase Equilibria 1986, 25 31.
38. Johnston, K.P.; Kim, S.; Wong, J.M. Fluid Phase Equilibria 1987, 38, 39.
39. Yonker, C.R.; Smith, R.D. J. Phys. Chem. 1988, 92, 2374.

RECEIVED May 2, 1989

PHASE BEHAVIOR

Chapter 6

Partition Coefficients of Poly(ethylene glycol)s in Supercritical Carbon Dioxide

Manouchehr Daneshvar and Esin Gulari

Chemical and Metallurgical Engineering Department, Wayne State University, Detroit, MI 48202

The coexisting phase compositions of poly(ethylene glycol) - carbon dioxide mixtures have been measured. Three polymer samples with average molecular weights of 400, 600, and 1000 were studied. The measurements were conducted at 313 and 323 K over a pressure range up to 28 MPa in a high pressure apparatus with countercurrent circulation. At equilibrium, both phases were sampled simultaneously and their compositions were determined. Poly(ethylene glycol) (PEG) amounts were measured by a colorimetric technique and molecular weight distribution of extracts were determined by Fast Atom Bombardment Mass Spectrometry (FAB-MS). It was observed that PEGs have appreciable solubilities in supercritical CO_2. At a given pressure and temperature, the lower molecular weight PEGs exhibited much higher solubility than the ones with higher molecular weights. The overall partition coefficients of the samples with different molecular weights indicated that the relative amounts of different fractions in a given mixture can be varied by changing the operating conditions. The individual n-mer partition coefficients for a sample of PEG(600) were calculated by combining the results from FAB-MS and colorimetric techniques. Two different regions of yield and selectivity with respect to the degree of polymerization were identified.

The polymers of ethylene oxide with the general formula of $H\text{-}(OCH_2CH_2)_n\text{-}OH$ are divided into two catagories based on their molecular weights. The polymers with an average molecular weight in the range 200-20,000 are referred to as PEGs. The ethylene oxide polymers or poly(ethylene oxide) resins are the higher molecular weight members of this series. PEGs are soluble in water and they are used in cosmetics, lubricants and pharmaceuticals depending on their molecular weights.

0097–6156/89/0406–0072$06.00/0

In general, synthetic polymers exhibit a wide molecular weight distribution. Their separation into narrower molecular weight fractions is a fairly difficult task. Distillation is not an effective separation technique for these materials because of their low vapor pressures. The solubility of different fractions of a parent polymer in conventional liquid solvents are normally too high which makes the liquid extraction a nonselective technique for these materials. Fractionation of polymers by using supercritical fluids presents unique advantages and has attracted the attention of several investigators in the past few years (1-5).

Supercritical fluid (SCF) extraction is a separation technique in which the extractant is a dense gas at temperatures and pressures above its critical point. The widespread interest in supercritical fluid processing is primarily due to the unique characteristics of the solvent. The solvation power of the solvent is controlled by its density; in the vicinity of the critical point, slight changes in pressure or temperature can produce large changes in density. In addition to the control of solvent power, there are other advantages in using a SCF as solvent. High diffusivity and low viscosity of a SCF cause it to penetrate into a polymer phase more effectively than a liquid solvent. The separation of the solute from the solvent can be achieved by pressure reduction or cooling or both. Therefore, it is possible to achieve a very selective and efficient fractionation.

The feasibility of a supercritical polymer fractionation is determined by the variations of the solubility of a homologous series of compounds. Qualitatively, the solubility of a polymer decreases as the degree of polymerization increases. This concept has been illustrated by Krukonis (2) in supercritical fluid fractionation of some heat-labile, low molecular weight polymers. Another observation made by Kumar et. al. (3) is that, at a given pressure and temperature, the partition coefficient of an n-mer between the polymer phase and the SCF phase depends only on its chain length and doesn't depend on the compositions of polymers of different chain lengths in the two phases at equilibrium. This is an important observation because it implies the possibility of achieving optimum conditions for the separation of a given molecular weight fraction.

In the literature, very little data on the solubility of polymers in supercritical fluids have been reported. The data are limited because experiments require precise detections of composition and molecular weight distribution. The study of the liquid solutes is even more complicated than solid solutes. For solid solutes, the solid phase can be assumed to remain pure and only the supercritical fluid phase is then sampled. In case of liquid solutes, the supercritical fluid is appreciably soluble in the solute and therefore both phases must be sampled.

In this paper, the coexisting phase compositions of different average molecular weight PEG - carbon dioxide systems are presented. The equilibrium compositions have been measured at 313 and 323 K over a pressure range up to 28 MPa. The partition coefficients based on the average molecular weights of the polymer are reported. PEG with an average molecular weight of 600 is used to study the mass based partitioning of individual n-mers between the SCF phase and the polymer phase. The experimental set up designed for sampling of both phases is described. The calibration data are reported for determining very small amounts of PEG in water by adapting a colorimetric technique. The effect of the molecular weight distribution of the parent polymer on the solubility data are discussed.

Apparatus and Procedure

The apparatus was designed to obtain vapor-liquid equilibrium data at temperatures up to 373 K and pressures up to 35 MPa. A schematic diagram of the equipment is shown in Figure 1. A high pressure metering pump introduces CO_2 to the system from a supply cylinder equipped with a siphon tube. Equilibrium is achieved by circulating both the liquid and vapor phases at approximate flow rates of 8 ml/min. The SCF phase is drawn from the top and driven to the bottom of the vessel by a reciprocating plunger pump while the polymer phase is drawn from the bottom and driven to the top by a second pump. The apparatus includes several quick-connects around the equilibrium vessel. These are helpful for loading the polymer and cleaning the system. The 150 ml vessel has an electric heating mantle with a temperature controller. The temperature is measured with a thermocouple installed inside the vessel to within $\pm$0.3 K. The circulation lines are maintained at the vessel temperature with heating tapes and controllers which are run off the thermocouples attached to different locations on the lines. The pressure is measured by a pressure transducer installed on the vessel to within $\pm$0.003 MPa. Two high pressure, six-port switching valves with sample loop volumes of 0.5350 (for the upper phase) and 0.1086 ml (for the lower phase) are installed on the circulation lines and are used for sampling.

After loading the polymer, the entire system is purged of air and then charged with CO_2. Once the desired CO_2 loading is attained, the CO_2 feed line is shut off and the circulation pumps are started. At equilibrium, a sample from each phase is taken by switching the sample loop out of the system while the circulation is routed through a bypass. A typical equilibration time is approximately 30 min and the sampling is performed after allowing atleast an hour.

The CO_2 content of each phase is determined in a different manner. The details around the two sampling valves are indicated on the insets of Figure 1. When the sample loops are switched out of the system, the samples expand between two valves and the polymer precipitates in the line. The amount of CO_2 in the supercritical phase is determined by throttling the expanded sample through valve 3 which is connected to a precision pressure gauge (Texas Instruments, Model 145). The gauge has a quartz Burdon tube element and measures pressure to a precision of 0.01%. The volume is known and the temperature of the sample is also measured after equilibration. In the polymer phase, the amount of CO_2 is determined by carefully opening valve 4 and measuring the volume of CO_2 corresponding to the displacement of water level in an inverted graduated tube which is immersed in CO_2 saturated water. Over the range of data the volume of CO_2 in the polymer phase measured at ambient conditions varies from 2 to 30 ml. The CO_2 amounts in both phases are then determined from PVT data.

The polymer content of each phase is analyzed by flushing the sample lines with 50 ml of water. The concentration of the polymer in water is then determined by a colorimetric technique.

Colorimetric Detection of PEG

A quantitative colorimetric detection of aqueous PEG mono-oleate solution was

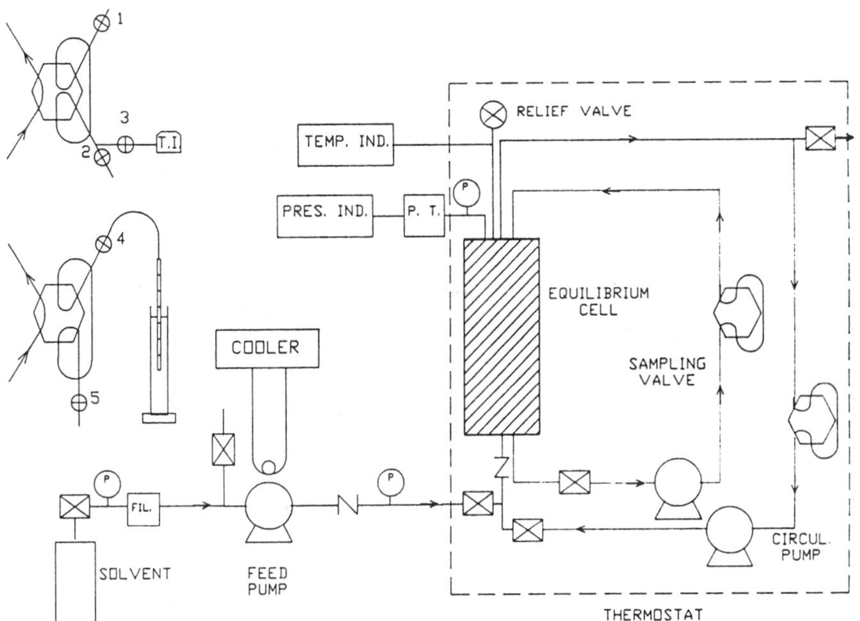

Figure 1. Schematic diagram of the equipment and details of the sampling valves.

described by Brown and Hayes (6). Crabb and Persinger (7) have also described a similar procedure for determining the concentration of polyoxyethylene nonionic surfactants in parts per million concentration range. We adapted a procedure, similar to that of ref. (3) with small modifications. This procedure is based on the formation of a blue complex between PEG and ammonium cobaltothiocyanate solution. The ammonium cobaltothiocyanate solution is prepared by dissolving 15 gr of cobalt nitrate hexahydrate and 100 gr of ammonium thiocyanate in 500 ml distilled water. In a 60 ml separating funnel, 10 ml of PEG solution is reacted with 20 ml of ammonium cobaltothiocyanate solution at 20 °C.

The complex is then extracted into 25 ml of chloroform in several repeated steps and the optical density of the blue chloroform solution at 320 nm is measured with a scanning diode array spectrophotometer (HP 8452A). Figure 2 shows a typical absorption spectra of PEG(600) at two different concentrations. There are two absorption peaks at 320 nm and 622 nm. The optical density of the peak at 320 nm is about six times larger than the one at 622 nm. Therefore, the sensitivity of 320 nm peak is much greater than that at 622 nm and it is used in our analysis.

It is reported that the higher the degree of polymerization, the less is the amount of polyglycol required to form the blue complex (7). In fact, at least six ethylene oxide units are required in a chain for the color development. This implies that the absorbance varies with the length of PEG chain. Therefore, different calibration curves for each compound with different nominal molecular weight must be developed.

Calibration data for the three polymer samples, PEG(400), PEG(600) and PEG(1000) were obtained using the colorimetric technique. In Figure 3, absorbance versus concentration is plotted for the three samples. The calibration data were fitted to polynomials of the form $y = a + bx + cx^2 + dx^3$ where y is concentration in (mg of PEG/50 ml H_2O) and x is absorbance in AU. The calibration curves are summarized in Table I.

Table I. Calibration data for poly(ethylene glycol)s with different molecular weights

Compound	Conc. Range	Coefficients			
	(mg/50 ml)	a	b	c	d
PEG(400)	0 - 10	0	55.27	900.9	5723
PEG(400)	10 - 50	0	103.23	287.6	2624
PEG(600)	1 - 14	-0.535	24.01	0	0
PEG(1000)	0 - 1.2	-0.080	10.834	0	0

FAB - MS

Fast Atom Bombardment Mass Spectrometry was used to obtain the molecular weight distribution of PEGs. FAB-MS like any mass spectrometry technique

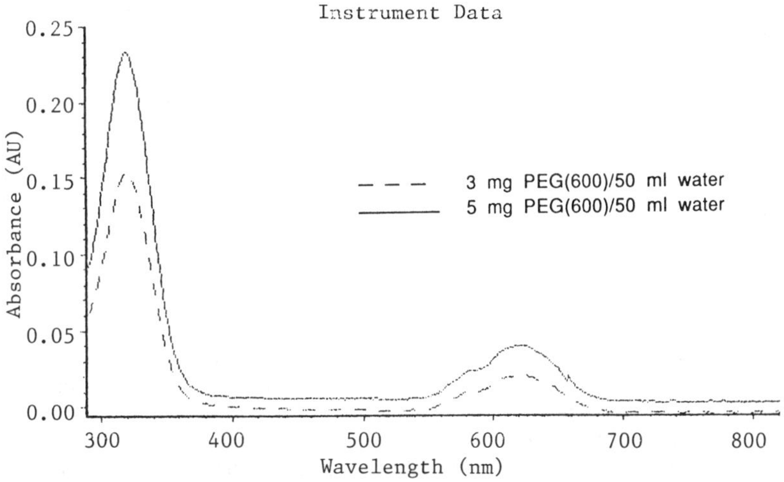

Figure 2. Absorption spectra of PEG(600) - Cobaltothiocyanate complex in Chloroform for two different concentrations.

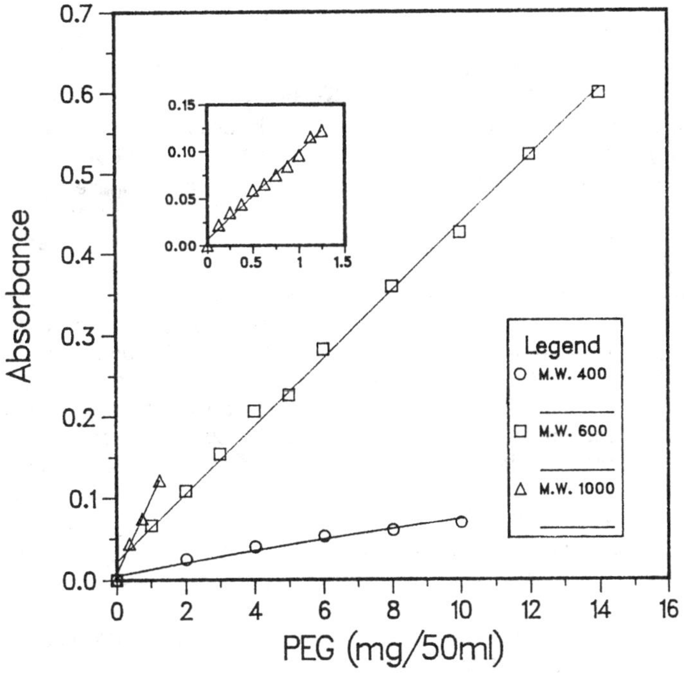

Figure 3. Calibration curves for poly(ethylene glycol) - cobaltothiocyanate complex at 320 nm.

utilizes the effects of electromagnetic fields to control the trajectory of isolated ions and thereby measures their mass to charge (m/z) ratio. In this work the formation of ions are accomplished by bombardment of polymer sample with a xenon gun. The samples are prepared by blending 1 μl of 1 wt% water solution of PEG with 0.5 μl of glycerol on the tip of a stainless steel probe. The xenon atoms striking the sample surface generate quasimolecular ions $[M+H]^+$ resulting in a series of peaks, 44 units apart.

Results and Discussion

The equilibrium phase compositions for PEG - carbon dioxide systems were experimentally measured. The data for PEG(400) - CO_2 and PEG(600) - CO_2 systems were obtained at 313 K. The PEG(400) - CO_2 and PEG(1000) - CO_2 systems were also studied at 323 K. The pressure versus composition diagrams are presented in Figures 4 and 5. The equilibrium compositions are reported as weight percent of the PEG.

Appreciable solubilities of PEGs in supercritical carbon dioxide were observed. It was found that the solubility is a strong function of molecular weight of the polymer. Clearly, there is a minimum value of pressure above which the solubility of a given molecular weight polymer is detected. The minimum pressure increases with molecular weight along isotherms; the minimum pressure corresponding to the solubility limits are about 10 MPA for PEG(400) and about 15 MPA for PEG(600).

At pressures above these limits, the solubility of CO_2 in the polymer phase remains relatively constant as seen on the right hand branches of Figures 4 and 5. This condition affects the distribution of CO_2 and polymer between the two phases. When the composition of the polymer phase is almost constant, a preferential partitioning of CO_2 into the SCF phase drives a certain amount of polymer from the polymer phase into the SCF phase based on the criterion of phase equilibria. This effect together with the solvent density increase due to pressure cause the enhancement of the solubility of polymer in the SCF phase as observed on the left hand branches of Figures 4 and 5.

It was observed that, the solubility drops significantly with molecular weight at a given pressure and temperature. The difference in the relative solubility of PEG(400) and PEG(600) at 313 K is about an order of magnitude and that for PEG(400) and PEG(1000) at 323 K is more than an order of magnitude.

The enhancement of solubility of PEG samples with increasing pressure is a complicated function of the molecular weight distribution. A sample with a very narrow molecular weight distribution is expected to exhibit a sharper solubility enhancement with pressure. The polydispersity or M_w/M_n of the PEG samples are typically high; value of 1.5 is quoted by the supplier. The measured solubility is a composite solubility with contributions from different molecular weight fractions. The composite solubility builds up slowly because the lowest molecular weight fraction starts dissolving at a certain threshold pressure and as pressure is increased, the next higher molecular weight fraction starts dissolving while a new equilibrium is established for the first fraction.

It is evident from Figures 4 and 5 that, the solubility of CO_2 in the polymer

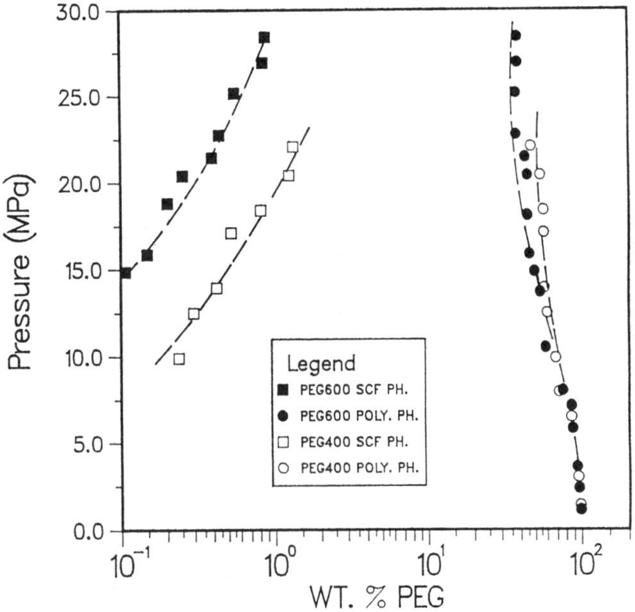

Figure 4. Pressure-Composition diagrams for poly(ethylene glycol) - carbon dioxide systems at 313 K.

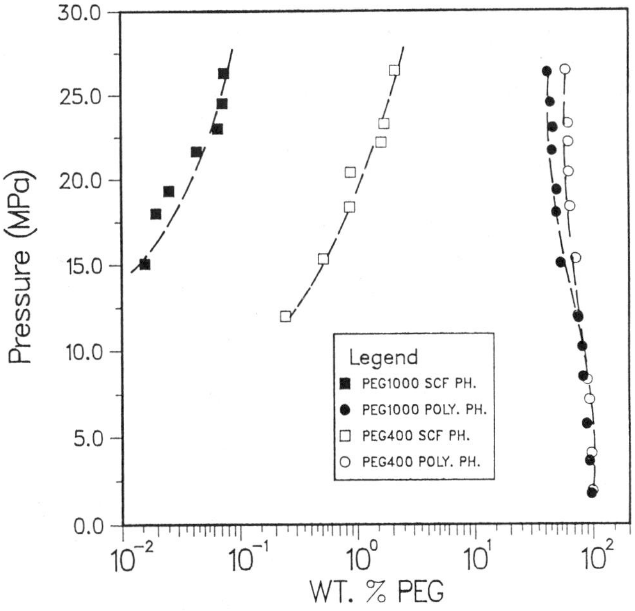

Figure 5. Pressure-Composition diagrams for poly(ethylene glycol) - carbon dioxide systems at 323 K.

phase displays two types of behavior along an isotherm at low and high pressures. At low pressures, there is a linear relation between CO_2 amount dissolved in the polymer and the overall pressure. The slope of a linear fit (wt% CO_2/ MPa) in the low pressure region was found to be 2.04 for PEG(400) at 313 K, 2.13 for PEG(600) at 313 K, 1.00 for PEG(400) at 323 K and 2.08 for PEG(1000) at 323 K. At high pressures, the solubility of CO_2 in polymer phase remains almost constant.

The effect of temperature on the solubility of PEG(400) in the SCF phase and CO_2 in the polymer phase is shown in Figures 6 and 7 respectively. In the SCF phase (Figure 6), a temperature change of 10 °C dose not affect the solubility of PEG(400) in CO_2. This observation suggests that the effects of the vapor pressure of solute and the density of the solvent are to some extent compensating. In the polymer phase (Figure 7), the solubility of CO_2 drops with temperature because CO_2 is very volatile and evaporates out of the liquid phase very effectively when temperature is increased from 313 to 323 K.

If the PEGs are assumed to be monodisperse, then a partition coefficient can be calculated by dividing the weight fraction of PEG in SCF phase by the weight fraction of PEG in polymer phase. Figure 8 shows plots of these partition coefficients as a function of pressure and temperature for PEGs with different molecular weights. The behavior of the partition coefficient as a function of pressure displays interesting features for a separation process. Depending on the choice of pressure, the concentration of PEGs with different molecular weights can be varied relative to each other. If the partition coefficient of an n-mer does not depend on the compositions of other chain lengths, as claimed in ref.(2), the yield as well as selectivity of a multistage operation can be controlled by pressure programming.

Polymer samples are inherently polydisperse. Molecular weight distribution of parent PEG(600) as measured by FAB-MS is shown in Fig. 9. For this nominal molecular weight PEG, the degree of polymerization ranges from n=8 to n=18. This level of polydispersity is quite typical of PEGs. In order to optimize a fractionation process, it is imperative to detect the overall solubility as well as n-mer distribution in the two coexisting phases.

The molecular weight distribution of polymer samples isolated from the coexisting phases for PEG(600) at T=313K and P=27.03 MPa were measured by FAB-MS. When the overall PEG solubility was weighted by the measured distribution, the solubility of each n-mer was calculated at the specified conditions. The individual n-mer partition coefficients are then defined as the weight fraction of the n-mer in the SCF phase divided by the weight fraction of the same n-mer in the polymer phase. Figure 10 shows a plot of the logarithm of the individual n-mer partition coefficients with respect to molecular weight. Two distinct relations between the partition coefficient and the molecular weight are observed. At a given temperature and pressure, there is a critical molecular weight below which the partition coefficients of the neighboring n-mers do not depend on molecular weight. Above this critical molecular weight, the partition coefficients of neighboring n-mers decrease exponentially with molecular weight. This behavior is expected to change with pressure. In the low molecular weight region, the relative yields are high while the selectivity is low. In the high

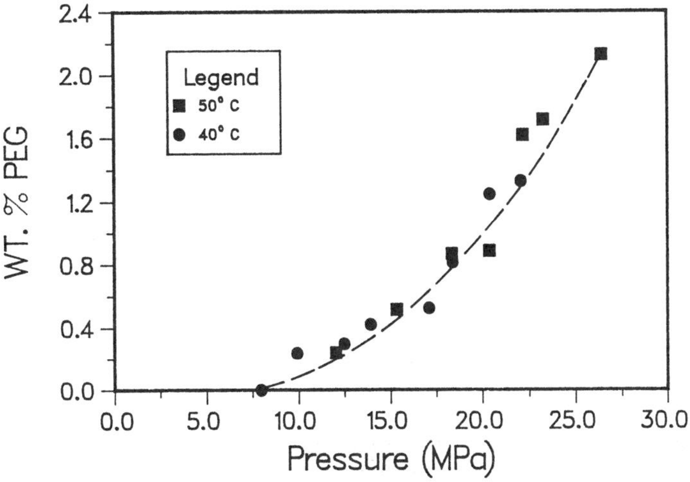

Figure 6. Solubility of PEG(400) in supercritical CO_2 as a function of pressure at 313 and 323 K.

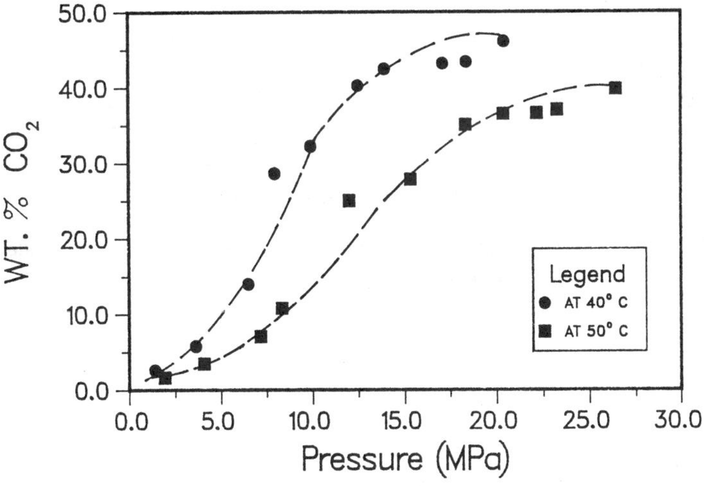

Figure 7. Solubility of CO_2 in PEG(400) as a function of pressure at 313 and 323 K.

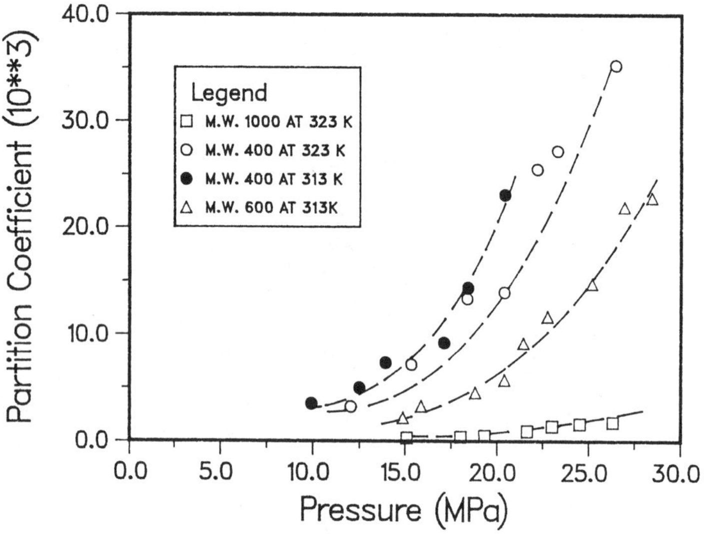

Figure 8. Partition coefficients for poly(ethylene glycol)s with different average molecular weights.

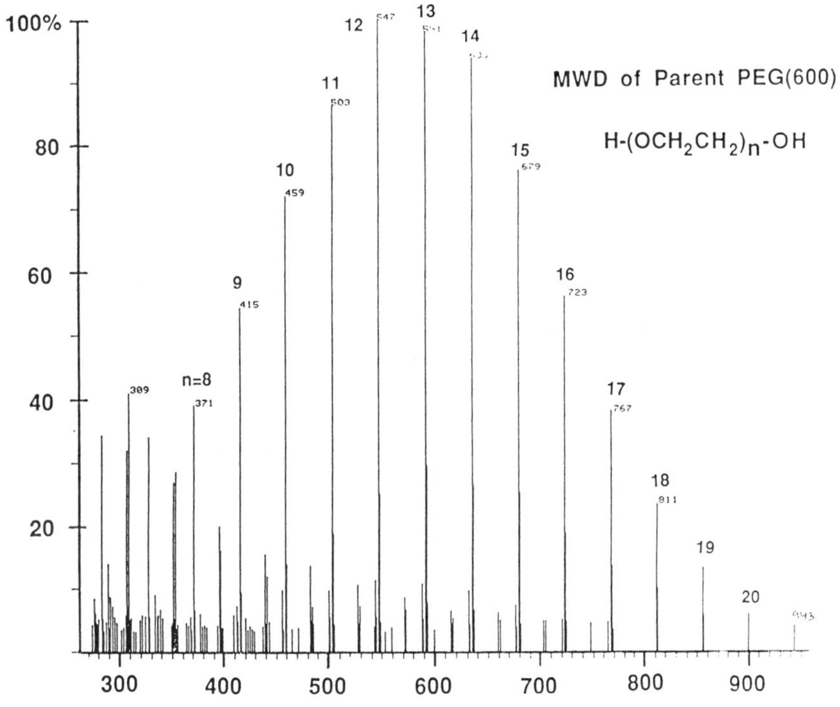

Figure 9. FAB mass spectrum of parent PEG(600).

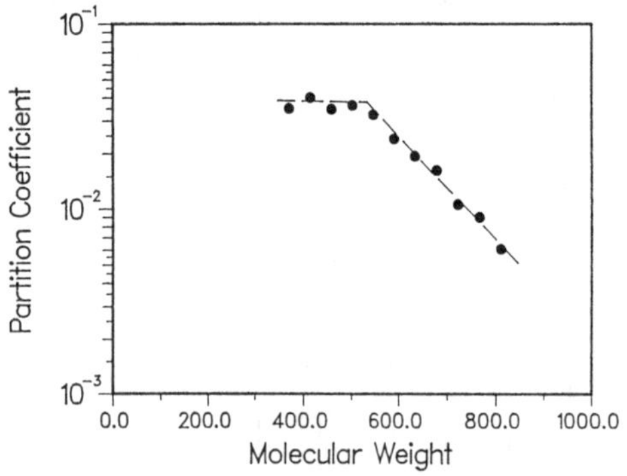

Figure 10. Individual n-mer partition coefficients of PEG(600) at 313 K and 27.03 MPa.

molecular weight region, the selectivity is high while the yield is low. The n-mer partition coefficients can provide means for optimizing a supercritical fractionation process in terms of yield and selectivity.

Conclusions

The equilibrium solubility data have been presented for poly(ethylene glycol)s - carbon dioxide systems at 313 and 323 K over a pressure range up to 28 MPa. The solubility of lower molecular weight PEGs in supercritical CO_2 were found to be appreciable. At a fixed temperature, there is a minimum pressure above which a given molecular weight PEG is soluble in CO_2. PEG solubility in supercritical CO_2 drops significantly with molecular weight. The enhancement of PEG solubility occurs over a pressure range when the amount of CO_2 soluble in the polymer phase becomes almost constant. The partitioning of individual n-mers between the SCF phase and the polymer phase exhibits two types of behavior in terms of yield and selectivity. The higher molecular weight neighboring n-mers exhibit high selectivity and relatively lower yield while the lower molecular weight neighboring n-mers show almost no selectivity and higher yield.

Acknowledgement

The authors gratefully acknowledge the financial support from the National Science Foundation Grant No. RII-8503643 and the University Science Partners, Inc..

Literature Cited

1. Kumar, S. K.; Suter, U. W.; Reid, R. C. Fluid Phase Equilibria 1986, 29, 373.
2. Krukonis, V. Polym. News 1985, 11, 7.
3. Kumar, S. K.; Chhabria, S. P.; Reid, R. C.; Suter, U. W. Macromolecules 1987, 20, 2550.
4. Yilgor, I.; McGrath, J. M. Polymer Bulletin 1984, 12, 499.
5. Scholsky, K. M.; O'Connor, K. M.; Weiss, C. S. J. Applied Polym Sci. 1987, 33, 2925.
6. Brown, E. G.; Hayes, T. J. Analyst 1955, 80, 755.
7. Crabb, N. T.; Persinger, H. E. J. Am. Oil Chem. Soc. 1964, 41, 752.

RECEIVED June 9, 1989

Chapter 7

Experimental Measurement of Supercritical Fluid–Liquid Phase Equilibrium

Huazhe Cheng[1], John A. Zollweg, and William B. Streett

School of Chemical Engineering, Cornell University, Ithaca, NY 14853

An analytic method has been used to produce pVTxy measurements for binary systems containing methyl oleate and supercritical solvents. A micro dual-sampling system has been added to our apparatus for taking vapor and liquid samples. The systems ethane-methyl oleate and carbon dioxide-methyl oleate were studied along isotherms at 313.15 K and 343.15 K up to pressures substantially greater than the critical pressures of the pure solvents. Comparisons are made between the experimental data and predictions using the Peng-Robinson equation of state.

More than one hundred years ago, certain fundamental principles in supercritical extraction had already been known, but viable processes for using this technique developed slowly. In the past two decades, process engineers in several industries have been interested in using supercritical fluids to extract soluble nonvolatile components from mixtures. One of many examples is enhanced oil recovery using carbon dioxide. Another is the fractionation of cod-liver oil using supercritical ethane (1).

Because of high energy costs and increasing regulation of solvents used in conventional distillation, alternative extraction and crystallization methods are now more frequently used. Furthermore, supercritical extraction combines the characteristics of both distillation and extraction; it is efficient, clean, and is able to extract substances which are difficult to separate by other unit operations. Thus it has recently become an attractive possibility for many separations.

Since the first large-scale supercritical extraction process was commercialized for the decaffeination of green coffee with carbon dioxide a decade ago, scientists and engineers in the food industry have been paying considerable attention to this technique for similar separations, *i.e.*, removal of cholesterol from butter, removal of cocoa butter from cocoa beans, and extraction of hops, spices, and

[1]Current address: Beijing Research Institute of Chemical Industry, Beijing, China

0097–6156/89/0406–0086$06.00/0

nicotine from raw materials (2-4). A related application is the separation and rearrangement of the fatty acids of food products. This is desirable because there is conclusive evidence relating the level of consumption of various fatty acids with the risk of developing coronary heart disease (5). Unfortunately, only a few such processes have been commercialized so far. The main impediment to further application of this technology is the lack of reliable data for the industry to use and reliable methods for handling heavy nonvolatile compounds. So far, no scientific basis exists for selecting the best operating conditions for the extraction of a multicomponent mixture with a supercritical solvent. There is an urgent need for accurate equilibrium and mass transfer data to support these new methods if they are to gain widespread industrial application. Of particular difficulty is the adaptation of the traditional cubic equations of state to deal with systems containing high molecular weight, relatively involatile components.

In this paper, we describe the apparatus we use to make phase equilibrium measurements on mixtures of components with greatly differing volatilities, putting particular emphasis on recent improvements over the previous version (6-7). We also describe quantitative measurements of the solubility of methyl oleate in supercritical fluids which can provide a basis for choosing a solvent to separate fatty acids in edible oils. In the following paper (8) we explore the utility of cubic equations of state to describe the results of supercritical fluid - liquid phase equilibrium measurements. Some additional experimental results on the mutual solubility of methyl linoleate and carbon dioxide are presented there also.

Apparatus and Methods

A diagram of the apparatus is shown in Figure 1. An analytic method is used to produce pVTxy measurements for systems containing methyl oleate and supercritical fluids. A micro dual-sampling system has been added to our VLE apparatus (6), which is of the dual-recirculation type, built by Adams in 1984. The heart of the apparatus is an optical sapphire pressure vessel (SPV) through which both liquid and vapor phases are continuously recirculated by two magnetic pumps (GRP, LRP). High pressure micro sample-injection valves (RV1, RV2), connected in the two recirculation loops, are used for sampling both the vapor and the liquid phases individually. A switch valve (RV3) is used for changing the direction of flow of the carrier gas (He) which transports these samples through a separately heated zone (HZ) (for total vaporization) to the gas chromatograph. Two hand-operated piston-screw pumps (PG1, PG2) shown on the far right side of the diagram in Figure 1 are used to generate pressure in the system. The pressure is measured with any accuracy of 0.15% using a stain gauge transducer. Temperature is controlled using a proportional controller and is measured to .01 K in two locations with thin-film platinum resistance thermometers. The apparatus enables vapor-liquid phase equilibrium experimentation in the temperature range 300 to 400 K at pressures up to 40 MPa.

The densities of the samples are determined by first measuring the response of the system (in terms of chromatograph peak area) to pure carbon dioxide at various pressures and to pure methyl oleate, and then using the integrated peak areas on the gas chromatograph to

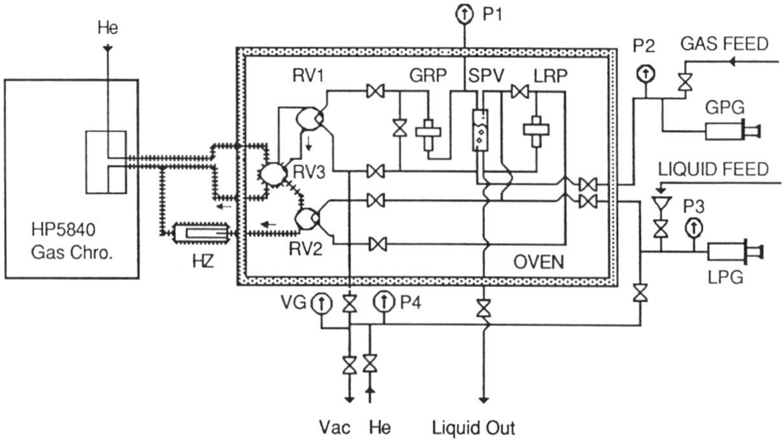

Figure 1. Schematic Diagram of the Dual-Sampling Vapor-Liquid Equilibrium Apparatus. Components are labeled as follows: GPG, Gas pressure generator; GRP, Gas recirculation pump; HZ, Heated zone; LPG, Liquid pressure generator; LRP, Liquid recirculation pump; P, Pressure gauge; RV, High pressure sampling valve; SPV, Sapphire pressure vessel; VG, Vacuum gauge. Hatchmarks on the lines in the sampling section of the apparatus indicate where heaters have been wrapped on them. The path of vapor samples is indicated by light arrows and the dark arrow shows the path of liquid samples.

determine the total quantity of material in the samples. The calibration of quantity of material *vs.* peak area for carbon dioxide is slightly nonlinear. The volume of sample trapped in the sampling valves is implicit in this calibration; thus, the two sampling valves have different calibrations. Density uncertainties are less than one percent.

The experimental procedure is as follows. The liquid to be studied is charged to the apparatus with the hand-operated pump, and liquid in the cell is degassed by evacuating. Sufficient solvent is then injected using the other piston-screw pump to give the desired pressure. The two magnetic pumps are operated until equilibrium is reached, as indicated by constant pressure readings. Equilibration times range from 2-4 hours at lower pressures to 4-5 hours near the critical pressure. After reaching equilibrium, liquid and vapor samples are taken every 15 minutes.

Methyl oleate is a nonvolatile component whose boiling point of 635.4 K (calculated by the group contribution method) differs strongly from those of the supercritical solvents ethane and carbon dioxide. Most of the experimental problems are associated with transferring a sample of the equilibrated mixture from the equilibrium cell to an appropriate analytical instrument without composition change. Although several new micro sampling techniques (9-11) have been reported and microchromatographic equipment is available, difficult obstacles still remain. These problems are especially acute for mixtures of compounds with widely different volatilities, as is true for supercritical fluid-liquid systems. Reliable measurement of supercritical fluid-liquid equilibria is also complicated by entrainment of small droplets of one phase in the other and the adsorption of the 'heavy component' (methyl oleate) on the walls of the sample lines. In order to ensure complete transfer of the equilibrium samples to the analytical instrument, the following alterations to the earlier appartus were made:

1. The heated zone and lines for vaporizing and transferring samples are glass-lined and are as short as possible.
2. Separate sampling valves are used for taking the liquid and vapor samples to prevent entrainment of small droplets and cross-contamination of samples.
3. The transfer lines and heated zone are well-insulated to ensure uniformity in temperature during sampling.
4. Provision is made for injecting a solvent into the sampling lines for clean-up of the sampling valves without contaminating the mixture under study.

There is no guarantee that adsorption of the methyl oleate on the sampling valve itself does not distort the results. In fact, that may be one of the principal sources of scatter in our results, which have an uncertainty of 2 %.

The carbon dioxide was supplied by Air Products and Chemicals, Inc., with a stated purity of 99.99% and research grade ethane was obtained from MG Industries. Crude methyl oleate (KODAK T.G. containing 65%-75% methyl oleate) was purified to 98%-99% (including methyl linoleate) using vacuum distillation.

Results

The experimental vapor and liquid phase compositions and molar volumes for the carbon dioxide + methyl oleate and ethane + methyl oleate systems at 313.15 K and 343.15 K are presented in Tables I and II, respectively. The pVTxy data span the pressure range from about 2 MPa up to nearly the critical pressure of the mixture at this temperature.

The vapor and liquid phase compositions for carbon dioxide + methyl oleate are shown graphically in Figure 2, and for ethane +

Table I. Equilibrium Phase Properties of Carbon Dioxide/Methyl Oleate

T = 313.15 K

Pressure (MPa)	Mole fraction Carbon dioxide		Volume (cm^3/gmole)	
	vapor	liquid	vapor	liquid
2.02		0.33		387.
3.61	0.9950		592.	
3.73		0.53		227.
5.28	0.9869		376.	
5.51	0.981	0.67	296.	201.
6.15	0.982		276.	
7.16		0.89		191.
7.25	0.9922	0.94	199.	184.
8.12		0.94		148.
9.87	0.9960	0.91	78.0	129.
11.87		0.94		108.
12.45	0.9952		75.3	
12.52	0.9972	0.96	73.2	102.

T = 343.15 K

Pressure (MPa)	Mole fraction Carbon dioxide		Volume (cm^3/gmole)	
	vapor	liquid	vapor	liquid
2.45	0.9997	0.25	1130.	295.
4.15	0.9999	0.33	651.	241.
5.59	0.9996	0.43	473.	212.
6.76	0.9996		365.	
8.46	0.99995	0.55	269.	187.
9.76	0.99997	0.53	215.	155.
11.01	0.99990	0.55	175.	150.
11.99	0.99998		136.	
13.81	0.99993	0.63	105.	123.
15.72	0.99985		78.5	
16.95	0.99980	0.89	66.7	98.5
17.88	0.99987	0.93	54.7	79.4
19.33	0.99984	0.95	40.7	89.6
19.68	0.9991		51.9	

Table II. Equilibrium Phase Properties of Ethane/Methyl Oleate

T = 313.15 K

Pressure (MPa)	Mole fraction Carbon dioxide		Volume (cm^3/gmole)	
	vapor	liquid	vapor	liquid
2.36	0.9966		1642.	
2.40		0.55		343.
2.70	0.9919		1632.	
3.55		0.76		345.
4.22	0.966		570.	
4.91	0.953	0.76	437.	188.
5.68	0.979	0.73	253.	226.
6.21	0.992		90.8	
6.60		0.76		170.
7.70	0.987		82.5	
7.77		0.88		93.
8.08	0.984		94.8	
8.15		0.95		110.

T = 343.15 K

Pressure (MPa)	Mole fraction Carbon dioxide		Volume (cm^3/gmole)	
	vapor	liquid	vapor	liquid
2.65	0.9916		921.	
3.38	0.9938		709.	
3.52	0.9992		680.	
4.11	0.9992		558.	
4.54		0.64		291.
4.77	0.9974		478.	
4.97	0.9987		451.	
5.55	0.9949	0.70	398.	258.
6.01		0.76		238.
6.43	0.988	0.82	323.	195.
7.09	0.984	0.83	277.	148.
8.82	0.988		204.	
9.87	0.9975		175.	
9.92	0.9987	0.75	199.	193.
10.88	0.9997		172.	
11.05	0.9974		167.	
12.24	0.9999	0.80	159.	205.

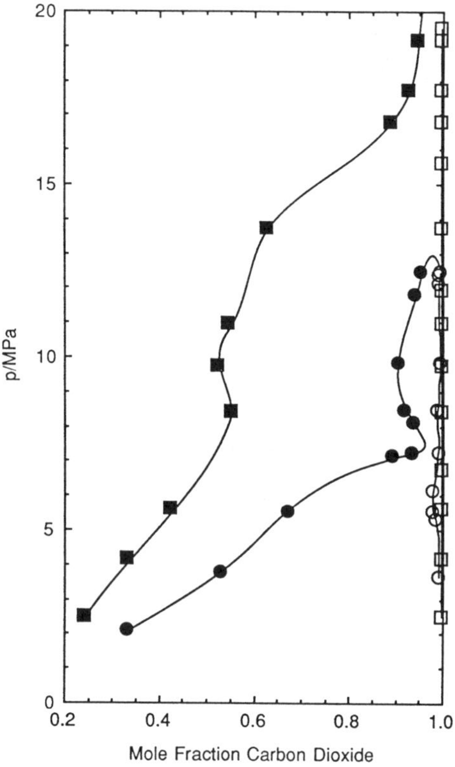

Figure 2. Pressure *vs.* composition for CO_2 + Methyl Oleate. Open symbols: vapor phase; closed symbols: liquid phase; ●, 313.15 K; ■, 343.15 K.

methyl oleate in Figure 3. Note that the isotherms have a characteristic "waisted" shape. The pressures at the narrow parts of the diagrams correspond to the critical pressures of the pure solvents. These phase diagrams appear to belong to class 4 in the phase diagram classification of van Konynenberg and Scott (12). Such waisted shapes have been observed for carbon dioxide + hexadecane (13) and carbon dioxide + 2,6,10,15,19,23-hexamethyltetracosane. It seems reasonable to have this kind of phase behavior for carbon dioxide + methyl oleate, since methyl oleate is a fatty acid ester which contains a total of 19 carbons with only one double bond, and thus is of similar size to the hydrocarbons which show this behavior.

Cubic equations of state have been modified in many ways to describe the behavior of fluids and their mixtures, but many of these modifications were found unsuitable as they could not predict the molar volumes of compounds with high molecular weight. Although more complicated equations with two or three dozen constants have been utilized for correlating high-precision PVT measurements, they are not preferred for involved thermodynamic calculations because they require tedious programming and long computing times for lengthy iterative calculations. Simpler semi-empirical cubic equations of state such as the Soave-Redlich-Kwong (SRK) (14) and Peng-Robinson (PR) (15) are thus used to compare with experimental phase equilibrium and molar volume data. We have used the Peng-Robinson equation for comparisons which are shown in Table III and Figure 4. It is necessary to have estimates of the critical pressure and temperature of methyl oleate, as well at its Pitzer acentric factor. Since the critical properties of methyl oleate are not directly available, they

Table III. Comparison of Experimental and Calculated Molar Volumes for Carbon Dioxide + Methyl Oleate at 313.15 K

Pressure (MPa)	Measured volume (cm^3/gmole)		Calculated volume (cm^3/gmole)	
	vapor	liquid	vapor	liquid
3.615	592.		594.	255.
3.726		227.		
5.515	296.	201.	335.	193.
7.158	214.	191.	211.	148.
7.251	199.	184.	206.	145.
8.124	163.	148.	153.	125.
9.119			91.1	107.
9.873	78.0	129.	72.8	99.4
11.867	76.4	108.	67.4	86.8
12.453	75.3		68.3	83.2
12.516	73.3	102.		
13.434			73.1	73.1

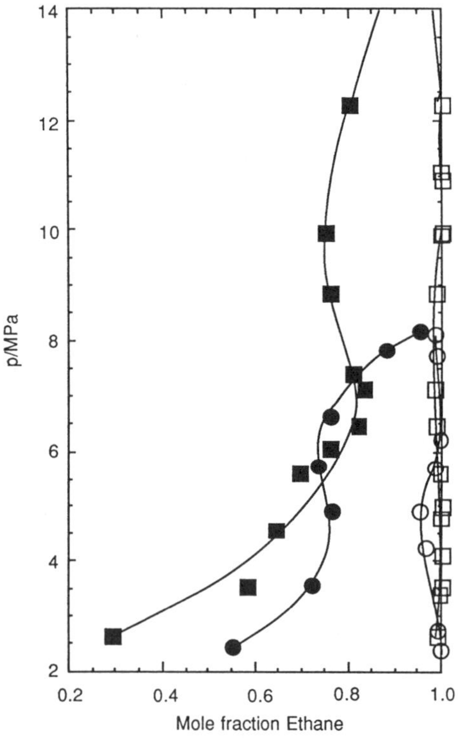

Figure 3. Pressure *vs.* composition for C_2H_6 + Methyl Oleate. Open symbols: vapor phase; closed symbols: liquid phase; ●, 313.15 K; ■, 343.15 K.

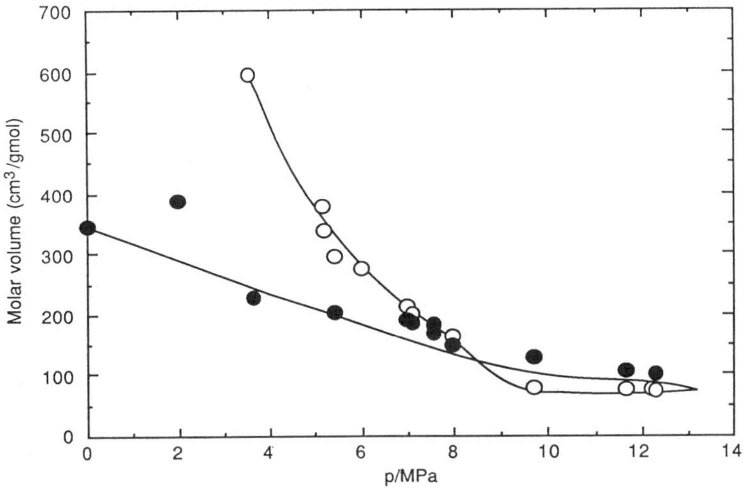

Figure 4. Mixture volumes versus pressure at 313.15 K. Points are experimental values. Curves are calculated using the Peng-Robinson equation. Open symbols: vapor phase; closed symbols: liquid phase.

are estimated by means of the Lydersen group-contribution method (16-18). The interaction parameter between carbon dioxide and methyl oleate in the attraction term was chosen to be 0.050. This choice led to good agreement between the calculated and measured critical pressure on this isotherm. The vapor phase molar volumes are also in good agreement with experiment, which can be seen in Table III and in Figure 4. More detailed comparisons between cubic equations of state and these data are presented in the following paper.

Discussion

Fluid mixtures may exhibit very different and complicated types of phase behavior. Many of them are described in the literature (13, 19, 20). The isotherms in Figures 2 and 3 take on a characteristic "waisted" shape, with a narrow upper branch and a wide lower branch fused together to form the "waist". The pressure at the waists roughly corresponds to the critical pressure of the pure solvents: ethane and carbon dioxide, respectively. These figures, together with the visual observation of two liquid phases at lower temperatures, imply that the p-T projections of these binary systems belong to class 4, according to the classification scheme of van Konynenberg and Scott (12).

Both the experimental and calculated molar volumes show that the molar volume of the vapor phase becomes smaller than that of the liquid at higher pressures. This seemingly anomalous result is a consequence of the very large difference in molecular weights of the solute and solvent. We saw no evidence, such as a phase inversion, that the mass densities of the liquid and vapor become equal except at the critical point on the isotherm.

We wish to emphasize that the design of experimental apparatus for pVTxy measurements of supercritical fluid-liquid equilibrium in fatty ester systems has to be undertaken with special care to avoid adsorption of the fatty acid on the sampling apparatus. Since the equilibrium fluid mixture passes through the sampling valve in our design, it is neither necessary nor advisable to use very small diameter lines.

Acknowledgment

This work was supported by the United States National Science Foundation under grant CBT-8704581.

Literature Cited

1. Zosel, K. In Extraction with Supercritical Gases; Schneider, G. M.; Stahl, E.; Wilke, G., Eds.; Verlag Chemie: Weinheim, 1980; pp 1-23.
2. Rizvi, S. S, H.; Benado, A. L.; Zollweg, J. A.; Daniels, J. A. Food Technology 1986, 40(6), 55-65.
3. Rizvi, S. S. H.; Daniels, J. A.; Benado, A. L.; Zollweg, J. A. Food Technology 1986, 40(7), 57-64.
4. Hubert, P.; Vitzthum, O. G. In Extraction with Supercritical Gases; Schneider, G. Schneider, G. M.; Stahl, E.; Wilke, G., Eds.; Verlag Chemie: Weinheim, 1980; pp 25-43.
5. Wood, D. A.; Butler, R. A.; Riemersma, R. A.; Thomson, M.; Oliver, M. F. Fulton, M.; Birtwhistle, A.; Elton, R. Lancet 1984 ii, 117-121.

6. Adams, W. R. Ph.D. Thesis, Cornell University, Ithaca, NY, 1986.
7. Adams, W. R.; Zollweg, J. A.; Streett, W. B.; Rizvi, S. S. H. AIChE J. 1988, 34, 1387-1391.
8. Zou, M.; Lim, S.B.; Rizvi, S. S. H.; Zollweg, J. A., This symposium.
9. King, M. B.; Alderson, D. A.; Fallah, F. H.; Kassim, D. H.; Kassim, K. M.; Sheldon, J. R.; Mahmud, R. S. In Chemical Engineering at Supercritical Fluid Conditions; Paulaitis, M. E.; Penninger, J. M. L.; Gray, R. D.; Davidson, P. D., Eds.; Ann Arbor Science: Ann Arbor, MI, 1983; pp 31-80.
10. Anderson, F. E.; Prausnitz, J. M. Fluid Phase Equilibria 1986, 32, 63-76.
11. Hsu, J. J-C.; Nagarajan, N.; Robinson, Jr, R. L. J. Chem. Eng. Data 1985, 30, 485-491.
12. van Konynenberg, P. H.; Scott, R. L. Phil. Trans. Roy. Soc. 1980, A298, 495-540.
13. Schneider, G. M. In Extraction with Supercritical Gases, Schneider, G. M.; Stahl, E.; Wilke, G., Eds.; Verlag Chemie: Weinheim, 1980, pp 45-81.
14. Soave, G. Chem. Eng. Sci. 1972, 27, 1197-1203.
15. Peng, D-Y.; Robinson, D. B. Ind. Eng. Chem. Fundam. 1976, 15, 59-64.
16. Chopey, N P.; Hicks, T. G., Eds. Handbook of Chemical Engineering Calculations; McGraw-Hill, New York, 1984; p 1-4.
17. Lydersen, A. L.; Greenkorn, R. A.; Haugen, O. A. Generalized Thermodynamic Properties of Pure Fluids Univ. Wis. Coll. Eng. Expt. Stn. Rep. 4, Madison, WI, 1955.
18. Lange's Handbook of Chemistry, John A. Dean, Ed..; McGraw-Hill: New York, Thirteenth Edition, 1985; p 10-57.
19. Streett, W. B. In Chemical Engineering at Supercritical Fluid Conditions; Paulaitis, M. E.; Penninger, J. M. L.; Gray, R. D.; Davidson, P. D., Eds.; Ann Arbor Science: Ann Arbor, MI, 1983; pp 3-30.
20. McHugh, M. A.; Krukonis, V. J. Supercritical Fluid Extraction; Butterworths: Boston, 1986; pp 23-67.

RECEIVED May 2, 1989

Chapter 8

Vapor–Liquid Equilibria of Fatty Acid Esters in Supercritical Fluids

M. Zou[1], S. B. Lim[1], S. S. H. Rizvi[1], and John A. Zollweg[2]

[1]Institute of Food Science, Cornell University, Ithaca, NY 14853
[2]School of Chemical Engineering, Cornell University, Ithaca, NY 14853

Several cubic equations of state such as Redlich-Kwong, Soave-Redlich-Kwong, and Peng-Robinson have been used to calculate vapor-liquid equilibria of fatty acid esters in supercritical fluids. Comparisons are made with experimental data on n-butanol, n-octane, methyl oleate, and methyl linoleate in carbon dioxide and methyl oleate in ethane. Two cubic equations of state with a non-quadratic mixing rule were successful in modeling the experimental data.

In recent years, a great deal of interest has been paid to supercritical fluid extraction (SFE) processes. They are especially suitable for the separation of substances with low volatility which decompose before reaching their normal boiling points. These processes are based on the phenomenon that the dissolving power of a solvent, as a first approximation, changes greatly with its density.

One of the recent major advances made in solvent extraction is the application of supercritical carbon dioxide (SC-CO_2) in the food and beverage industry for the extraction and concentration of natural products and flavorings (1-2). Several actual or potential commercial applications of this method include the extraction of fragrances and flavors from liquids, decaffeination of coffee beans, deodorization of oils, extraction of oil seeds, fractionation of highly unsaturated methyl esters derived from fish oil triglycerides, and separation of organic materials from water. Additionally, there is increasing interest in separating and rearranging the fatty acids of food materials to formulate new products. However, in order to establish commercial SFE processes which involve vapor phase extraction, it is important to have reliable equilibrium data and methods for predicting phase equilibrium behavior.

0097–6156/89/0406–0098$06.00/0

For solid solubilities in SC-fluids, numerous experimental data are reported in the literature (3). At moderate pressure, equilibrium solubilities can be calculated from the truncated virial equation of state since reliable methods are available for estimating virial coefficients for such mixtures; at higher pressure, the Peng-Robinson (PR) equation of state represents the solubility behavior quite well. The Soave-Redlich-Kwong (SRK) equation of state as well as other simple cubic equations of state would give comparable results. In the case of supercritical fluid-liquid equilibria, it is particularly difficult to adapt the traditional cubic equations of state to systems containing components which have high molecular weight and are relatively non-volatile. The additional complexity in the equilibrium calculations introduced by the solvent dissolving in the solute makes these calculations much more difficult.

The objective of the present paper is to describe the behavior of supercritical fluid-liquid mixtures by using simple equations of state (EOS) with different mixing rules.

Experimental

The apparatus and methods are described in detail in the previous paper (4) and elsewhere (5-6). It is of the dual recirculation-type and is comprised, in part, of a central pressure vessel through which both liquid and vapor phases are continuously recirculated. Samples of the liquid and vapor phases are removed from the circulation loops and analyzed by gas chromatography. The uncertainty in measured compositions is 0.02 mole fraction for the liquid and 0.001 for the vapor.

The densities of the samples are determined by first calibrating the sampling system using pure materials, and then using the integrated peak areas on the gas chromatograph to determine the total quantity of material in each sample. The calibration curve is slightly nonlinear. In order to obtain good results, separate sampling valves are used for taking the liquid and vapor samples so as to prevent cross-contamination of samples. Measured phase equilibrium data are shown in Table I.

The manufacturer's stated mole percent purities of the compounds were 99.98 for carbon dioxide (MG Industries) and 99.0 for methyl linoleate (Sigma Products).

Vapor-Liquid Equilibrium calculation:

For the vapor-liquid equilibrium calculations, at the equilibrium state the fugacities for all species i must be the same in all phases, namely

$$f_i^V = f_i^L \qquad \text{or} \qquad \phi_i^V y_i = \phi_i^L x_i \tag{1}$$

where f_i is the fugacity and ϕ_i is the fugacity coefficient. To calculate fugacity coefficients, equations of state which are

Table I
Experimental Data
Carbon dioxide (1)/Methyl linoleate(2)
at 343.15K

Pressure (bar)	x_1 (liquid phase)	y_1 (vapor phase)	liquid volume (cm^3/g-mol)	Vapor volume (cm^3/g-mol)
21.3	0.29	0.9989	295	1022
35.5	0.43	1.0000	220	567
53.6	0.48	0.9987	209	345
71.9	0.56	0.9987	207	235
88.4	0.58	0.9987	193	177
95.8	0.55	0.9996	180	150
140.3	0.62	0.9989	189	90

valid for both the vapor phase mixture and the liquid phase mixture were used (7).

Cubic equations of state (EOS) such as the Redlich-Kwong (RK), Soave-Redlich-Kwong and Peng-Robinson equations of state have become important tools in the area of phase equilibrium modeling, especially for systems at pressures close to or above the critical pressure of one or more of these system components. The functional form of the Soave-Redlich-Kwong and Peng-Robinson equations of state can be represented in a general manner as shown in Equation 2:

$$p = RT/(v-b) - a/(v^2 + uvb + wb^2) \tag{2}$$

where u and w are numerical constants. For the Soave-Redlich-Kwong equation of state, u = 1, w = 0; for the Peng-Robinson equation of state, u = 2, w = -1.

For simple mixtures, the parameters a and b are related to the pure component parameters and composition through the following mixing rules:

$$a_m = \sum_i^n \sum_j^n x_i x_j a_{ij} \tag{3}$$

$$b_m = \sum_i^n \sum_j^n x_i x_j b_{ij} \tag{4}$$

In these equations, a_{ii} and b_{ii} are parameters corresponding to pure components; while a_{ij} and b_{ij} ($i \neq j$) are called the unlike interaction parameters. It has been customary to relate the unlike interaction parameters to the pure component parameters by combining rules, such as the following:

$$a_{ij} = (a_{ii} a_{jj})^{1/2} (1 - k_{ij}) \tag{5}$$

$$b_{ij} = (b_{ii}+b_{jj})/2 \qquad (6)$$

In Equation 5, k_{ij} is called a binary interaction parameter. It is calculated from experimental binary phase equilibrium data on a given isotherm by regression.

For the calculations presented in this paper, we first elected to use three simple cubic equations of state: PR-EOS; SRK-EOS; and RK-EOS. For the pure components, critical properties (P_c, T_c) and Pitzer's acentric factor (ω) are needed to obtain a_i and b_i. Critical properties have been measured for most of the low molecular weight components and are reported by Reid et al. (8). For biomaterials that are thermally unstable and decompose before reaching the critical temperature, several estimation techniques are available. We have used the Lydersen group contributions method (8). Other techniques available for predicting critical properties have been reviewed and evaluated by Spencer and Daubert (9) and Brunner and Hederer (10). It is also possible to determine the EOS parameters from readily measurable data such as vapor pressure, and liquid molar volume instead of critical properties (11). We used the Lydersen method to get pure component parameters because the vapor compositions we obtained were in closer agreement with experiment than those we got from pure component parameters derived by Brunner's method. The critical properties we used for the systems we studied are summarized in Table II.

Table II Critical Properties and Pitzer's Acentric Factor for the Working Materials

Component	T_c/K	P_c/bar	ω
Carbon dioxide	304.2	73.82	0.225
Ethane	305.4	48.84	0.098
n-Butanol	562.93	44.12	0.59
n-Octane	568.8	24.82	0.394
Methyl oleate	785.99	12.83	0.9835
Methyl linoleate	786.88	13.06	0.9869

The binary interaction parameter, k_{ij}, is initially assumed to be zero, and a modification of the Levenberg-Marquardt algorithm (MINPACK) is applied to minimize the sum of the squares given by Equation (1). This calculation was applied to the following systems at the indicated temperatures:

Carbon dioxide (1)/n-Butanol(2) at 40°C and 110°C (12)
Carbon dioxide (1)/n-Octane(2) at 40°C and 110°C (12)
Carbon dioxide(1)/Methyl oleate(2) at 40°C and 70°C (4)
Ethane(1)/Methyl oleate(2) at 40°C and 70°C (4)
Carbon dioxide(1)/Methyl linoleate(2) at 70°C (this work)

The optimum binary interaction parameters are shown in Table III. An example of the results is shown in Figure 1 for the PR-EOS applied to carbon dioxide/methyl oleate at 70°C. Comparing the results of those three simple equations of state, the Redlich-Kwong equation of state gave the poorest prediction.

There has been criticism directed toward the oversimplicity of the cubic equation form, especially in the modeling of supercritical vapor-liquid equilibrium. Nevertheless, this representation does describe at least qualitatively all the important characteristics of vapor-liquid equilibrium behavior. Alternative equations of state have been suggested, but none have been widely used and tested. Also, other EOS are significantly more complex and bring with them additional parameters which must be evaluated by regression from experimental data.

It is our opinion that the key to success in employing the cubic equations of state at high pressure to model phase equilibrium with supercritical fluids is in the choice of the mixing and combining rules and in keeping the EOS in the simplest form with the fewest interaction parameters.

A number of suggestions for the improvement of mixing rules, some of which show promise, have evolved from recent work in this area (13-16). Based on statistical mechanical theory, the following mixing rules have been derived (17):

$$\sigma^3 = \sum_i^n \sum_j^n x_i\, x_j\, \sigma_{ij}^3 \tag{7}$$

$$\epsilon\sigma^3 = \sum_i^n \sum_j^n x_i\, x_j\, \epsilon_{ij}\, \sigma_{ij}^3 \tag{8}$$

where ϵ_{ij} is the interaction energy parameter between molecule i and j, and σ_{ij} is the intermolecular interaction distance between the two molecules. Knowing that the coefficients a and b of the cubic equations of state are proportional to ϵ and σ according to the following expressions,

$$a \propto N_o\, \epsilon\sigma^3 \tag{9}$$

$$b \propto N_o\, \sigma^3 \tag{10}$$

where N_o is Avogadro's number, one can derive the mixing rules for the cubic equations of state (17-18).

Redlich-Kwong:

$$a = \{\sum_i \sum_j x_i\, x_j\, a_{ij}^{2/3}\, b_{ij}^{1/3}\}^{3/2} / \{\sum_i \sum_j x_i\, x_j\, b_{ij}\}^{1/2}$$
$$b = \sum_i \sum_j x_i\, x_j\, b_{ij} \tag{11}$$

Table III
Binary Interaction Parameters for Different Simple Cubic Equations of State

Component i	Component j	EOS	T (C°)	k_{ij}	AAD in x_1	AAD in y_1
Carbon dioxide	n-Butanol	PR	40	0.09741	0.0042	0.0031
		SRK		0.09540	0.0041	0.0032
		RK		0.18257	0.0117	0.0172
		PR	110	0.07716	0.0112	0.0175
		SRK		0.07742	0.0117	0.0192
		RK		0.22451	0.0129	0.0316
Carbon dioxide	n-Octane	PR	40	0.07806	0.0127	0.0011
		SRK		0.08582	0.0127	0.0015
		RK		0.12685	0.0146	0.0067
		PR	110	0.04936	0.0138	0.0076
		SRK		0.05997	0.014	0.0078
		RK		0.18032	0.0166	0.0185
Carbon dioxide	Methyl oleate (MO)	PR	40	0.02294	0.0237	0.0079
		SRK		0.04504	0.0231	0.0072
		RK		0.09021	0.0160	0.0096
		PR	70	0.07644	0.0313	0.0014
		SRK		0.08661	0.0325	0.0011
		RK		0.21525	0.0234	0.0030
Ethane	Methyl oleate (MO)	PR	40	0.00080	0.0553	0.0041
		SRK		-0.00121	0.0571	0.0032
		RK		0.06519	0.0483	0.0148
		PR	70	-0.00338	0.0454	0.0075
		SRK		-0.00571	0.0466	0.0072
		RK		0.12366	0.0249	0.0091
Carbon dioxide	Methyl linoleate	PR	70	0.00397	0.1042	0.0008
		SRK		0.01195	0.1051	0.0007
		RK		0.1194	0.0863	0.0039

with the following combining rules:

$$a_{ij} = (1-k_{ij})(a_{ii}a_{jj})^{1/2}$$
$$b_{ij} = [(b_{ii}^{1/3} + b_{jj}^{1/3})/2]^3 \tag{12}$$

Peng Robinson:

$$a = \sum_{i}\sum_{j} x_i\, x_j\, a_{ij}$$
$$b = \sum_{i}\sum_{j} x_i\, x_j\, b_{ij} \tag{13}$$
$$c = \sum_{i}\sum_{j} x_i\, x_j\, c_{ij}$$

with the following combining rules:

$$a_{ij} = (1-k_{ij})\, b_{ij}\left(\frac{a_{ii}\, a_{jj}}{b_{ii}\, b_{jj}}\right)^{1/2}$$
$$b_{ij} = (1-l_{ij})[(b_{ii}^{1/3} + b_{jj}^{1/3})/2]^3 \tag{14}$$
$$c_{ij} = (1-m_{ij})[(c_{ii}^{1/3} + c_{jj}^{1/3})/2]^3$$

Kwak and Mansoori (17) tested these mixing rules through the prediction of solubility of high molecular weight solids in supercritical fluids. They showed that these mixing rules can predict supercritical solid solubilities more accurately than the conventional mixing rules for the Redlich-Kwong and Peng-Robinson equations of state.

Mansoori and co-workers also tested the conformal solution mixing rules with other equations of state on systems containing a high molecular weight liquid in a supercritical fluid mixture. They showed that the Peng-Robinson equation of state using mixing rules based on conformal solution theory can predict the fluid phase equilibrium of high molecular weight liquids in supercritical fluids more accurately than others (18,19).

The Panagiotopoulos-Reid mixing rule (P & R Mixing Rule) was developed by making the normal single binary interaction parameter, k_{ij}, composition dependent (20,21). Two binary interaction parameters k_{ij} and k_{ji} ($k_{ij} \neq k_{ji}$), are determined from regression of experimental data. The "effective" interaction parameter between component i and j approaches k_{ij} as x_i approaches zero and approaches k_{ji} as x_i approaches unity. Application of this mixing rule for the calculation of the mixture parameter a_m results in a cubic expression for the mole fraction dependence, instead of the conventional mixing rule, which is a quadratic expression for a_m. The form of the empirical modifications of the mixing rules and combining rules and the resulting expressions for the fugacity coefficient in a mixture for the case of a general cubic EOS are given below:

$$a_m = \sum_{i}^{n}\sum_{j}^{n} x_i x_j a_{ij} \quad (15)$$

$$a_{ij} = \sqrt{(a_{ii}a_{jj})}[(1-k_{ij} + (k_{ij}-k_{ji})x_i)] \quad (16)$$

$$\ln \phi_k = (B_k/B)(Z-1) - \ln(Z-B) + ((\sum_i x_i (a_{ik}+a_{ki}) - \sum_i\sum_j x_i{}^2 x_j (k_{ij}-k_{ji})\sqrt{(a_{ii}\ a_{jj})} + x_k \sum_i x_i (k_{ki}-k_{ik})\sqrt{a_i\ a_k})/a_m - B_k/B)* a_m/(\sqrt{(u^2-4w)}b_m RT)*\ln(2v+b_m(u-\sqrt{u^2-4w})/(2v+b_m(u+\sqrt{u^2-4w}))) \quad (17)$$

where, $B = b_m P/RT$.

Results

The principal motivation behind the development of the new mixing rules has been the representation of phase equilibrium in systems that contain a supercritical component and one biomaterial. Both the Mansoori and the Panagiotopoulos & Reid mixing rules were tested for improving the prediction of equilibrium phase behavior. For the biomaterials methyl oleate and methyl linoleate in supercritical fluids, the Panagiotopoulos & Reid mixing rules give better results. It might be anticipated that the Panagiotopoulos and Reid mixing rules would give better agreement with experiment than the Mansoori rules with the Redlich-Kwong type equations because they have two parameters instead of one. However, in the case of the Peng-Robinson equation the two parameter Panagiotopoulos & Reid mixing rules still give superior results despite there being three adjustable parameters in the Mansoori mixing rules for this equation. The Mansoori mixing rules were also successful in predicting fluid phase equilibrium, but only in the moderate pressure range (22).

The inadequacy of the conventional method for such systems is demonstrated in Figures (1-2) that show calculated values of liquid and vapor compositions using both one-parameter and two-parameter mixing rules. It is clear that when the adjustable parameter in the single-parameter correlation is fitted to the composition of one phase, the results for the other phase are very poor. In contrast, the agreement between the model predictions using two-parameters and experiment is better. Both the Peng-Robinson and Soave-Redlich-Kwong equations of state have been used with the Panagiotopoulos & Reid mixing rules, giving similar results. The pure component parameters for the working materials are shown in Table IV. The optimum values of the two parameters, k_{ij} and k_{ji}, for phase equilibrium prediction are shown in Table V.

It is interesting to note that for the systems carbon dioxide with n-butanol and with n-octane, the optimal values of the two-parameters k_{ij} and k_{ji} are quite close to each other. Use of the conventional mixing rules for these simple systems would result in almost as good agreement between experiments and predictions as for the two-parameter correlation. However, for biomaterials, the Panagiotopoulos & Reid mixing rules improves the prediction of the phase behavior appreciably.

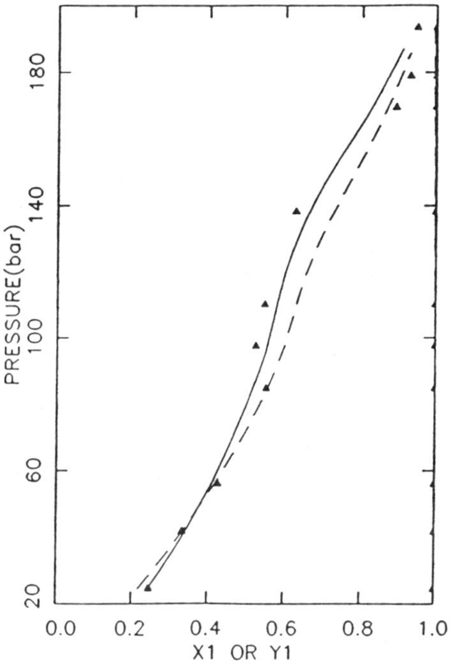

Figure 1 VLE data of methyl oleate in SC-CO_2 at 70°C as calculated with Peng-Robinson equation of state with conventional and Panagiotopoulos and Reid mixing rules compared with experimental data.

- - - k_{ij} = 0.07644 (PR-EOS),

—— k_{ij} = -0.01513 and k_{ji} = 0.1284 (PR-EOS with P & R mixing rule).

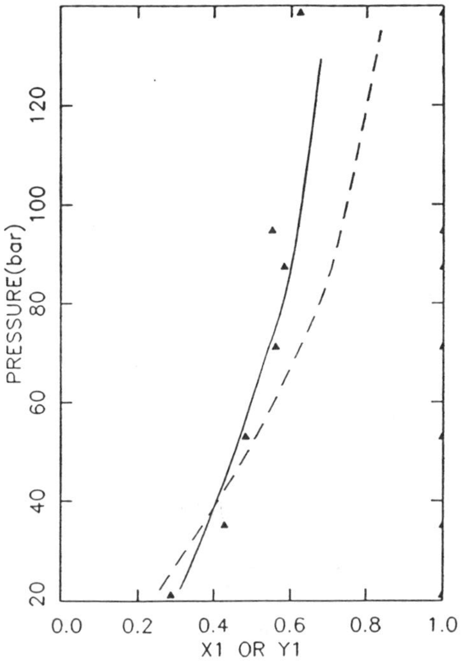

Figure 2 VLE data of methyl linoleate in SC-CO_2 at 70°C as calculated with Peng-Robinson equation of state with conventional and Panagiotopoulos and Reid mixing rules compared with experimental data.

- - - k_{ij} = 0.00397 (PR-EOS),

——— k_{ij} = -0.23581 and k_{ji} = 0.11142 (PR-EOS with P & R mixing rule).

Table IV
Pure Component Parameters for the Working Materials

Component	Temp (°C)	a[atm(liter/mol)^2] PR-EOS	a[atm(liter/mol)^2] SRK-EOS	b(liter/mol) PR-EOS	b(liter/mol) SRK-EOS
Carbon dioxide	40	3.8293	3.5676	0.0267	0.0297
	70	3.5736	3.2898	0.0267	0.0297
	110	3.2621	2.9539	0.0267	0.0297
Ethane	40	5.8782	5.4799	0.0404	0.0450
	70	5.5884	5.1521	0.0404	0.0450
N-Butanol	40	38.0053	37.7222	0.0825	0.0919
	110	32.7045	31.9697	0.0825	0.0919
N-Octane	40	62.7682	61.9394	0.1482	0.1650
	110	55.5096	54.0169	0.1482	0.1650
Methyl oleate	40	384.9416	398.9905	0.3961	0.4411
	70	362.1318	373.4044	0.3961	0.4411
Methyl linoleate	70	357.5449	368.7952	0.3897	0.4340

Table V
Optimum Values of Two Interaction Parameters
For Phase Equilibrium Predictions

Systems	°C	k_{ij} PR-EOS	k_{ij} SRK-EOS	k_{ji} PR-EOS	k_{ji} SRK-EOS
CO_2(1) /n-Butanol(2)	40	0.10034	0.09833	0.09532	0.09334
	110	0.06676	0.06527	0.09203	0.09436
CO_2(1) /n-Octane(2)	40	0.09679	0.10545	0.07166	0.07916
	110	0.02766	0.03864	0.06001	0.07000
CO_2(1) /Methyl oleate(2)	40	-0.09295	-0.11297	0.03950	0.05375
	70	-0.01513	-0.00887	0.12840	0.13987
Ethane (1) /Methyl oleate(2)	40	-0.3383	-0.35403	0.04158	0.04095
	70	0.09363	0.0874	-0.00056	-0.00310
CO_2(1) /Methyl linoleate(2)	70	-0.23581	-0.24088	0.11142	0.12404

Although the phase equilibrium in binary mixtures of simple molecules and biomolecules has been modeled fairly successfully, a general extension of this approach to more complex, multi-component systems remains a challenge.

Conclusions

A two-parameter mixing rule is used with several cubic equations of state and is shown to be relatively successful in correlating the phase equilibrium behavior of biomolecules that cannot be correctly represented by conventional one-parameter mixing rules. The modification is related to the idea of local composition, which has been shown to improve the representation of the phase equilibrium in asymmetric mixtures. However, further improvement is still needed.

The use of such a model may substantially facilitate the task of process design and optimization for separations that utilize supercritical fluids.

Acknowledgements

Financial support from the National Science Foundation (CBT-8704581) is gratefully acknowledged. The authors also thank Professor Panagiotopoulos for helpful discussion.

Legend of Symbols

AAD	absolute average deviation
a,b	equation of state parameters
f	fugacity
N_o	Avogadro's number
P	pressure
R	gas constant
T	absolute temperature
v	molar volume
x,y	mole fraction
Z	compressibility factor
ϕ	fugacity coefficient
ω	Pitzer's acentric factor
σ	molecular interaction distance
ϵ	energy parameter

Subscripts

c	critical property
i,j	components
m	mixture

Superscripts

L	liquid phase
V	vapor phase

Literature Cited

1. Rizvi, S.S.H.; Benado, A.L.; Zollweg, J.A.; Daniels, J.A. Food Tech. 1986, 40(6),55-65.
2. Rizvi, S.S.H.; Daniels, J.A.; Benado, A.L.; Zollweg, J.A. Food Tech. 1986, 40(7),57-64.
3. Paulaitis, M.E.; Krukonis, V.J.; Kurnik, R.T. Chem. Eng. 1982, 181-240.
4. Cheng, H.; Zollweg, J.A.; Streett, W.B. This Symposium.
5. Adams, W.R.; Zollweg, J.A.; Rizvi, S.S.H.; Streett, W.B.; AIChE J. 1988, 34, 1387-91.
6. Cheng, H.; Zollweg, J.A.; Streett, W.B.; Paper Presented at International Symposium on Thermodynamics in Chemical Engineering & Industry, Beijing, China. 1988.
7. Prausnitz, J.M.; Lichthentiler, R.N.; Azeudo, E.G.; In Molecular Thermodynamics of Fluid Phase Equilibria. 2nd Edn. Prentice Hall, Inc.
8. Reid, R.C.; Prausnitz, J.M.; Poling, B.E.; In The Properties of Gases and Liquids 1987, 4th Edn. McGraw Hill Book Company.
9. Spencer, C.F.; Daubert, T.E.; AIChE J. 1973, 19(3),482-86.
10. Brunner, G.; Hederer, H.; High Pressure Science and Technology 1979, j-3, 527-534.
11. Panagiotopoulos, A.Z.; Kumar, S.K.; Fluid Phase Equilibria 1985, 22,77-88.
12. Paulaitis, M.E.; Penninger, J.M.; Gray, R.D., Jr.; Davidson, P.; In Chemical Engineering at Supercritical Fluid Conditions 1983. Ann Arbor Science Publishers.
13. Huron, M.J.; Vidal, J.; Fluid Phase Equilibria 1977, 3,255-71.
14. Mollerup, J.; Fluid Phase Equilibria 1981, 7,121-38.
15. Vidal, J.; Ber. Bunsenges. Phy. Chem. 1984, 88,784-91.
16. Mansoori, G.A.; Ely, J.F.; J. Chem. Phys. 1985, 82,406-13.
17. Kwak, T.Y.; Mansoori, G.A.; Chem. Eng. Sci. 1986, 41(5),1303-09.
18. Park, S.J.; Kwak, T.Y.; Mansoori, G.A.; International J. Thermophysics. 1987, 8(4),449-71.
19. Benmekki, E.H.; Mansoori, G.A.; Fluid Phase Equilibria, 1987, 32,139-149.
20. Panagiotopoulos, A.Z.; Reid, R.C.; ACS Symposium Series No. 300; Equation of State, Theories and Application, p 571. 1986.
21. Panagiotopoulos, A.Z.; Reid, R.C.; ACS Symposium Series No. 329; p 115. 1987.
22. Rizvi, S.S.H.; Zou, M.; Kashulines, P.; Benkrid, H.; 1988. International meeting at Porto, Portugal, Oct. 17.

RECEIVED May 2, 1989

Chapter 9

Four-Phase (Solid–Solid–Liquid–Gas) Equilibrium of Two Ternary Organic Systems with Carbon Dioxide

Gary L. White and Carl T. Lira

Department of Chemical Engineering, Michigan State University, East Lansing, MI 48824–1226

Melting temperatures of organic solids may be depressed significantly when contacted with supercritical fluids. In this work, P-T traces are reported for the high temperature branch of the three phase (S-L-G) line for the naphthalene-CO_2 and phenanthrene-CO_2 systems and the four phase (S-S-L-G) lines beginning at the solid-solid eutectic for the naphthalene-phenanthrene-CO_2 and naphthalene-biphenyl-CO_2 systems.

Competent design of chemical processes requires accurate knowledge of such process variables as the temperature, pressure, composition and phase of the process contents. Current predictive models for phase equilibria involving supercritical fluids are limited due to the scarcity of data against which to test them. Phase equilibria data for solids in equilibrium with supercritical solvents are particularly sparse. The purpose of this work is to expand the data base to facilitate the development of such models with emphasis on the melting point depressions encountered when solid mixtures are contacted with supercritical fluids.

Previous workers (1-7) have demonstrated that the melting points of solids in binary systems with ethylene, ethane, or carbon dioxide at elevated pressures may be significantly reduced. This phenomenon is due to the solubility of the light component in the liquid phase. Temperature minimums have been observed along the S-L-G three phase line for some systems (4-7). For these systems, the three phase line intersects the critical locus for the mixture at the upper and lower critical end points (UCEP and LCEP) as shown in Figure 1. A similar phenomenon is also observed for the eutectic melting point of certain binary mixtures of solids in contact with gases at elevated pressures as illustrated in Figure 2. This could potentially lead to a shift in the ratio of the molten solids in the eutectic liquid.

0097–6156/89/0406–0111$06.00/0

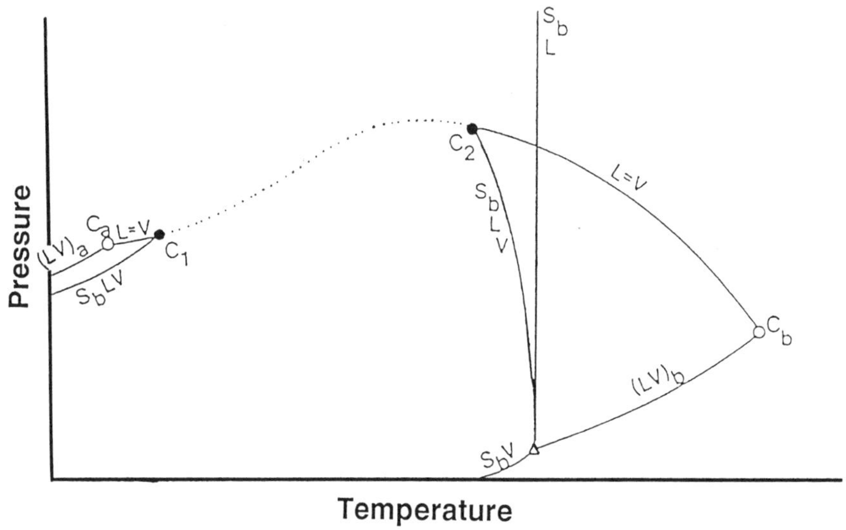

Figure 1. P-T projection for a binary system with a discontinuous critical mixture line.

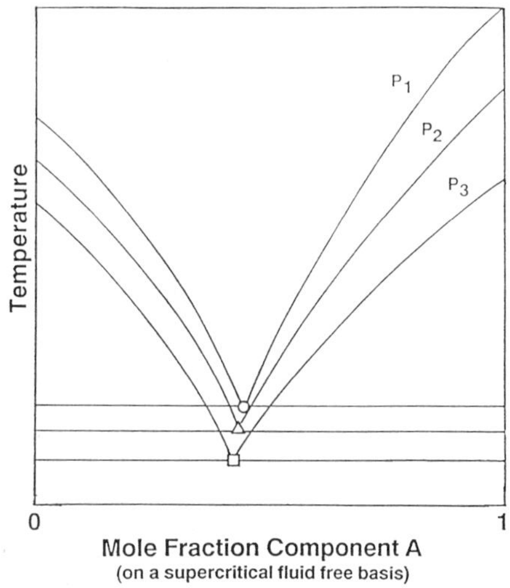

Figure 2. Depression of eutectic melting point by a supercritical fluid in an A-B-SCF system, where A,B are immiscible solids and $P_1 < P_2 < P_3$. The upper lines represent first freezing and the lower lines represent first melting (o - P_1, Δ - P_2, □ - P_3).

In work with a ternary mixture of ethylene, naphthalene, and hexachloroethane, van Gunst et. al.(8) observed an analogous temperature depression of the four phase (S-S-L-G) minimum melting solid line. As shown in Figure 3, such system exhibited two ternary critical end points, designated as the "p" (lower temperature) and "q" (higher temperature) points. Ternary systems may display such an interruption of the critical locus if the binaries mixtures of the individual solids with the solvent gas also have interrupted critical loci. The existence of the UCEP and LCEP for each of the binaries does not, however, mandate such behavior for the ternary critical locus.

For the binary systems, only the P-T trace extending to the UCEP is sought in this work. The LCEP is typically very near the critical temperature and pressure for the pure supercritical fluid. Similarly, for the ternary systems, only the q point branch of the four phase line is sought in this work. Systems studied here exhibit solid-solid immiscibility.

Experimental Apparatus

A schematic of the experimental apparatus is shown in Figure 4. Solvent gas from a supply cylinder is compressed with an air operated gas booster (Haskell model AC-152) to a pressure above that neede in the view cell. A High Pressure Equipment Company reactor vessel is used as a reservoir for the compressed gas. Flow of the solvent gas from the reservoir is controlled by a shutoff valve. The inlet and outlet from the valve are to .03 inch I.D. tubing, which permits incremental pressure increases in the cell of as little as 2 psi. Pressure in the view cell is measured with a Bourdon tube gauge (Heise model CMM-63457). The temperature within the cell is measured with a calibrated thermistor (Omega Engineering, model THX-400-GP) which passes through a compression fitting into the cell. Temperature control is achieved with an insulated water bath. A 600 watt copper tubing base heater and a Bayley Instruments model 123 temperature controller with a 1000 watt quartz bayonet makeup heater provide heat to the bath. Tap water flowing through coiled copper refrigeration tubing provides cooling for the bath.

Figure 5 shows a cross section of the view cell. This cell is fabricated from 316 stainless steel. The interior of the cell is illuminated through the window at the top of the cell by a fiberoptic light. The states of the contents are observed through the side window by means of a closed circuit color camera connected to a borescope. Although view of the upper region of the cell is restricted, any additional phase formed in this region must have a mass density less than the mass density of the supercritical phase. Such behavior is not anticipated with the systems studied here. The windows are 3/4 inch x 3/4 inch quartz. A triangular magnetic stir bar at the bottom of the cell stirs the lower phase in the cell. A rectangular stainless steel wire mesh "flapper" on a shaft set into the magnetic stir bar stirs the upper phase. The solid sample rests on a stainless steel wire mesh platform at the level of the side window of the view cell.

Method

First melting and first freezing have been used to study melting point melting point depressions. Both methods may be used in

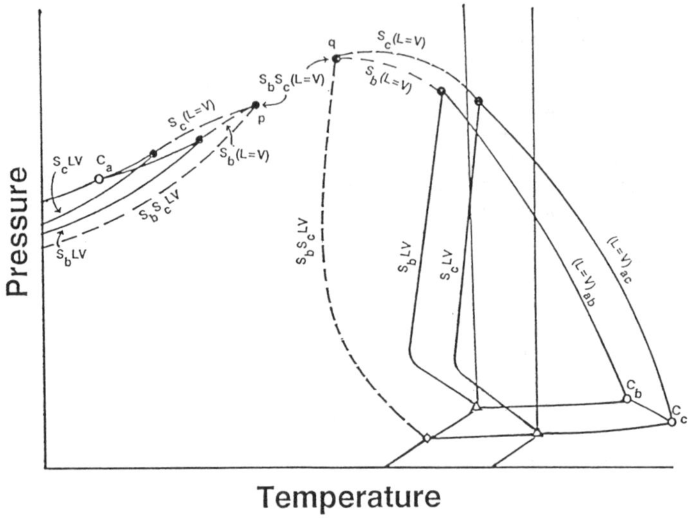

Figure 3. P-T projection for a ternary system with discontinuous four phase line.

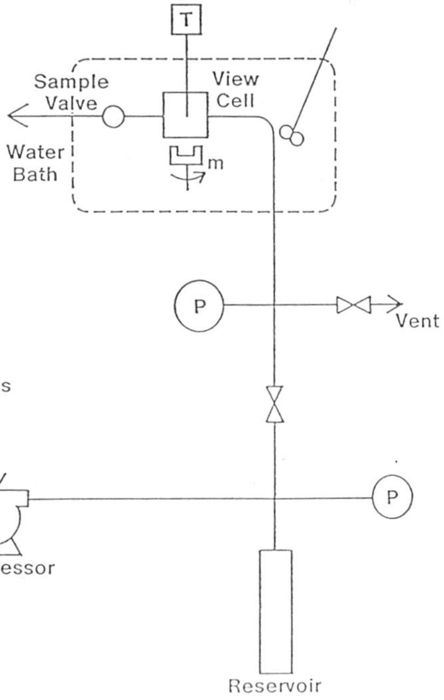

Figure 4. Schematic of the experimental apparatus. (P - pressure gauge, T - thermistor, m - magnetic stirrer for view cell).

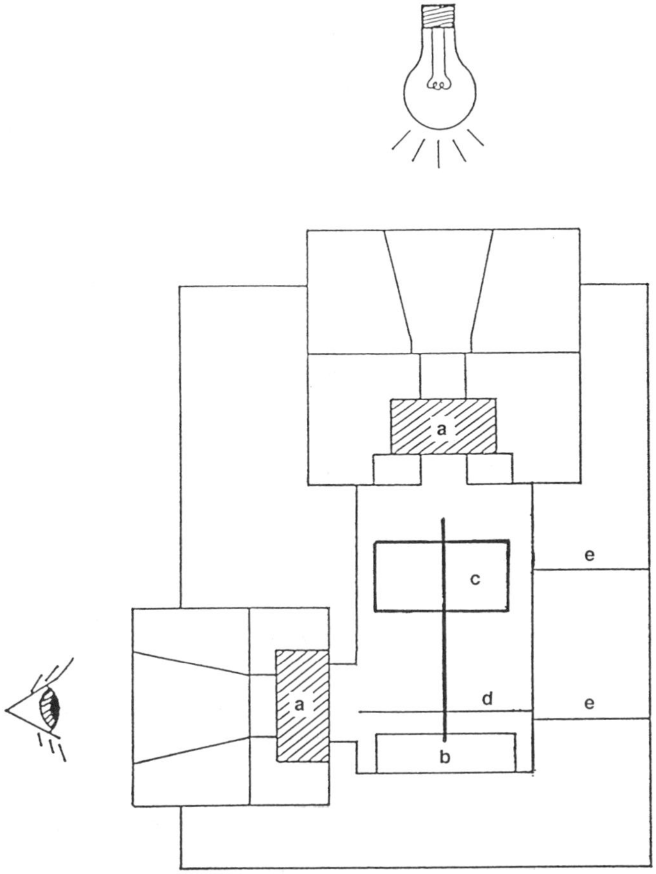

Figure 5. Diagram of the view cell. (a - quartz windows, b - magnetic stir bar, c - wire mesh flapper, d - sample platform, e - sampling ports).

binary systems, but only the first melting method is suitable for ternary systems (see Figure 2). Sample preparation is similar for both the binary and the ternary systems. For binary system measurements, some of the solid is melted and drawn into a section of capillary tubing. A 3/4 inch section of this filled tubing is placed on the wire platform and serves to hold the a portion of the sample being studied in a constant position and prevent it from draining away when melting occurs.

For ternary system measurements, the two solids to be studied are first mixed to provide intimate contact. The minimum melting point for the solid-solid binary at atmospheric conditions occurs at the eutectic composition. A solid mixture of this composition is prepared in the expectation that a similar ratio of the components will be present in the liquid phase at the first melting point in the ternary mixture. The solid mixture is melted and stirred to obtain a homogenous liquid. Some of the liquid is drawn into a capillary tube. The remaining liquid is poured out onto a clean sheet of aluminum foil. After the binary liquid cools and solidifies, the thin sheet of solid material is broken up into "flakes" for loading into the view cell.

Once the sample is prepared, adequate solid is loaded into the view cell to ensure the presence of excess solid at all conditions to be studied. The capillary tube is placed on the wire mesh platform as close as possible to the side window and parallel to it. With the sample loaded, the cell is sealed, placed in the temperature control bath, and connected to the high pressure gas reservoir. The cell is purged with the solvent gas and then filled with enough of the solvent fluid to bring it to near the desired final pressure. The bath is also brought up to near the desired temperature.

Measurements are made by two methods. Method A is used in the initial lower pressure region where the decrease in melting point with increasing pressure is most rapid. The temperature of the bath is held constant and the pressure in the cell is varied by adding and releasing small amounts of the gas or vapor phase. The pressure at which the first melting occurred within the capillary tube (near the ends is recorded as the melting point. Accuracies for these measurements are ± 5 psia and ± .05 °C.

In the higher pressure region where the P-T curve becomes almost parallel to the pressure axis, method B is used. The pressure of the cell is raised to near the desired level and then the temperature is raised slowly until the first melting is observed. Method B corresponds to the method used by McHugh (4-5) to determine similar P-T traces. For this method, both the pressure and the temperature are changing with time. Accuracies for these measurements are ± 1 psi and ± .2 °C.

Materials

Chemicals used in this study are listed in Table I. All chemicals were used without further purification. The purity of the naphthalene and phenanthrene were verified by measuring the melting point range of each at atmospheric pressure.

Results

The P-T data obtained in this study are listed in Table II and Table III. The P-T trace of the S-L-V line of naphthalene-CO_2 was

Table I. Purity of Materials

Chemical	Supplier	Purity
Naphthalene	Aldrich	99+%
Biphenyl	Aldrich	99%
Phenanthrene	Aldrich	98+%
CO_2	Linde Co.	Bone dry grade

determined and compared to literature data to validate the method. Data obtained in this study for the naphthalene-CO_2 system are plotted in Figure 6 with data from McHugh (4-5) and Cheong et. al. (6) for comparison.

Figure 7 illustrates the P-T trace of the ternary system napthalene-biphenyl-CO_2. The P-T traces of the constituent binary systems are included for comparison. Biphenyl-CO_2 data shown in Figure 7 for are those reported by McHugh (4-5) and Cheong et. al. (6). The three phase P-T traces for both the naphthalene-CO_2 and the biphenyl-CO_2 binaries lie above the critical temperature of carbon dioxide. For the ternary mixture, however, the four phase line runs into the subcritical region for carbon dioxide. The characteristics of the phase transition change at slightly above the last reported data point (870 psia and 23.1 °C). In this region, attempts to extend the S-L-V line by increasing the pressure cause a CO_2 rich liquid phase to begin forming in the bottom of the cell. The crystals on the platform appear to be unchanged during this phase transition. Further experiments are in progress to study the phase behavior of this system.

Figure 8 shows the data obtained for the naphthalene-CO_2 and phenanthrene-CO_2 binaries and the naphthalene-phenanthrene-CO_2 ternary system in this study. In the ternary, the four phase line falls above the critical conditions for pure carbon dioxide but significantly below the three phase P-T line for each of the single solid-CO_2 binaries. Based on data taken, the ternary critical end point q point) for the for this system lies between 32.0 and 34.2 °C and between 1335 and 1415 psia. Outside this region, no melting of the solids is observed with increasing temperature, but decreasing the pressure from 1415 psia while holding the temperature constant at 34.2 °C causes a liquid to condense out which, when the pressure is further decreased, completely

Table II. P-T Data for the Naphthalene-Biphenyl-CO_2 System

Naphthalene/CO_2		Naphthalene/Biphenyl/CO_2	
P(psia)	T(°C)	P(psia)	T(°C)
15	80.2	15	39.4
420	74.1	210	34.3
480	73.2	250	33.4
810	68.8	390	30.4
960	67.0	560	26.9
1090	65.6	650	25.5
1240	63.8	735	24.5
1480	61.65	870	23.1
1950	58.5		
2470	58.6		
2890	59.5		

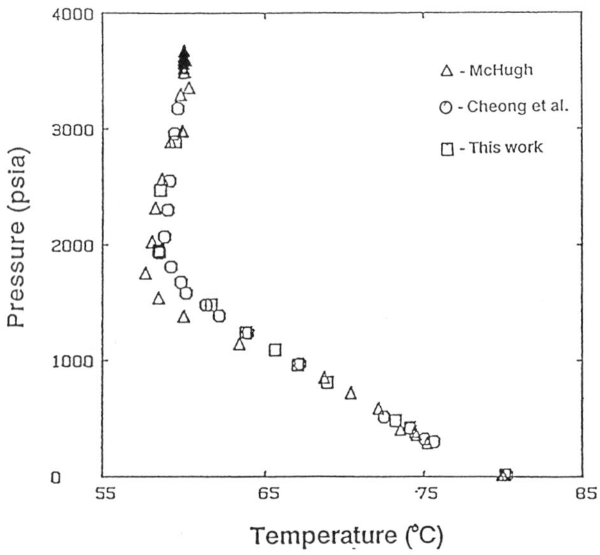

Figure 6. P-T projection for the naphthalene-CO_2 system. (Data from Ref. 4-6 and this work).

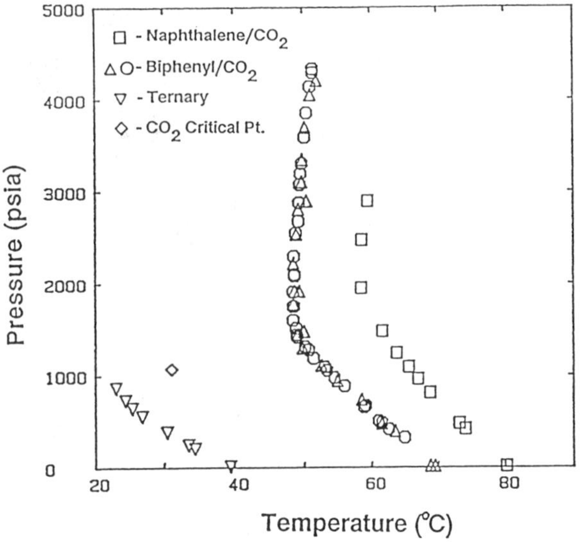

Figure 7. P-T projections for naphthalene-CO_2, biphenyl-CO_2, and naphthalene-biphenyl-CO_2 systems. (Biphenyl data from Ref. 4-6).

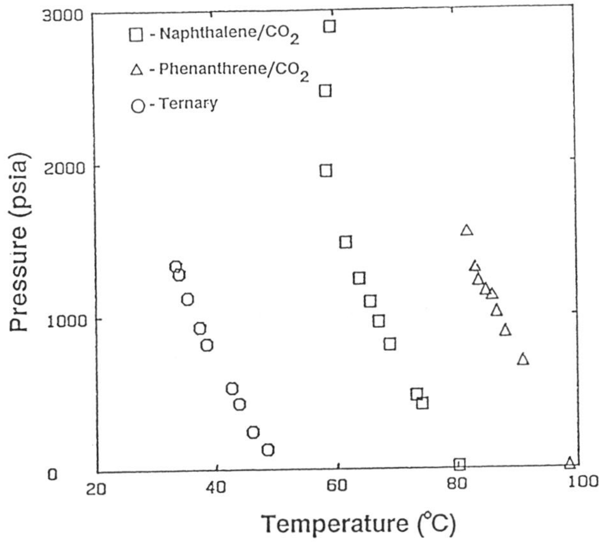

Figure 8. P-T projections for naphthalene-CO_2, phenanthrene-CO_2, and naphthalene-phenanthrene-CO_2 systems.

Table III. P-T Data for the Naphthalene-Phenanthrene-CO_2 System

Phenanthrene/CO_2		Naphthalene/Phenanthrene/CO_2	
P(psia)	T(°C)	P(psia)	T(°C)
15	98.5	130	48.4
700	91.0	245	46.0
890	88.1	430	43.8
1020	86.7	535	42.6
1110	86.5	820	38.5
1160	84.9	930	37.4
1225	83.7	1120	35.4
1310	83.9	1280	34.0
1540	82.5	1335	33.5
1542	82.6		

solidifies at the expected pressure for the four phase P-T line. Further work will be required to locate the q point of this system more accurately.

Conclusions

New data are presented for the P-T traces of the four phase line representing the melting point depressions in two ternary systems. These data are valuable for testing and development of predictive models of melting point depressions in multicomponent systems. This work demonstrates that binary data alone do not necessarily indicate some important aspects of the phase behavior, such as melting point depressions of multicomponent systems involving high pressure gases and supercritical fluids.

Acknowledgments

Partial support for this work by NSF grant No. CBT 86 10705 is gratefully acknowledged.

Literature Cited

1. van Welie, G. S. A.; Scheffer, F. E. C. J. Am. Chem. Soc. 1948,70, 4081-4085.
2. van Gunst, C. A.; Scheffer, F. E. C.; Deipen, G. A. M. J. Phys. Chem. 1953, 57, 578-581.
3. van Welie, G. S. A.; Diepen, G. A. M. Rec. Trav. Chim. 1961, 80,659-680.
4. McHugh, M. A. Ph.D. Dissertation, University of Delaware, Newark, DE 1981.
5. McHugh, M. A.; Yogan, T. J. J. Chem. Eng. Data 1984, 29, 112-115.
6. Cheong, P. L.; Zhang, D; Ohgaki, K.; Lu, B. C. Y. Fluid Phase Equilibria 1986, 29, 555-562.
7. Zhang, D.; Adachi, Y.; Lu, B. C. Y. Proc. Int. Symp. Supercrit. Fluids, Nice, France, October 17-19, 1988, Tome 1, 19-26.
8. van Gunst, C. A.; Scheffer, F. E. C.; Deipen, G. A. M. J. Phys. Chem. 1953, 57, 581-583.

RECEIVED May 2, 1989

SURFACTANTS, GELS, AND POLYMERS

Chapter 10

Direct Viscosity Enhancement of Carbon Dioxide

Andrew Iezzi[1], Robert Enick[1], and James Brady[2]

[1]Department of Chemical and Petroleum Engineering, 1249 Benedum Engineering Hall, University of Pittsburgh, Pittsburgh, PA 15261
[2]Department of Chemistry, 905A Chemistry Building, University of Pittsburgh, Pittsburgh, PA 15260

A high pressure, visual, falling-cylinder viscometer has been designed for use over a wide range of temperature and pressure, including near-critical and supercritical conditions. The viscometer consists of a precision-bore glass capillary which is contained within a high pressure windowed cell. The terminal velocity an aluminum cylinder falling through this tube was used to determine fluid viscosity. A vent at the bottom of the tube eliminated any pressure differential across the tube wall.

This viscometer was used in the evaluation of several additives which were thought to have the potential of enhancing the viscosity of dense carbon dioxide when present in dilute concentrations. Solutions of liquid CO_2 which were saturated with several surfactants did not have viscosities significantly different from pure CO_2, even when water was introduced into the system to stabilize and swell any micelles. Tri-n-butyltin fluoride forms weakly associating, linear polymers in light alkanes if its concentration is greater than about 0.3 weight percent. Although tri-n-butyltin fluoride is very soluble in pentane (29 weight percent) at ambient temperature, the use of pentane as a cosolvent did not increase the solubility of this compound in CO_2 (0.13 weight percent). High pressure mixtures of the semi-fluorinated alkanes in CO_2 did result in the formation of stable gels due to the microfibrillar morphology of the small excess amount of the semi-fluorinated alkane in the saturated liquid. This class of compound does, therefore, have potential for applications in which dramatic increases in CO_2 viscosity are desired.

0097–6156/89/0406–00122$06.00/0

The miscible displacement of oil from porous media is most effective when the mobility of the displacing fluid, defined as the ratio of the relative permeability of the fluid to its viscosity, is less than that of the fluid it is displacing. Although carbon dioxide is an effective fluid for recovering oil, its low viscosity results in an unfavorably high mobility in porous media. This leads to the 'fingering' of CO_2 through the oil-bearing formation, leaving much of the oil unrecovered. The direct viscosity enhancement of CO_2 to a level comparable to the oil it is displacing would decrease its mobility, thereby inhibiting this channeling and improving the volumetric sweep efficiency of this process.

Many of these formations consist of layers of porous media with varying permeability. If a zone is very permeable relative to the rest of the formation, most of the injected fluids will enter it. Extremely viscous CO_2 could be used as a diverting agent which would tend to minimize flow through a highly permeable layer. In this application a gel or very viscous liquid would be formed near the wellbore in this layer and remain immobile. Subsequently injected fluids would then flow into the other layers.

The direct viscosity enhancement of CO_2 would also improve its performance as a fracturing fluid in low permeability formations. A more viscous fluid would be able to transport the proppant particles into the fracture more effectively. These proppants hold the high permeability fracture open, increasing the rate of fluid recovery from the well.

LITERATURE REVIEW

Heller and Taber (1) were the first to study and report data involving the use of a direct thickener for dense carbon dioxide. Their efforts focused on commercially available polymers that would be sufficiently soluble in CO_2 to increase its viscosity a factor of 20. From their study, the authors found that none of the commercially available polymers were able to increase the viscosity of CO_2 to the desired level. However, they were able to make certain generalizations on the influence of various polymer properties on their solubility in liquid or supercritical CO_2.
Heller and Taber (2) then began pursuing three different approaches. The first approach was the synthesis of amorphous and preferably atactic polymers of varying molecular weights and with side chains which vary in carbon number. The goal was to create a large proportion of disorder and irregularity in such a multicomponent polymer so that, when combined with CO_2, it would impart high entropy to the system and become soluble. Some of the polymers prepared were found to be soluble in CO_2 but did not increase its viscosity. Secondly, the synthesis of ionomers from higher alpha-olefin terpolymers was then considered. However, since the work in the first approach had net yet reached the level needed to study this approach, no conclusions could be made concerning these compounds. The third approach dealt the the synthesis of organometallic compounds, specifically, tri-alkyltin fluorides. These compounds, when dissolved in non-polar solvents, form high molecular weight polymers by transient associations between adjacent molecules. A number of such compounds were prepared and were found useful in increasing the viscosity of dense butane and propane.

Heller, Kovarik, and Taber (3), continuing along the same lines of their previous work, reported that additional multicomponent polymers had been synthesized but their molecular weights were too high to attain sufficient solubilities in CO_2. They also began synthesis of diene containing terpolymers. Furthermore, they found that none of the organometallic compounds prepared were soluble enough in CO_2 to increase its viscosity.

Terry et al.(4) presented a novel idea to increase CO_2 viscosity. They attempted to increase its viscosity by insitu polymerization of CO_2 miscible monomers. The authors found that light olefins can be polymerized in such an environment using commonly available initiators. However, no apparent viscosity increases were detected since the polymers were insoluble in CO_2 and precipitated.

The use of entrainers to improve CO_2 mobility control for enhanced oil recovery was investigated by Llave, Chung, and Burchfield (5). The candidate entrainers were selected from high molecular weight alcohols and hydrocarbons and ethoxylated compounds including n-decanol, isooctane, 2-ethylhexanol, and an ethoxylated alcohol. They found that these entrainers were appreciably soluble in the CO_2 phase and effectively increased the phase viscosity and density. However, the viscosity increase could only be attained after the introduction of relatively large amounts of the entrainer.

In an attempt to improve the fracturing capabilities of low temperature liquid CO_2, Lancaster et al. (6) investigated compounds such as polymers, fumed silica, and amines, which are known to increase the viscosity of liquid alkanes. None of these compounds induced any viscosity change in CO_2.

Recently, Heller, Kovarik, and Taber (7) reported that the closest approaches to synthesizing practical direct thickeners for CO_2 had occurred in the organotin fluorides. Although the solubility of these compounds in CO_2 was not high enough to cause any significant change in solution viscosity, the authors felt that this type of associating compound offered a number of promising directions. For example, the synthesis of compounds with various types of groups attached to the tin molecules which may increase the solubility of the associating compound in CO_2, such as tris-(trimethylsilylpropyl) tin fluoride. The authors also reported that they had apparently exhausted all possibilities in their search for atactic, straight chain hydrocarbon polymers with varying side chains that would be soluble enough in CO_2 to increase its viscosity sufficiently.

EXPERIMENTAL APPARATUS AND PROCEDURE

Initial low pressure screening of the proposed "thickeners" was performed by using a falling ball viscometer to measure the viscosity of dilute mixtures of these compounds in liquids. The liquids chosen to simulate high pressure carbon dioxide were isoctane, pentane, and perfluorinated hexane. These compounds have similar densities, dipole moments, solubility parameters, dipolarity/polarizability (8-12) as dense CO_2 (Table I). At typical reservoir conditions, the CO_2 density ranges between .4 and .7 grams per cubic centimeter. Based on the solubility parameter, dipole moment, and the polarizability/dipolarity parameter, the perfluorinated hexane should be the best screening compound. The density of the compound, 1.7 grams per cubic centimeter, is about 3 times larger than that of CO_2. At higher denisites, .8 - 1.0 grams per cubic centimeter, the light alkanes should be the better screening compound based on all of the aforementioned parameters. Compounds which exhibit solubility in these liquids and an increase in solution viscosity were then evaluated in CO_2.

Table I. Characteristics of CO_2 and Several Liquids Suggested for Low-Pressure Screening

Compound	Density $\rho(g/cm^3)$	Solubility Parameter $(cal/cm^3)^{.5}$	Dipole Moment μ (debye)	Polarizability/ Dipolarity π^*
CO_2	.4	3.1	0	-.25*
	.5	3.8	0	-.18*
	.6	4.7	0	-.12*
	.7	5.2	0	-.08*
	.8	6.0	0	-.04*
	.9	6.7	0	-.01*
	1.0	7.4	0	0.0*
n-C_5H_{12}	.626**	7.0	0	-.08
n-C_6H_{14}	.684**	7.4	0	-.04
i C_8H_{18}	.692**	6.9	0	-.04
C_6F_{14}	1.69**	5.9	0	-.40
C_7F_{16}	1.73***	6.0	0	-.39

*at 50 °C from reference 8. Over this range, values increase with density. Reported values from other references (9,10) increase from -.6 to 0 over this density range.

**at 20 °C ***at 25 °C

A novel, high pressure, visual falling cylinder viscometer developed by Barrage (13) was employed to measure the viscosity of the solution formed when these compounds were combined with dense CO_2. The apparatus is shown in Figure 1. The two major components are an autoclave mixing chamber and a viscometer. After the "thickener" was charged into the system, CO_2 was compressed in until the desired pressure, 1000-2500 psia, was attained. Mixing and rapid equilibration was accomplished by the rotating impellers of the mixing cell. Any compound not solubilized into the dense phase settled into a trap in the bottom of the mixing cell.

A high pressure, visual rotameter has been modified to become a viscometer by replacing the original tapered tube and ball with a straight, precision bored tube and aluminum cylinder (see Figure 2). A pressure vent at the bottom of the tube permits pressure drop across the tube wall to be eliminated. The pressure distortion characteristic of sapphire crystal viscometers, which have high pressure CO_2 on the inside of the tube and ambient pressure on the outside, is thereby eliminated. There is not a vent at the top of the tube, therefore flow occurs only through the inside of the tube. The cylinder was machined from aluminum to minimize the density difference between the CO_2 and the cylinder, thereby decreasing the terminal velocity. Extremely small gap sizes could also be attained (0.113 mm) in order to decrease the terminal velocity. This enabled precise measurements of low viscosity fluids to be attained.

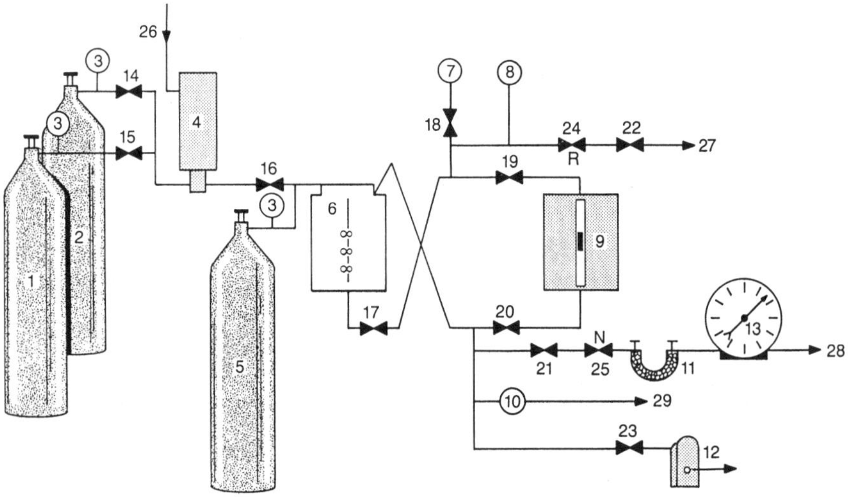

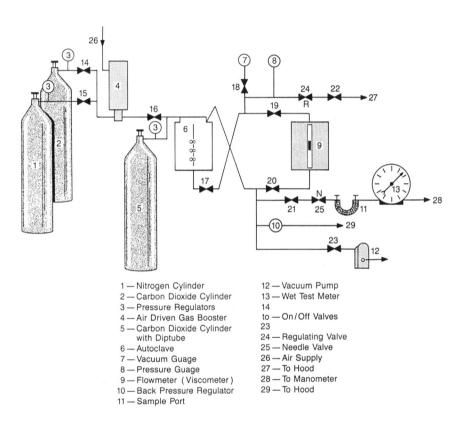

1 — Nitrogen Cylinder
2 — Carbon Dioxide Cylinder
3 — Pressure Regulators
4 — Air Driven Gas Booster
5 — Carbon Dioxide Cylinder with Diptube
6 — Autoclave
7 — Vacuum Guage
8 — Pressure Guage
9 — Flowmeter (Viscometer)
10 — Back Pressure Regulator
11 — Sample Port
12 — Vacuum Pump
13 — Wet Test Meter
14 to 23 — On/Off Valves
24 — Regulating Valve
25 — Needle Valve
26 — Air Supply
27 — To Hood
28 — To Manometer
29 — To Hood

Figure 1. Experimental Apparatus

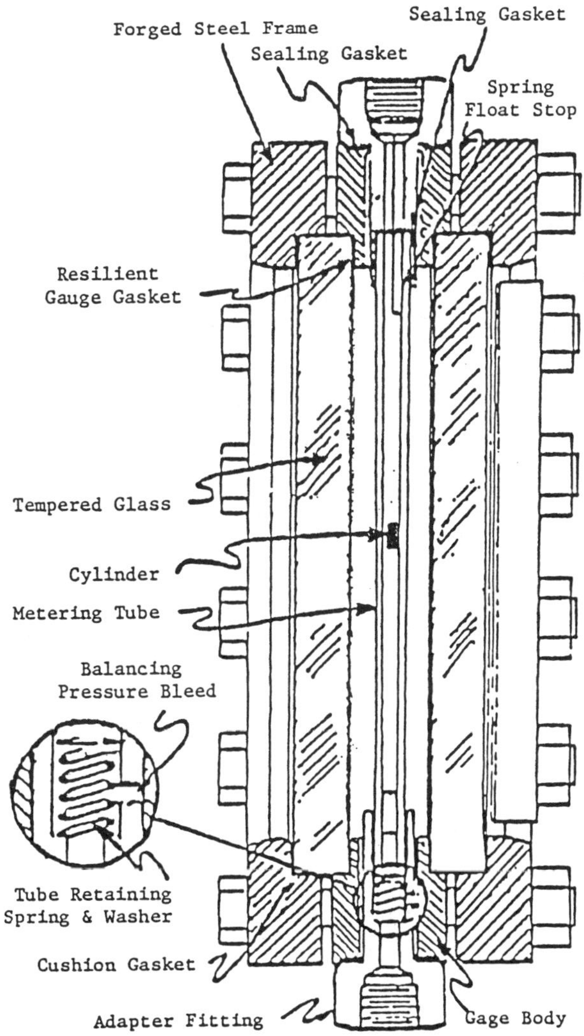

Figure 2. High Pressure, Visual, Falling Cylinder Viscometer

The impellers were also used to develop fluid flow through the viscometer. The upward flow created by the impellers forced the small aluminum cylinder to the top of the pressure equilibrated tube. Flow through the viscometer was then terminated by closing the valves leading to it. The cylinder then fell through a stationary column of the solution. Viscosities were calculated from the terminal velocities. The viscometer was calibrated with a fluids of known density and viscosity. This particular geometry, a coaxial cylinder and tube, also allowed the analytical derivation of the calibration constants (13) since the fluid displaced by the falling cylinder flowed through the annulus between the cylinder and tube wall. The specific parameters of the viscometer were r_o=2.000 mm, r_t= 1.887 mm and ρ_c=2.7 g/cm^3. Note that neither the length of the cylinder, nor the tube length, affected the viscosity measurements as long as the tube was long enough for terminal velocity to be attained. End effects were negligible since the ratio of cylinder length to annulus gap width was 70:1.

A comparison of experimental CO_2 viscosities obtained over a wide range of temperatures and pressures with that of previously reported viscosites (14) taken over that same range is illustrated in Figure 3. The three different experimental curves shown correspond to a particular calibration fluid, water or CO_2, or the constant derived from the Navier-Stokes equation.

$$\mu = K \left\{ \frac{(\rho_C - \rho_f)(t_2 - t_1)}{(l_2 - l_1)} \right\} \tag{1}$$

where

$$K = \frac{r_c\, g\, A}{-4r_c - 2B\{-(2r_c/B) - (r_t^2 - r_c^2) + A\}} \tag{2}$$

and

$$A = (r_t^2 - r_c^2)\,[\ln(r_c/r_t)] + (r_t^2 - r_c^2) \tag{3}$$

$$B = \frac{1}{r_c \ln(r_c/r_t)} \tag{4}$$

The curve which corresponds to the calibration based on CO_2 at 1000 psia was, as expected, the most accurate of the three with an absolute average percent deviation (AAPD) of 3.84. The constant derived from the Navier-Stokes equation was found to be extremely sensitive to the gap size between the aluminum cylinder and tube wall. For example, changing the tube diameter to values within the specified tolerance (+/- 0.0002 inches), changed viscosity calculations by as much as 17.60 percent.

CO_2 "THICKENING" CANDIDATES, RESULTS

This paper will examine three possible methods of enhancing the viscosity of dense carbon dioxide. Each of these is capable of increasing the viscosity of liquids which exhibit similar properties as CO_2.

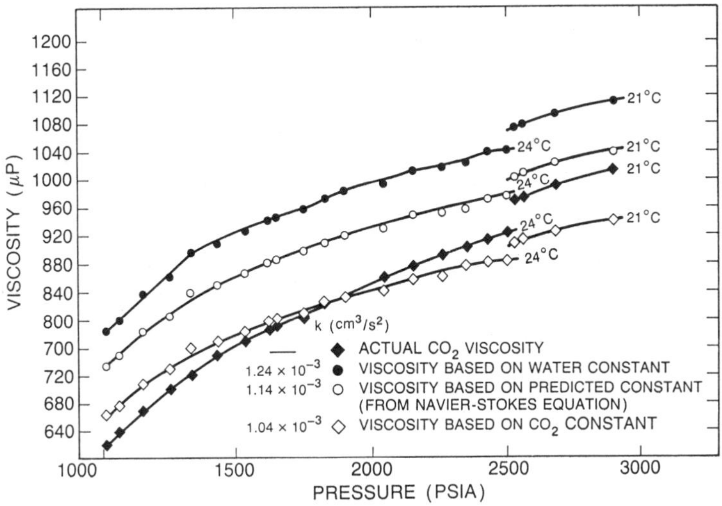

Figure 3. Viscometer Calibration

Reverse Micelles. Reverse Micelles in supercritical fluids are currently being studied for several distinct applications (15-18). Normal micelles and microemulsions in aqueous solutions are known to be capable of increasing solution viscosity in several applications, including the surfactant flooding of petroleum reservoirs.(19) If reverse micelles or microemulsions can be formed in CO_2, an increase in solution viscosity could possibly occur. The surfactants chosen as candidates for CO_2 flooding application should be characterized by low water solubility and a strong CO_2 solubility, minimal adsorption onto the porous media, and stability at reservoir conditions. (20)

In general, nonionic oil soluble surfactants were chosen for a study since they would satisfy these conditions, although many ionic surfactants with low HLB's were also evaluated. Pentane was used to simulate dense carbon dioxide. In the initial, low pressure evaluation of eighty-four commercially available surfactants which statisfied most of the criteria listed previously. Thirty-two of the samples exhibited at least 0.5 weight percent solubility in pentane (13). At weight concentrations of 0.5 - 4.0 percent surfactant (as is), none of these surfactants induced an increase in the solution viscosity above 10.0 percent.

Table II lists the surfactants which exhibited the highest degree of solubility in pentane. Saturated solutions of these surfactants in CO_2 and 23°C and 1000 -2500 psia were tested in the high pressure viscometer, with and without water present. However, none of these saturated solutions or a solution of CO_2 and Aerosol OT, which is known to form reverse micelles in supercritical light alkane systems, (15) resulted in viscosity increases relative to pure CO_2. It must be noted that experimental verification of reverse micelle formation was not performed. This method of increasing the CO_2 viscosity was not pursued because we felt that (1) the low viscosity of the continuous phase, CO_2, would not promote the formation of viscous, micellar solutions, and (2) the surfactants listed in Table II were the most likely to satisfy our requirements. Their inability to increase the viscosity of CO_2, whether micelles formed or not, did not lead us to believe other surfactants would yield dramatically better results.

Table II. Surfactants Used in Dense CO_2

Surfactant	Highest Degree* of Solubility (wt%)	Manufacturer
Pluronic L121	4.0	BASF Wynadote Corporation
Trylox 5900 CO-5	1.0	Emery Industries, Inc.
Trylox 6960 NP-1	1.0	Emery Industries, Inc.
Pluronic L101	1.0	BASF Wyandote Corporation
Witcamide 511	1.0	Witco Chemical Corporation
Petrosul 750	1.0	Penreco

*in pentane

Cosolvents and Tri-n-butyltin Fluoride. Dunn and Oldfield (15) first reported the ability of tri-n-butyltin fluoride to dramatically increase the viscosity of a relatively light alkane, namely hexane. Tri-n-butyltin fluoride exists in the form of a penta co-ordinate species, and is thereby able to form linear polymer chains by weak dipole-dipole interactions between the fluorine and tin of adjacent molecules. The fluid in which the tri-alkyltin fluoride is dissolved must have no dipole moment in order not to disrupt the transient associations which form the long chain polymer. As stated earlier, this compound was only very slightly soluble in CO_2 (less than two weight percent) and no viscosity increase was observed. Heller et at. (2,3,7) have conducted extensive research on these tri-alkyltin fluorides varying the alkyl groups in an attempt to enhance their solubility in CO_2.

An attempt was made to increase the solubility of commercially available tri-n-butyltin fluoride (TBTF) in liquid CO_2 using a cosolvent. The use of a small amount of a second component in dense CO_2 has been used extensively in supercritical extraction studies in an attempt to enhance the solubility of slightly polar compounds, heavier compounds, or compounds which are difficult to separate from a mixture (16,21). Significant increases in solubility (over an order of magnitude) have been achieved with as relatively small concentrations of cosolvent (less than 5%). The obvious cosolvents for enhancing the dissolution of TBTF in CO_2 were the C_3-C_6 n-alkanes. (TBTF was essentially insoluble in the C_7-C_8 alkanes at room temperature.) These alkanes are completely miscible in CO_2 at reservoir conditions.

The determination of the optimum cosolvent was conducted by comparing solubility of TBTF in various solvents. Figure 4 illustrates the solubilities and viscosities of mixtures of TBTF with four light alkanes. TBTF exhibited its greatest solubility, 29 weight percent, in pentane at room temperature, therefore it was chosen as the cosolvent. The high pressure data for propane and butane was measured and reported by Heller (3).

Viscous solutions of TBTF in pentane, ranging from 5 to 15 grams of TBTF per liter of pentane, were prepared by slowly adding the TBTF, which is a white powder, to the pentane and stirring for 24 hours. These solutions were then introduced into the mixing chamber. CO_2 was then introduced into the system until the desired pressure was attained. Temperatures varied between 23^oC and 24^oC for these tests. By using more concentrated solutions of TBTF in pentane, smaller amounts of the cosolvent could be employed for a given overall weight percentage of TBTF. By using larger amounts of any particular viscous solution, the effect of increasing the amount of TBTF in solution could be determined at a constant ratio of TBTF to pentane. The determination of the number of phases present, one or two (liquid - solid), was achieved by increasing the impeller rotational speed. This increased the fluid flow rate through the visual viscometer, enabling the observation of any entrained TBTF particles. The mole fraction of pentane in CO_2 were estimated using the Peng -Robinson equation of state (22) with the interaction parameter adjusted to -.2 to match the single liquid phase density of known composition. The amount CO_2 compressed into the apparatus was measured after the experiment by venting the CO_2 through the wet test meter.

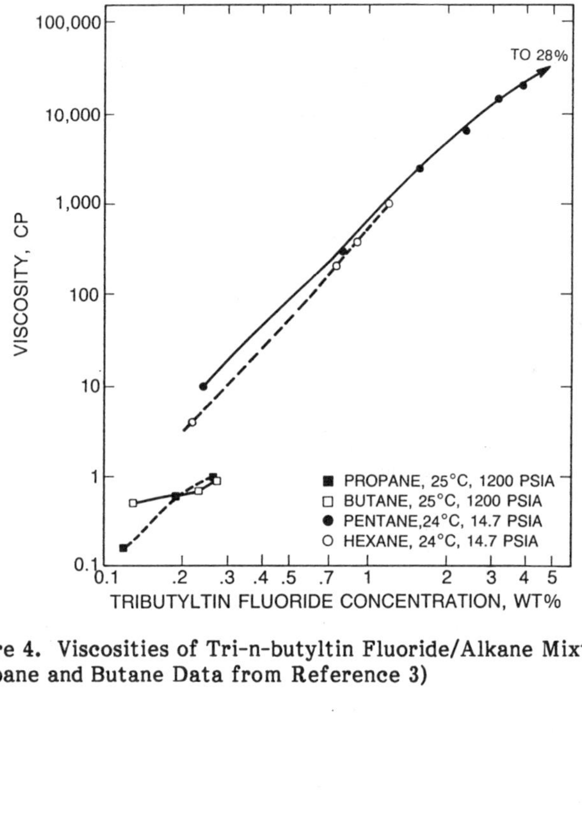

Figure 4. Viscosities of Tri-n-butyltin Fluoride/Alkane Mixtures (Propane and Butane Data from Reference 3)

Pentane was not an effective cosolvent. Although the solubility of TBTF in CO_2 was increased to levels as high as 0.8 weight percent, nearly all of the single phase solutions were metastable (see Figure 5). After several minutes to several hours, or after the rotational speed of the impeller was increased, the TBTF precipitated throughout the solution (like snow). High viscosity solutions (values between 100 and 1800 cp at TBTF concentrations between 0.6 and 0.8 weight percent were recorded) quickly deteriorated to solutions whose viscosity was characteristic of pure CO_2 after the precipitation of the excess TBTF. The metastable results indicated, however, that some other type of organotin fluoride molecule, more stable in CO_2, may be an effective viscosity enhancing agent.

Viscous, single phase solutions could probably be obtained by using a low ratio of TBTF to pentane, such as 5.0 gram per liter. These solutions would contain extremely large amounts of pentane (approximately 50.0 mole percent), which is impractical for application.

Semi-fluorinated Alkanes. Semi-fluorinated alkanes have been shown to form gels when dissolved in alkanes such as decane and octane, due to a microfibrillar morphology. This gel phase occurs when the mixture is heated above the melting point of the semi-fluorinated alkane, and then allowed to cool (23). Upon cooling from above the melting point, the semi-fluorinated alkane solidified in the form of needles in the presence of the remaining solution.

Micrographs of the crystallizing solution show extremely long needles which become interlocked in disarray (24). Thus, the gels are interdigitated crystallites which enclose large amounts of solvent in the cavities formed during crystallization (24). Semi-fluorinated alkanes ($F(CF_2)_n(CH_2)_mH$), denoted as F_nH_m, with varying ratios (n/m) and lengths (n + m) were synthesized and investigated in order to determine if a stable, viscous fluid or gel could be formed in liquid CO_2. The $F_{12}H_{8-20}$ series has been shown to form gels in decane, and $F_{10}H_{12}$ has formed gels in octane. Low pressure screening can be performed with the observation of gel formation in light liquid alkanes and perfluorinated alkanes.

Several semi-fluorinated alkanes (F_nH_m) were synthesized (25) from the two starting compounds, a 1-alkene and an iodoperfluorinated alkane. The solubilities of these F_nH_m compounds in the screening liquids were determined visually by mixing small amounts of the solid, white F_nH_m into a known volume of liquid. When a small excess of the F_nH_m appeared in the solution, the mixture was considered to be saturated. Several additional solutions with an even greater excess of the F_nH_m were then prepared. In order to form the gel phase, these saturated solutions containing an excess of the solid F_nH_m were slowly heated until the solid F_nH_m melted (25) and dissolved. This single phase solution was then allowed to cool to room temperature. As the temperature of the mixture decreased, the solution became supersaturated. The excess F_nH_m did not precipitate to the bottom of the flask, but instead formed a network of microfibers as indicated in Figure 6. If an excess of only 1.0 - 2.0 weight percent of F_nH_m was present, the entire mixture formed a gel. These gels were stable, remaining at the bottom of an inverted flask for weeks. When the gels were compressed, the microfiber network remained intact and the solvent dropped out, much like squeezing a sponge saturated with water.

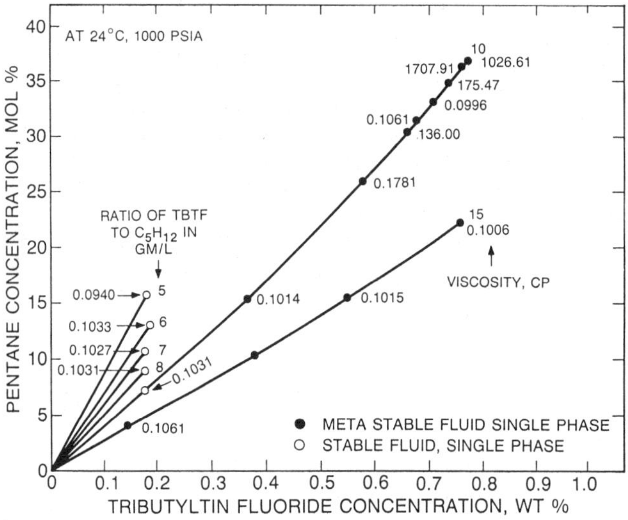

Figure 5. Viscosities of CO_2/Pentane/Tri-n-butyltin Fluoride Mixtures

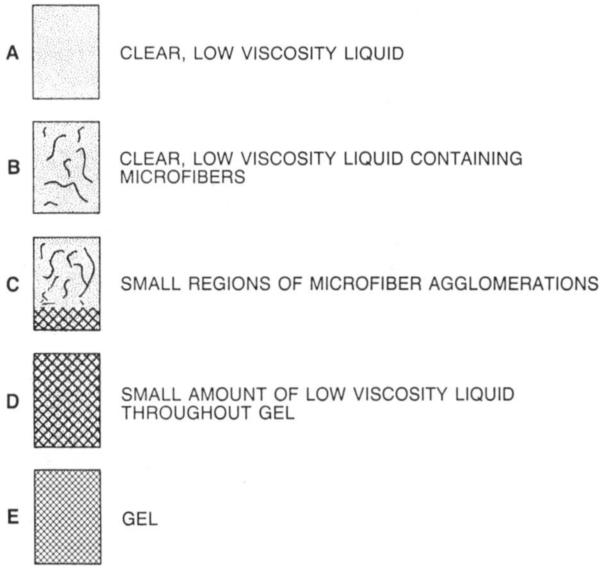

Figure 6. Low Pressure Gel Formation Observed in Solutions of Semi-Fluorinated Alkanes and Liquids

Table III lists the concentrations of F_nH_m required to obtain the phase behavior illustrated in Figure 6. It is evident that the amount of F_nH_m required to form a gel decreases dramatically with the length of the perfluorinated portion of the F_nH_m. Unfortunately, the largest commercially available iodoperfluorinated alkane is iodo-perfluorododecane. A similar, but less marked, trend is observed with the length of the hydrocarbon portion of the compound. Therefore, two high pressure CO_2 tests using the largest compounds available, $F_{12}H_{12}$ and $F_{12}H_8$, were conducted.

Table III. Stage Transition Concentrations of Figure 7 for Light Alkanes and Perfluorinated Alkanes

F_nH_m	Solvent	A	B	C	D	E
F_8H_{10}	1	87	89	91	93	95
	2	89	91	93	95	97
	3	74	76	78	80	82
F_8H_{12}	1	86	88	90	92	94
	2	87	89	91	93	95
	3	48	50	52	54	56
$F_{10}H_{10}$	1	65	67	69	71	73
	2	64	66	68	70	72
	3	16	18	20	22	24
$F_{12}H_8$	1	15	17	19	21	23
	2	14	16	18	20	22
	3	9	19	11	12	12
$F_{12}H_{12}$	1	13	15	17	19	21
	2	14	16	18	20	22
	3	1	2	3	4	5

All concentrations in weight percent F_nH_m

Solvents: 1 - Pentane
2 - Isooctane
3 - Perfluorohexane

Several grams of the semi-fluorinated compound were placed in a windowed Jergeson Cell. The cell was placed in a horizontal position and the F_nH_m distributed on the window so that visual observations could easily be made. CO_2 was then compressed into the windowed cell until the F_nH_m dissolved completely. In order to obtain a supersaturated liquid phase, the vapor phase above the solution was vented. The progression from liquid to gel occurred in both of the compounds tested as shown in Figure 7. Unfortunately, not enough of the F_nH_m was available to conduct PVT

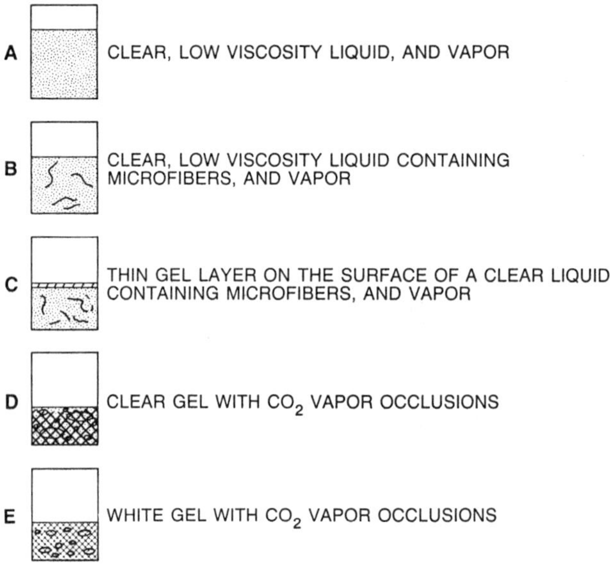

Figure 7. High Pressure Gel Formation Observed in Solutions of Semi-Fluorinated Alkanes and Liquid CO_2

experiments which could accurately determine the weight fraction of the F_nH_m required to form the gel.

Large lots of $F_{12}H_{12}$ and $F_{12}H_{20}$ (the largest, easily synthesized F_nH_m) are currently being synthesized in order to perform a detailed PVT study. The solubilities of the F_nH_m compounds in CO_2, which also corresponds approximately to the amount required to form a gel, will be determined in a visual cell over the 25 - 125°C, 1000 - 3000 psia range.

CONCLUSIONS

A novel, high pressure, visual, falling cylinder viscometer has been developed which can be used effectively for high pressure fluids, such as dense carbon dioxide (CO_2).

Neither the dissolution of surfactants into CO_2 nor the addition of pentane into CO_2 in order to increase tri-n-butyltin fluoride (TBTF) concentration resulted in any significant increase in CO_2 viscosity.

Some semi-fluorinated alkanes, such as $F_{12}H_8$ and $F_{12}H_{12}$, formed gels in CO_2 by forming a microfibrillar network in a slightly supersaturated solution. The size of the hydrocarbon and fluorocarbon portions of the semi-fluorinated alkane will be increased in an attempt to reduce the amount required to form a gel.

ACKNOWLEDGEMENTS

This work was supported by three initiation grants: AIChE, Engineering Foundation RI-A-86-12; American Chemical Society ACS-PRF-G 19986-G7; and National Science Foundation CBT-8808443.

LEGEND OF SYMBOLS

- A constant related to the viscometer constant, cm^2
- B constant related to the viscometer constant, cm^{-1}
- g gravitational acceleration, cm/sec^2
- K viscometer constant, cm^3/sec^2
- l_1 first position (at terminal velocity), cm
- l_2 second position (at terminal velocity), cm
- r_c radius of the cylinder, cm
- r_t radius of the tube, cm
- t_1 time when the cylinder is at l_1 sec
- t_2 time when the cylinder is at l_2 sec

Greek Symbols

- μ dipole moment, debye
- μ viscosity, g/cm sec
- ρ_c density of the cylinder, g/cm^3
- ρ_f density of the fluid, g/cm^3

LITERATURE CITED

1. Heller, J.P.; Taber, J.J. Development of Mobility Control Methods to Improve Oil Recovery by CO_2 - Final Report, U.S. Dept. of Energy Report DOE/MC/10689-17, November 1983.
2. Heller, J.P.; Taber, J.J. Improvement of CO_2 Flood Performance, Annual Report 1985," U.S. Dept. of Energy Report DOE/MC/21136-6, June 1986.
3. Heller, J.P.; Kovarik, F.S; Taber J.J. Improvement of CO_2 Flood Performance, Annual Report 1986," U.S. Dept. of Energy Report DOE/MC/21136-10, May 1987.
4. Terry, R.E.; Zaid, A; Angelos, C; Whitman, D.L. Polymerization in Supercritical CO_2 to Improve CO_2/Oil Mobility Ratios, paper SPE 16270, 1987.
5. Llave, F.M.; Chung, F.T.H; Burchfield, T.E. The Use of Entrainers in Improving Mobility Control of Supercritical Carbon Dioxide, paper SPE 17344, 1988.
6. Lancaster, G.W.; Barrientos, C.; Li, E.; Grrenhorn, R.C. High Phase Volume Liquid CO_2 Fracturing Fluids, CIM 87-38-71, presented at the Petroleum Society of CIM 38th Annual Meeting, Calgary, Canada, June 7-10, 1987.
7. Heller, J.P.; Kovarik, F.S.; Taber, J.J. Improvement of CO_2 Flood Performance, Annual Report 1987, PRRC Report 87-9, June 1988.
8. Yonker, C.R.; Frye, S.L.; Kalkwarf, D.R.; Smith, R.D. J. Phys. Chem; 1986, 90, 3022-26.
9. Sigman, M.E.; Lindley, S.M.; Leffler, J.E. J. Am. Chem. Soc., 1985, 197, 1471.
10. Hyatt, J.A. J. Org. Chem., 1984, 49, 5097.
11. Brady, J.E.; Carr, P.W. J. Phys. Chem., 1985, 89, 1813-22.
12. Kamlet, M.J.; Abboud, J.M.; Abraham, M.H.; Taft, R.W. J. Org. Chem., 1983, 48, 2877.
13. Barrage, T.C. M.S. Thesis, Dept. of Chemical and Petroleum Engineering, University of Pittsburgh, 1987.
14. Reid, R.C; Prausnitz, J.M.; Poling, B.E. The Properties of Gases and Liquids, 3rd Edition, McGraw-Hill, New York, 1986, 426-30.
15. Gale, R.W.; Fulton, J.L.; Smith, R.D. J. Am. Chem. Soc., 1987, 109, 920-21.
16. Johnston, K.P.; Lemert, R.M.; McFann, G. in Supercritical Fluid Science and Technology, ACS Symposium Series 1989.
17. Olesik, S.V.; Hedrich, D. Reverse Micelles in Supercritical Fluids, ARO/ONR Workshops on Supercritical Fluid Technologies, Seattle, WA, May 6-8, 1987.
18. Randolph, T.W.; Prausnitz, J. Enzymatic Conversation of Steroids in a Supercritical Fluid, ARO/ONR Workshop on Supercritical Fluid Technologies, Seattle, WA, May 6-8, 1987.
19. Shah, D. Fundamental Aspects of Surfactant - Polymer Flooding Process, proceedings of the 3rd European Symposium on EOR, Bournemoth, U.K., September 21-23, 1981.
20. Dunn, P.; Oldfield, D. J. Macromol. Sci. Chem., 1970, A4, 1157-68.
21. Sunol, A.K.; Hugh, B.; Chen, S. Entrainer Selection in Supercritical Extraction, in Supercritical Fluid Technology, Elsevier, New York 1985, 125-145.
22. Peng, D.Y.; Robinson, D.B. Ind. Eng. Chem. Fundam., 1976, 15, 59.
23. Tweig, R.J.; Russell, T.P.; Siemens, R.; Rabolt, J.F. Macromolecules, 1986, 18 No.6, 1361.

24. Pugh, C.; J. Hopken; M. Moller, Amphiphilic Molecules with Hydrocarbon and Fluorocarbon Segments," ACS Polymer Preprints, 1988, No.1, 460-1.
25. Rabolt, J.F.; Russell, T.P.; Tweig, R.J. Macromolecules, 1984, 17, 2786-94.

RECEIVED May 2, 1989

Chapter 11

Pressure Tuning of Reverse Micelles for Adjustable Solvation of Hydrophiles in Supercritical Fluids

Keith P. Johnston, Greg J. McFann, and Richard M. Lemert

Department of Chemical Engineering, The University of Texas, Austin, TX 78712

The spectroscopic probe pyridine-N-oxide was used to characterize polar microdomains in reverse micelles in supercritical ethane from 50 to 300 bar. For both anionic and nonionic surfactants, the polarities of these microdomains were adjusted continuously over a wide range using modest pressure changes. The solubilization of water in the micelles increases significantly with the addition of the cosolvent octane or the co-surfactant octanol. Quantitative solubilities are reported for the first time for hydrophiles in reverse micelles in supercritical fluids. The amino acid tryptophan has been solubilized in ethane at the 0.1 wt.% level with the use of an anionic surfactant, sodium di-2-ethylhexyl sulfosuccinate (AOT). The existence of polar microdomains in aggregates in supercritical fluids at relatively low pressures, along with the adjustability of these domains with pressure, presents new possibilities for separation and reaction processes involving hydrophilic substances.

Supercritical fluids (SCFs) such as carbon dioxide have a "hydrocarbon-like" solvent strength at typical conditions, so that they are appropriate solvents for lipophilic substances. The solvent strength may be raised significantly by the addition of small amounts of cosolvents such as ethanol to increase solubilities of moderately polar substances selectively(1), sometimes by several hundred percent(2,3,4). The solvent and cosolvent form clusters about solutes, in which the cosolvent concentrations are enhanced significantly(5,6). The present objective is to explore the effects of considerably more powerful solvent additives, that is surfactants. Since very little is known about surfactants in SCFs, spectroscopic probes were used to measure polarities inside the reverse micelles. Polarity is a key indicator of the ability of a reverse micelle to solvate a hydrophile. Using the

0097–6156/89/0406–0140$07.25/0

polarity data, surfactant /co-surfactant systems were designed for the purpose of solubilizing hydrophilic substances.

Reverse micelles are thermodynamically stable aggregates of amphiphilic molecules in non-aqueous solvents. The hydrophilic heads form a core, and the lipophilic tails extend into the oil continuous phase, as shown in Figure 1. If enough water is present it collects as a "water pool" in the micelle core. It is difficult to distinguish between the micelle core and the interfacial region containing the surfactant head groups, so we will refer to both as the "polar microdomain." In reverse micelles the driving forces for aggregation are the ionic interactions among the head groups and counterions, as well as dipolar forces and hydrogen bonding (7,8). Aggregation is opposed by the entropy gain in dispersing the surfactant and by steric repulsion between the tails. A change in the solvent can promote or oppose aggregation as discussed below. The forces that promote aggregation in reverse micelles are weaker than the hydrophobic effect, which can cause aggregation numbers of over 100 for normal micelles in a water continuous phase. Thus the aggregation numbers for reverse micelles are small, often no more than 20.

The magnitude of solvent effects on reverse micelles is dependent upon the nature of the surfactant. It is important to make the distinction between the solvent effect on [1]aggregation number for dry reverse micelles, and [2]the ability of reverse micelles to solubilize water. These are two very different phenomena. AOT is a good example for highlighting this difference, as the solvent effect is small for the former and large for the latter. For AOT, the maximum amount of water solubilization varies over a wide range for a series of hydrocarbon solvents(9). This means that the micelle size depends upon the solvent, since size is related directly to W_0 (molar ratio of water to surfactant) (10). The optimal hydrocarbon solvent for solubilizing water in AOT is octane. A model has been developed to relate the alkane carbon number of the optimal hydrocarbon solvent to the solvent-solvent, solvent-tail, and tail-tail interactions(11). For systems without added water, very little data are available. The solvent effect on the aggregation number is pronounced for sodium dinonylnaphthalenesulfonate (7), but small for AOT (12).

The cohesive energy density of a SCF solvent may be adjusted by a change in the pressure or temperature. Therefore, it is likely that pressure may be used to adjust the size and/or the polarity for reverse micelles. It is possible to vary the cohesive energy density of a SCF over a wider range than available for a series of hydrocarbon liquid solvents. Previous studies lend support to this hypothesis of significant pressure effects. It is already well-known, for example, that pressure may be used to manipulate chemical potentials of solutes, with partial molar volumes reaching thousands of mL/mole negative (13). Randolph, et al., (14), found striking changes in the size of small cholesterol aggregates in near-critical CO_2. As pressure is increased from 81 to 104 bar, the cholesterol suddenly begins to aggregate, reaches a maximum size, then begins to dissociate.

Reverse micelles have been investigated in SCF solvents only recently. It was observed qualitatively that Cytochrome-c forms a colored solution for AOT in

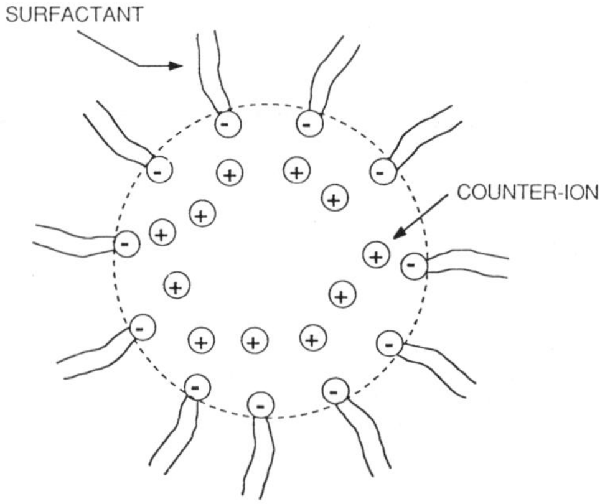

Figure 1. Structure of reverse micelles of AOT(sodium di-2-ethylhexyl sulfosuccinate) in nonaqueous solvents

propane, although concentrations were not reported(15). At 240 bar, AOT in SCF ethane solubilizes three moles of water per mole of surfactant ($W_o = 3$) for surfactant concentrations up to 0.15 M (16). Extensive visual experiments have been performed for AOT in ethane, propane, and butane to construct a generalized density versus temperature phase behavior map (17). By using dynamic light scattering, it was shown that the apparent hydrodynamic radius of AOT reverse micelles increases 20% as the pressure is decreased from 340 to 220 bar in ethane(18). The increase was attributed to micelle-micelle interactions, which become more pronounced with the approach to the two-phase region. These studies were performed at high pressures, typically above 200 bar ($P_r = 4$) for ethane (to remain in the one-phase region), or in liquid propane at 25 C ($T_r = 0.8$). In both cases, the isothermal compressibilities are relatively small compared with ethane at 35 C and 80 bar. Hence the pressure effects were modest, yet larger than in conventional liquid solvents.

The present study focuses on the behavior at lower pressures, particularly in the region where the SCF is highly compressible and thus has adjustable properties. Here surfactant and water solubilities are relatively small, so that many of the experiments were performed in the two-phase liquid-SCF region. Although this two-phase region is more complex than the one-phase region, it is interesting for several reasons. The properties of the aggregates are likely to be much more adjustable in this region. In an extraction process, it is likely that pressure adjustments could be used to recover selectively certain products and/or the surfactant. Finally, capital costs would be reduced significantly at these lower pressures.

The first objective is to explore the possibility of pressure tuning of the polarity in the reverse micelles(or aggregates) over a wide range in a single SCF. Polarities (solvent strengths) have been measured extensively using solvatochromic probes to predict solvent effects on a wide variety of chemical phenomena (19), including surfactant aggregation in non-aqueous liquid solvents (20,21). Pyridine-N-oxide was chosen as the probe since it partitions into more hydrophilic regions than most other indicators, because of its large dipole moment - 4.3 Debye. Our second objective is to form highly polar microdomains in ethane even below 100 bar. Co-solvents such as octane and co-surfactants such as octanol play an important role in achieving these pressure reductions. The final objective is to measure quantitatively the solubilities of hydrophiles, that is hydroquinone and tryptophan, in reverse micelles of AOT in SCF ethane.

Shield et al. (22) have demonstrated that reverse micelles can be used in organic solvents to recover proteins selectively from aqueous solutions. Protein denaturation can occur, however, during recovery from the organic phase, which requires changes in pH or ionic strength. Supercritical fluid solvents offer the potential advantage that proteins could be recovered simply by changing the pressure. Additional potential applications of surfactants in supercritical fluids include separations of other types of hydrophilic substances, enzymatic catalysis(14), and mobility control in enhanced oil recovery(23).

Experimental

Spectroscopic and phase behavior studies were conducted using a fixed path length cell. Pressure was increased by adding pure solvent to the cell, so molar concentrations were constant while mole fractions varied. Additional phase behavior studies and solubility studies were conducted in a variable volume view cell, based on an existing design(24), in which mole fractions were held constant but molarities varied.

AOT (Fluka, 98%) was purified according to Kotlarchyk's method (25). The purity of the final product was checked using HPLC. Purified AOT was stored in a desiccator, and AOT solutions were stored with molecular sieves to minimize hydration. The nonionic surfactant $C_{11-14}EO_5$ was provided by Shell Development Co. Brij 56 surfactant was from Fluka. Pyridine-N-Oxide (Aldrich 13,165-2) was dried prior to use. Purified de-ionized water was used at all times.

Reverse micelle solutions of known concentration were prepared in 2 mL constant volume 2.5 inch o.d. by 5/8 inch i.d. stainless steel cells fitted with 1" diameter x 3/8" thick sapphire windows. The path length was 1 cm. The cell was equipped with cartridge heaters and thermostated to $\pm$ 0.1 C by means of a platinum resistance thermometer and a temperature controller. A 60 mL syringe pump pressurized the system, and pressure was controlled to $\pm$ 0.2 bar. The surfactant and pyridine-N-oxide were introduced into the static cell as solutions in volatile solvents, so that concentrations were known to $\pm$2 %. The solvent was removed by volatilization and water, if needed, was added with a syringe. Experiments were always performed in the order of increasing pressure so that the overall molarities of surfactant, water, and pyridine-N-oxide were constant.

The contents of the cell were equilibrated rapidly using a magnetic stir bar which was small enough to rest below the light beam of the spectrophotometer. Repeated UV measurements over time showed that any suspended drops settled in less than about 10-20 minutes after stopping the stir bar. On one occasion the cell was allowed to rest overnight and no change in λ_{max} was seen. The cell was removed periodically from the spectrophotometer to ensure that there was no cloudiness in the SCF phase or deposits on the windows. The λ_{max} was determined by fitting the absorbance band using a cubic polynomial. The typical uncertainty in λ_{max} was $\pm$0.2 nm.

The magnetically-stirred variable volume view cell apparatus, which was used to measure solubilities, is shown in Figure 2. The 2 in. o.d. x 5/8 in. i.d. 304 ss cell (28 mL usable volume) contained a 1 in. diameter by 3/8 in. thick sapphire window. A piston with two 90-durometer buna-N o-rings separated the experimental fluid from the pressurizing fluid, CO_2. Pressure was controlled by a 175 mL Lee Scientific model 501 computer controlled syringe pump. The view cell

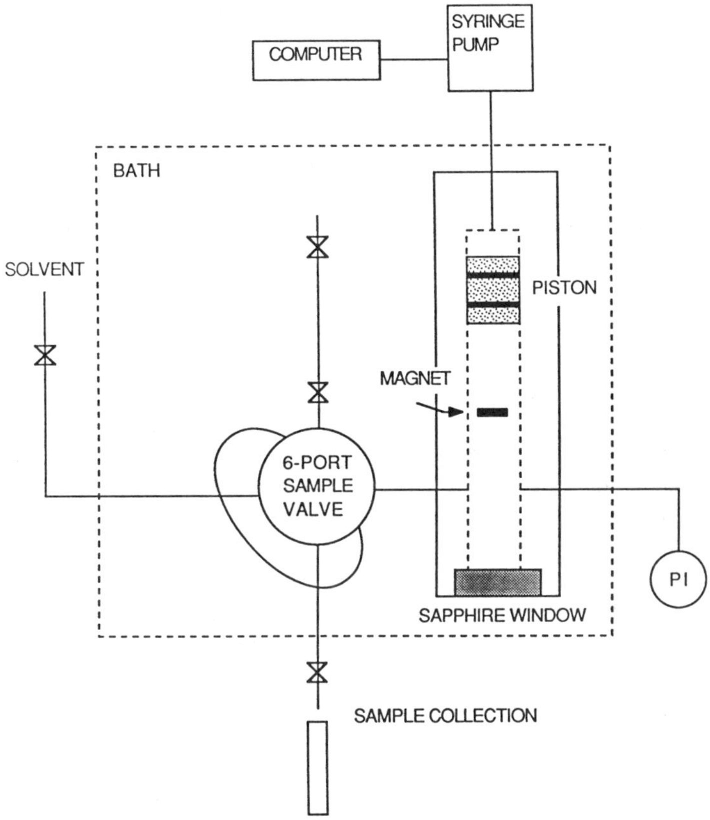

Figure 2. Variable-volume view cell apparatus with microsampling

was placed in a polycarbonate water bath which served also as a safety shield, and the temperature was controlled to within 0.1°C. Pressures on the sample side of the piston were measured to within ±0.1 bar with a strain gauge transducer. The cell was loaded at atmospheric pressure with the various components to within 0.1 mg. Ethane was condensed into the cell from a 300 cm^3 transfer cylinder, which was weighed to within 0.1 g.

Samples were obtained from the variable volume view cell by displacing a controlled volume of solution through a Valco (C6U) six-port HPLC sampling valve equipped with a 100 microliter sample loop. The samples were collected by discharging the contents of the loop through a suitable liquid solvent, then rinsing the loop with additional solvent. Instantaneous pressure drops in the cell during the sampling were typically below 10 bar, and can be reduced further in the future by adding more ballast. The cell contents were stirred at least twenty minutes to equilibrate at a new pressure. To purge the small dead volume between the cell and the sample loop, the first sample taken at each pressure was discarded. The procedure was tested by measuring the solubility of anthracene in CO_2 and was found to give results within 5% of the literature values (1). The samples were analyzed with UV spectroscopy in a Cary 2290 spectrophotometer. Tryptophan was measured at 280 nm.

Calibration of the solvatochromic probe in liquid solvents

Before using a solvatochromic probe in the SCF state, it is important to perform calibrations in well-defined one-phase liquid systems. A variety of solvatochromic indicators have been used previously to probe AOT reverse micelles at atmospheric pressure. For example El Seoud et al. (26) investigated malachite green and thymol blue, which are hydrophobic. They are insoluble in heptane despite their hydrophobicity. Accordingly it was found that these probes are absorbed at the surfactant/oil interface, but do not partition into the water pool. The λ_{max} and UV absorbance for thymol blue change significantly for W_o below 6, but become constant for W_o from 8.3 to 22.2. UV-VIS and positron annihilation studies showed that even less hydrophobic probes such as nitrophenols are also solubilized in the surfactant/oil interface.

Pyridine-N-oxide is one of the smallest and most hydrophilic (μ = 4.3 D) probes that has been studied. Consequently, it is located at the surfactant/water interface or, if sufficient water is present, in the water pool itself (20). We chose to study pyridine-N-oxide since it partitions into more hydrophilic regions of reverse micelles than thymol blue and malachite green. All of these probes are complementary.

Pyridine-N-oxide is a "blue shift" indicator in that λ_{max} shifts to shorter wavelengths as the solvent polarity increases. The λ_{max} decreases from 281.7 nm for isooctane to 254.4 nm for water (27) Although it is only sparingly soluble in a hydrocarbon solvent such as cyclohexane, its λ_{max} can be measured at a concentration as low as 10^{-5} M because the extinction coefficient is 9000. Thus solvatochromic data can be obtained for a wide variety of solvents, from supercritical ethane to water. In the figures below, the characteristic values of λ_{max} in liquid solvents will be indicated to place the polarities in perspective.

Figure 3 shows solvatochromic shifts for pyridine-N-oxide in AOT with the solvent cyclohexane, without added water. At extremely low AOT concentrations the probe senses a nonpolar environment that has a solvent strength (polarity) equivalent to that of pure cyclohexane. The pronounced change in slope at 0.001 M is the "apparent critical micelle concentration" where the average size begins to increase significantly. This apparent cmc should not be considered to be a sharp dividing line between surfactant monomer and large micelles, as in the case of aqueous systems (11). The average micelle size increases with surfactant concentration until the largest size that can be supported by the solvent is attained. For AOT this occurs at approximately 0.01 M in agreement with an earlier study(10). Here the pyridine-N-oxide senses a polarity equivalent to that of methanol. These results for AOT, which are in accord with those of earlier investigators (10,28,29), further substantiate the reliability of pyridine-N-oxide as a probe.

The effect of adding water to the system is shown in Figure 4 for a series of AOT concentrations. As W_o is increased from 1 to 6, the change in λ_{max} is large. It becomes less pronounced at higher values of W_o as the water pools are formed . As W_o reaches 20, λ_{max} approaches that in pure water. These results indicate that the probe is useful, as it is sensitive to both AOT concentration and W_o over a wide range.

To further test pyridine-N-oxide as an indicator, the solvatochromic data are correlated versus micelle radius in Figure 5. The λ_{max} values for pyridine-N-oxide were determined in solutions of .025 M AOT in two solvents, n-octane and n-hexane. For a given W_o the values are very similar in the two solvents. The micelle radii are from the photon correlation experiments of Zulauf and Eicke (10), at the same AOT concentration, for a similar solvent, isooctane. It is widely accepted that W_o is a good indicator of the size of reverse micelles, as is evident in the relationship for the two horizontal axes in the figure. There is a relatively linear relationship between micelle size and λ_{max} for a W_o up to 15. The λ_{max} at this point approaches that of pure water. This is in accord with Eicke and Kvita (28), who indicate that at a W_o of approximately 15, the water pool has the characteristics of free water. These results supply additional evidence of pyridine-N-oxide's hydrophilic nature and utility as an indicator.

In order to make sure that pyridine-N-oxide does not perturb the system, its concentration was varied for AOT in n-octane as shown in Table I. Consider the case for AOT without added water. Assuming an average aggregation number of 25 (12), the number of pyridine-N-oxide molecules per micelle may be estimated. At 0.1 molar AOT, as the pyridine-N-oxide/micelle ratio increases from 0.025 to 0.125 (corresponding to pyridine-N-oxide concentrations of .0001 M and .0005M,

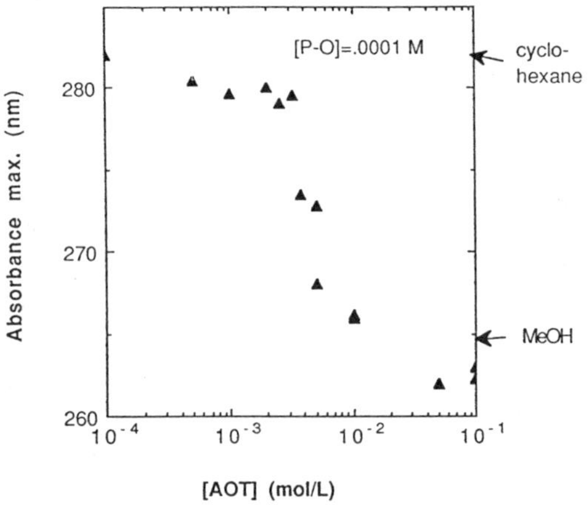

Figure 3. Calibration of pyridine-N-oxide as a solvatochromic probe in cyclohexane- effect of [AOT]

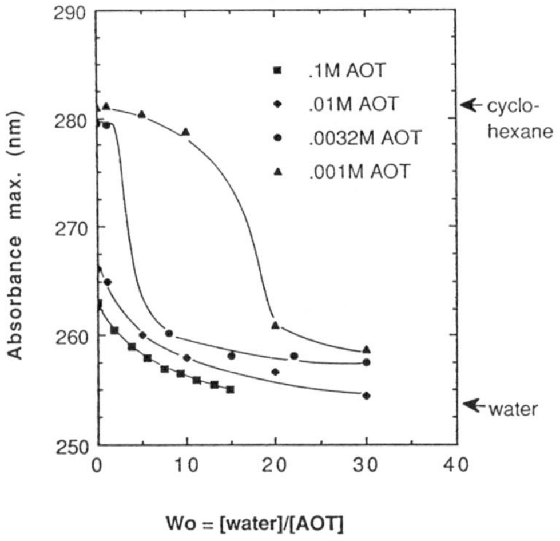

Figure 4. Pyridine-N-oxide as a solvatochromic probe in cyclohexane- effect of W_o = [water]/[AOT]

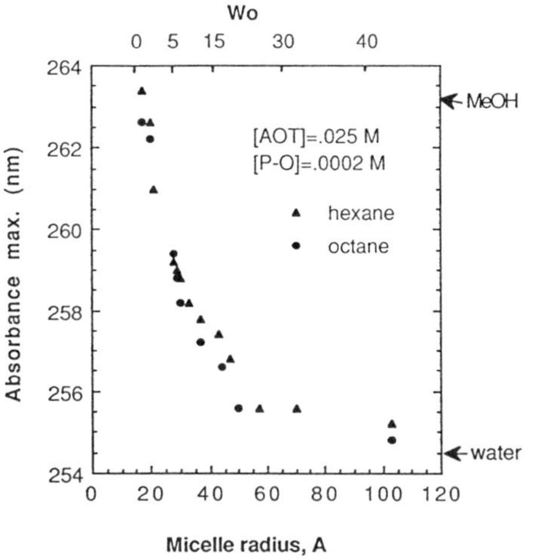

Figure 5. Correlation between λ_{max} for pyridine-N-oxide in AOT and micelle radius. (Micelle radius data taken from ref. 10.)

respectively), the change in λ_{max} is quite small, usually less than 2 nm. Even at 0.025 M AOT, where the pyridine-N-oxide/micelle ratio increases to 0.2, there is little change with probe concentration. The changes in λ_{max} with probe concentration are also small for systems with added water. It can be concluded that pyridine-N-oxide does not perturb the system significantly. This is a useful result for studies in the two-phase region, where pyridine-N-oxide partitions between the liquid and fluid phases. Here, changes in pyridine-N-oxide concentration in the fluid phase with pressure should not affect the λ_{max} significantly.

Table I λ_{max} (in nm) of pyridine-N-oxide in AOT solutions in liquid octane with varying amounts of water at 25 C

[AOT] = .025 M

W_o	pyridine-N-oxide concentration	
	0.1 mM	0.2 mM
0	263.2	263.4
1	263.0	262.6
2	261.0	261.0
4	258.7	259.2
5	259.0	259.0
6	258.6	258.8
8	258.7	258.2
10	257.2	257.8
14	256.0	257.4
18	255.9	256.8
25	254.8	255.6
30	254.5	255.6
40	254.1	255.2

[AOT] = .1 M

W_o	pyridine-N-oxide concentration		
	0.1 mM	0.3 mM	0.5 mM
0	264.6	263.6	263.0
1	266.2	262.8	262.5
4	264.8	260.2	259.7
5	258.0	258.8	258.5
6	256.8	258.6	258.2
10		257.4	
15		256.2	256.0
20	255.6	255.6	255.5
40	256.6	254.8	255.0

Pressure tuning of the polarity of reverse micelles in supercritical ethane

Anionic Surfactant: AOT has been studied extensively since it forms reverse micelles readily in a variety of organic solvents, even without a co-surfactant(28). Figure 6 shows the solvatochromic shifts of 0.0002 M pyridine-N-oxide in solutions of AOT in SCF ethane at 345 bar. No water was added to the system, but it is likely that W_o was 1 given the difficulty of completely dehydrating AOT(29). The pressure was fixed at 345 bar so that all of the AOT solutions would be in the one-phase region(16). Notice how closely the results for ethane match those for

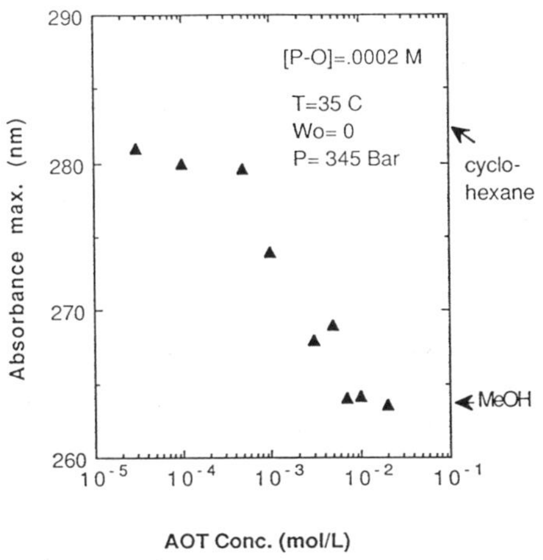

Figure 6. λ_{max} for pyridine-N-oxide as an indicator of aggregation for AOT in ethane without added water

cyclohexane shown in Figure 3. For low AOT concentrations the polarity(actually the polarizability per volume) of the solution is similar to that of cyclohexane, as would be expected for SCF ethane. As seen above for cyclohexane, aggregation commences at an AOT concentration of approximately 0.001 M, and at higher concentrations the polarity approaches that of methanol. These results are novel in that they indicate that polar microdomains may be formed even in a supercritical fluid with a low cohesive energy density.

The concept of pressure tuning of the polarity inside reverse micelles was explored at a constant AOT concentration of 0.01 M. The results are shown in Figure 7 for three values of W_o at 35 C. The system is in the two phase region except for the highest pressure data point at $W_o = 1$. The light beam was transmitted through the SCF phase only and did not contact the small liquid pool on the bottom of the cell. The concentrations listed in the figure refer to overall concentrations loaded into the cell, and not those for each phase. For $W_o = 1$ ("dry" AOT) there is an enormous increase in the polarity with pressure, which is equivalent to a change from liquid chloroform to ethanol. It is likely that much of this change may be explained by a simple mechanism. A pressure increase causes more AOT to go into solution, which drives the equilibrium towards larger aggregates. AOT concentrations were estimated from in-situ UV absorbance measurements at 230 nm, and the concentrations at each pressure agreed fairly well with those measured previously (16). For each pressure, the value of λ_{max} was consistent with the expected value based on the AOT solubility in accordance with Figure 6. Further evidence for this solubility mechanism is that solvent effects on aggregation are small for liquids in the one-phase region as discussed above.

Additional insight into the increase in aggregation with pressure may be obtained from studies of water uptake by AOT in a series of liquid alkanes. These solvents are characterized commonly by the alkane carbon number (ACN), which is simply the number of carbon atoms. Obviously, the apparent ACN must vary with pressure for a SCF. The effect of ACN on aggregation number involves competing interactions: surfactant tail-tail interactions, solvent-solvent interactions, and solvent-surfactant interactions(11). For methane, the tail-tail interactions are much stronger than the solvent-surfactant interactions, so that AOT is essentially insoluble. As the ACN increases, the water solubilization goes through a maximum at n-octane and decreases for larger values of ACN where solvent-solvent interactions become dominant(9). In SCF ethane, as pressure increases, the "apparent ACN" (which is related to the cohesive energy density) increases. Although water solubilization increases with pressure, it remains far below the optimum value in octane, even at 300 bar. This ACN effect is consistent with observations for the series ethane, propane, and butane (17), as the pressure required to dissolve a given amount of water per surfactant is lower for butane than ethane.

For systems with added water (Figure 7, $W_o = 12.5$ and 20), an extremely polar microdomain was observed with a polarity close to that of bulk water at only 60 bar. The existence of such a microdomain is novel and interesting; however, the

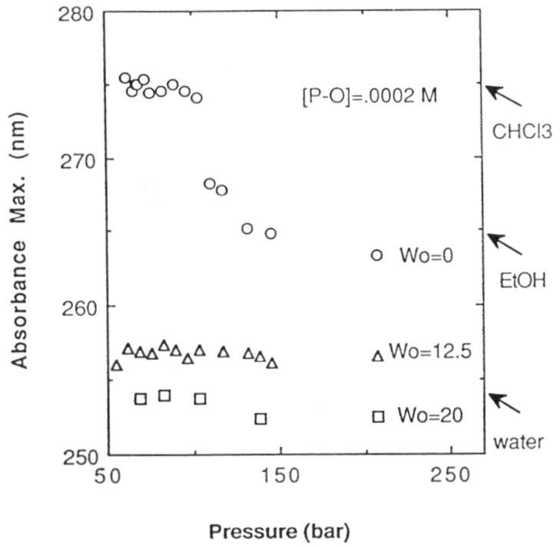

Figure 7. Variable polarities of microdomains in reverse micelles of anionic surfactant (0.01 M AOT) in ethane (T = 35 C)

structure of the microdomain and the amount of solubilized water are unknown since the system is in the two-phase region. For a one-phase system of AOT in SCF ethane the maximum W_o is no more than 3 or 4 (17). Therefore, the probe is not sensing bulk water, but a combination of surfactant and water that is much more polar than either one alone in ethane. Measurements of the actual water content, perhaps by FTIR or Karl Fischer titration, are needed to learn more about the size of these small polar microdomains.

Experiments were also performed without surfactant present to rule out the possibility of large effects from water that is simply dissolved in ethane. Table II shows that water has little effect on λ_{max} for pyridine-N-oxide in liquid hexane. The same result is observed for SCF ethane in the one-phase region from 58 to 285 bar. This indicates that water must be present in conjunction with surfactant to give highly polar microdomains seen in Figure 4.

Table II λ_{max} of pyridine-N-oxide in systems containing water without surfactant

$[H_2O]$ M	Solvent	
	n-hexane	SCF ethane (58-285 bar)
0	281.5	281.0
0.005	281.5	281.0
0.01	281.5	-

Additional experiments were performed in ethane over a wide range of AOT concentration (10^{-4} to 10^{-1} M), W_o (1 to 26) and pressure (70 to 480 bar) in order to further understand the aggregation process. The pyridine-N-oxide absorption peaks were broader in SCF ethane than in cyclohexane, yet the shapes were nearly Gaussian, so that accurate values of λ_{max} could be obtained. The peak broadening in SCF ethane suggests several possibilities. There may be a broader range of aggregate sizes in the SCF solvent. Alternatively, there may be a greater amount of exchange of pyridine-N-oxide between micelles, exposing it to a wider variety of environments. It would be interesting to pursue these proposed ideas using other techniques such as nmr, FTIR, and tracer diffusion.

The effect of AOT concentration, pressure, and W_o on peak shape, in addition to λ_{max}, will be described further to understand the environment of the probe. Again, many of the data were obtained in the two-phase region, so that concentrations refer to overall values for both phases in the cell. At high AOT concentrations above 0.01 M where the polarity is high, the peak widths are the smallest. At high Wo's above 15 the peak heights become small (due to pyridine-N-oxide partitioning into the excess water phase). At low AOT concentrations below 0.005 M, two peaks are formed, each with a λ_{max} which indicates a low polarity. Typical wavelengths for the two peaks are 279 and 274 nm. This suggests that the probe partitions between bulk ethane (or surfactant monomer) and very small aggregates which do not have a very polar microdomain. Since there is little difference in the polarities of the two locations, there is not a strong preference for one versus the other. As the AOT concentration is increased and reverse micelles are formed, pyridine-N-oxide partitions into the polar micellar interior and the second peak for the ethane environment disappears.

Nonionic surfactant The nonionic surfactant $CH_3(CH_2)_{10-13}(OCH_2CH_2)_5OH$, which is designated C_{11-14} EO_5, was studied in order to investigate the generality of the above observations of pressure tunability and water-like polarities at low pressures. The hydrophilic moiety consists of five ethoxy groups, while the hydrocarbon chain length varies from 11 to 14 carbon atoms. The pressure effect is just the opposite that for AOT as shown in Figure 8. These data are also in the two phase region, except for the highest pressure data point at $W_o = 0$. At the lowest pressure for $W_o = 0$, the polarity is comparable to that of ethanol, but the surfactant concentration is low as indicated by the low absorbance intensity for pyridine-N-oxide. As pressure increases, the solubility of the surfactant increases significantly, but the polarity actually decreases, indicating a decrease in aggregation. The pressure effect is enormous as an increase of about 10 bar changes the polarity from a value equivalent to that of ethanol to that of hexane. At 300 bar, nearly all of the surfactant goes into solution. Here pyridine-N-oxide senses a highly nonpolar environment, which indicates a solution of monomers or very small oligomers. These results suggest that a strong solvent effect opposes the increase in aggregation which would be expected from an increase in surfactant concentration in the fluid phase.

Again, an examination of water solubilization into micelles in various solvents helps explain the differences for this surfactant and AOT. The uptake of water into 0.1 M C_{11-14} EO_5 was found to increase as ACN decreased from 10 (decane) to 5 (pentane). We do not expect that this increase would continue all the way to ethane. However, we do find that the optimal ACN for water uptake into this surfactant is less than 8, the value for AOT. The lower ACN for C_{11-14} EO_5 versus AOT could be a factor in the explanation of their different behavior in ethane.

An examination of solvent effects on reverse micelles without added water provides even greater insight into the results for C_{11-14} EO_5. The solvent effect on the aggregation number of dry alkali dinonylnaphthalenesulfonates was studied in a classic paper by Little and Singleterry (7). For sodium as the counterion, the

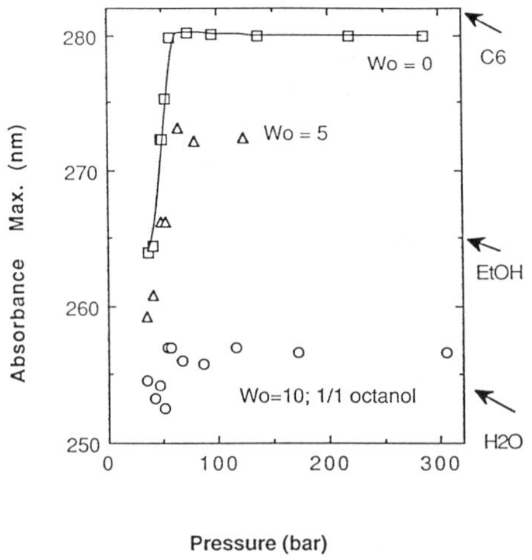

Figure 8. Variable polarities of microdomains in reverse micelles of nonionic surfactant (0.01 M Shell C_{11-14} EO_5) in ethane (T = 35 C)

aggregation number increases from 3 to 15, as the solubility parameter of the solvent decreases from 10 to 6.5 $(cal/cm^3)^{1/2}$. As the solubility parameter decreases, the ability of the solvent to shield head groups of monomers decreases, thus promoting aggregation. The reduction in aggregation with pressure for C_{11-14} EO_5 is consistent with this mechanism, although other unknown factors are likely to be present.

Pressure tuning of reverse micelles was also accomplished in systems with added water. Again, we must emphasize that solvent effects on aggregation are known to be very different for dry systems compared with those containing water. The addition of water, at levels of $W_o = 5$, and $W_o = 10$(with an octanol to surfactant ratio of 1/1), raised the polarity in the SCF phase tremendously, even though only part of the water dissolved. For $W_o = 10$, polarities equivalent to those of pure water were obtained even at 50 bar. Even at this value of W_o, λ_{max} is still adjustable as it shifts 5 nm with an increase in pressure. This change of 5 nm is significant as it is comparable to gradients in polarity which are used commonly to achieve large changes in retention times in reverse phase HPLC. The same experiment was carried out with another polyethylene oxide nonionic surfactant, Brij 56, $CH_3(CH_2)_{15}(OCH_2CH_2)_{12}OH$, which has many more EO groups. Although it was much less soluble than C_{11-14} EO_5, the trends in λ_{max} with pressure were the same. With these two nonionic surfactants, pressure tunability was even more pronounced than in AOT, and polar microdomains were also present at low pressures.

Cosolvent and co-surfactant effects on phase behavior and polarity

Below 300 bar, the cohesive energy density of ethane is too low for the formation of AOT reverse micelles which would be sufficiently large to solubilize large amounts of water(17). Again, octane is the optimal solvent for water solubilization. We have blended the cosolvent octane with ethane to attempt to overcome this limitation. We have also used an amphiphilic cosolvent, octanol, which will be called a co-surfactant. The length of octanol is consistent with AOT's preference for an alkane carbon number of 8. It can intermingle with AOT in the interfacial region and stabilize larger aggregates.

The influence of octane on solubilization of water and solvatochromic polarity is pronounced as shown in Table III. Two combinations of octane concentration and W_o were investigated. In both cases, large amounts of water are solubilized at low pressures. The polarities do not change with an increase in pressure as they are already comparable with that of bulk water. At these high concentrations of octane, the fluid is much less compressible than ethane. In the future, we will explore lower concentrations of octane to determine if pressure could be used to tune the polarity in the one phase region.

Table III. Addition of octane to SCF ethane to swell reverse micelles of AOT with water at modest pressures at 35 C

[octane] M	W_o	P (bar)	λ_{max} (nm)
0	3	300	260
1.2	20	42	257
2.2	10	41	254

[AOT] = 0.084 M, [pyridine oxide] = 0.0002 M

The influence of octanol is shown in Table IV for a constant AOT concentration of 0.01 M. Pressures are indicated which are required to dissolve the specified amounts of water so that the systems become one phase. The existence of multiple phases is common for these systems, particularly at lower pressures. For 0.5 M octanol, and W_o= 3, four phases are present at 52 bar.

Table IV. Addition of octanol to SCF ethane to swell reverse micelles with water at modest pressures at 35 C

[octanol] M	W_o	P (bar)	λ_{max} (nm)
0.	3	300	260
0.01	11	345	255
0.5	19	117	262

[AOT] = 0.01 M, [pyridine oxide] = 0.0002 M

The addition of 0.01 M octanol increases significantly the solubilization of water from a W_o of 3 to 11. The polarity increases to approach the pure water value of 254 nm, but it was not adjustable with pressure in the one phase region. As the octanol concentration is increased by a factor of 50, a large amount of water is solubilized at only 117 bar. Here the ratio [octanol]/[AOT] = 50 and [octanol]/[water] = 2.5. Compare this case with the one above it. Even though a larger amount of water is solubilized, the polarity has shifted significantly towards that of pure octanol where λ_{max} is 266 nm. Similar behavior was obtained in liquid octanol, as the polarity varied only slightly for a water concentration of 0 to 0.4 M.

Based on these results, we explored the possibility of using octanol as an amphiphilic cosolvent without the need for AOT. With the addition of octanol at a concentration of 0.5 M, it becomes possible to solubilize water at 0.05 M at only 80 bar. The λ_{max} is close to that of pure octanol, so that it is unlikely that water pools were formed. The cosolvent octanol could play an important role in SCF technology by increasing the solubilization of water.

Use of AOT and octanol to solubilize hydrophilic substances

Hydroquinone The ability of AOT/co-surfactant to solubilize non-ionic hydrophiles in ethane was tested using 1,4-dihydroxybenzene(hydroquinone, HQ). To improve aggregation, octane and octanol were utilized based on the above results. Before measuring solubilities, phase boundaries were determined using the variable volume view cell. Octanol was more effective than octane for forming a one-phase system (not counting the solid phase). For $W_o = 10$, the system became one-phase at 145 bar for 3.6 mole % octanol and at 195 bar for 6.2 mole % (AOT = 0.62 mole %). At low concentrations, the octanol acts as a co-surfactant and favors water solubilization, but it causes a new liquid phase to form at high concentrations.

In cases where only a single fluid phase is present (in addition to the solid), its appearance usually did not change with pressure, but an interesting effect was observed in the ethane - 0.6% AOT - 6.0% octanol system with $W_o = 10$. At high pressures well removed from the dew point, the fluid was clear. As pressure was lowered, the fluid very slowly became gold in color. Within twenty to thirty bar of the dew point, the intensity of the color increased rapidly, and at the dew point, the fluid was practically opaque. Possible explanations include light scattering due to coalescence of micelles(18), or to critical phenomena.

The solubility of HQ is shown in Table V for various octanol concentrations at a pressure of 300 bar. For all of the data listed, there is only one fluid phase in equilibrium with the solid. In this solid-fluid region, solubilities did not change significantly with pressure. Without any AOT present, octanol is an excellent cosolvent. At a concentration of 1 M, it raises the solubility by over two orders of magnitude to reach 17 mM(about 1 wt.%). At this octanol concentration, the addition of 0.62 mole % AOT raises the solubility by a factor of 1.4 with no water added and by 2.2 with a W_o of 10. For HQ, high solubilities can be achieved using octanol without AOT, but in the next section a system will be discussed where both AOT as well as octanol are required for solubilization.

Table V. Effect of AOT and octanol on the solubility of hydroquinone (in mM) in SCF ethane at 35 C and 300 bar

[Octanol] (mole %)	[Octanol] (M)	0% AOT	0.62% AOT* (approx. 0.1M)
0	0	<0.2	-
0.77	0.1	-	s-l-f region
1.95	0.3	2.4	-
3.6	0.6	-	28
6.3	1	17	23 (W_o = 1)
6.0	1	-	38

*W_o = 10 unless indicated otherwise

Tryptophan In order to focus on the role of surfactant on solubilization, we chose to study an ionic solute which is insoluble in ethane, even when doped with several percent octanol. Amino acids, which are hydrophilic zwitterions, are not soluble in ethane and other hydrocarbons. Tryptophan was selected because its strong chromophore simplified analysis. At 37 C, its solubility is below the detection limit, 0.2mM, in a solution of 6 mole % octanol in ethane at 325 bar.

Figure 9 shows that tryptophan can be solubilized with the addition of AOT, octanol, and water. At lower pressures, the surfactant partitions mostly into the liquid phase, and tryptophan is only sparingly soluble. At pressures above 140 bar, the liquid phase disappears as the AOT, octanol, and water form micelles in the fluid phase. The micelles cause the solubility of tryptophan to increase dramatically. The solubility becomes well above 0.1 wt.%, which is quite sufficient for practical applications. At pressures above 200 bar in the solid-fluid region, solubilities vary little with pressure, which is consistent with the relatively constant polarities shown in Figure 7 for similar values of W_o. This ability to adjust solubilities of ionic species at modest temperatures and pressures opens up the possibility of interesting new practical applications.

In this study no attempt was made to control the pH of the water. In liquid reverse micelle systems, the pH of the aqueous phase has a significant effect on the

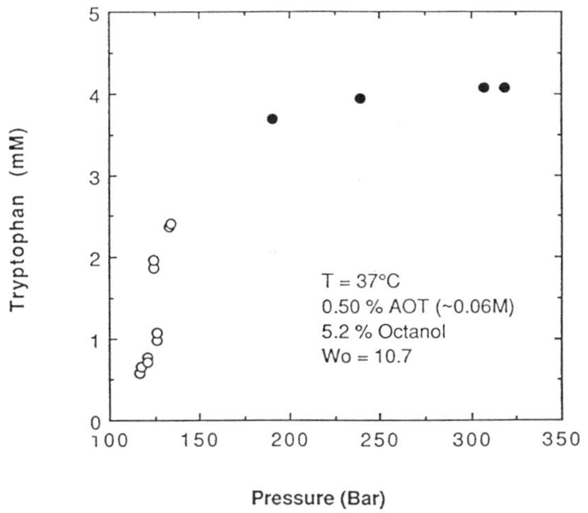

Figure 9. Solubilization of tryptophan in ethane using AOT (open circles: solid-liquid-fluid region, closed circles: solid-fluid region)

extraction of proteins from water, based on the isoelectric point(30,31) . Future studies of the role of pH and of salt effects in SCF's, as Leodidis and Hatton (32) have done recently for liquid solvents, would be very useful.

Conclusions

Highly polar microdomains exist in reverse micelles of AOT and nonionic polyethylene oxide surfactants in ethane, even below 100 bar, both with and without cosolvents. Without cosolvents these domains are likely very small since values of W_0 are small. The addition of the cosolvent octane provides a means to take up large amounts of water over a wide pressure range. The polarities in the interior of the micelles approach that of bulk water. The existence of polar microdomains in supercritical fluids at relatively low pressures presents an opportunity for new separation and reaction processes involving hydrophilic substances.

Pressure affects aggregation by influencing both the partitioning of water and surfactant between the liquid and fluid phases, as well as the surfactant-solvent interactions. For nonionic surfactants, the latter effect is dominant as polarities in the micelles (aggregation) decrease with pressure. For dry AOT, the pressure effect is consistent with the change in AOT solubility. Other spectroscopic techniques would be useful for understanding these large pressure effects more fully, particularly in the highly compressible region of ethane.

An amino acid, tryptophan, was dissolved in SCF ethane for the first time at the 0.1 wt.% level using AOT and octanol. Solubilities are highly adjustable in the solid-liquid-fluid region, where the partitioning of water and surfactant are variable with pressure. It is possible that this type of adjustability could be used to selectively extract and recover hydrophilic substances such as proteins. It could be used also as a means to recover surfactant for recycle, but further work is needed to understand these complex systems.

The solubility behavior is related strongly to the polarity in the reverse micelles. The pressure effects on reverse micelles are very different in the two-phase liquid-fluid region (where the fluid phase is highly compressible), versus in the one-phase fluid region (at higher pressures where the compressibilities are much smaller). In the two-phase region, it was discovered that polarities may be adjusted continuously over a wide range using small changes in pressure. Here, the solubilities of hydrophiles are also highly adjustable. In the one-phase region, both polarities and solubilities are relatively independent of pressure since the density of the SCF is relatively constant. This is consistent with the modest pressure effect on the size of reverse micelles in ethane in the one-phase region above 200 bar(18). This raises an interesting question of whether it is possible to formulate one-phase solutions with large pressure effects, for example using co-surfactants. To attempt to do this, we are continuing to explore regions where the solvent is highly compressible.

Acknowledgment

Acknowledgement is made to the Camille and Henry Dreyfus Foundation for a Teacher-Scholar Grant, and the Separations Research Program at the University of Texas. We thank Bob Schechter and Ted Randolph for many stimulating discussions, and Rob Fuller, Vikas Grover, and Caron Arnold for assistance in performing spectroscopic analysis.

Literature Cited

1. Dobbs, J.M.; Johnston, K.P. Ind. Engr. Chem. Res. 1987, 26, 1476.
2. van Alsten, J. G. Ph.D. Diss., Univ. Illinois, 1986.
3. Dobbs, J. M.; Wong J.M.; Lahiere R.J.; Johnston, K.P. Ind. Eng. Chem. Res. 1987, 26, 56.
4. Schmitt, W.J.; Reid, R.C. Fluid Phase Equilibria 1986, 32, 77.
5. Kim, S.; Johnston, K.P. AIChE J. 1987, 33, 1603.
6. Yonker, C.R.; Smith, R.D. J. Phys. Chem. 1988, 92, 2374.
7. Little, R. C.; Singleterry C.R. J. Phys. Chem. 1964, 68, 3453.
8. Ruckenstein, E.; Nagarajan, R. J. Phys. Chem. 1980, 84, 1349.
9. Middleton, M.A.; Schechter, R.S.; Johnston, K.P. Langmuir 1989, submitted.
10. Zulauf, M.; Eicke, H. F. J. Phys. Chem. 1979, 83, 480.
11. Bourell, M.; Schechter R.S. Microemulsions and Related Systems, Surfactant Science Series V. 30, 1988, Marcel Dekker, N.Y.
12. Peri, J. B. J. Coll. Interface Sci. 1969, 29, 6.
13. Eckert, C. A.; Ziger, D. H.; Johnston, K. P.; Kim, S. J. Phys. Chem. 1986, 90, 2738.
14. Randolph T.W.; Clark, D.S.; Blanch, H.W.; Prausnitz, J.M. Science 1987, 238,387.
15. Gale, R. S.; Fulton, J. L.; Smith, R. D. J. Amer. Chem. Soc. 1987, 109, 920.
16. Fulton, J. L.; Smith, R. D.; J. Phys. Chem. 1988, 92, 2903.
17. Steytler, D.C.; Lovell, D.R.; Moulson, P.S.; Richmond, P.; Eastoe, J.; Robinson, B.H. International Symp. Supercritical Fluids, Soc. Francaise de Chemie. 1988, 67.
18. Blitz, J.P.; Fulton, J.L.; Smith, R.D. J. Phys. Chem. 1988, 92, 2707.
19. Reichardt, C. Solvent Effects in Organic Chemistry, 1988, Verlag Chemie, Weinheim.
20. Menger F. M.; Donohue J. A.; Williams R.F. J. Am. Chem. Soc. 1973, 95, 286.
21. Fendler J. H.; Liu L. J. J. Am. Chem. Soc. 1975, 97, 999.
22. Shield, J.W.; Ferguson, H.D.; Bommarius, A.S.; Hatton T. Ind. Eng. Chem. Fundam. 1986, 25, 603

23. Iezzi, A.; Enick, R., this symosium.
24. McHugh, M. A.; Krukonis, V. J. Supercritical Fluid Extraction; Principles and Practice, 1986, Butterworths; Boston, Mass.
25. Kotlarchyk, M.; Chen, S.-W.; Huang, J. S.; Kim, M. W. Phys. Rev. A 1984, 29, 2054.
26. El Seoud, O. A.; Chenoletto, A. M.; Shimiz, M. R.. J. Coll. Interface Sci. 1982, 88, 420.
27. Kosower, E.M. J. Am. Chem. Soc. 1958, 80, 3253.
28. Eicke, H. F.; Kvita, P. Reverse Micelles, 1984, Luisi, P. L.; Straub, B. E.,eds; Plenum, New York.
29. Kotlarchyk, M.; Huang, J. S. J. Phys. Chem. 1985, 89, 4382.
30. Luisi, P.; Henninger, F.; Joppich, M. Biochem. Biophys. Res. Comm. 1977, 74, 1384.
31. Goklen, K. E.; Hatton, T. A. Sepn. Sci. Tech. 1987, 22, 831.
32. Leodidis, E. B.; Hatton T.A. Langmuir 1989, submitted.

RECEIVED May 2, 1989

Chapter 12

Structure of Reverse Micelle and Microemulsion Phases in Near-Critical and Supercritical Fluid as Determined from Dynamic Light-Scattering Studies

Richard D. Smith, Jonathan P. Blitz, and John L. Fulton

Chemical Methods and Separations Group, Chemical Sciences Department, Pacific Northwest Laboratory, Richland, WA 99352

Dynamic light scattering methods were used to study reverse micelle and microemulsion phases formed in liquid and supercritical alkane continuous phases. These reverse micelle or microemulsion (w/o) phases were formed in the alkanes from ethane through decane using the surfactant aerosol-OT (AOT) with variable amounts of water. A high-pressure cylindrical sapphire cell, capable of operation at pressures up to 500 bar and over 80°C, was used for measurements of both single-phase and the upper (alkane continuous) phase of two-phase systems. Measured changes in hydrodynamic diameter of reverse micelle or microemulsion phases can be largely attributed to attractive micelle-micelle interactions. Such interactions become increasingly evident at solution pressures approaching the phase boundary between the one-phase and two-phase systems and are also enhanced at higher surfactant and water concentrations. It is shown that reverse micelles exist in the upper fluid phase of two-phase systems, and have sizes (or W values) which are highly pressure dependent.

Several years ago we reported initial observations of reverse micelles and microemulsions in supercritical fluid solvents (1). These studies suggested the possibility of creating a previously unsuspected broad range of organized molecular assemblies in dense gas solvents. Such systems are of interest due to potential applications which exploit the readily variable properties of supercritical fluids as well as the unique solvent environments of reverse micelles and microemulsions. These initial studies showed that even gram quantities of proteins, such as Cytochrome-c (Mwt. 12,842 dalton) could be solvated in a liter of supercritical ethane or propane due to the microemulsion solvent environment, something which is not achievable with "conventional"

0097-6156/89/0406-0165$06.00/0

supercritical fluids due to the limited polarity and practical temperature constraints.

The formation of reverse micelles and water-in-oil (w/o) microemulsions in liquid hydrocarbons using the surfactant sodium bis(2-ethylhexyl) sulfosuccinate (AOT) has been widely studied (2,3). In nonpolar liquid solvents, these molecular aggregates generally consist of 3- to 20-nanometer-diameter, roughly spherical shells of surfactant molecules surrounding a polar core, which is typically an aqueous solution. This combination of hydrophilic, hydrophobic, and interfacial environments in one solvent has created potential applications in separations (4,5), chromatography (6), and catalytic reactions (7).

In conventional liquid microemulsions, the properties of the surfactant interfacial region are of primary importance in determining the size and shape of surfactant aggregates. The ionic interactions between surfactant head groups and interactions with their counter ions, as well as the hydrogen bonding within the aqueous core, are important factors which lead to aggregation. The equilibrium structure of the interfacial region is also determined by a delicate balance of several additional factors. Small changes in the surfactant's hydrocarbon tail structure (8) as well as small changes in the composition of the nonpolar solvent (8,9) cause profound changes in the aggregate structure or size as evidenced by the large changes in the amount of water or surfactant (and other substances) which can be solubilized by the microemulsion. Previous results have indicated that aggregation or clustering of two or more micelles can also occur when the micelles are of sufficient size or at high enough concentrations (10). Indeed, the phase behavior of certain micellar solutions is known to resemble that of a simple molecular fluid which has liquid-gas phase equilibria and a well-defined critical point (11,12).

The ability of surfactants such as AOT (usually with water) to form micelle and microemulsion phases in supercritical fluids (dense gases) (1,13) opens up a range of potential new applications. A supercritical fluid is a substance above its critical temperature and pressure which has properties that are highly dependent on pressure due to the proximity to the critical point. In supercritical fluids density, dielectric constant and viscosity, as well as other properties, can be continuously varied between the gas and liquid phase limits by manipulating pressure. Fluids, in which AOT can be used to create micelles and which are supercritical at moderate temperatures and pressures include ethane (T_c = 32.2°C, P_c = 48.8 bar), xenon (T_c = 16.6°C, P_c = 58.4 bar), and propane (T_c = 96.7°C, P_c = 42.4 bar). Microemulsions formed in supercritical fluids have been previously characterized by light scattering (14,15), spectroscopic methods (15), conductivity, density, and phase behavior (13). These surfactant/supercritical fluid systems have potential applications in enhanced oil recovery (16), reaction processes (17-19), chromatography (20,21), and bulk separations processes (22) where the high diffusivities of solutes in the fluid continuous phase may greatly increase reaction or extraction rates.

The properties of a supercritical fluid continuous phase provide a useful tool with which to study surfactant aggregation. Haydon and coworkers (23) have shown that a low molecular weight liquid alkane, e.g., butane, penetrates and solvates the hydrocarbon tail region of the surfactant interface to a much greater extent than higher molecular weight liquids, e.g., hexadecane. Smaller molecules, such as xenon or ethane, might then be expected to possess the capability of even greater solvation of this hydrocarbon tail region (15). It is also expected that the degree of solvation of the hydrocarbon tails by the fluid will be strongly density dependent; it is well established that the solvation of simple nonpolar, higher molecular weight organic substances in supercritical fluids is highly density dependent (24). On a macro-scale, the degree of solvation of the micelle species will also be dependent upon pressure. By changing the pressure at constant temperature, relatively large changes in the solvating power of the continuous phase occur, and the attractive micelle-micelle interactions can be strongly affected. We believe that as pressure is lowered, so as to approach the phase boundary, one finds that these micelle-micelle interactions become increasingly important, ultimately leading to large-scale aggregation of micelles and phase separation (22).

In this paper we use dynamic light scattering (DLS) methods to examine micelle size and clustering in (1) supercritical xenon, (2) near-critical and supercritical ethane, (3) near-critical propane as well as (4) the larger liquid alkanes. Reverse micelle or microemulsion phases formed in a continuous phase of nonatomic molecules (xenon) are particularly significant from a fundamental viewpoint since both theoretical and certain spectroscopic studies of such systems should be more readily tractable. Diffusion coefficients obtained by DLS for AOT microemulsions for alkanes from ethane up to decane are presented and discussed. It is shown that micelle phases exist in equilibrium with an aqueous-rich liquid phase, and that the apparent hydrodynamic size, in such systems is highly pressure dependent.

Experimental

The surfactant AOT ("purum" grade, Fluka) was purified as described by Kotlarchyk (25). The AOT solution was filtered through a 0.2-μm Millipore filter prior to drying in vacuo for eight hours. The AOT was stored in a desiccator over anhydrous calcium sulfate. The molar water-to-AOT ratio (W) was assumed to be 1 in the purified, dried solid (25). Water was distilled and filtered through a Millipore Milli-Q system. Ethane, propane ("CP" grade, Linde), and xenon (Research grade, Linde) were used as received. The alkanes had a reported purity of >99% (Aldrich) and were used as received.

A cross-sectional view of the high-pressure light scattering cell is shown in Figure 1. The cylindrical light scattering cell window was a high-precision sapphire tube with an inside diameter of 1.9 cm and an outside diameter of 3.2 cm. To achieve the

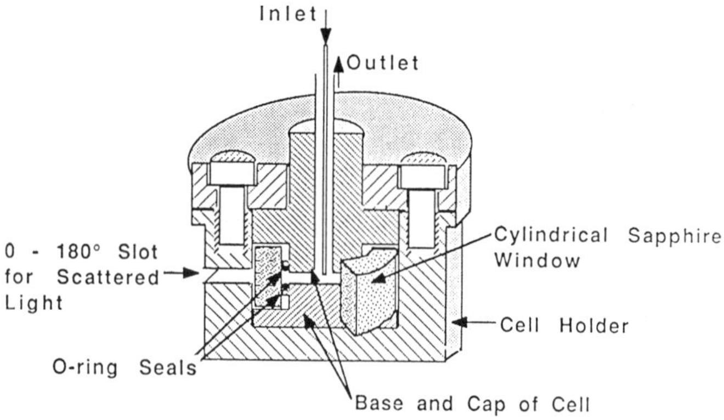

Figure 1. High pressure cell and holder for dynamic light scattering studies.

necessary alignment of the cell, the inside and outside surfaces of the sapphire tube are round and concentric to within 0.0005 cm. The axis of this tube was located within 0.002 cm of the axis of rotation of the goniometer. The cell was placed in an 9.25-cm diameter, thermostated quartz vat filled with toluene. The physical dimensions of the scattering sample cell were minimized to eliminate convection of the low-conductivity, low-viscosity fluid. The total cell volume was 1.5 cm^3. The alignment of the instrument was confirmed using 58-nm polystyrene latex spheres dispersed in H_2O, and the measured size of this standard was found to agree within 5% of the reported value. In addition, our measurement of micelle size for the H_2O/AOT/iso-octane system was within 5% of Zulauf and Eicke's (26) measurement for that system and showed the same functional dependence upon the W.

Experiments with supercritical xenon and ethane were done by adding 0.10 g of AOT (150 mM) into the scattering cell with either 16 µl water (W = 5) or no added water (W = 1). A miniature magnetic stir bar was also placed directly into the scattering cell. A Varian 8500 syringe pump filled with pure xenon or ethane was connected directly to the scattering cell. Upon pressurizing with pure continuous phase, the solution was mixed for 15 min. and allowed to equilibrate for one hour prior to data acquisition. The outlet of the scattering cell was connected to a pressure transducer (Setra Systems, No. 300C). Experiments with the other alkanes were conducted by filling the syringe pump with the filtered micellar solution and then pumping this solution directly into the scattering cell. The concentration of the alkane solutions was 0.015 mole of AOT/mole of alkane with a water-to-surfactant ratio of 5.

A Malvern PCS-100 spectrometer (Malvern Instruments, Malvern, England) equipped with a 5 W Ar^+ laser (488 nm) was used. The spectrometer and laser were mounted on a vibration-free optical table (Technical Manufacturing Corp.). All measurements were taken at a constant scattering angle of 54°. The signal from the photomultiplier was processed on a 128 channel, real time digital correlator (K7032-OS) using either a 50 or 100 nanosecond sample time. Temperatures were maintained with a Malvern temperature controller at 25 ± 0.1°C for experiments conducted in xenon and the alkane liquids, and at 37 ± 0.1°C for experiments in supercritical ethane. The photon autocorrelation function was analyzed by the method of second order cumulants (27) in which the logarithm of the normalized autocorrelation function, $G^{(2)}(q,t)$, was fitted to a polynomial equation by using a nonlinear least-squares fitting routine,

$$\frac{1}{2} \ln [G^{(2)}(t)] = \Gamma_o - \Gamma_1 t + \frac{1}{2} \Gamma_2 \frac{t^2}{2} \quad (1)$$

where Γ_o is ideally zero. Γ_1, the first cumulant, is equal to the diffusion coefficient, D_T, by

$$\Gamma_1 = D_T q^2 \quad (2)$$

where q is the scattering vector. The magnitude of the scattering vector, q, is given by

$$q = \frac{4\pi \sin \frac{1}{2}\theta}{\lambda/n} \tag{3}$$

where θ is the scattering angle, λ is the wavelength of the incident beam in vacuum, and n is the index of refraction of the scattering medium. In the absence of interparticle interactions the second-order cumulant, Γ_2, is related to the variance of the particle size distribution. The contribution to the measured autocorrelation function from the pure xenon or ethane fluid was determined to be negligible because of the high signal-to-noise ratio (the scattered intensity at 54° of the xenon/AOT mixture was 10 times greater than for pure xenon). In addition, the measured autocorrelation function of pure xenon or ethane showed no autocorrelation on the time scale used to examine the micelle solutions.

The mean (apparent) hydrodynamic diameter, d_H, was calculated from D_T using the Stokes-Einstein relation for spherical particles

$$d_H = \frac{kT}{3\pi \eta D_T} \tag{4}$$

where k is Boltzmann's constant, T is the absolute temperature, and η is the viscosity of the solvent. The mean, ±1 standard deviation of five replicate measurements, is reported in all cases. It must be remembered that the reported hydrodynamic diameters may reflect differences from the actual micelle diameters due to micelle-micelle interactions. Large changes in the viscosity of near critical and supercritical fluids occur with moderate changes in density. Since diffusion coefficients will be strongly effected by the fluid viscosity it is easier to interpret fluid structural changes by consideration of the data in terms of the apparent hydrodynamic diameter.

To calculate micelle size and diffusion coefficient, the viscosity and refractive index of the continuous phase must be known (equations 2 to 4). It was assumed that the fluid viscosity and refractive index were equal to those of the pure fluid (xenon or alkane) at the same temperature and pressure. We believe this approximation is valid since most of the dissolved AOT is associated with the micelles, thus the monomeric AOT concentration in the continuous phase is very small. The density of supercritical ethane at various pressures was obtained from interpolated values (28). Refractive indices were calculated from density values for ethane, propane and pentane using a semi-empirical Lorentz-Lorenz type relationship (29). Viscosities of propane and ethane were calculated from the fluid density via an empirical relationship (30). Supercritical xenon densities were interpolated from tabulated values (31). The Lorentz-Lorenz function (32) was used to calculate the xenon refractive indices. Viscosities of supercritical xenon (33), liquid pentane, heptane, decane (34), hexane and octane (35) were obtained from previously determined values.

Results and Discussion

Micelle-Micelle Interactions. Solvatochromic studies have recently demonstrated for one-phase xenon and ethane systems that the aqueous core solvent environment of the reverse micelles or microemulsion droplets undergoes only very small changes for pressures from 200 bar (the approximate minimum pressure for formation of a one-phase system at [AOT]>10 mM) up to 1500 bar (15). Since we expect that the micelle size and the core solvent environment to be strongly correlated, we conclude that the physical size of the surfactant/water assembly is largely independent of pressure. This is supported by spectroscopic studies of the iso-octane/AOT/water systems, in which a direct correlation was established between micelle size, as determined by neutron and light scattering (26,36), and the measured solvatochromic shift (15). The electrostatic interactions of the head groups and hydrogen bonding within the aqueous core control the size of the nanometer-sized droplet, and not the properties of the nonpolar continuous phase solvent (except for relatively small changes). This hypothesis is further supported by studies in liquid alkanes and cyclohexane which showed little or no effect of the continuous phase solvent on droplet size (36).

The bulk of the surfactant in a reverse type microemulsion forms aggregate structures while a small amount of the surfactant, equal to the critical micelle concentration (cmc), exists as dissolved monomer or small clusters in the continuous phase. In liquid systems, such as iso-octane (36), the cmc is approximately equal to 6 X 10^{-4} M. At similar molecular densities, iso-octane will be a better solvent for surfactant monomer than either ethane or propane because iso-octane has a much higher polarizability. In supercritical fluid microemulsions the cmc (if such a discrete value even exists) has not been determined. However, since the cmc is equal to the solubility of surfactant monomer in the continuous phase, it is expected that the cmc will be dependent on the fluid density (i.e., solvent strength) (13). Changes in the cmc will lead to changes in the size of surfactant aggregates. In this study, where the surfactant concentration is greater than 10^{-2} M, these changes in aggregate size would be negligible unless the cmc was unexpectedly much larger than 10^{-4} M.

The general observation from DLS studies is that the apparent hydrodynamic diameter increases as the pressure is decreased towards a phase boundary (where surfactant and water will precipitate to form a second phase). Figures 2 and 3 show DLS results for AOT/water micelles in supercritical xenon (at 25°C) and ethane (at 37°C), respectively. Results are presented for $[H_2O]/[AOT]$ molar ratios (W) of 1 (a) and 5 (b). All measurements were obtained in single-phase systems at constant W. The apparent hydrodynamic micelle diameter decreases with increasing pressure or density of the continuous phase in both fluids. The second cumulant in Equation 1, which is a qualitative measure of the polydispersity of the system, is very close to zero for all conditions of this study. There is no statistically

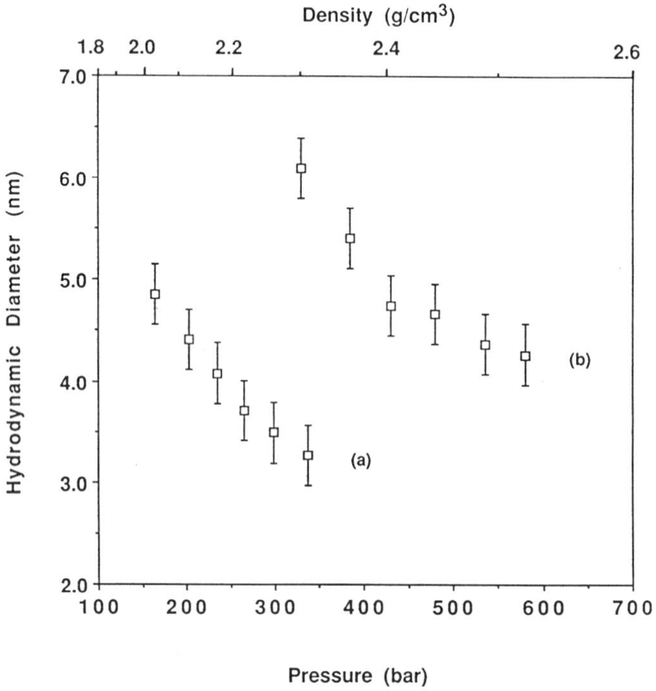

Figure 2. Apparent hydrodynamic diameters of AOT reverse micelles in supercritical xenon as a function of pressure and density (of the pure fluid) at 25°C, with (a) W = 1 and (b) W = 5. [AOT] = 150 mM.

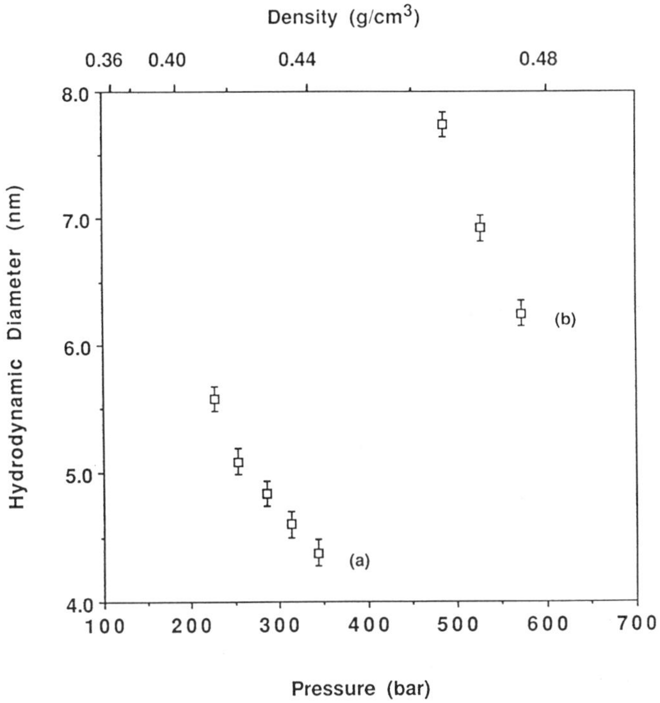

Figure 3. Apparent hydrodynamic diameters of AOT reverse micelles in supercritical ethane as a function of pressure and density (of the pure fluid) at 37°C, with (a) W = 1, and (b) W = 5. [AOT] = 150 mM.

significant effect of pressure on the second cumulant for the single phase microemulsions formed in xenon and ethane over the pressure ranges from 150 to 560 bar and W values from 1 to 5. The measured autocorrelation function of supercritical xenon/AOT/water mixtures suggests that within the limits of detectability of the DLS technique, that a monodisperse micelle phase exists in supercritical xenon above 200 bar.

The apparent hydrodynamic diameter of the AOT reverse micelles in supercritical xenon is less than that found in supercritical ethane at equal pressures, and the effect of pressure on micelle size is less than that in ethane. This is expected since the two-phase boundary for xenon is at lower pressures than for ethane. The limited data obtained from the supercritical ethane micelle solution (37°C) at W = 5 is due to the pressure constraints of our experimental apparatus and the relatively high pressure at which this solution ([AOT] = 150 mM) becomes one phase (450 bar). For both fluids the size appears to increase as the two-phase boundary is approached. For W = 1 and W = 5 in xenon, the asymptotic values for the apparent hydrodynamic diameter at high pressure are approximately 3.0 and 4.3 nm; smaller than the size of AOT reverse micelles in liquid iso-octane (7) (3.8 and 5.6 nm) at the same water contents, but consistent with our DLS results for other liquid alkanes presented later. Observations for the xenon system are complicated due to formation of the gas hydrate (clathrate) at higher pressures (15), which may result in the actual micelle (fluid phase) W value at higher water content being slightly lower than expected. (The clathrates are observed as a solid mass at the top of the high pressure cell, and we have found no evidence for corresponding structures dissolved in the fluid phase which might interfere with the present measurements.) Because the apparent micelle size in ethane changes sharply as the pressure is increased, a similar asymptotic value for micelle size is not reached even at the highest pressures examined. Solvatochromic probe studies of the aqueous core solvent environment show small changes in micelle size as a function of pressure which are due perhaps to the extent of solvent penetration into the surfactant tails (15). However, these changes represent only a small fraction of the size changes observed by DLS.

The most reasonable explanation for the increase in apparent hydrodynamic diameter measured by DLS is the enhanced micelle-micelle interactions as the boundary of a two-phase system is approached (i.e., the pressure is lowered). Figure 4 illustrates this concept of micelle-micelle interactions, which is manifested as aggregation (or clustering) of the reverse micelle or microemulsion droplets. Since the solvent environment is essentially unchanged by this "macromolecular aggregation" (15) we exclude the possibility of (other than transitory) micelle-micelle coalescence to form stable, larger micelles. The micelles may coalesce briefly to form transitional species (which might be a "dumbbell" or more cylindrical structures), in which the water cores collide and intermix.

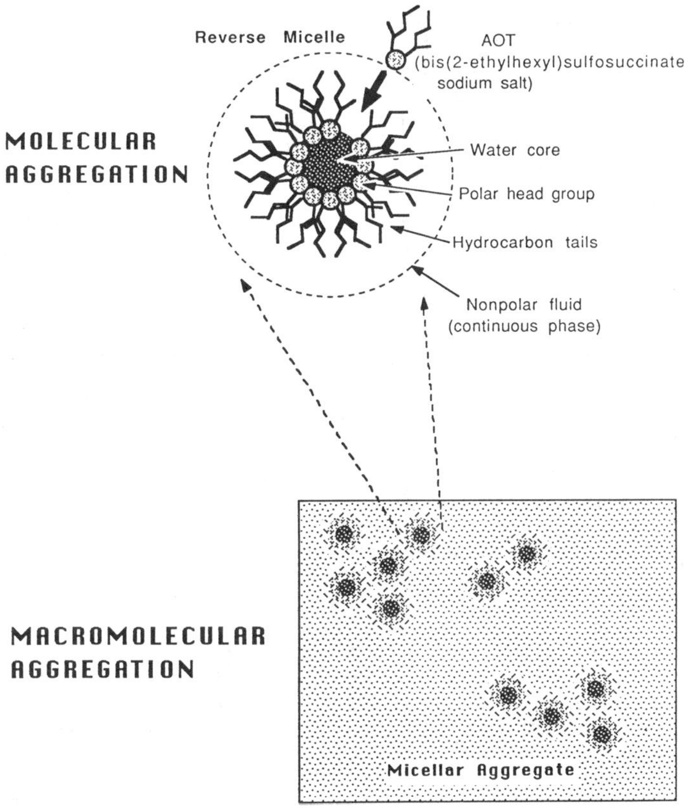

Figure 4. Idealized AOT reverse micelle or microemulsion structure and a proposed aggregation (or clustering) mechanism which maintains the distinct solvent environments for the reverse micelle components.

Diffusion Coefficients of Reverse Micelles in Liquid and Supercritical Alkanes. Figure 5 compares diffusion coefficients for reverse micelles (W = 5) in alkanes ranging from propane to decane at 25°C and pressures up to 600 bar. The diffusion coefficients measured in liquids generally decrease by 10 to 15 percent as pressure is increased to 400 bar, and show the expected systematic increase as alkane length decreases (and viscosity increases).

Diffusion coefficients of micelles in near critical propane (25°C) and supercritical ethane contrast with the larger alkanes by showing an initial increase as pressure increased. This is ascribed to the strong micelle-micelle interactions in these fluids at lower pressures. The interactions are relatively small for propane (but still much greater than for the larger alkanes), resulting in a maximum for the diffusion coefficient in propane at ~100 bar due to the opposing effect of increased viscosity at higher pressures. The supercritical ethane reverse micelles shown in Figure 5, have diffusion coefficients which increase with pressure and due to the pressure limitations of our instrumentation the expected maximum was not observed. These results, again attributed to micelle-micelle interactions, suggest that significantly improved mass transport properties for these systems are often obtained at higher pressures even though the viscosity (of the fluid continuous phase) is higher.

Figure 6 shows hydrodynamic diameters for the reverse micelles in liquid alkanes at 25°C and supercritical ethane at 37°C. The results show that micelle diameter is generally in the 4 to 5 nm range for the larger alkanes, although slightly larger diameters were observed for decane. Generally, the larger alkanes show little or no change in hydrodynamic diameter with pressure, although the larger diameters for decane (and possibly heptane) may suggest some micelle-micelle attractive or more complex steric interactions. In contrast, propane and ethane show hydrodynamic diameters which decrease substantially as pressure is increased, due to decreased micelle-micelle interactions.

Hydrodynamic Diameters of Reverse Micelles-Fluid Phases in Equilibrium with Aqueous Phases. The formation and properties of reverse micelle and microemulsion phases in equilibrium with a second predominantly water continuous phase is of practical interest for extraction processes. Figure 7 compares apparent hydrodynamic diameters observed in the ethane/AOT/water system at 37°C for values of 1, 3 and 16. In single phase systems at W = 1 (a) and 3 (b) the apparent hydrodynamic diameter decreases with increased pressure due to decreased micelle-micelle interactions as the solvent power increases. In contrast for a system with an overall W = 16 (c), where a second aqueous phase exists, hydrodynamic diameter increases continuously with pressure.

Corresponding data for the propane/AOT/water system at 25°C are presented in Figure 8 for W = 1, 5, and 20. In a single phase at W = 1 (a) hydrodynamic diameter is nearly invarient with pressure (3.8 ± 0.3 nm) with a slight increase suggested at the very lowest pressures. In a single phase system at W = 5 (b),

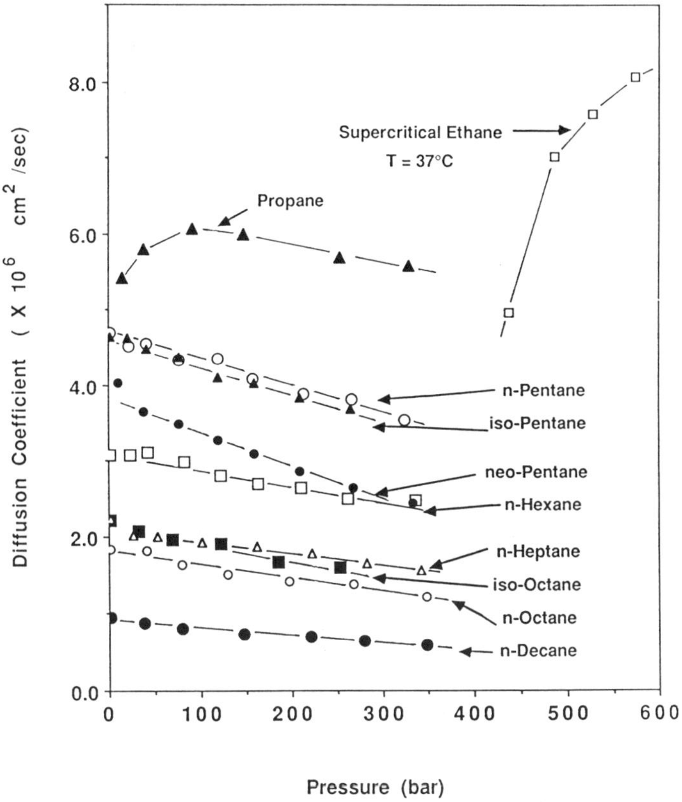

Figure 5. Micelle diffusion coefficients for various alkanes as a function of pressure measured by DLS. T = 25°C, W = 5, Y_{AOT} = 0.015 (mole fraction.)

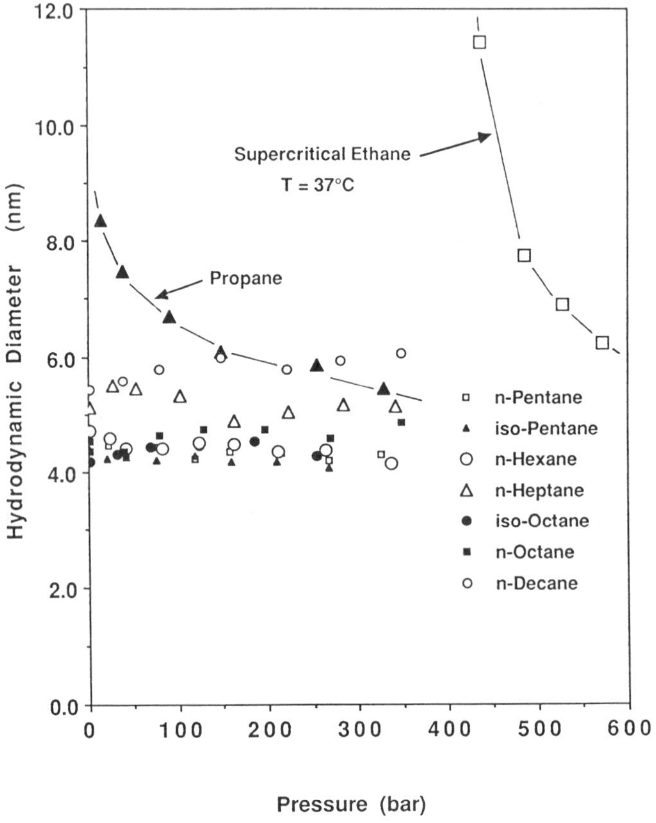

Figure 6. Hydrodynamic diameters measured by DLS for alkane/AOT/water solutions. T = 25°C, W = 5, Y_{AOT} = 0.015 (mole fraction.)

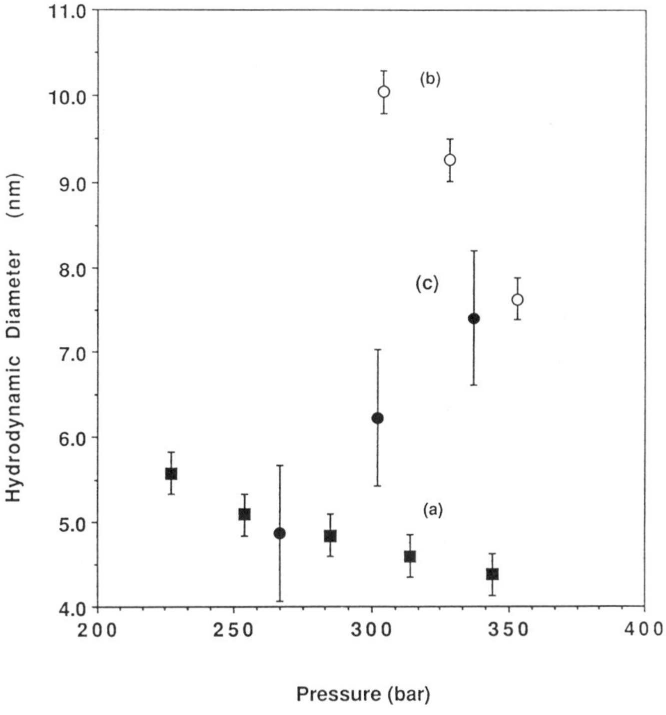

Figure 7. Hydrodynamic diameters for supercritical (37°C) ethane/AOT/water solutions as a function of pressure for W values of the overall system of 1 (a - ■), 3 (b - ○), and 16 (c - ●).

greater micelle-micelle attractive interactions are evident for these micelles with larger water cores leading to the expected attractive increases at lower pressures. As shown previously in Figure 6, the hydrodynamic diameter is larger than the actual physical diameter (which we assume to be approximately equal to micelle size in the larger alkanes).

The measured hydrodynamic diameters in propane at W = 20 (Figure 8, c) show a maximum at ~80 bar corresponding to the phase boundary for formation of a two-phase fluid-liquid system. At lower pressures, the liquid propane is in equilibrium with a lower predominantly water phase. Thus, as the phase boundary is approached from higher pressures, micelle-micelle interactions become increasingly important. As the phase boundary is approached hydrodynamic diameters increase exponentially and substantially increased light scattering is observed at the detector. At the phase boundary the attractive interactions cause a phase change where portions of both the AOT and water precipitate to form a predominantly water liquid phase. Importantly, hydrodynamic diameters are substantial at pressures as low as 25 bar. While the actual micelle size (W) has not yet been determined, it is apparent that the water to surfactant ratio will vary continuously through the two-phase region. Further measurements of the micelle number density and physical diameters (W for the upper phase) are required to understand microemulsion formation and diameter, and hence solvent properties, in the two-phase region.

Conclusions

The DLS results, taken in conjunction with previous solvatochromic probe studies (15), show the important role of micelle-micelle interactions in determining both mass transport properties and phase behavior for reverse micelle and microemulsion systems. Clearly micelle-micelle attractive interactions of the London-van der Waals type are of much greater importance in near-critical and supercritical fluids than in liquid solvents under the conditions studied. At low continuous phase densities, solvent-micelle attractive forces are lower and the lower dielectric continuous phase can less effectively "shield" micelle-micelle attractive forces. These attractive forces lead to clustering and, at low pressures, eventual coalescence of micelles to form either a second reverse micelle phase or a second aqueous phase. In the two-phase region similar considerations apply, but here both micelle number density and $[H_2O]/[AOT]$ (W) in the upper phase are variable with pressure. It is apparent that the attractive interactions, which cause the apparent hydrodynamic diameter to be larger than the physical diameter, are increased at both higher surfactant concentrations and larger W (at constant surfactant concentration). Experimental results, to be reported later (37), are consistent with this expectation. These results suggest that for fluids such as ethane and propane that the size of the water core has a much greater effect upon micelle clustering than the micelle concentration. Current research is aimed at a

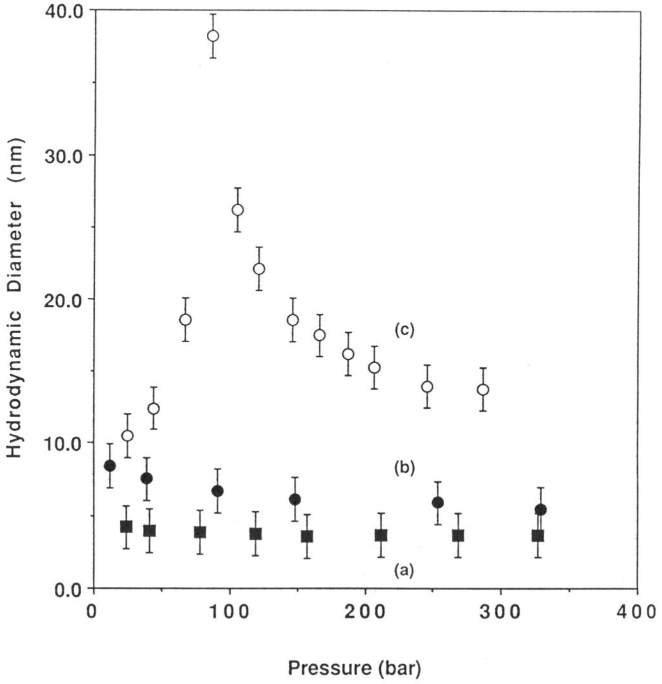

Figure 8. Hydrodynamic diameters for liquid "near-critical" (25°C) propane/AOT/water solutions as a function of pressure for W values of 1 (a - ■), 5 (b - ●) and 20 (c - ○).

quantitative understanding of the factors governing micelle number density and W in the two-phase region and development of a predictive model for phase behavior in these systems.

Acknowledgment

Work supported by the Director, Office of Energy Research, Office of Basic Energy Sciences, Chemical Sciences Division of the U. S. Department of Energy (DOE) under contract DE-AC06-76RLO 1830. Pacific Northwest Laboratory is operated for the DOE by Battelle Memorial Institute.

Literature Cited

1. Smith, R. D.; Fulton, J. L.; Gale, R. S. J. Am. Chem. Soc. 1987, 109, 920-921.
2. Clarke, J.H.R.; Brown, D. J. Phys. Chem. 1988, 92, 2881-2888.
3. Wong, M.; Thomas, J. K.; Nowak, T. J. Am. Chem. Soc. 1977, 99, 4730-4736.
4. Fletcher, P.D.I.; Parrott, D. J. Chem. Soc., Faraday Trans 1 1988, 84, 1131-1144.
5. Hatton, T. A.; Goklen, K. E. Separation Science and Technology, 1987, 22, 831-824.
6. Dorsey, J. G.; Hernandez-Torres, M. A.; Landy, J. S. Anal. Chem. 1986, 58, 744-747.
7. Fendler, J. H. In Reverse Micelles, Luisi, P. L., Straab, B. E., Eds.; Plenum: New York, 1984; 305-322.
8. Evans, D. F.; Sen, R.; Warr, G. G. J. Phys. Chem. 1988, 92, 774-783.
9. Eicke, H. F.; Hilfiker, R.; Kim, V. J. Colloid Interface Sci. 1988, 121, 579-584.
10. Eicke, H. F.; Hilfiker, R. J. Chem. Soc., Faraday Trans. 1 1987, 83, 1621-1629.
11. Huang, J. S. J. Chem. Phys. 1985, 82, 480-484.
12. Roux, D.; Bellocq, A. M. In Surfactants in Solution, Mittal, K. L., Lindman, B., Eds.; Plenum: New York, 1984; 1247-1261.
13. Smith, R. D.; Fulton, J. L. J. Phys. Chem. 1988, 92, 2903-2907.
14. Blitz, J. P.; Fulton, J. L.; Smith, R. D. J. Phys. Chem. 1988, 92, 2707-2710.
15. Fulton, J. L.; Blitz, J. P. Tingey, J. M. Smith, R. D. J. Phys. Chem., in press.
16. Smith, R. D.; Fulton, J. L. In: Surfactant-Based Mobility Control, ACS Symposium Series, 373, Smith, D. H., Ed.; American Chemical Society: Washington,D. C., 1988.
17. Prausnitz, J. M.; Randolph, T. W.; Clark, D. S.; Blanch, H. W. Science 1988, 238, 387-390.
18. Beckman, E. J.; Smith, R. D. J. Phys. Chem. in press.
19. Matson, D. W.; Fulton, J. L.; Smith, R. D. Mat. Lett. 1987, 6, 31-33.
20. Gale, R. W.; Fulton, J. L.; Smith, R. D. Anal. Chem. 1987, 59, 1977-1979.

21. Smith, R. D.; Gale, R. W.; Fulton, J. L. LC•GC 1987, 6, 134-142.
22. Smith, R. D.; Fulton, J. L.; Blitz, J. P.; Tingey, J. M. J. Phys. Chem., submitted.
23. Gruen, D.W.R.; Haydon, D. A. Pure & Appl. Chem. 1980, 52, 1229-1240.
24. Schmitt, W. J.; Reid, R. C. J. Chem. Eng. Data 1986, 31, 204-212.
25. Kotlarchyk, M.; Chen. S.; Huang, J. S.; Kim, M. W. Phys. Rev. A 1984, 29, 2054-2069.
26. Zulauf, M.; Eicke, H. F. J. Phys. Chem. 1979, 83, 480-486.
27. Koppel, D. E. J. Chem. Phys. 1972, 57, 4814-4815.
28. Younglove, B. A.; Ely, J. F. J. Phys. Chem. Ref. Data 1987, 16, 577.
29. Hardich, J. J. Chem. Phys. 1976, 64, 2265-2266.
30. Reid, R. C.; Prausnitz, J. M.; Sherwood, T. K. In The Properties of Gases and Liquids, 3rd ed.; McGraw Hill: New York, 1977, p. 426.
31. Michels, A.; Wassenaar, T.; Louwerse, P. Physica XX 1954, 99-106.
32. Smith, B. L.; Parpia, D. Y. J. Phys. C: Solid St. Phys. 1971, 4, 2251-2257.
33. Thodos, G.; Shimotake, H. A.I.Ch.E Jour. 1958, 4, 257-262.
34. Stephen, K.; Lucas, K. in Viscosity of Pure Fluids, Plenum Press: New York, 1979.
35. Brazier, D. W.; Freeman, G. R. Can. J. Chem. 1969, 47, 893-899.
36. Kotlarchyk, M.; Huang, J. S.; Chen, S. H. J. Phys. Chem. 1985, 89, 4382-4386.
37. Fulton, J. L.; Smith, R. D. J. Phys. Chem., to be submitted.

RECEIVED May 1, 1989

Chapter 13

Inverse Emulsion Polymerization of Acrylamide in Near-Critical and Supercritical Continuous Phases

Eric J. Beckman, John L. Fulton, Dean W. Matson, and Richard D. Smith

Chemical Methods and Separations Group, Chemical Sciences Department, Pacific Northwest Laboratory, Richland, WA 99352

The inverse microemulsion polymerization of water-soluble acrylamide monomers within near-critical and supercritical alkane continuous phase provides a potential route for production of polymers with novel physical properties and at high reaction rates. In order to define conditions for a model polymerization process, the phase behavior of a nonionic surfactant/acrylamide/water system in near- and supercritical mixtures of ethane and propane was examined. Results show that in mixtures of ethane and propane the continuous-phase density determines the phase behavior. Results also show that acrylamide acts as a co-surfactant with $C_{16}E_2/C_{12}E_4$ (Brij) surfactant blend used for these experiments. Surprisingly, increasing the total dispersed-phase volume fraction lowers the density (and consequently pressure) required to form a stable microemulsion. Dynamic light scattering results suggest the presence of strong micelle-micelle interactions, or clustering, the extent of which increases rapidly as the phase boundary is approached. Initial polymerization results indicate possible dependencies of both the polymerization rate and the molecular weight on continuous-phase density and/or the degree of micelle-micelle clustering, suggesting that the monomer may not be as accessible to a growing chain as in a classical emulsion polymerization.

Emulsion polymerization is an important commercial process because, in contrast to the same free-radical polymerization performed in the bulk, molecular weight and reaction rate can be increased simultaneously (1-3). Furthermore, the lower viscosity of an emulsion system compared with that of the corresponding bulk process provides better control over heat transfer. Commercial emulsion processes usually use a surfactant/water/monomer system

0097-6156/89/0406-0184$06.75/0

that is stabilized by vigorous stirring. The dispersed phase contains micelles, approximately 10 to 50 nm in diameter, as well as monomer droplets, which can be 10 to 100 times larger than the micelles. In the absence of agitation, these monomer droplets will coagulate and separate as a second phase. If, as is the usual practice (1), a continuous-phase soluble initiator is used, polymerization commences at the micelle interface and proceeds within the micelles. During the reaction, monomer diffuses from the large droplets into the micelles. Exhaustion of these monomer reservoirs signals the end of the polymerization. In contrast to the emulsion system described above, a microemulsion is thermodynamically stable, and thus one-phase and optically clear in the absence of agitation. Microemulsion polymerization has been used to produce stable latices with a very fine (approx. 50 nm) particle size (4).

Microemulsion polymerization in a supercritical fluid may provide some significant advantages compared with the same reaction in a conventional liquid. Removal of the continuous phase following polymerization would certainly be faster and easier than removal following a similar reaction carried out in a conventional liquid. The ability to remove the continuous phase without the formation of a liquid-vapor meniscus and its accompanying strong surface forces could allow production of polymer with a very fine particle size.

In this article we describe the phase behavior of a microemulsion system chosen for the free radical polymerization of acrylamide within near-critical and supercritical alkane continuous phases. The effects of pressure, temperature, and composition on the phase behavior all influence the choice of operating parameters for the polymerization. These results not only provide a basis for subsequent polymerization studies, but also provide data on the properties of reverse micelles formed in supercritical fluids from nonionic surfactants.

In addition, we present some initial results on the effect of pressure, temperature, and the various composition variables on the rate of polymerization and the molecular weight of the polymer formed.

Experimental

Materials. Nonionic surfactants Brij 52 (B52) and Brij 30 (B30) were obtained from the Sigma Chemical Company and used as received. These surfactants are ethoxylated alcohols with the nominal structures $C_{16}E_2$ and $C_{12}E_4$, respectively, where E represents the number of ethylene oxide units. Acrylamide was obtained from the Aldrich Chemical Company (Gold Label 99+%) and recrystallized twice from chloroform. Azo bis(isobutyrnitrile) (AIBN), obtained from the Alfa Products Division of Morton Thiokol, was recrystallized from methanol. Water was doubly deionized. Propane obtained from Union Carbide Linde Division (CP Grade) and ethane from Air Products (CP Grade) were used without further purification.

Phase Behavior. Phase transitions were observed visually using a high-pressure view cell (volume = 47 cm^3), capable of pressures to 600 bar, whose design has been previously described (5). Material was introduced to the magnetically stirred cell, which was then sealed and pressurized with the fluid of choice using a Varian 8500 syringe pump. Gas mixtures were prepared by weight (composition ±0.25%) in a 400 cm^3 lecture bottle, stirred for 15 minutes, then transferred to the syringe pump. Temperature in the cell was controlled to within 0.1°C using an Omega thermocouple/temperature programmer. Pressure was measured using a Precise Sensor 0- to 10,000-psi transducer, and readout was calibrated to within ±10 psi using a Heise Bourdon-tube gauge.

Quasi-Elastic Light Scattering. Dynamic light scattering measurements were made using a Malvern PCS-100 spectrophotometer modified for high-pressure work. The conventional scattering cell was replaced by a high-precision sapphire tube, described elsewhere (6), which allowed measurements at pressures up to 500 bar. Temperature in the cell, whose volume is approximately 1.5 cm^3, is maintained via a thermostated (±0.1°C) toluene bath. The spectrophotometer used a 5-W argon laser (488 nm), and the signal from the photomultiplier was processed on a 128-channel, real-time digital correlator (K7032-OS) with a 50-ns sample time. The instrument alignment was checked using 58-nm polystyrene latex spheres dispersed in water; the measured size was within 5% of the reported value.

The photon autocorrelation function was analyzed by the method of cumulants (7), in which the logarithm of the normalized autocorrelation function, $G^{(2)}(q,t)$, is fit to a polynomial using a nonlinear, least squares routine,

$$\frac{1}{2}\ln\left[G^{(2)}(t)\right] = \Gamma_\phi - \Gamma_1 t + \frac{1}{2}\Gamma_2\frac{t^2}{2} \quad (1)$$

where Γ_ϕ is ideally zero. Γ_1, the first cumulant, is related to the diffusion coefficient, D_t, by

$$\Gamma_1 = D_t q^2 \quad (2)$$

where q is the scattering vector. In the absence of interparticle interactions, the second cumulant, Γ_2, can be related to the variance of the distribution of diffusion coefficients. The mean hydrodynamic radius of the particles can be calculated from the diffusion coefficient using the Stokes-Einstein relation for spherical particles:

$$R_H = \frac{kT}{6\pi\eta D_t} \quad (3)$$

where k is Boltzmann's constant, T is the temperature, and h is the viscosity of the continuous phase.

Because the solubility of acrylamide (8), water (9), and the surfactants in ethane or propane is low, the viscosity of the continuous phase was taken to be that of the pure fluid. The viscosity of the various ethane/propane mixtures was calculated using a reduced-density correlation developed by Dean and Stiel (10), which is reported to be accurate to within 2 to 4% for light hydrocarbon mixtures. The density of the ethane/propane mixtures was either calculated via a modified Benedict-Webb-Rubin equation of state (11) or, in some cases, measured using a Mettler-Paar DMA-512 vibrating tube densimeter. The densimeter was thermostated via a circulating water bath to within ±0.01°C, and calibrated using water and propane at the temperatures of interest.

The refractive index of the ethane/propane mixture, needed for calculation of the scattering vector, was determined from experimental density measurements and an empirical Lorentz-Lorenz relationship (12). Incorporating the errors from the viscosity and refractive index calculations into the Stokes-Einstein relation results in a maximum error in the hydrodynamic radius of approximately 5%.

Polymerization. The hydrodynamic radius of the micelles both before and during polymerization was followed using the apparatus shown in Figure 1. First, the surfactant, acrylamide, and water are added to the view cell. Next, while stirring, the ethane/propane continuous phase is charged to the system via the syringe pump. After setting the temperature and pressure of the view cell-scattering cell system to the desired point, the circulating pump is operated for 30 minutes to thoroughly homogenize the system. The circulating pump is shut off, and an additional 30 minutes is allowed to elapse to allow equilibrium to be established and any dust to settle out in the scattering cell. Micelle size measurements are then made as a function of temperature or pressure while the phase behavior is observed via the view cell. For polymerization studies, following the initial micelle size measurement at the pressure and temperature of interest, a known volume of initiator (AIBN/toluene, 4% by weight) solution is injected using the manual syringe pump (which had been calibrated previously). The recirculating pump is again operated for 30 minutes, followed by 30 minutes settling time. Micelle size measurements are then made every 20 minutes at constant temperature and pressure for six hours.

At the end of six hours, the ethane/propane is vented and a 3:1 chloroform/methanol solution is injected to the system, dissolving the surfactant, water, and unreacted monomer. The polymer, which is insoluble in the chloroform/methanol mixture, is collected by filtration, washed with additional chloroform/methanol and acetone, dried under vacuum, and weighed. The average molecular weight of the polymer (M_w) is measured using static light scattering [see for example (13)] of dilute solutions of the polymer in water.

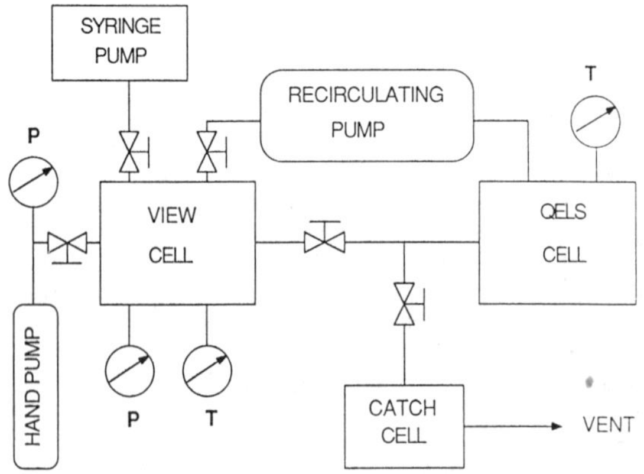

Figure 1. Apparatus for high-pressure dynamic light scattering of microemulsions.

Results and Discussion

Surfactant Selection. It was desired that the polymerization be conducted in a stable microemulsion at temperatures between 50-70°C, thus permitting relatively rapid thermally induced radical production from the initiator, AIBN. For any ethoxylated alcohol nonionic surfactant of the general structure C_iE_j in a water/oil (w/o) ternary system, lowering temperature induces the micellar interface to curve towards the oil phase, creating a w/o emulsion [14]. Because the w/o situation was required for this work, it was therefore necessary to choose a surfactant system that would show a phase inversion temperature (the point at which micellar curvature changes direction) which is below room temperature, thus providing the largest possible temperature range to conduct our research. The empirical hydrophilic-lipophilic balance (HLB) method (15,16) was used to guide the selection of a C_iE_j surfactant system, although there is considerable disagreement over the ideal value for inverse microemulsions. HLB, a means by which to categorize nonionic surfactants, normalizes the weight fraction of hydrophilic groups in a molecule to a 0 to 20 scale. According to Griffin (17), a w/o microemulsion requires an HLB of approximately 3 to 6, whereas Moore and Bell (18) claim an HLB of 7-8 is more effective. Moore and Bell's results agree with those of Kane and Shields (19), who found the HLB range of 7 to 9 to be useful for forming stable w/o emulsions. In a study of a nonionic surfactant/water/acrylamide inverse emulsion in an alkane continuous phase, Candau et al. (3) found the most effective HLB range to be 9 to 9.5. This large range of "acceptable" HLB values is due to the fact that although HLB is often calculated from the surfactant structure alone, the behavior of a surfactant is governed by the continuous phase composition, the concentration of other components (e.g., the monomer and water) (20,21), and of course, the temperature. This is because emulsion stability depends on the proper balance of lipophile-oil and hydrophile-water interactions. Preliminary screening of mixtures of the Brij 52 and Brij 30 (B52/B30) nonionic surfactants (which cover an HLB range of approximately 5 to 10) in a pentane/acrylamide/water system at atmospheric pressure showed that an 80/20 ratio of the surfactants, a calculated HLB value of approximately 7.5, produced a stable microemulsion with a phase inversion temperature of approximately 18-20C and relatively high acrylamide content (acrylamide/surfactant molar ratio greater than 2.0). Consequently, the 80/20 mixture of B52/B30 was used throughout this work.

Effect of Continuous-Phase Composition. In order to determine the temperature range in which acrylamide microemulsion polymerization could be conducted within a supercritical alkane continuous phase, the phase behavior of the Brij mixture/water/acrylamide system in mixtures of propane (T_c = 97°C) and ethane (T_c = 32°C) was investigated. In addition, employing a mixture of ethane and propane allows us to examine the effect of the continuous-phase density on the polymerization at constant pressure and temperature. In this series of experiments the water concentration was fixed at W=5.0 and that for the acrylamide at 1.0. (Water and acrylamide concentrations are reported as molar ratios to the surfactants; the nominal molecular weights of 330 for Brij 52 and 360 for Brij 30 were used. Henceforth all water and acrylamide concentrations will be given in terms of such molar ratios.) The volume fraction of the dispersed phase (volume of surfactants + water + acrylamide divided by the total system volume) in this series of experiments was 0.136.

The phase behavior for various percentages of ethane is given in Figure 2 in terms of clearing points, or cloud points, where the one-phase region is above each line. As the pressure in the view cell is raised to the clearing point, the B52/B30/acrylamide/water system turns from opaque to a transparent reddish-purple color. As pressure is increased still further, the color of transmitted light changes progressively to red-orange to orange to yellow (the color changes are reversible). These color changes occur regardless of the ethane concentration in the mixtures; their implications for the structure of the microemulsion will be discussed in a later section.

The cloud point data in Figure 2 reveal a series of curves that are essentially parallel and shifted to higher pressures as the amount of ethane in the mixture increases. When these data are replotted as cloud point *density* versus temperature (Figure 3), the central role of fluid density upon phase behavior becomes apparent. (Densities for pure ethane and propane were taken from the literature, those for the 80.4/19.6 mixture were measured, and those for the other mixtures were calculated.) The coalescence of the data (when plotted versus density) is not surprising since continuous-phase density affects the chemical potential of the fluid, which in turn determines the degree of solvation of the surfactant tails and, potentially, the structure of the interfacial region. Increasing the degree of solvent penetration or the relative rigidity of the interface may increase the spontaneous radius of curvature and consequently allow for a larger amount of water and acrylamide to be solubilized (16). In addition, at low fluid densities the micelles will display a much greater affinity for each other than for the continuous phase. Increasing the density of the continuous phase effectively screens the micelle-micelle interaction, allowing greater micelle number densities or larger micelles to exist without phase separation. The density effect can also be examined in terms of liquid-liquid phase equilibria thermodynamics; increasing the continuous phase density increases the entropy of mixing by lowering the free volume difference between the constituents. Thus the various two-phase envelopes (water-oil, oil-surfactant, and water-surfactant) that combine to affect the emulsion stability will shrink upon increasing density (20).

Figure 3 also suggests that increasing the temperature increases the stability of the emulsion since the continuous phase density at the clearing point decreases. The surfactant will

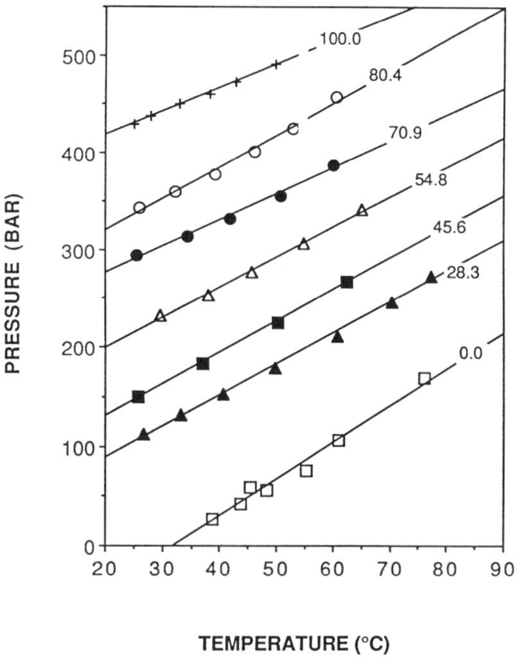

Figure 2. Cloud point curves of Brij 52/Brij 30 80/20 mixture with a water/surfactant ratio of 5.0, acrylamide/surfactant ratio of 1.0, total dispersed-phase volume fraction of 0.136, and seven continuous-phase ethane concentrations (weight %).

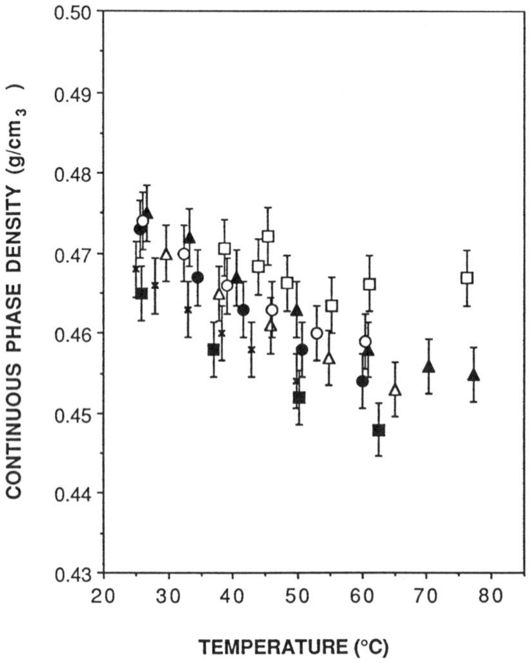

Figure 3. Data from Figure 2 replotted as density of continuous phase at cloud point versus temperature; symbols same as in Figure 2.

become less hydrophilic as temperature increases, decreasing the pressure required to solubilize a given amount. However, as the hydrophilicity of the surfactant decreases, the equilibrium water content will also decrease. Consequently, above 72°C this system becomes turbid.

Effects of Acrylamide and Water Concentration. To maximize the amount of product from this inverse emulsion polymerization, it would be desirable to solubilize as much acrylamide as possible in the precursor microemulsion, while using the minimum amount of surfactant. Because water is merely a solvent that must eventually be removed from the product, the ideal water content of the microemulsion for our purposes would be zero. However, without some water, it is likely that micelles formed from ethoxylated alcohols will be quite small, approximately 5-10 molecules each (22). Thus the effect of water and acrylamide levels on the phase behavior was examined.

The solubility of either acrylamide or water in ethane/propane mixtures is extremely low (8,9). The solubility of the surfactant mixture B52/B30 is also quite low (23). Adding acrylamide to the B52/B30 blend allows significantly larger amounts of both components to be solubilized in the alkane continuous phase, suggesting that acrylamide is a co-surfactant in this system. However, the B52/B30 mixture will solubilize acrylamide only up to an [acrylamide]:[surfactant] molar ratio of 1:4; larger amounts of acrylamide lead to precipitation of a solid phase.

The addition of water significantly increases the amount of acrylamide that can be solubilized by the B52/B30/ethane/propane system (see Figure 4). Accurate determination of cloud point curves for microemulsions with acrylamide levels higher than 2.0 is difficult since the reddish-purple color that is evident upon clearing (see previous section) darkens significantly as the acrylamide level increases. Adding acrylamide increases the effective hydrophilicity of the surfactant system which in turn leads to an increase in the size of the two-phase region in that part of the phase diagram rich in alkane. Therefore, greater pressure (density) is required to increase the entropy of mixing sufficient to offset the effect of raising the hydrophilicity of the surfactant.

Although water allows for greater uptake of acrylamide by the microemulsion, water alone ([acrylamide]=0) will not produce a one-phase system with the Brij 52/30 blend in an ethane/propane continuous phase. As postulated earlier, acrylamide is a co-surfactant with the B52/B30 blend, as evidenced by the results in Figure 5. The existence of the maximum in the allowable water as a function of [AM] has been observed in other micelle systems where acrylamide behaves as a co-surfactant (16). When more than the maximum allowable water level is added at a particular acrylamide content, the system becomes turbid, followed by the appearance of what appears to be a solid second phase. That acrylamide behaves as a co-surfactant is possibly due to its

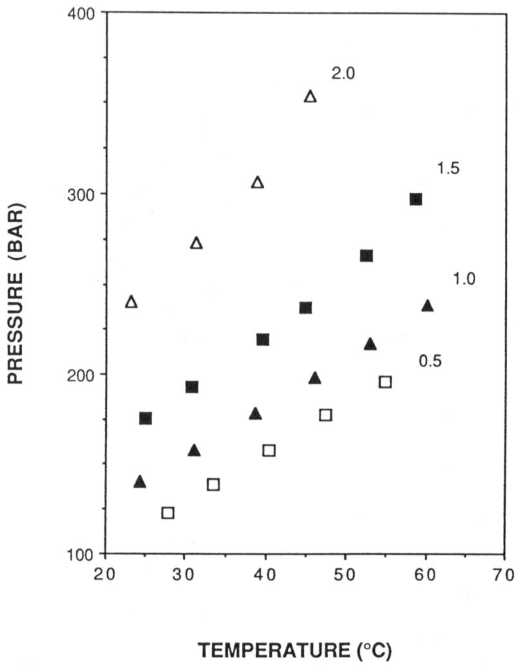

Figure 4. Cloud point curves of B52/B30 in a 41.9/58.1 w/w ethane/propane mixture with a water/surfactant ratio of 5.0 and four acrylamide/surfactant ratios.

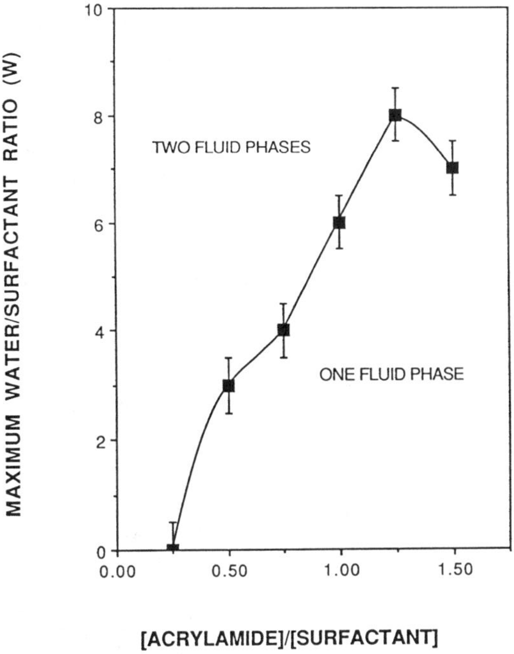

Figure 5. Maximum water/surfactant ratio of B52/B30 in 80.4/19.6 w/w ethane/propane at 30°C and 500 bar versus acrylamide/ surfactant ratio.

ability to coordinate and hydrogen bond to the surfactant ethylene oxide units, as is the case for low molecular weight alcohols or amines (24,25). This coordination of the EO units allows the water to self-associate in the core region, which is thermodynamically favorable. Given an EO coordination number of 0.5 (24) and an average structure of our surfactant blend of $C_{15.3}E_{2.4}$, complete coordination of the EO units by acrylamide corresponds to an [AM] ratio of 1.2, which agrees well with the maximum in the curve in Figure 5.

Effect of Dispersed Phase Concentration. In any application of these microemulsion systems to polymerization processes, it would be desirable to maximize polymer yield. This can be accomplished by maximizing the acrylamide ratio at constant surfactant loading or, of course, the acrylamide ratio (as well as the water ratio) to the surfactant can be fixed and the total amount of surfactant in the system increased. Therefore, the effect of total dispersed phase concentration on the phase behavior was investigated. The dispersed phase concentration was defined as a volume fraction equalling the total volume of surfactants + acrylamide + water divided by the total volume.

Results shown in Figure 6 reveal that increasing the dispersed phase volume fraction significantly reduces the pressure required to form a stable microemulsion. At a volume fraction of 0.09, a one-phase system will not form at any pressure up to 550 bar, whereas increasing volume fraction to 0.15 will produce a stable microemulsion at less than 300 bar. This decrease in clearing pressure as the volume fraction dispersed phase is increased may be due to the stabilizing effect of surfactant tail overlap, or micelle-micelle clustering, as the number density of the micelles increases. It is interesting to note that the point at which the phase boundary (in Figure 6) becomes independent of pressure is close to the volume fraction of 14% which Cazabat et al. (26) report as the point where geometrical percolation (formation of transient, very large clusters) occurs in hard sphere systems. As before, we can also explain this phenomenon in light of the effects of composition on liquid-liquid phase behavior. It can be shown that increasing the surfactant concentration at constant [H_2O] and [AM] moves the system away from a phase boundary, thus allowing a stable system of lower pressure (23).

As mentioned earlier, if the temperature is increased to a certain point, these microemulsion systems will remain turbid regardless of the pressure at which they become one phase. This ceiling temperature decreases as the dispersed-phase volume fraction increases (see Figure 7).

Effect of Pressure on Apparent Micelle Size. The pressure-induced color changes that are observed in the one-phase region of the Brij-based microemulsion systems suggest that the size of the organized assemblies in the ethane/propane mixtures is changing as the pressure is changed. Initial considerations of dynamic light

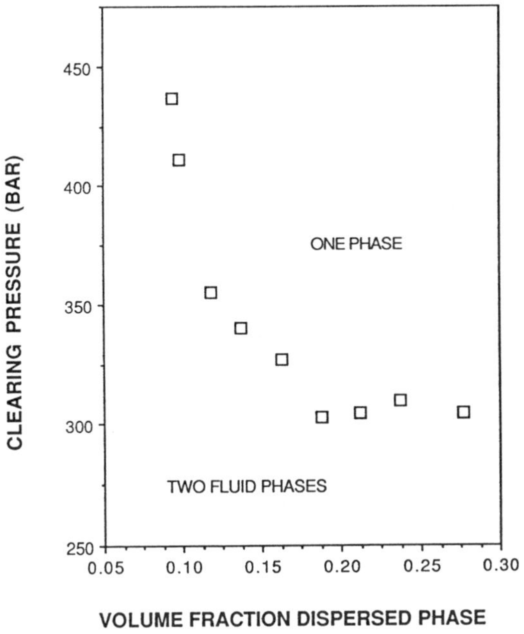

Figure 6. Clearing pressure (cloud point) of B52/B30 in 80.4/19.6 w/w ethane/propane with a water/surfactant ratio of 5.0 and an acrylamide/surfactant ratio of 1.0 versus dispersed-phase volume fraction.

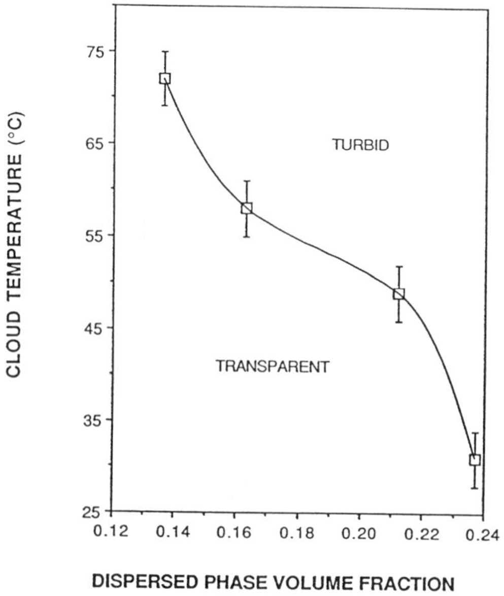

Figure 7. Cloud point temperatures of B52/B30 in 80.4/19.6 w/w ethane/propane with a water/surfactant ratio of 5.0 and an acrylamide/surfactant ratio of 1.0 versus dispersed-phase volume fraction.

scattering results tended to support this hypothesis (see Figure 8). As the pressure is increased above the cloud point, the apparent hydrodynamic radius (as calculated from the measured micelle diffusion constant and the calculated continuous-phase viscosity) appears to decrease dramatically. It is likely, however, that this apparent size change is due to a decrease in micelle-micelle interactions and, consequently, a declustering as the pressure increases, rather than an actual shrinkage of the micelles (27). Calculations based on literature values (28-29) for the length and coverage surface area of ethoxylated alcohols like the Brij surfactants, plus the amount of water and acrylamide in the system, suggest a micelle radius of 5 to 7 nm. The much larger apparent radii observed in Figure 8, as well as the results from the previous section, support the notion that the micelles tend to cluster in this system, and that the extent of clustering increases rapidly as the phase boundary is approached.

Polymerization Results. Preliminary polymerization runs were conducted to evaluate the effect of initiator concentration, temperature, and continuous-phase density on the rate of reaction as well as the ultimate molecular weight of the polymer. Continuous-phase density could be varied in two ways: 1) by varying the pressure at constant temperature and ethane/propane ratio, and 2) by varying the ethane/propane ratio at constant temperature and pressure. In all of these polymerizations, the acrylamide ratio was 1.0, water was 3.5, and the total dispersed-phase volume fraction was 0.16.

Somewhat surprisingly, increasing the concentration of AIBN did not decrease the polymer molecular weight, as is observed in conventional emulsion polymerizations (see Figure 9). This type of behavior has been observed in microemulsion polymerizations (30), and has been attributed to the difficulty of the initiator to penetrate the interface because of the larger amount of surfactant in microemulsion systems than in conventional emulsion polymerizations.

The results presented in Table 1 are representative of the general trends that were observed in a number of polymerizations. First, by simply lowering the continuous-phase density (by increasing the ethane composition) at constant pressure, temperature and [AIBN], the molecular weight of the polymer increased by as much as a factor of two. Our research has also shown (31) that lowering the continuous-phase density at constant temperature and pressure increases the apparent cluster size. However, increased clustering should not affect the molecular weight if, as is usually assumed, the monomer passes freely from micelle to micelle during collisions. It may be that because the monomer is a co-surfactant in our system, and therefore closely coordinated about the ethylene oxide units of the surfactant, it may not be as free to migrate to a reacting center as in the case where it resided in the micelle core. High-pressure NMR or fluorescence studies would be required to investigate this hypothesis in detail.

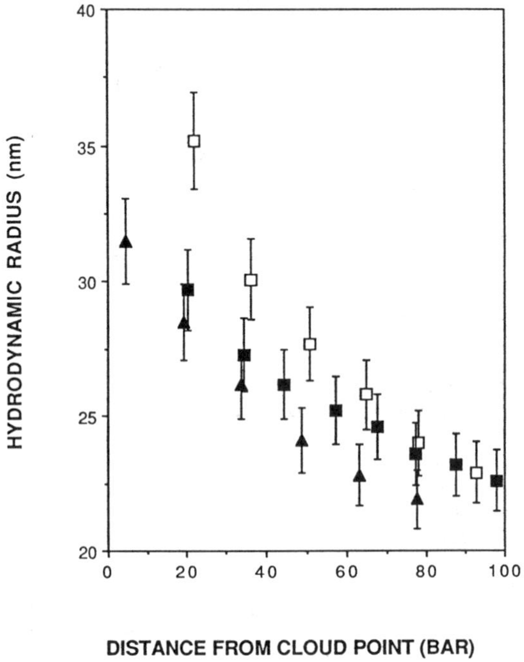

Figure 8. Hydrodynamic radius for system from Figure 7 plotted versus distance from cloud point at 30 °C (□); 40 °C (▲); and 50 °C (■).

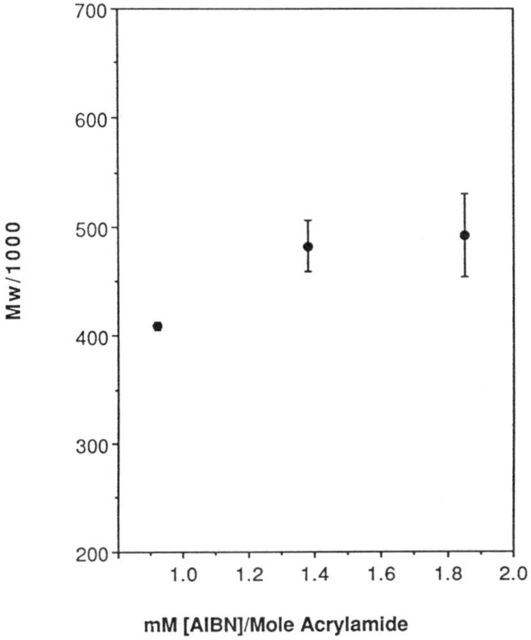

Figure 9. Effect of initiator concentration on molecular weight of polyacrylamide after six hours reaction time in microemulsion system with acrylamide/surfactant = 1.25, water/surfactant = 3.5, dispersed-phase volume fraction = 0.16, P = 362 bar, and T = 60 °C.

Table 1. Polymerization Results

Run No.	Propane Wt. %	[AIBN]	ρ (g/cc)	Yield (%)	MW
1	50[a]	1.35	.453	<5	483,000
2	35[a]	1.35	.439	<5	979,000
3	50[a]	1.85	.453	<5	492,000
4	35[a]	1.85	.439	<5	629,000
5	57.5[b]	1.85	.454	>95	560,000
6	42.2[b]	1.85	.440	<5	986,000
7	42.2[c]	1.85	.454	20-30	864,000

a Temperature = 60°C, pressure = 362 bar
b Temperature = 65°C, pressure = 362 bar
c Temperature = 65°C, pressure = 440 bar

The rate of polymerization appears to be both temperature and density dependent. At 60°C it is likely that the rate of radical formation by the initiator is too low to permit complete reaction in a reasonable amount of time (1). Raising the temperature to 65°C induced nearly complete conversion (see run 5); yet at the same temperature, lowering the density (in run 6) resulted in very little polymer produced over the same time period. Raising the pressure on the system in run 6 in an attempt to equal the density of the system in run 5 increased the rate of reaction, but not as fast as that in run 5. Again, the extent of clustering may be influencing the reaction rate. Decreasing clustering by raising the density will increase the micelle surface area, which may explain the rise in reaction rate.

Finally, contrary to expectations, the apparent micelle radius decreases (by approximately 10 to 15%) as the polymerization progresses (see Figure 10). This trend was observed for all of the runs shown in Table 1. Not until almost total conversion, when coagulation and phase separation were observed in the view cell, was any increase in the apparent micelle size observed during polymerization. However, the growing polymer molecules could be in a highly collapsed state such that they fit inside a micelle core. (A one-million molecular weight molecule in its bulk state, density of approximately 1g/cc, would require a sphere of approximately 7 to 8 nm in radius).

Clearly, much work remains to be done to fully explain the effects of continuous-phase density and the extent of micelle clustering, on both the rate of polymerization and the molecular weight of the polymer formed via this process.

Conclusions

Investigation of the phase behavior of the Brij-based microemulsion system in ethane/propane mixtures defines the operating conditions for the polymerization process and provides evidence of formation of stable microemulsions in supercritical

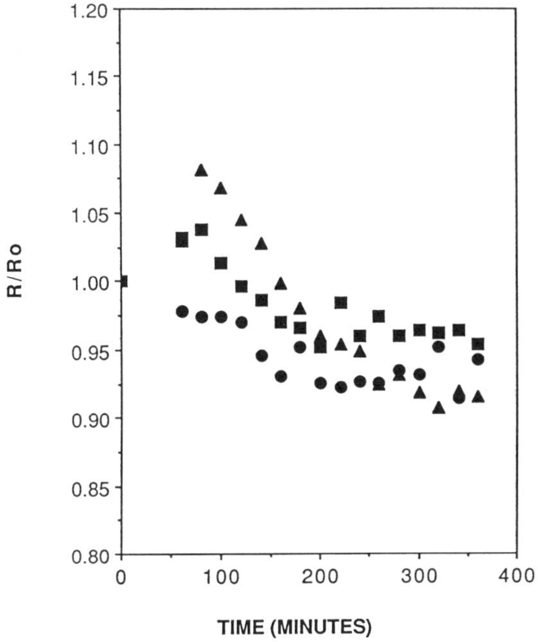

Figure 10. Micelle hydrodynamic radius versus reaction time for the three initiator concentrations from Figure 9; [AIBN] = .92 mMole (■); 1.35 mMole (▲); 1.85 mMole (●).

fluids using nonionic surfactants. Several important observations were made concerning the effects of the various composition variables on the stability of the system.

First, it is apparent that the density of the ethane/propane continuous phase, rather than the molecular composition, determines the stability of the microemulsion. Stable microemulsions can be prepared in mixtures of ethane and propane over the entire concentration range. This allows examination of the effect of continuous-phase density on reaction rate, etc., while temperature and pressure remain constant.

Second, acrylamide behaves as a co-surfactant in this microemulsion system. This implies that as the acrylamide is consumed during the polymerization, the microemulsion may become unstable. This has indeed been observed in subsequent polymerizations that have been carried out to high conversion. Third, increasing the total dispersed-phase volume fraction tends to increase the stability of the microemulsion and, thus, lowers the pressure required to maintain a single-phase system. This result may be due to the stabilizing effect of micelle-micelle clustering. Clustering is also indicated by dynamic light scattering results, which show apparent micelle radii 5 to 10 times larger than would be expected according to molecular considerations, and which appear to increase dramatically near a phase boundary.

Continuous-phase density and the extent of micelle-micelle clustering also appear to play important roles in determining the rate of polymerization and the ultimate molecular weight in this system. Because the monomer in our system is a co-surfactant, it may be less mobile and accessible to reaction than in a classical emulsion polymerization. In this case the greater extent of micelle clustering found in our system at lower densities would produce polymer with higher molecular weight, as we have observed. Further work is necessary to shed more light on these hypotheses. These results thus provide a foundation for those interested in this novel type of polymerization or for anyone interested in the effect of continuous-phase density on phase-transfer reactions in micellar systems.

Acknowledgment

We wish to thank the U.S. Army Research Office for their support of this work through Contract No. DAAL03-87-K-0137. We also appreciate the helpful discussions and suggestions of E. W. Kaler (University of Washington).

Literature Cited

1. Odian, G. Principles of Polymerization; John Wiley & Sons: New York, 1981.
2. Candau, F.; Leong, Y.S. J. Polym. Sci. - Polym. Chem. Ed. 1985, 23, 193.

3. Candau, F.; Zekhnini, Z.; Durand, J.-P. J. Coll. Int. Sci. 1986, 114, 398.
4. Leong, Y.S.; Candau, S.J.; Candau, F. In Surfactants in Solution; Mittal, K.L., Lindman, B., Eds.; Plenum Press: New York 1983; Vol. 3, p 1897
5. Blitz, J.P.; Fulton, J.L.; Smith, R.D. submitted to J. Phys. Chem.
6. Blitz, J.P.; Fulton, J.L.; Smith, R.D. J. Phys. Chem. 1988, 92, 2707.
7. Koppel, D.E. J. Chem. Phys. 1972, 57, 4814.
8. Windholz, M.; Budavari, S., Eds., The Merck Index, Tenth Ed., Merck and Co.: Rahway, N.J., 1983.
9. Parrish, W.R.; Pollin, A.G.; Schmidt, T.W. Proc. Sixty-First Ann. Conv., Gas Proc. Assoc., 1982, 164.
10. Dean, D.E.; Stiehl, L.I. AIChE J. 1965, 11, 526.
11. Reynolds, W.C. Thermodynamic Properties in SI, Dept. of Mech. Eng., Stanford Univ., Stanford, CA 1979
12. Hadrich, J. J. Chem. Phys. 1976, 64, 2265.
13. Billmeyer, F. W. Textbrook of Polymer Science, 2nd Edition. Wiley-Interscience: New York, 1971.
14. Marszal, L. In Nonionic Surfactants-Physical Chemistry; Schick, M. J., Ed.; Marcel Dekker, Inc: New York; Chapter 9, 1987.
15. Beerbower, A.; Hill, M.W. Am. Cosmet. Perfumery 1972, 87, 85.
16. Hou, M-J.; Shah, D.O. Langmuir 1987, 3, 1086.
17. Griffin, W.C. J. Soc. Cosmet. Chem. 1949, 1, 311.
18. Moore, C.D.; Bell, M. Soap, Perfum. Cosmet. 1956, 29, 893.
19. Kane, J.; Shields, J. U.S. Patent No. 3 997 492, 1975.
20. Kahlweit, M.; Strey, R. Angew. Chem. Int. Ed. Engl. 1985, 24, 654.
21. Becher, P. In Surfactants in Solution; Mittal, K.L., Lindman, B., Eds.; Plenum Press: New York, 1983, Vol. 3, p. 1925.
22. Ravey, J. C.; Buzier, M.; Picot, C. J. Coll. Int. Sci. 1984, 97, 9.
23. Beckman, E. J.; Smith, R. D., submitted to J. Phys. Chem.
24. Kon-No, K.; Kitahara, A.; El Seoud, O. A. In Nonionic Surfactants-Physical Chemistry; Schick, M. J., Ed.; Marcel Dekker, Inc.: New York; Chapter 4, 1987.
25. Kumar, C.; Balasubramanian, D. J. Phys. Chem. 1980, 84, 1985.
26. Cazabat, A.M.; Chatenay, D.; Guering, P., Langevin, D.; Meunier, J.; Sorba, O. In Surfactants in Solution; Mittal, K.L., Botheral, P., Eds.; Plenum Press: New York, 1986; Vol. 3, p 1737.
27. Smith, R. D.; Fulton, J. L.; Blitz, J. P.; Tingey, J. M. submitted to J. Phys. Chem.
28. Triolog, R.; Magid, L. J.; Johnson, J. S.; Child, H. F. J. Phys. Chem. 1982, 86, 3689.
29. Ravey, J-C. J. Coll. Int. Sci. 1983, 94, 289.

30. Jayakrishnan, A.; Shah, D. O. J. Polym. Sci., Polym. Lett. Ed. 1984, 22, 31.
31. Beckman, E. J.; Smith, R. D. submited to J. Phys. Chem.

RECEIVED May 1, 1989

Chapter 14

Interaction of Polymers with Near-Critical Carbon Dioxide

A. R. Berens[1] and G. S. Huvard[2]

BFGoodrich Company, Research and Development Center, Brecksville, OH 44141

The kinetics and equilibria of carbon dioxide transport in a wide variety of polymers have been studied at pressures up to the saturated vapor pressure at 25°C. Gravimetric data were obtained by rapid weighing of polymer films of varied thickness during desorption at atmospheric pressure, following exposures to compressed CO_2 for various times and pressures. Transport kinetics are Fickian, and diffusivity increases with CO_2 concentration. The solubility of CO_2 generally increases with increasing content of polar groups in the polymer. For several glassy polymers, the isotherms plotted vs. CO_2 activity are sigmoid in shape, combining dual-mode character at low activity with Flory-Huggins form at high activity. The assembled evidence shows that near-critical CO_2 behaves as a polar, highly volatile organic solvent, rather than as a simple gas, in its interactions with polymers.

The transport of carbon dioxide in polymers has historically been analyzed in the same manner as other simple gases (1). Recent studies have shown, however, that the effects of CO_2 on polymers include some features commonly associated with organic solvents, including swelling (2-5), and depression of glass transition temperatures, i.e., plasticization (6-8). Moreover, CO_2 can be handled as a liquid at room temperature under rather moderate pressures; its critical temperature is 31°C and its saturated vapor pressure at 25°C is 64.6 atm (950 psi). For these reasons it seems appropriate to consider near-critical CO_2 as a highly volatile solvent, rather than as a gas, in its interactions with polymers.

[1]Current address: R.D. No. 2, Box 3510, Middlebury, VT 05753
[2]Current address: E. I. du Pont de Nemours and Company, P.O. Box 27001, Richmond, VA 23261

0097–6156/89/0406–0207$06.00/0

This paper describes a simple new technique for obtaining both kinetic and equilibrium data on the transport of CO_2 in polymers, and compares results for several polymers with the behavior of organic vapor/polymer systems. In subsequent publications, we will discuss transport data for additional polymer/CO_2 binary systems and for ternary systems containing an added low molecular weight component. It now appears that the high diffusivity, solublity and plasticizing action of compressed CO_2 in polymers makes this gas uniquely useful in promoting the impregnation of many polymers with a wide variety of additives (9).

Experimental

The general experimental method used here involves sorption of CO_2 into polymer film or sheet samples in a simple pressure vessel, followed by rapid venting and transfer of the samples to a balance for recording weight changes during desorption. This gravimetric method does not require a balance capable of operation under high pressure, but can provide kinetic data in both sorption and desorption, as well as equilibrium solubilities, through suitable experimental procedures and data analyses.

Procedures. Experiments in this study involved sorption of both gaseous and liquid CO_2 and hence covered pressures up to the saturated vapor pressure of CO_2. All measurements were carried out at 25°C. Polymer samples included a wide variety of glassy, rubbery, and semicrystalline materials. To avoid use of solvents, the polymers were formed into films 0.1 to 1 mm. thick by compression-molding from the melt; a sample about 1 x 2 cm. was used in each experiment.

The procedure for high pressure sorption experiments is illustrated schematically in Figure 1. A polymer sample, of 20 - 200 mg. dry weight, w_o, was placed in a 100 ml. pressure vessel fitted with a pressure gauge, valve and a screw closure which could be opened quickly. The vessel was evacuated, then filled to the desired sorption pressure, P_s, from a cylinder of liquid CO_2 and left at this pressure for an appropriate sorption period, t_s. Because of the relative sizes of sample and vessel, sorption caused no appreciable pressure drop. For each polymer studied, experiments were run at various pressures, sorption times, and sample thicknesses, l. At the end of the sorption period, the CO_2 pressure was rapidly vented to atmospheric, the vessel opened, and the sample quickly placed on the pan of a fast-response electronic digital balance readable to 0.00001 g (Mettler AE163). With a computerized data acquisition system, sample weights during desorption at atmospheric pressure were recorded as a function of desorption time, t_d, at intervals as short as 5 seconds beginning within 10 to 20 seconds after venting the vessel.

Supplemental gravimetric sorption/desorption data for CO_2 pressures below atmospheric were obtained by conventional procedures using a Cahn recording vacuum microbalance, for greater precision in the low solubility range.

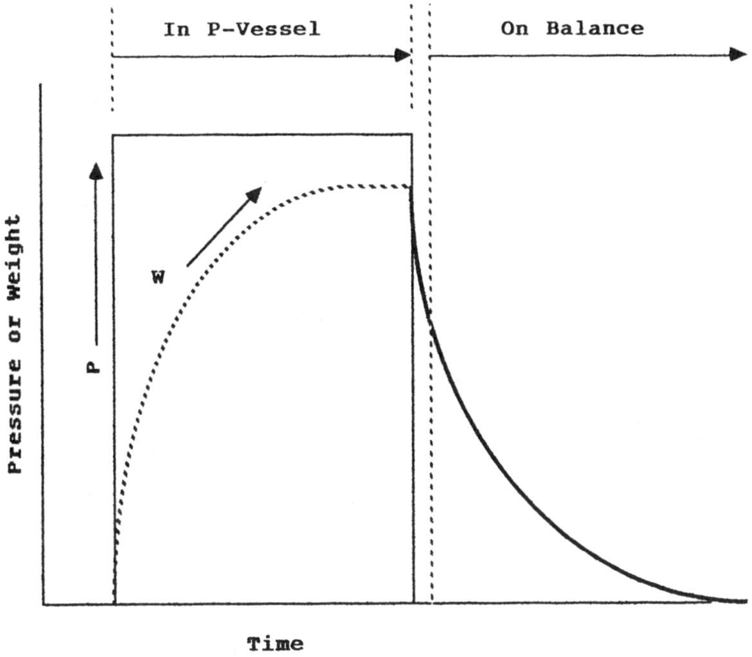

Figure 1. Experimental procedure for following desorption from polymer samples after exposure to high-pressure CO_2. (Schematic).

Data analysis. Using only sample weights, w_t, recorded during desorption, it is possible to develop a rather complete picture of sorption equilibria as well as the kinetics of both absorption and desorption. First, $M_{t,d} = (w_t - w_o)/w_o$ is plotted vs. the square root of t_d; for Fickian diffusion from a plane sheet, this plot should be initially linear. Figure 2, typical of our results on glassy polymers, demonstrates this behavior. Extrapolation to $t_d = 0$ gives $M_{t,s}$, the weight of CO_2 in the sample at the end of the sorption period t_s. The equilibrium CO_2 solubility is established by running several samples at a given P_s, for successively longer t_s, until a constant value of $M_{t,s}$ establishes the equilibrium uptake, M_∞. The equilibrium sorption isotherm is determined from solubility measurements at various pressures. From the initial slope of $M_{t,d} / M_\infty$ vs. $\sqrt{t_d} / l$ for samples sorbed to equilibrium, one calculates D_d, the mean diffusivity for desorption over the concentration interval of that experiment (1). The precision of $M_{t,s}$ (intercept) and D_d (slope) determinations depends upon the rate of desorption, and hence on the polymer type and sample thickness. In most of the experiments reported here, at least six points were obtained in the linear region of $M_{t,d}$ vs $\sqrt{t}$; the precision of $M_{t,s}$ is generally within ±1 g /100 g polymer, and of $\log D_d$, within ±0.2.

While each experimental run gives a complete set of weight vs. time data for desorption, it provides only one point on the weight vs. time curve for absorption. To define the kinetics of absorption, $M_{t,s}$ is determined for several samples exposed to a given CO_2 pressure for varied t_s shorter than the equilibration time, as schematically suggested in Figure 3. The mean diffusivity in sorption, D_s, is obtained from the initial slope of $M_{t,s} / M_\infty$ vs $\sqrt{t_s} / l$.

This analysis of the desorption data implicitly assumes that the polymer sample maintains its plane sheet geometry throughout the experiment. For most of the glassy polymers studied, this assumption seems valid, as no visible change in shape or appearance was observed upon release of the CO_2 pressure or during desorption. A number of rubbery or highly plasticized polymers, however, expanded and foamed quite noticeably when the pressure was released. Nonetheless, the amount of CO_2 absorbed could be estimated by quickly weighing the sample upon removal from the pressure vessel. The determination of equilibrium sorption, and particularly of diffusivity, from the intercept and slope of the weight vs. square-root time plots, however, is clearly imprecise for samples which show such geometric changes.

Results

Transport kinetics. An example of our kinetic results is shown in Figure 4, as M_t vs $\sqrt{t} / l$ plots for poly(methyl methacrylate) (PMMA) samples of two different thicknesses exposed to liquid CO_2 at 25°C. The superposition of data for different

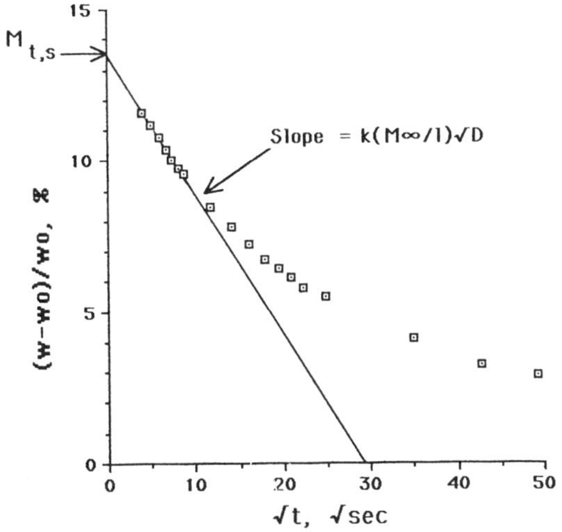

Figure 2. Representative CO_2 desorption data: Weight loss of 0.3 mm cellulose acetate film after 17 hours exposure to CO_2 at 500 psi.

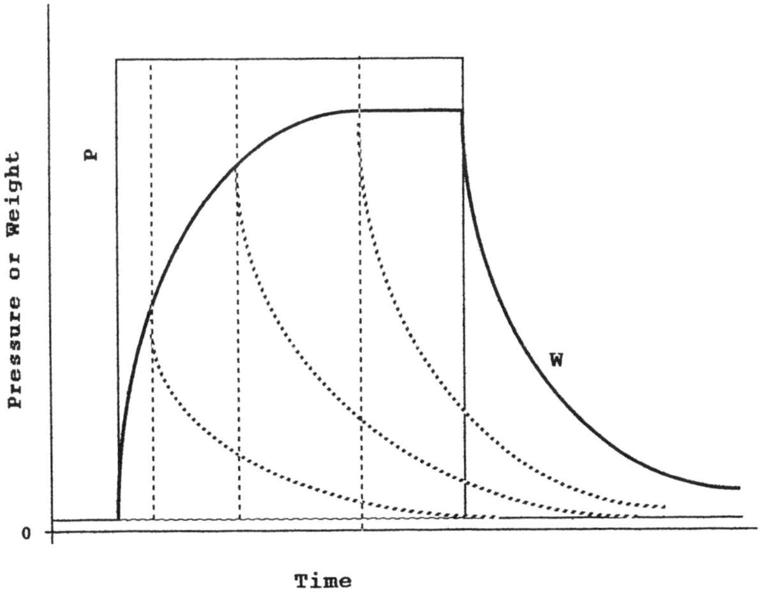

Figure 3. Determination of adsorption kinetics from desorption runs after varied adsorption periods. (Schematic).

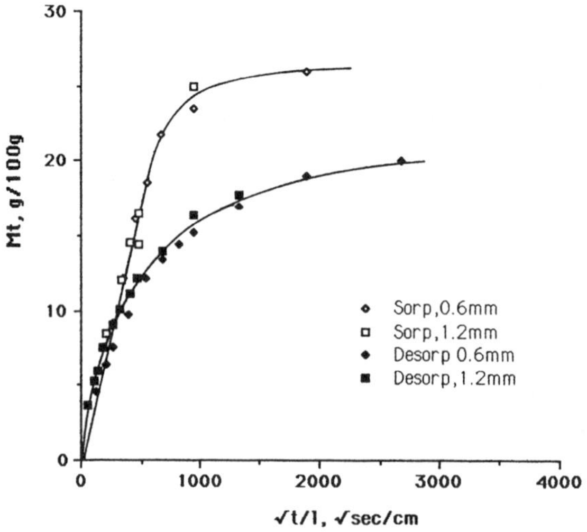

Figure 4. Adsorption and desorption kinetics for CO_2 in PMMA films of two thicknesses at 25°; sorption at 950 psi., desorption to atmospheric pressure.

sample thickness indicates that the transport kinetics are Fickian. The relative behavior of the sorption and desorption curves depends on the form of the concentration dependence of diffusivity. The divergence of the sorption and desorption curves, with the desorption curve falling well below the absorption curve at longer times, indicates that the diffusion coefficient is an increasing function of CO_2 pressure or concentration (1). Crank (10) shows examples of curves resembling Fig. 4 when there is a maximum in the diffusivity-concentration relationship, as may be suggested by some of our results.

More extensive and quantitative data on the variation of diffusivity with concentration has been obtained from desorption runs following equilibration of polymer samples at different CO_2 pressures. In Figure 5, D_d values obtained in this way for four glassy polymers (PMMA, poly(vinyl chloride) (PVC), cellulose acetate (CA), and polystyrene (PS)) are plotted against the CO_2 concentrations extrapolated to zero desorption time. The order of CO_2 diffusivities, PS >> PVC > PMMA , is the same as previously found for organic vapors (11). The points at lowest CO_2 concentration for each polymer were obtained by conventional gravimetric sorption/desorption experiments using a Cahn vacuum microbalance, and agree well with results of permeation experiments for PS (1), PVC (12), and PMMA (13). The extrapolations of our high-pressure results are quite consistent with these low pressure data, attesting to the validity of the simple new technique described here. Unfortunately, few diffusivity data obtained by other methods are available for direct comparison with our high-pressure kinetic results.

With increasing concentration, the diffusivity of CO_2 shows parallel increases for the four polymers, and approaches the range of 10^{-6} to 10^{-7} cm^2/sec. which is typical for CO_2 diffusivity in rubbers (1). This evidence from transport kinetic data is consistent with other indications that CO_2 has a substantial plasticizing effect upon glassy polymers (6-8). The downward curvature of the **log** D vs C plots is similar to that reported for several organic vapor/polymer systems at temperatures not far above T_g (1). The apparent leveling off or maxima of the curves may also be a result of the opposing effects of increasing CO_2 concentration and increasing hydrostatic pressure (14).

Sorption equilibria. Equilibrium solubilities of CO_2 in a variety of different polymers at 25°C have been determined by our gravimetric desorption method . At higher pressures, the solubilities are quite substantial and vary markedly with the polymer type. Equilibrium uptakes of liquid CO_2 range from about 3 g/100 g in polyethylene to over 50 g/100g in poly(vinyl acetate); other high values have been found for PMMA (27 g/100), ethyl cellulose (30 g/100), and cellulose acetate (27 g/100).

To help elucidate the solvent action of CO_2, the effects of some systematic variations of polymer structure have been investigated. Informative trends have been observed for several series of copolymers, as illustrated in Figure 6. The equilibrium sorption of liquid CO_2 in butadiene/acrylonitrile (BAN) copolymer rubbers increases

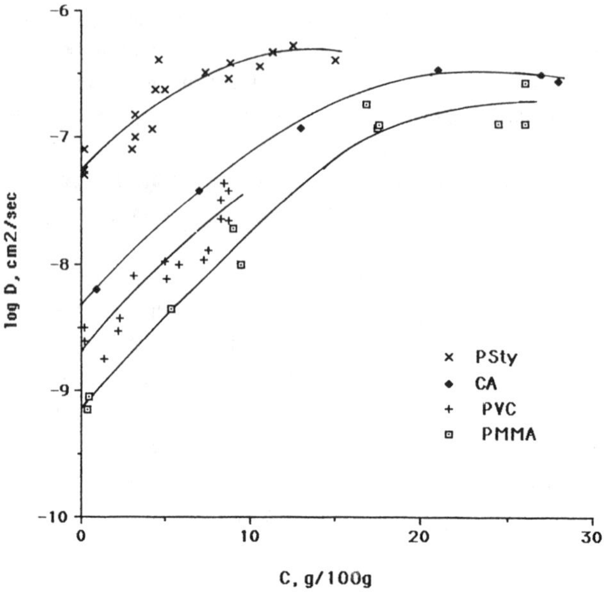

Figure 5. Diffusivity of CO_2 vs. CO_2 concentration, for four polymers, at 25°, from desorption data.

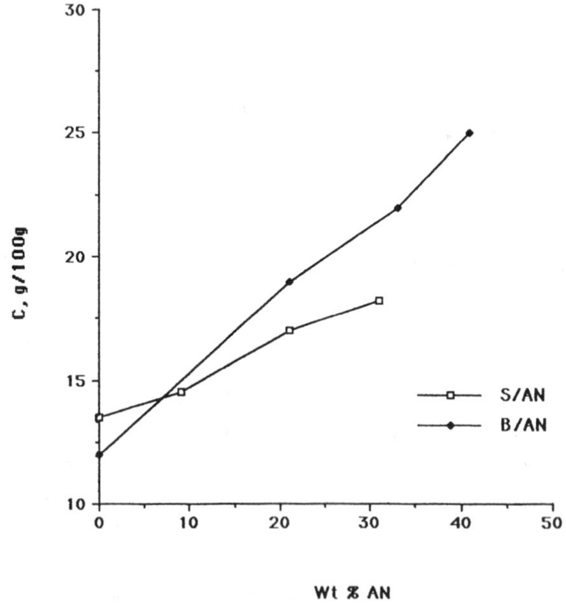

Figure 6. Solubility of liquid CO_2 at 25° vs. acrylonitrile content of styrene (S/AN) and butadiene (B/AN) copolymers.

regularly with increasing AN content; the same trend was observed for CO_2 at atmospheric pressure in the early studies of van Amerongen (15). Figure 6 also shows a similar trend for a series of glassy styrene/acrylonitrile (SAN) copolymers, hence this effect seems independent of the polymer state. Ethylene/vinyl acetate (EVA) copolymers also show a steady increase in CO_2 sorption with increasing content of the more polar monomer, vinyl acetate. These trends suggest that relatively high sorption values may result from specific polar interactions of CO_2 with carbonyl or nitrile groups in the polymers.

The solvent interaction of liquid or dense gaseous CO_2 with various organic solutes is a topic of continuing research interest (16-18). The low value of its solubility parameter, 6.0 $(cal/cc)^{1/2}$ (19), suggests that liquid CO_2 should behave like a non-polar hydrocarbon. Our results for polymers, however, show behavior resembling a somewhat polar organic solvent; perhaps the quadrupole moment, or the H-bonding basicity (17) of CO_2 may account for its unexpectedly high solubility in polar polymers. Additional data and interpretation of our results will be topics of subsequent publications.

Sorption Isotherms. The dependence of equilibrium sorption upon CO_2 pressure has been studied for several glassy polymers; representative data are shown in Figure 7, for PVC, polycarbonate (PC), PMMA, and poly(vinyl acetate) (PVA). These 25°C sorption isotherms show features well-known for both gases and organic solvents in glassy polymers: At low pressures or low concentrations, the curvature is concave downward, as is typical for gases in glassy polymers and generally described as "dual-mode" behavior. At high concentrations, the curvature is upward, in the nature of the Flory-Huggins isotherms typical for vapors of swelling solvents in rubbery polymers. In the case of PMMA, the isotherm seems to show an inflection, changing from dual-mode to Flory-Huggins form with increasing pressure.

Sigmoidal sorption isotherms combining dual-mode behavior at low concentrations with Flory-Huggins form at higher concentration were first reported for the system vinyl chloride monomer (VCM)/PVC (20); examples are reproduced here in Figure 8. It was suggested that this behavior is related to the depression of the glass transition temperature, T_g, by the dissolved VCM. It was later demonstrated (21) that the isotherm inflection, which occurs at lower concentrations as temperature is increased, coincides with independent measurements of T_g in the VCM/PVC system. Dual-mode sorption thus was shown to be characteristic of the glassy state, and the Flory-Huggins form, of the rubbery state. Observations of similar sigmoidal isotherms for several other polymer-vapor systems have prompted the recent suggestion that this may be the general form describing sorption of penetrants in glassy polymers whenever a sufficiently broad range of concentrations is covered (22).

In view of the other similarities between CO_2 and organic vapors in their effects on glassy polymers, further investigation of the applicability of the generalized sigmoidal isotherm to CO_2/polymer systems seems appropriate. The conventional and

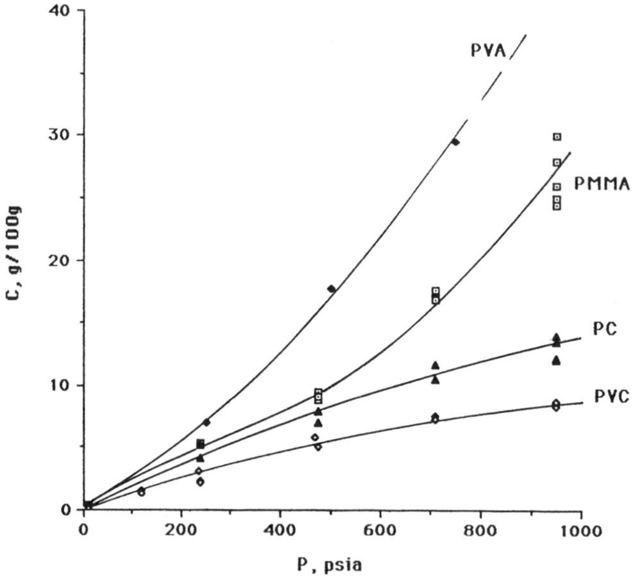

Figure 7. Solubility of CO_2 at 25° in four glassy polymers vs. CO_2 pressure.

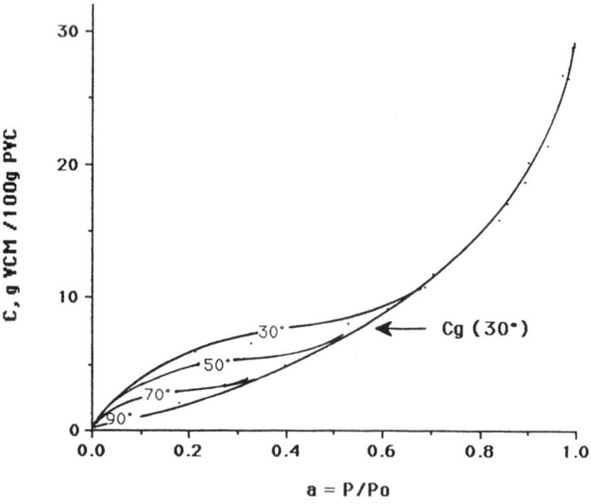

Figure 8. Sorption isotherms for vinyl chloride in PVC (20).

thermodynamically sound practice for polymer/vapor systems is to plot equilibrium solubility against thermodynamic activity, rather than against pressure as gas/polymer isotherms are generally published. While activity of organic vapors is adequately approximated by the ratio of actual partial pressure to saturated vapor pressure, p/p_o the high-pressure non-ideality of CO_2 requires use of fugacities to accurately express activity.

We have used the following approach in calculating CO_2 activities: At temperatures below T_c, the reference state is taken to be liquid CO_2 at p_o. The reference state at higher temperatures is selected by extrapolation of p_o into the supercritical region; the linear relation between p_o and $1/T$ is valid for this extrapolation up to about $T = 2T_c$ (19). For each temperature, the fugacity at either the actual or extrapolated p_o is taken as the reference state fugacity, f_o, and evaluated by interpolation from published f/P data (23). Using these f_o values and the published f/P data, a table of activities, f/f_o, vs. P is calculated for T at 10K intervals. Activities at each experimental P and T are then obtained by double parabolic interpolation from this activity table.

In exploring the sorption isotherms, we have utilized both our own data for gas and liquid CO_2 solubility at 25°C and published data at other temperatures for a number of glassy polymers. Figures 9-11 show a few examples, plotted both against pressure and against activity.

Isotherms for CO_2 in poly(vinyl benzoate) (PVBz), replotted from the data of Kamiya et al. (24), are shown in Figure 9. Conversion from pressure to activity coordinates has the effects of compressing the higher temperature isotherms along the activity axis, increasing the upward curvature, and superimposing the isotherms at higher activities. The resulting set of isotherms is remarkably similar to the VCM/PVC curves of Figure 8. In both cases, the isotherms show dual-mode curvature at lower activity, then appear to converge to a common curve of Flory-Huggins form at higher activity, as indicated by the dashed line in Figure 9b. With increasing temperature, the apparent Langmuir or hole-filling portion of the dual-mode isotherm diminishes, and the apparent inflection shifts to lower concentration or activity. As in the VCM/PVC case, the inflections presumably represent the glass transitions of the CO_2/(PVBz) system; Kamiya (24) has applied the term glass composition, or C_g, to the penetrant concentration which lowers T_g to a specified temperature.

In Figure 10, data for CO_2 solubility in polycarbonate (PC) from the present work and two published studies (2, 25) are plotted against both pressure and activity. Here, the sigmoidal form, not apparent in the pressure plot,is revealed by conversion to the activity coordinate. Data in the low-activity, dual-mode range show both a decreasing solubility with increasing temperature and a sorption/desorption hysteresis (25). The higher pressure data are brought nearly into coincidence when replotted against activity.

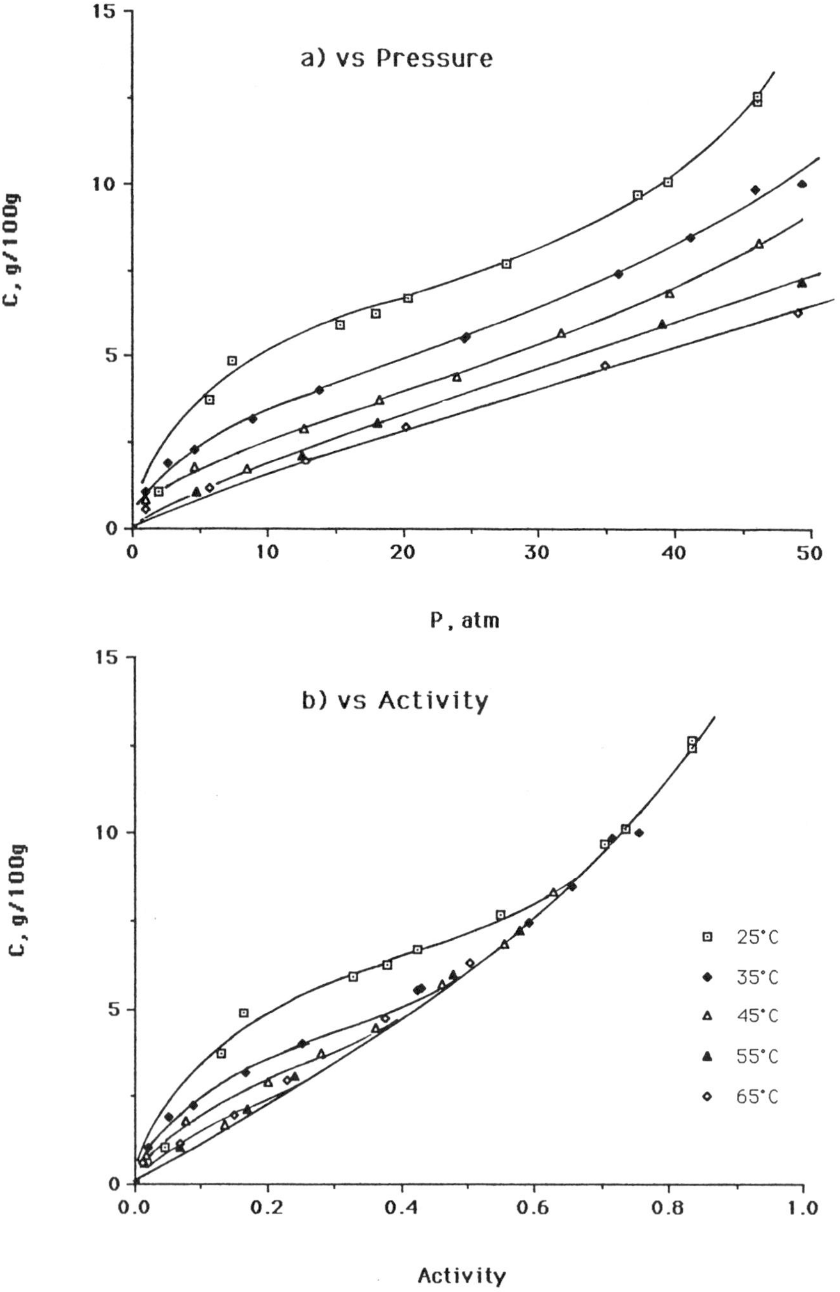

Figure 9. Sorption isotherms for CO_2 in poly(vinyl benzoate); data from Kamiya, et al. (24).

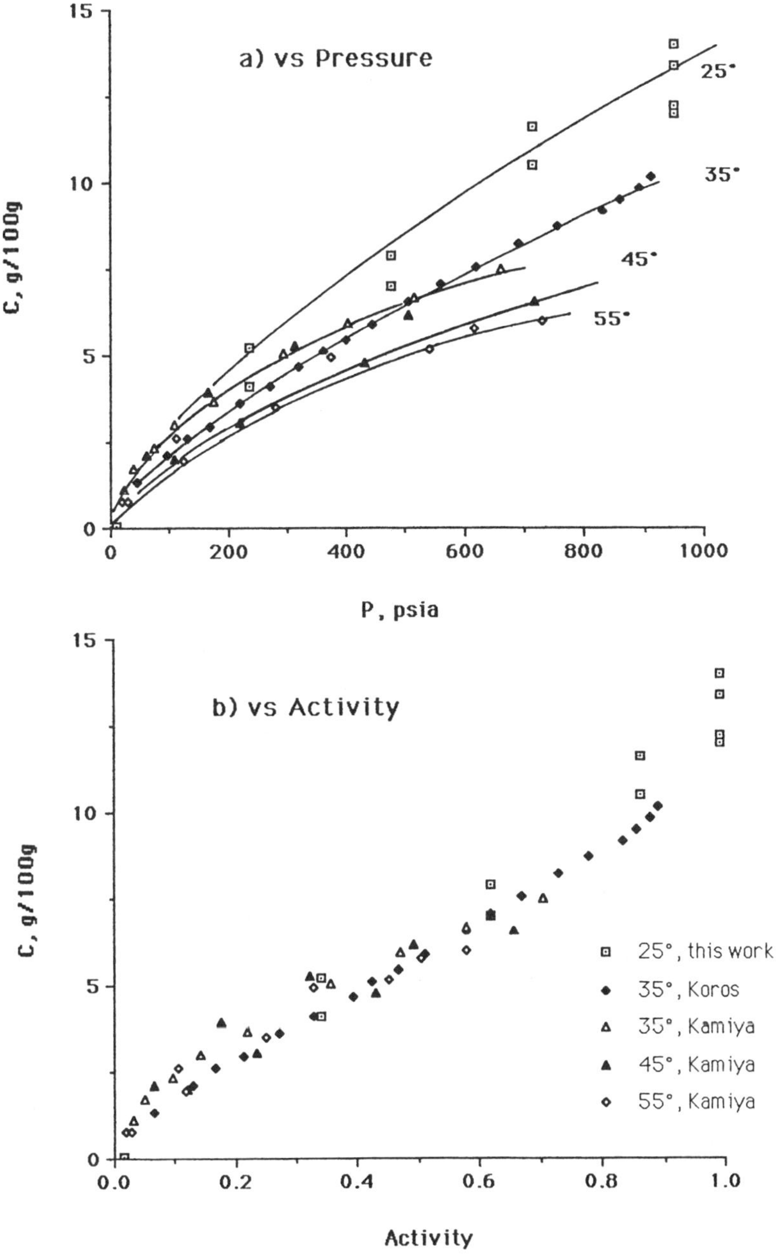

Figure 10. Sorption isotherms for CO_2 in polycarbonate; data from Fleming and Koros (2), Kamiya et al. (25), and this study.

Equilibrium solubility data for the CO_2/PMMA system are assembled in Figure 11; as in the above cases, results of our gravimetric desorption experiments are again in line with published data obtained by other methods (5, 26). The sorption of CO_2 in PMMA, at equal activity, is substantially greater than in PVBz or PC. Transformation of the abscissa from pressure to activity results in close superposition of the PMMA data over a 55° temperature range. Over most of the activity range, the isotherms show upward curvature, yet inflections are apparent in the data for the lower temperatures. The identification of the inflections with the glass transition seems quite consistent with experimental and calculated values of T_g in the PMMA/CO_2 system (7). At corresponding temperatures, C_g may be somewhat higher for PMMA than for PVBz because of the difference in T_g of the pure polymers (105°C for PMMA, 65.5° for PVBz (24)), but occurs at lower activity in PMMA because of the higher solubility of CO_2. Dual-mode behavior in PMMA appears relatively less pronounced than in PVBz or PC, probably reflecting a greater contribution of the Henry's Law mode, relative to the Langmuirian portion, in PMMA compared to the other two polymers.

These three examples, and others to be included in subsequent publications, seem clearly to demonstrate that the sorption isotherms for CO_2 follow the general sigmoid form observed for organic vapors in glassy polymers. This behavior seems to be a reasonable consequence of the high solubility and plasticizing action of CO_2 at high pressures. At sufficiently high pressures, this gas indeed produces many of the same effects as an organic solvent. Use of the activity scale, rather than pressure, accentuates the similarity between CO_2 and conventional solvents and demonstrates the continuity of behavior below and above the critical temperature. This continuity allows estimation of polymer behavior in supercritical CO_2 at high pressures from data obtained at subcritical temperatures and much lower pressures.

The close superposition of CO_2/polymer isotherms at various temperatures, when plotted vs. activity, suggests that most of the apparent temperature dependence of the isotherms plotted vs. pressure is related to the activity change of CO_2 with temperature. At constant activity, the actual mixing of CO_2 with PMMA, PC, or PVBz appears to be nearly athermal; i.e., the energy of interaction of CO_2 with these polymers seems to be essentially that associated with the compression of the gas to its molar volume in the sorbed state. This aspect of polymer interactions with CO_2 will also be considered further in forthcoming publications.

The effects of high-pressure CO_2 upon polymers, as demonstrated in this study, seem relevant in several practical applications. Because of its plasticizing action and its rapid absorption and desorption, compressed CO_2 dramatically accelerates the transport of other small molecules in glassy polymers. This effect has recently been applied to the impregnation of glassy polymers with a wide variety of additives (9). The same action of CO_2 also may play a large role in its effectivness in supercritical extractions and fractionations of polymers, and in the resistance of barrier polymers to attack by swelling agents during service in high-pressure CO_2 environments.

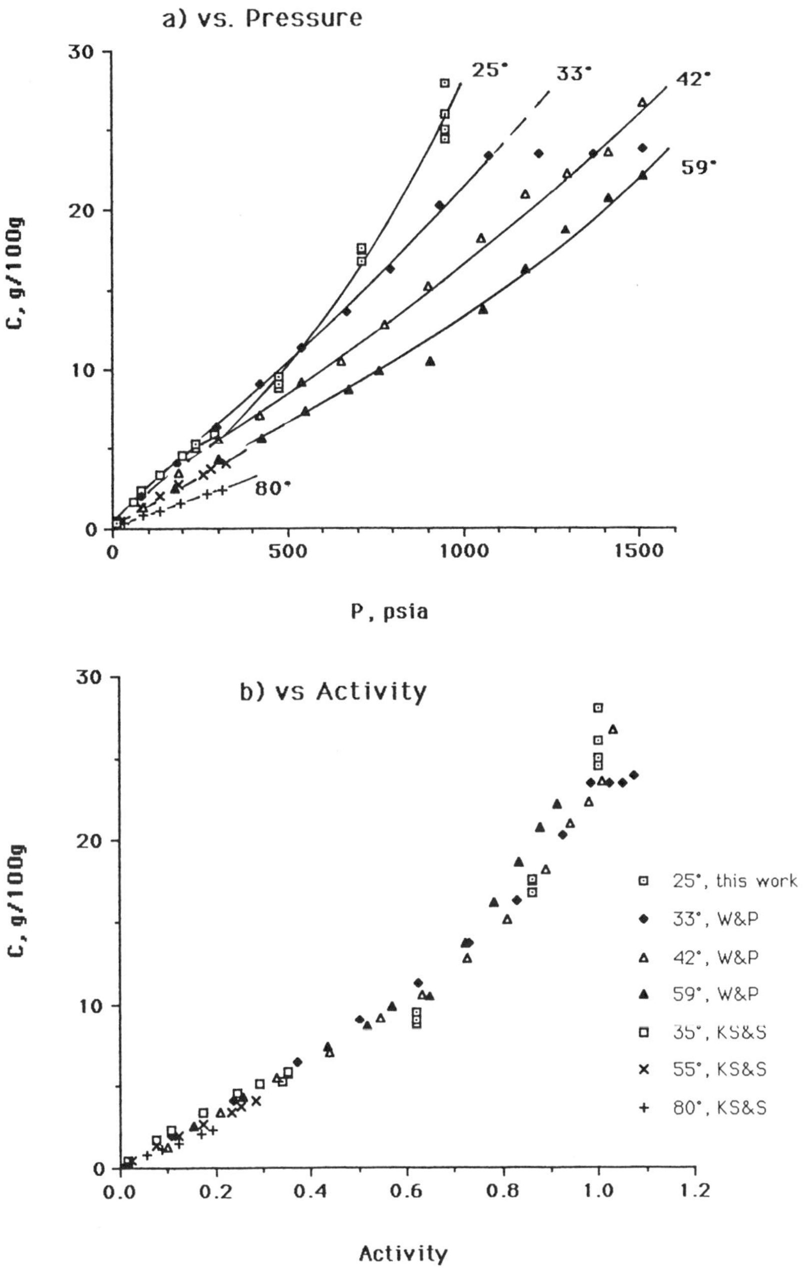

Figure 11. Sorption isotherms for CO_2 in PMMA; data from Wissinger and Paulaitis (5), Koros, Smith and Stannett (26), and this study.

Conclusions

1. Useful data on the kinetics and equilibria of CO_2 transport in polymers can be obtained by rapidly weighing film or sheet samples during desorption after exposure to liquid or gaseous CO_2 at high pressure.

2. The diffusion of CO_2 in polymers follows Fickian kinetics over the entire pressure range; diffusivity increases with concentration, in line with other evidence of the plasticizing action of CO_2.

3. Liquid CO_2 is a swelling agent for a wide variety of polymers; equilibrium solubilities at 25°C range from 3 to at least 50 g per 100 g polymer. The variation of swelling with polymer type and polarity indicates that CO_2 behaves as a somewhat polar, rather than hydrocarbon-like, solvent.

4. Sorption isotherms plotted against CO_2 activity in several glassy polymers show the sigmoid form observed for a number of glassy polymer/organic vapor systems. Isotherm curvature is downward (dual-mode form) at low concentrations, as is typical of the glassy state, and upward (Flory-Huggins form) at higher concentrations, characteristic of the rubbery state. The inflection corresponds to a composition having its glass transition at the isotherm temperature.

5. For several polymers, replotting of CO_2 solubilities against activity, rather than pressure, nearly removes the temperature dependence of the isotherms above their inflection. This result indicates a near-zero heat of mixing; variation of CO_2 activity with temperature apparently accounts for most of the temperature dependence of the isotherms plotted vs pressure..

Acknowledgements: The authors appreciate the experimental assistance of F. W. Kunig and the permission of The BFGoodrich Company to publish this work.

Literature Cited

1. Crank, J.; Park, G. S. Diffusion in Polymers, Academic Press, London, 1968.
2. Fleming, G. K.; Koros, W. J. Macromolecules 1986, 19, 2285.
3. Sefcik, M. D. J. Polym. Sci, Polym. Phys. 1986, 24, 935.
4. Hirose, T.; Mizoguchi, K.; Kamiya, Y. J. Polym. Sci, Polym. Phys. 1986, 24, 2107.
5. Wissinger, R. G.; Paulitis, M.E. J. Polym. Sci., Polym. Phys. 1987, 25, 2497.
6. Wang, W. V.; Kramer, E. J.; Sachse, W. H. J. Polym. Sci, Polym. Phys. 1982, 20, 1371.
7. Chiou, J. S.; Barlow, J. W.; Paul, D. R. J. Appl. Polym Sci. 1985, 30, 2633.

8. Sefcik, M. D. J. Polym. Sci, Polym. Phys. 1986, 24, 957.
9. Berens, A. R.; Huvard, G. S.; Korsmeyer, R. W. AIChE National Meeting, Washington, DC, November 28 - December 2, 1988 (to be published).
10. Crank, J. The Mathematics of Diffusion, Clarendon Press, Oxford, 1956; p. 283.
11. Berens, A. R.; Hopfenberg, H. B. J. Membrane Sci.,1982, 10, 283.
12. Tikhomirov, B. P.; Hopfenberg, H B.; Stannett, V.; Williams, J. L. Makromol. Chem. 1968, 118, 177.
13. Patel, V. M., et al. Makromol. Chem. 1972, 158, 65.
14. Rogers, C.E., in Polymer Permeability, Comyn, J., Ed.; Applied Science Publishers: London, 1984.
15. van Amerongen, G. J. J. Polym. Sci. 1950, 5, 307.
16. Francis, A. W. J. Phys. Chem. 1954, 58, 1099.
17. Hyatt, J. A. J. Org. Chem. 1984, 19, 5097.
18. Dandge, D. K.; Heller, J. P.; Wilson, K. V. Ind. Eng. Chem. Prod. Res. Dev. 1985, 24, 162.
19. Prausnitz , J. M.; Shair, F. H. A.I.Ch.E. Journal 1961, 7, 682.
20. Berens, A. R. Angew. Makromol. Chem. 1975, 47, 97.
21. Berens, A. R. Polym. Eng. Sci., 1980, 20, 95.
22. Connelly, R. W.; McCoy, N. R.; Koros, W. J.; Hopfenberg, H. B.; Stewart, M. E. J. Appl. Polym. Sci., 1987, 34, 703.
23. International Thermodynamic Tables of the Fluid State: Carbon Dioxide, Angus, S.; Armstrong, B.; de Reuck, K.M., Eds., Pergamon Press, Oxford, 1976.
24. Kamiya, Y.; Mizoguchi, K.; Naito, Y.; Hirose, T. J. Polym. Sci, Polym. Phys., 1986, 24, 535.
25. Kamiya, Y.; Hirose, T.; Mizoguchi, K.; Naito, Y. J. Polym. Sci, Polym. Phys., 1986, 24, 1525.
26. Koros, W. J.; Smith, G. N.; Stannett, V. T. J. Appl. Polym. Sci. 1981, 26, 159.

RECEIVED May 1, 1989

CHEMICAL REACTIONS

Chapter 15

Kinetic Elucidation of the Acid-Catalyzed Mechanism of 1-Propanol Dehydration in Supercritical Water

Ravi Narayan and Michael Jerry Antal, Jr.

Department of Mechanical Engineering and the Hawaii Natural Energy Institute, University of Hawaii at Manoa, Honolulu, HI 96822

Experimental data are presented which describe the acid catalyzed dehydration chemistry of 1-propanol and 2-propanol in supercritical water at 375°C and 34.5 MPa. The data for 1-propanol dehydration are kinetically consistent with the acid catalyzed E2 mechanism, but not consistent with the related E1 mechanism. Neither the Ad_E3 mechanism nor the Ad_E2 mechanism is able to mimic the kinetic behavior of 2-propanol formation. The steady state idealization of the E2 mechanism does not represent the true kinetic behavior of the E2 mechanism over the range of experimental conditions presented in this paper.

The mechanism by which proton acids catalyze the dehydration of primary and secondary alcohols in water is not perfectly well understood (1). There is universal agreement that the dehydration of tertiary alcohols can be explained by an E1 mechanism (1,2) involving either a Π complex (3) or a symmetrically solvated carbonium ion (4) as the key reaction intermediate. Although an occasional text (5) also describes the dehydration of primary alcohols by an E1 mechanism, authoritative reviews (3,4) conclude that a concerted E2 type mechanism is more probable. The dehydration behavior of secondary alcohols is presumed to be similar to primary alcohols (4). Discussions of the gas phase dehydration of alcohols by heterogeneous Lewis acid catalysts admit more possibilities. In their authoritative review Kut, et al. (6) consider E1-, E2-, and E1cB-like mechanisms, as well as the possible role of diethyl ether as a reaction intermediate, but they reach no conclusion concerning the relative importance of these mechanisms in the formation of olefins from alcohols.

Early work (7) in this laboratory established the heterolytic nature of ethanol dehydration in supercritical water. Trace (0.001 to 0.01 M) concentrations of strong mineral acids (such as H_2SO_4 and HCl) were found to catalyze significant conversions of ethanol to ethene in water at 385°C, 34.5 MPa after a few

0097–6156/89/0406–0226$06.00/0

seconds or more. Later work (8) established the inability of simple, single step rate laws to describe experimental measurements of the dependence of ethanol conversion on time, reactant and catalyst concentrations in supercritical water. However, experimental studies (8,9) of the acid catalyzed dehydration of 1-propanol in supercritical water did evidence first order behavior when the reactant concentration was low. At higher concentrations (typically > 0.5 M) departures from first order behavior were observed. In addition, significant concentrations of 2-propanol were detected. These observations prompted the kinetic analysis described in this paper of all our relevant experimental data on the acid (H_2SO_4) catalyzed dehydration of both 1-propanol and 2-propanol in supercritical water at 375°C, 34.5 MPa. The objectives of this kinetic analysis were to determine (1) if the data for each alcohol could be described by either an acid catalyzed E1 or E2 elimination mechanism, (2) if the data contained sufficient information to enable us to distinguish between the two mechanisms and thereby ascertain which governed the dehydration process, and (3) if the data contained sufficient information to enable us to identify values for the rate constants associated with each elementary step of the governing mechanism.

These objectives might prompt the casual reader to conclude that this paper is really about model discrimination using chemical kinetics, and has little to do with supercritical fluids. In fact, chemical kinetics, model discrimination and parameter estimation are likely soon to become foci of interest for many workers concerned with reaction chemistry in supercritical water. Why? There are both practical and fundamental explanations for this prognostication. It has been established (7) that supercritical water with ion constant $K_w > 10^{-14}$ behaves chemically as very hot liquid water, favoring heterolytic reactions involving charged species as intermediates. For example, at 375°C an acid catalyzed reaction whose rate doubles every 15°C will proceed 2^{18} ($= 2.6 \times 10^5$) times faster than at 100°C. This finding has important practical consequences for chemists concerned with reactions involving carbohydrates, since these reactions typically require very high concentrations of acids and long reaction times. A 10^5 increase in reaction rates permits acid catalyst concentrations to be reduced to 10^{-2} M (or less) and reaction times to be reduced to 10^{-2} h (or less). A chemical engineer cannot exploit the practical implications of these dramatic changes in reaction conditions without a detailed understanding of the governing chemical kinetics. An unexpected bonus of an engineering inquiry into reaction kinetics in supercritical water is the new light it sheds on mechanism. Earlier kinetic examinations of mechanism by chemists were largely hamstrung by very high acid and reactant concentrations, requiring the use of Hammett acidities and reactant activities to estimate reaction rate constants. In supercritical water the reactant and acid concentrations are so low that complete dissociation of H_2SO_4 to H^+ and HSO_4^- (which does not dissociate) may be assumed (Narayan, R.; Antal, Jr., M.J., submitted to J. Am. Chem. Soc.), and activity coefficients may be set equal to unity with no loss of accuracy. Thus the unique attributes of supercritical water as a solvent enable chemists to begin rigorous kinetic examinations of reaction mechanism. In this paper we illustrate the above through the development of a kinetic model for the acid catalyzed dehydration of 1-propanol which has both practical and mechanistic implications.

Apparatus and Experimental Procedures

Figure 1 is a schematic of one of the two supercritical flow reactors used in this work. The system is first brought up to the operating pressure by an air compressor. An HPLC pump forces the reactant solution through the reactor, the ten-port valve and dual-loop sampling system, and into the product accumulator, where the flow of products displaces air through a back-pressure regulator. The reactant inflow is rapidly heated to reaction temperature by an electric entry heater/water jacket combination, and maintained at isothermal conditions by a Transtemp Infrared furnace and an exit electric heater/water jacket combination. Product samples captured in 5.0 ml sample loops are collected in sealed, evacuated test tubes for qualitative and quantitative analysis. The weight of the reactant solution is continuously monitored on a Mettler E2000 balance and the flow rate is measured using a stopwatch. A more complete description of the reactor and its operation is given by Antal, et al. (7) and Ramayya, et al. (8).

The outer shell of the first reactor is a 4.7 mm ID Hastelloy C-276 tube, and the inner annulus is a 3.2 mm OD sintered alumina tube, giving the reactor a hydraulic diameter of 3.0 mm. The alumina tube accommodates a movable type K thermocouple along the reactor's axis, which provides for the measurement of axial temperature gradients along the reactor's functional length of approximately 0.46 m. Radial temperature gradients are measured as differences between the centerline temperature and temperatures measured at ten fixed positions along the outer wall of the reactor using type K thermocouples. Pressure in the reactor system is measured using an Omega PX176 pressure transducer with an accuracy of 0.2 MPa and calibrated by a Wika test gauge (NBS traceable) with an accuracy of 0.2 MPa. This reactor permits residence time studies from approximately 15 s to 100 s. The characterization of this reactor using non-dimensional numbers is fully described elsewhere (9). An analysis of characteristic times associated with these non-dimensional numbers reveals that the reactor performs as an ideal plug flow reactor (9-11).

The second reactor resembles the first, but it is fabricated from a 1.6 mm OD Hastelloy capillary tubing and lacks an inner annulus. It has a length of approximately 0.28 m and enables studies involving residence times below 10 s.

All reactant solutions were prepared using degassed, distilled water. Fisher certified grade 1-propanol and Fisher HPLC grade 2-propanol were used as the reactants. No impurities were detected in these reagents by HPLC or GC analyses. The sulfuric acid used was a Fisher certified grade 10N solution.

At each operating condition, triplicate samples of the reactor effluent were collected for analysis. Quantification of liquid products was accomplished by triplicate analysis of each of these samples using a Waters High Performance Liquid Chromatograph (Model

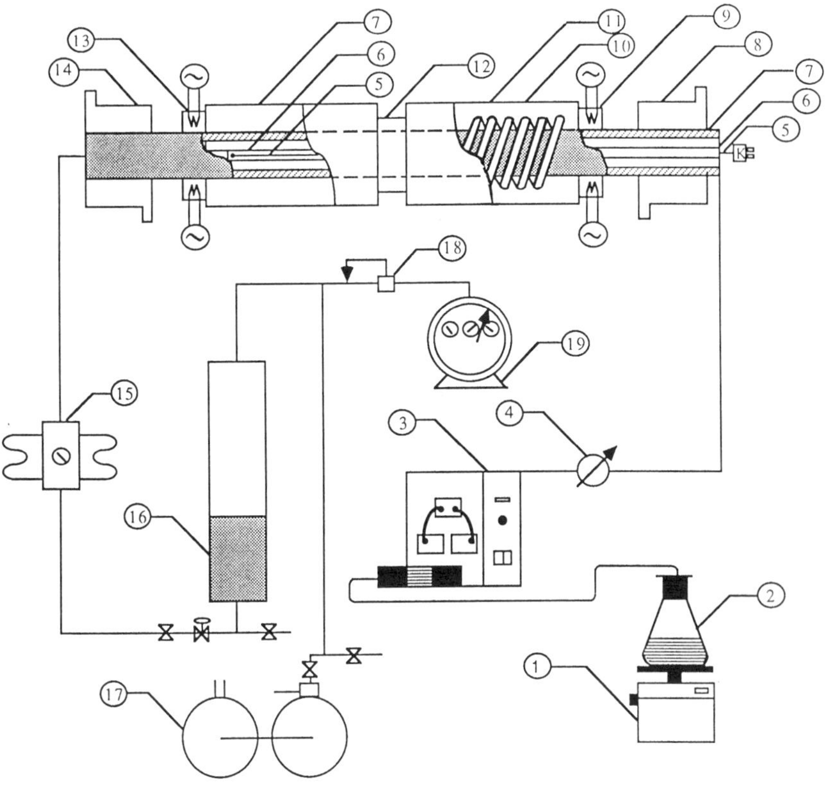

Figure 1. Supercritical flow reactor schematic:
1) Mettler balance, 2) Flask with filtered and deaerated distilled water, 3) HPLC pump, 4) Bypass (3 way) valve, 5) Probe thermocouple (type K), 6) Ceramic annulus, 7) Hastelloy C-276 tube, 8) Entrance cooling jacket, 9) Entrance heater, 10) Furnace coil, 11) Quartz gold plated IR mirror, 12) Window (no coils), 13) Guard heater, 14) Outlet cooling jacket, 15) Ten port dual loop sampling valve, 16) Product accumulator, 17) Air compressor, 18) Back pressure regulator, 19) Outflow measuring assembly (Wet test meter)

6000A solvent delivery system, Perkin Elmer LC 600 autosampler and a differential refractometer) and a Hewlett-Packard Model 3388A integrator. An Alltech C18 column was employed with degassed, distilled water as the solvent at a flow rate of 2 ml/min. Gaseous products were analyzed using a Hewlett-Packard Model 5840 Gas Chromatograph equipped with a flame ionization detector. A Poropak Q column operating at 200°C with 8.5% hydrogen in helium as the carrier gas was used to separate the gaseous products. 1-Propanol, 2-propanol, propene (99% pure Matheson C.P. grade) and air standards were used for calibration.

Kinetic Models and Parameter Estimation

Figures 2a and 2b display the acid catalyzed E2 and E1 mechanisms for the dehydration of 1-propanol and 2-propanol. Note that the E1 mechanism involves four more rate constants (kinetic parameters) than the related E2 dehydration mechanism. Chemists employ the terminology (1) Ad_E3 to describe the hydration mechanism which forms 2-propanol from propene in Figure 2a, and Ad_E2 to refer to the mechanism which forms 2-propanol from propene in Figure 2b. In this paper we do not distinguish between bare carbocations, Π-complexes, encumbered carbocations and symmetrically solvated carbocations, since these intermediates all manifest themselves similarly in the E1 kinetic model.

The coupled set of non-linear, ordinary differential equations governing the dehydration of 1-propanol to propene by the E2 mechanism (omitting for the moment the formation of 2-propanol) is given by

$$dy_1/dt = -k_1[H_3O^+]y_1 + k_2[H_2O]y_3 \quad (1)$$

$$dy_2/dt = k_3[H_2O]y_3 - k_4[H_3O^+][H_2O]y_2 \quad (2)$$

$$dy_3/dt = k_1[H_3O^+]y_1 - (k_2+k_3)[H_2O]y_3 + k_4[H_3O^+][H_2O]y_2 \quad (3)$$

where y_1 = [1-PrOH], y_2 = $[C_3H_6]$, y_3 = $[1\text{-PrOH}_2^+]$, and the k_i are rate constants. Values of $[H_3O^+]$ are calculated using the relationship $[H_3O^+] = [H_2SO_4]_o - [1\text{-PrOH}_2^+]$, where $[H_2SO_4]_o$ is the initial concentration of acid evaluated at reaction temperature and pressure (RTP). Note that earlier work (Narayan, R.; Antal, Jr., M.J., submitted to J. Am. Chem. Soc.) has established the effectively complete dissociation of H_2SO_4 to H_3O^+ and HSO_4^-, and the non-dissociation of HSO_4^- under the experimental conditions employed here. Values of $[H_2O]$ are obtained using tabular equation of state data. (12) Similar sets of coupled, non-linear, ordinary differential equations result from the E1 mechanism displayed in Figure 2b, and the related mechanisms for 2-propanol formation.

Solutions to the rate Equations 1-3 depend upon the kinetic parameters k_1, k_2, k_3, and k_4. If kinetic parameters can be identified which cause the value of a y_i (at a particular residence time, initial acid and reactant concentration) to be

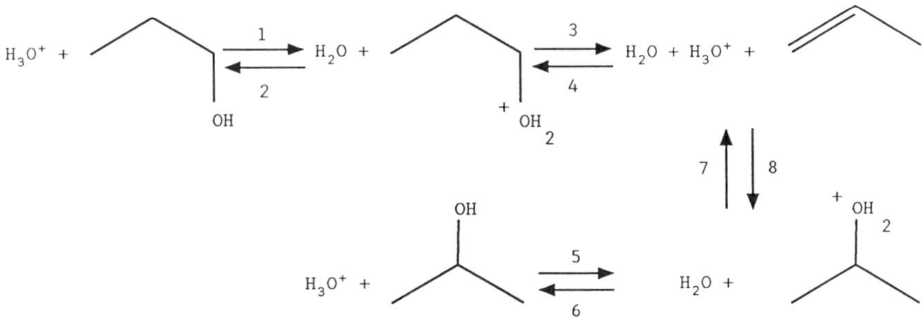

Figure 2a. Acid catalyzed E2 mechanism for 1-propanol dehydration and Ad_E3 mechanism for 2-propanol formation from propene.

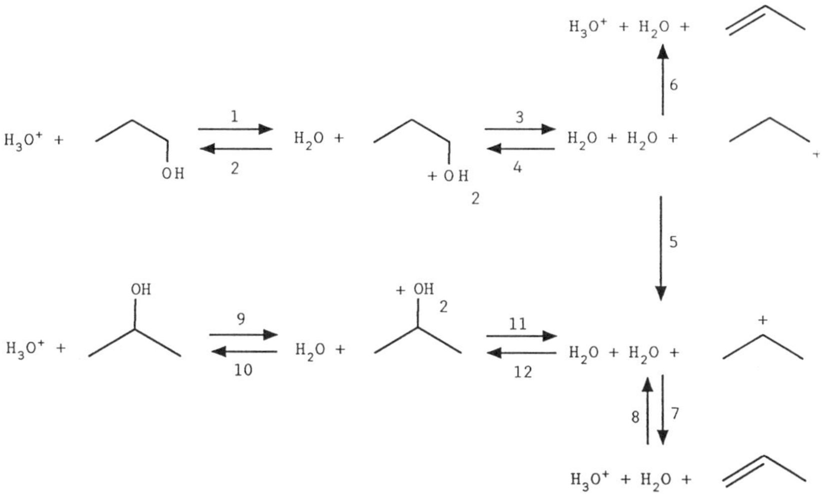

Figure 2b. Acid catalyzed E1 mechanism for 1-propanol dehydration and Ad_E2 mechanism for 2-propanol formation from propene.

within experimental error of an experimentally measured value, then the mechanism is consistent with that particular experimental data point.

More generally, for a given reaction network representing a particular mechanism of interest, mass action kinetics specify the governing ordinary differential equations

$$\dot{\underline{y}} = d\underline{y}/dt = \underline{f}\ (\underline{y};\underline{p}) \qquad (4)$$

where the vector $\underline{y}$ specifies the concentrations of the n species involved in the reaction network, $\underline{y}_o$ is given as an initial condition, and the vector $\underline{p}$ is composed of q rate constants which remain to be determined. If the m experimental measurements are designated as $(t_1, \underline{Z}_1)$, ..., $(t_m, \underline{Z}_m)$ where $\underline{Z}_j$ is a vector whose n components are experimental measurements at time t_j of the n species involved in the reaction network, then the inverse chemical kinetic problem (13-15) is to determine optimal values of the kinetic parameters $\underline{p}$ which minimize the l_2 - norm $S(\underline{p})$ of the residual vectors $\underline{s}(t_j) = \underline{y}(t_j;\underline{p}) - \underline{Z}_j$, where

$$S(\underline{p}) = \sum_{j=1}^{m} \underline{s}(t_j)^T D_j\ \underline{s}(t_j) \qquad (5)$$

and D_j is a diagonal (nxn) weight matrix. In this work the diagonal elements $D_{j,kk}$ of D_j satisfy $D_{j,kk} = 1/\sigma^2_{kk}$ where $\sigma_{kk} = 0.10 * Z_{j,k}$. This choice of σ_{kk} reflects our experience that the month to month reproducibility of the experimental values $Z_{j,k}$ as measured by the sample standard deviation is about 10% of the actual value (8,10). With this choice of D_j we have $\chi_\nu^2 = S(\underline{p}) / \nu$ where χ_ν^2 is the familiar chi squared statistic (16) and ν is the number of degrees of freedom.

We employ the IMSL (17) subroutine BCLSF to search for values $\tilde{\underline{p}}$ which minimize $S(\underline{p})$. BCLSF solves the nonlinear least squares problem given by Equations 4 and 5 subject to bounds on the parameters ($p_k \geq 0$ for $1 \leq k \leq q$) using a modified Levenberg-Marquardt algorithm and a finite difference Jacobian. BCLSF requires a value of $S(\underline{p})$ for each trial parameter set $\underline{p}$. This value is calculated by integrating the coupled set of stiff, non-linear, first order, ordinary differential equations which specify $\underline{y}$ using the IMSL subroutine IVPAG. IVPAG solves initial value problems, such as Equation 4, using an Adams-Moulton or Gear method. Employing this approach to the mathematics, we determine the optimal values of the q kinetic parameters composing $\tilde{\underline{p}}$, and then scrutinize the n components s_i of each of the m residual vectors $\underline{s}\ (t_j)$ to determine what fraction of their values exceed the 95% confidence interval (C.I.) associated with the experimental measurement. If a nominal 1 in 20 values of s_i exceed the C.I., then the mechanism is said to be consistent with the experimental data. If significantly more than 1 in 20 values exceed the C.I., then the mechanism is said to be inconsistent with the experimental measurements.

The Bodenstein steady state (SS) idealization is often used to simplify the coupled set of ordinary differential equations given by Equation 4. If species y_{k+1}, ... y_{k+r} are intermediates whose concentrations are thought to be very low during the course of the reaction, the SS idealization assumes

$$dy_i/dt = 0 \quad k+1 \leq i \leq k+r \tag{6}$$

to derive r algebraic expressions for the unknown concentrations $y_{k+1}, \ldots, y_{k+r}$, which are then substituted into the remaining equations in order to obtain a simplified set of n-r coupled, non-linear, ordinary differential equations which still must be solved using the computer. For example, application of the SS idealization to Equations 1-3, which describe the rate of disappearance of 1-propanol, results in the expression

$$\frac{dy_1}{dt} = \frac{-k_1k_3[H_2O]y_1 + k_2k_4[H_2O]^2y_2}{k_1y_1 + (k_2 + k_3)\,[H_2O] + k_4\,[H_2O]y_2}\,[H_2SO_4]_0 \tag{7}$$

With the present availability of very powerful software which runs on personal computers (such as the IBM PS2 (MOD 50) machine used in this work), the SS idealization no longer contributes significantly to visualizing the behavior of solutions to Equation 4. Nevertheless, we employ the SS idealization here in order to scrutinize its ability to mimic the actual solutions to Equation 4, and thereby gain insight into its real utility.

Equation 7 may be simplified further by recognizing that the concentration of the protonated intermediate must be small in order to satisfy the steady state approximation (i.e. $(k_2 + k_3)[H_2O] \gg k_1y_1 + k_4[H_2O]y_2$), and recalling from our earlier work (8) that equilibrium strongly favors the dehydration reaction (i.e. $k_2k_4[H_2O]^2y_2 \simeq 0$). With these additional assumptions equation 7 becomes the first order expression

$$\frac{dy_1}{dt} = \frac{-k_1k_3y_1}{k_2 + k_3}\,[H_2SO_4]_0 \tag{8}$$

$$= -k_{H,ss}y_1[H_2SO_4]_0 \tag{9}$$

Results and Discussion

Eighteen experimental measurements of product yields from the acid catalyzed dehydration of 1-propanol and 2-propanol in water (at 375°C and 34.5 MPa) are displayed in Table I. These data span a wide range of reactant (0.05 to 2.0 M) and catalyst (0.001 to 0.025 M) concentrations, and residence times (1.15 to 68.4 s). The error bars displayed in Table I reflect 95% confidence intervals based on yield variations detected during a single run. A typical sample standard deviation is almost 2% of the mean yield value for these data. Using these measurements, optimal values of the kinetic parameters associated with each trial mechanism were calculated.

TABLE I : Experimental and calculated fractional yields in water at 375°C, 34.5 MPa, using the E2AdE3 and E1AdE2 models

REACTANT /M	SULFURIC ACID/mM	RESIDENCE TIME/s	1-PROPANOL YIELD Expt*	1-PROPANOL YIELD Calc E2	1-PROPANOL YIELD Calc E1	2-PROPANOL YIELD Expt*	2-PROPANOL YIELD Calc E2	2-PROPANOL YIELD Calc E1
0.05 {1}	1.1	16.5	0.69±0.07	0.70	0.74	0.10±0.02	0.07	0.07
0.05 {1}	2.0	24.3	0.43±0.08	0.40	0.44	0.28±0.04	0.14	0.15
0.05 {1}	2.9	16.5	0.43±0.04	0.40	0.45	0.21±0.02	0.14	0.14
0.05 {1}	5.0	4.8	0.66±0.04	0.62	0.67	0.08±0.01	0.09	0.09
0.05 {1}	5.0	16.5	0.25±0.03	0.23	0.25	0.26±0.02	0.18	0.20
0.05 {1}	25.0	17.2	0.06±0.05	0.05	0.00	0.42±0.34	0.25	0.27
0.50 {1}	1.0	18.9	0.75±0.08	0.73	0.81	0.04±0.02	0.06	0.05
0.50 {1}	2.0	18.0	0.54±0.07	0.58	0.67	0.07±0.02	0.09	0.08
0.50 {1}	3.0	18.9	0.43±0.04	0.46	0.52	0.09±0.01	0.12	0.12
0.50 {1}	5.0	17.1	0.28±0.10	0.34	0.37	0.16±0.01	0.15	0.16
0.50 {1}	5.0	34.2	0.15±0.10	0.17	0.12	0.20±0.01	0.18	0.22
0.50 {1}	5.0	68.4	0.07±0.04	0.08	0.01	0.19±0.04	0.20	0.25
0.10 {1}	5.0	17.1	0.25±0.04	0.24	0.25	0.25±0.01	0.18	0.20
0.20 {1}	5.0	17.1	0.25±0.04	0.27	0.28	0.28±0.01	0.17	0.19
2.00 {1}	5.0	17.1	0.51±0.02	0.52	0.62	0.09±0.01	0.10	0.09
0.10 {2}	1.0	1.16	0.00±0.00	0.00	0.00	0.37±0.05	0.31	0.33
0.20 {2}	1.0	1.15	0.00±0.00	0.00	0.00	0.38±0.03	0.35	0.36
0.50 {2}	1.0	1.16	0.00±0.00	0.00	0.00	0.38±0.01	0.43	0.42

{1} : reactant is 1-propanol.
{2} : reactant is 2-propanol.

* ± indicates the 95% confidence interval for variations in yield during a single experimental run using student's t distribution.

Note that a total of 36 independent concentration measurements are available to estimate the values of eight to twelve kinetic parameters (i.e. the number of degrees of freedom ranges from 24 to 28). The trial mechanisms tested in this work include: (i) the E2 mechanism for 1-propanol dehydration combined with the Ad_E3 mechanism for 2-propanol formation from propene (E2AdE3), (ii) the E1 mechanism for 1-propanol dehydration and the Ad_E2 mechanism for 2-propanol formation (E1AdE2), (iii) the E2 mechanism for 1-propanol dehydration combined with the Ad_E2 mechanism for 2-propanol formation (E2AdE2), and (iv) the $E2Ad_E3$ mechanism modified by the steady state assumption (E2AdE3SS). Recall that the Ad_E3 mechanism for 2-propanol formation is the reverse of the E2 mechanism for 2-propanol dehydration, and the Ad_E2 mechanism is the reverse of the E1 dehydration mechanism. (1) Calculated best fit values for the yields of 1-propanol and 2-propanol according to the E2AdE3 and E1AdE2 models are also listed in Table I.

The yield data given in Table I for 1-propanol is summarized in Figure 3, which plots experimental vs calculated yields for the two models. In Figure 3 all the E2AdE3 yields fall well within the confidence envelope; whereas many of the E1AdE2 yields lie outside the confidence envelope. Clearly the E2AdE3 model is consistent with our experimental data; whereas the E1AdE2 model is not.

Unfortunately, the situation concerning 2-propanol is not so simple. Both the E1AdE2 and the E2AdE2 models fit the 2-propanol data equally well (see Table II), but a careful examination of the data in Table I reveals that the E1AdE2 fit of the 2-propanol yields is not acceptable. The fit of the E2AdE2 model to the experimental data is equally unacceptable. The fit of the remaining two models is worse. We believe that our inability to kinetically describe the formation of 2-propanol from 1-propanol results from scatter in the experimental data for the low concentrations of 2-propanol obtained in this work. To remedy this situation we are accumulating more data involving 2-propanol as a reactant. Results will be presented in a future paper.

TABLE II : Comparison of the various models for propanol dehydration

MODEL	OVERALL FIT (χ_ν^2)	1-PROPANOL FIT (L1 NORM)	2-PROPANOL FIT (L1 NORM)
E2AdE3	5.8	1.1	4.5
E2AdE2	5.2	1.4	3.8
E1AdE2	12.8	3.3	3.8
E2AdE3-SS	14.9	2.8	5.8

Table II also displays χ_ν^2 values for the various models. Clearly the E2 model for 1-propanol dehydration offers the best fit to the experimental data. The large values of χ_ν^2 for the two E2 models reflects their inability to fit the 2-propanol data. If the 2-propanol data is omitted from the χ_ν^2 evaluation, the value χ_ν^2 = 0.25 (14 degrees of freedom) is obtained for the E2AdE3 model optimized for 1-propanol disappearance. Such a low value for χ_ν^2

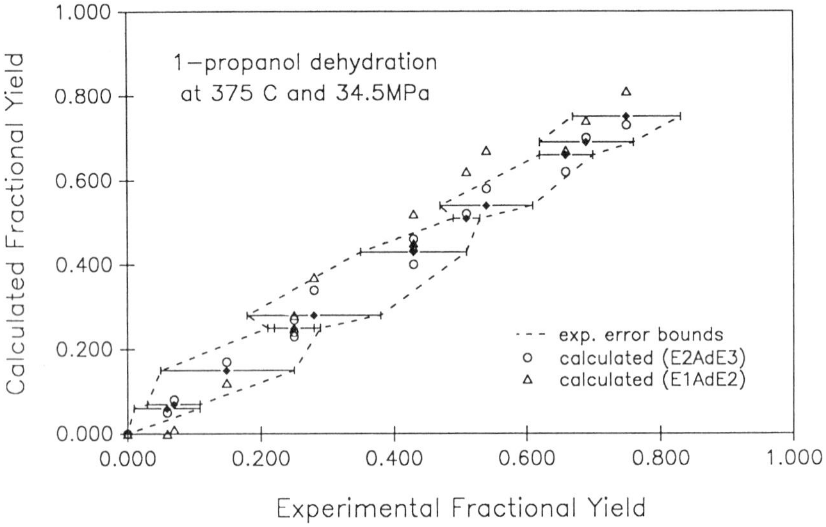

Figure 3. Calculated vs experimental yields of 1-propanol using the E2AdE3 and E1AdE2 models.

is consistent with the excellent agreement displayed in Figures 2a and 2b between the E2AdE3 model and the experimental data for 1-propanol disappearance.

Figure 4 displays representative, simulated fractional yields of 1-propanol and 2-propanol as a function of initial acid concentration (at NTP) using the E2AdE3 and E1AdE2 models which best fit our data. The agreement of the two models was even better when the simulated yields were plotted as a function of residence time or initial propanol concentration. The close agreement of these two models illustrates the difficulty in differentiating between two mechanistic models on the basis of kinetic information alone (1). We have not displayed experimental data in Figure 4 because only one data point is available for this particular set of conditions. Figures 2a and 2b and Table I offer the best comparison of the agreement of the two models with the experimental data.

Table III lists values of the k's for the E2AdE3 and the E1AdE2 models (see Figures 2a and 2b). These values minimized the objective function (given by Equation 5) for the experimental data displayed in Table I. The small value of k_2 is consistent with the fact that equilibrium strongly favors protonation of the primary alcohol. Further simulations indicated that some of the pairs of k's were not independent. For example, widely differing values of k_3 and k_4, satisfying $k_3/k_4 = 6.6$ (for the E2AdE3 model) result in the same minimum value of the objective function. Hence we conclude that some of the reactions are in equilibrium. For these reactions, the ratios of rate constants are listed in Table III.

TABLE III: Elementary rate constants for the E2AdE3 and E1AdE2 models

E2AdE3 model	E1AdE2 model
$k_1 = 36$	$k_1 = 31$
$k_2 = 0.54$	$k_2 = 0$
$k_3/k_4 = 6.6^*$	$k_3/k_4 = 0.14^*$
$k_5 = 3400$	$k_5 = 3000$
$k_6 = 4.8$	$k_6 = 8.0$
$k_7/k_8 = 0.16^*$	$k_7/k_8 = 2.4^*$
	$k_9/k_{10} = 140^*$
	$k_{11} = 790$
	$k_{12} = 2900$

* indicates equilibrium reactions.

Using values of k's for the E2AdE3 model in Equation 8 with the assumption $k_2/k_3 \simeq 0$, we obtain the steady state (SS) first order rate constant $k_{H,SS} = 36\ s^{-1}\ (mol/l)^{-1}$. This value is consistent with the first order rate constant $k_H = 31$ obtained in earlier work (Narayan, R.; Antal, Jr., M.J., submitted to J. Am. Chem. Soc.) from experimental data at low initial concentrations of 1-propanol reactant. Apparently the steady state approximation is valid at low reactant concentrations. Nevertheless, Figure 5

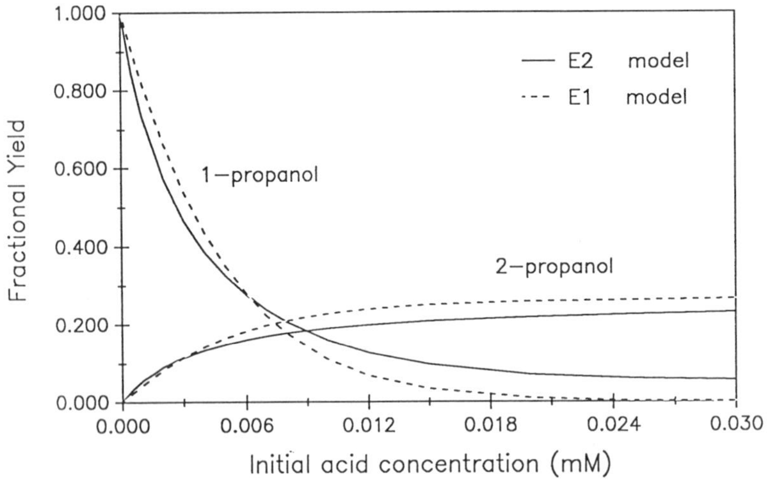

Figure 4. Comparison of the E2AdE3 and E1AdE2 models using the best-fit kinetic parameters (0.5M 1-propanol reactant, residence time = 18 s at 375°C, 34.5 MPa).

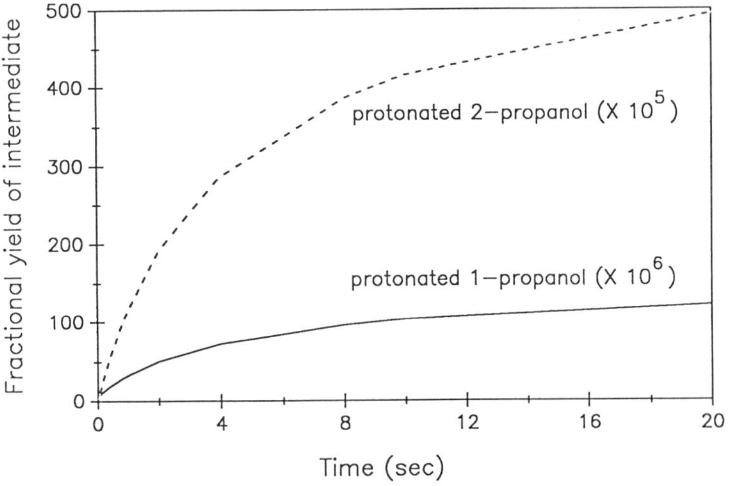

Figure 5. Fractional yields of intermediates for the E2AdE3 model (0.5M 1-propanol reactant, 5mM sulfuric acid catalyst at 375°C, 34.5 MPa).

reveals that the concentrations of the intermediates vary significantly during the typical course of a reaction. We also note that simulations using the steady state idealization of the E2AdE3 mechanism resulted in fits that were not acceptable. Hence we caution that the use of steady state idealization can be misleading.

Conclusions

1. The E2 mechanism for the acid catalyzed dehydration of 1-propanol is kinetically consistent with a wide range of experimental measurements of 1-propanol disappearance in supercritical water at 375°C and 34.5 MPa. The agreement of the calculated values of 1-propanol conversion with the experimental data is excellent ($\chi_\nu^2 = 0.25$).
2. The E1 mechanism is not consistent with these data.
3. Neither the AdE3 nor the AdE2 mechanism is consistent with available data concerning the acid catalyzed hydration of propene to 2-propanol in supercritical water at 375°C and 34.5 MPa. More data are being accumulated to sustain a rigorous kinetic examination of the hydration/dehydration mechanism.
4. The steady state "idealization" of the E2 mechanism does not adequately represent the true kinetic behavior of the E2 mechanism over the range of conditions presented in this paper.
5. The experimental data contain sufficient information to identify meaningful values of the individual rate constants (or their ratio in the case of an equilibrium reaction) associated with the E2 mechanism for 1-propanol disappearance. More work is needed to establish the confidence interval associated with each rate constant.
6. The use of kinetics to detail mechanism is a foundation-stone of modern chemistry. Nevertheless, many chemical engineers believe that with a sufficient number of free parameters, a "reasonable" model can be adjusted to fit any set of experimental data. The results of this paper (wherein a chemically motivated model with 12 free parameters could not fit the experimental data for 1-propanol disappearance; whereas an alternative model with only 8 parameters did fit the data) are in accord with the chemist's perspective that kinetics can be used to elucidate mechanism when sufficient data are available.

Acknowledgments

This work was supported by the National Science Foundation under grant CBT85-14867. The authors thank William Mok, Maninder Hunjan and Tongchit Leesomboon (University of Hawaii) for assistance with the experiments, Professor Donald G.M. Anderson (Harvard University), Professor Maitland Jones, Jr. (Princeton University), Professor Geoffrey Richards (University of Montana), Professor Jefferson W. Tester (M.I.T.) and Dr. Gabor Varhegyi (Hungarian Academy of Sciences) for many stimulating discussions concerning applied mathematics and reaction mechanisms, Dr. Duane Bruley and Dr. Maria Burka (NSF) for their continuing interest in this work.

Literature Cited

1. Lowry, T.H. and Richardson, K.S. Mechanism and Theory in Organic Chemistry 3rd Ed.; Harper & Row: New York, 1987; pp 567-620.
2. March, J. Advanced Organic Chemistry 3rd Ed.; J. Wiley & Sons, Inc.: New York, 1985; p 902.
3. Knozinger, H. In The Chemistry of the Hydroxyl Group; Patai, S., Ed.; pt 2; Interscience: New York, 1971; pp 641-718.
4. Saunders, W.H.; Cockerill, A.F. Mechanisms of Elimination Reactions; J. Wiley & Sons, Inc.: New York, 1973; pp 221-274.
5. Morrison, R.T.; Boyd, R.N. Organic Chemistry; Allyn and Bacon: Boston, 1973.
6. Kut, O.M.; Tanner, R.D.; Prenosil, J.E.; Kamholz, K. In Symposium on Catalytic Conversions of Synthesis Gas and Alcohols to Chemicals; Harman, R.G., Ed.; Plenum Press Inc.: New York, 1984.
7. Antal, M.J.; Brittain, A.; DeAlmeida, C.; Ramayya, S.; Roy, J.C. In Supercritical Fluids; Squires, T.G.; Paulaitis, M.E., Eds.; ACS Symposium Series 329; American Chemical Soc.: Washington, D.C., 1987; pp 77-87.
8. Ramayya, S.; Brittain, A.; DeAlmeida, C.; Mok, W.S.L.; Antal, Jr., M.J. FUEL 66 1987, pp 1364-1371.
9. Ramayya, S. MS thesis in Mechanical Engineering, University of Hawaii, 1987.
10. DeAlmeida, C. MS thesis in Mechanical Engineering, University of Hawaii, 1987.
11. Cutler, A.H.; Antal, Jr., M.J.; Jones, M. Ind. Eng. Chem. Res., 1988; Vol. 27, p 691
12. Haar, L.; Gallagher, J.S.; Kell, G.S. NBS/NRC Steam Table, 1984.
13. Bock, H.G. In Springer Series in Chemical Physics; Springer-Verlag: Berlin, 1981; Vol. 18, pp 102-125.
14. Bock, H.G. In Numerical Treatment of Inverse Problems in Differential and Integral Equations; Deuflhard, P.; Havier, E., Eds.; Birkhauser: Boston, 1982; pp 95-121.
15. Nowak, U.; Deuflhard, P. In Numerical Treatment of Inverse Problems in Differential and Integral Equations; Deuflhard, P.; Hairer, E., Eds.; Birkhauser: Boston, 1982.
16. Bevington, P.R. Data Reduction and Error Analysis for the Physical Sciences, McGraw Hill Book Co.: New York, 1969.
17. IMSL Inc., 2500 Park West Tower One, 2500 City West Blvd., Houston, TX 77042-3020.

RECEIVED May 1, 1989

Chapter 16

Chemistry of Methoxynaphthalene in Supercritical Water

Johannes M. L. Penninger[1] and Johannes M. M. Kolmschate[2]

[1]Akzo Salt and Basic Chemicals bv., 7550 GC Hengelo, Netherlands
[2]University of Twente, Enschede, Netherlands

Decomposition of methoxynaphthalene in supercritical water at 390 °C occurs by proton-catalyzed hydrolysis and results in 2-naphthol and methanol as main reaction products. The rate of hydrolysis is enhanced by dissolved NaCl. The dielectric constant and the ionic strength of supercritical water was found to affect the hydrolysis rate constant according to the "secondary salt effect" rate law, which commonly describes ionic reactions in liquid solvents. In subcritical water vapor the decomposition of the ether results in a mixture of cracking products and polycondensates, which is characteristic for a radical type thermolysis.

Decomposition of ethers in supercritical (SC) water is studied as a model for the conversion of natural raw materials, such as coal, shale and biomass, into chemicals and fuels by SC water exposure. The decomposition of such materials is proposed to occur by chemical action at hetero functional groups, in particular at oxygen-containing functional groups, such as ester and ether functions.

The dramatic change in product selectivity as compared to thermal pyrolysis has led to the hypothesis that SC water interacts chemically with those materials, e.g. subbituminous coal (1). Work with pure ethers (2-4) has shown that the ether reactivity is structure-dependent and that decomposition paths are affected by the density of SC water. Reactive in SC water are alkyl-alkylethers and alkyl-arylethers while aryl-arylethers, e.q. diphenylether, are inert. The change observed in product distribution with the density has been explained as a gradual transformation from a homolytic radical decomposition at zero or low water density to a hydrolytic decomposition at densities generally higher than the critical density. The same hypothesis has been forwarded for the decomposition of ethanol and 1,3 dioxalane (5) and for di-n-butylphthalate in SC water (6).

The current work focusses on the investigation of the reaction paths of a specific alkyl-arylether, viz. methoxynaphthalene, in

0097–6156/89/0406–0242$06.00/0

relation to the density of SC water. The principal objective is to present sound experimental evidence for hydrolysis chemistry at SC densities and to establish the reaction kinetics of such chemistry.

Experimental

The experiments were carried out in a 316 SS bomb with a volume of 64 cm^3. In each run 5.0 grams of methoxynaphthalene were held at 390 °C (T_r = 1.04) for 90 minutes in a quantity of water varying with each run. The density was thus varied from 0 gram/cm^3 (no water) to 0.51 gram/cm^3 (ρ_r = 1.57) by addition of 33 grams of water.

After the bomb was cooled to ambient temperature the gases formed were released in a gasometer over water and the volume measured at barometric pressure. Subsequently the bomb was opened and its content dissolved in acetone. This solution was transfered to a 50 ml volumetric flask and topped with additional acetone.

The gas phase collected in the gasometer was analyzed by GC (2 m x 1/8" 5 Å molsieves, in series with 4 m x 1/8" Porapak S, He, hot wire), to determine the absolute concentration of its constituents. The acetone solution was also analyzed by GC (CP-SIL-5 capillary 25 m x 0.32 mm, N_2, FID) and the concentration of its constituents determined relative to ethylnaphthalene, added as an internal standard.

This analytical procedure allowed for accurate mass balancing so that material prevailing as polycondensates could be quantified as the difference between consumed methoxynaphthalene and the sum of all products eluting from the GC column.

Results

Thermal Pyrolysis. As a base of reference, data are presented for the thermal pyrolysis of methoxynaphthalene, at a ρ_r of zero (Table I). The main reaction products are naphthols and naphthalenes as a result of reactions at the methoxy group; simultaneously gases such as CO and CH_4 and small quantities of CO_2 and H_2 are formed.

Methylation occurs preferentially with naphthalene, but also naphthol and methoxynaphthalene are partially methylated.
From the mass balance follows however that most of the ether is converted into non-identified polycondensates, amounting to 56 % of consumed naphthyl groups, 66 % of the oxygen and 44 % of the methyl groups.

Pyrolysis reactions proceed most commonly through radical intermediates; as a working hypothesis is therefore proposed the radical mechanism as is illustrated in Figure 1.
The pyrolysis is initiated by thermal fission of the CH_3-O bond (reaction 1). The resulting oxynaphthyl radical and methyl radical abstract H-atoms from the parent forming naphthol, CH_4 and the methylene-naphthylether radical (reactions 2 and 3).
The latter decomposes into CO and H_2 and a naphthylradical (reaction 4) which subsequently abstracts a H-atom from the parent molecule forming naphthalene and regenerating the methylene-naphthylradical (reaction 5).

$$ArOCH_3 \longrightarrow ArO\bullet + \bullet CH_3 \quad (1)$$

$$ArO\bullet + ArOCH_3 \longrightarrow ArOH + ArO\dot{C}H_2 \quad (2)$$

$$\bullet CH_3 + ArOCH_3 \longrightarrow CH_4 + ArO\dot{C}H_2 \quad (3)$$

$$ArO\dot{C}H_2 \longrightarrow Ar\bullet + CO + H_2 \quad (4)$$

$$Ar\bullet + ArOCH_3 \longrightarrow ArH + ArO\dot{C}H_2 \quad (5)$$

$$\bullet CH_3,\ Ar\bullet,\ ArO\bullet + H_2 \longrightarrow CH_4,\ ArH,\ ArOH + H \quad (6)$$

$$ArO\dot{C}H_2,\ ArO\bullet,\ Ar\bullet,\ \bullet CH_3 \longrightarrow \text{terminations} \quad (7)$$

	$ArOH/CH_4$	CO/ArH	$ArH/ArOH$	
model	1	1	1	if (6),(7) $\gg$ (4),(5)
			$\gg 1$	if (6),(7) $\ll$ (4),(5)
experiment	0.49/0.46	0.41/0.42	0.42/0.49	

Figure 1. Radical mechanism of methoxynaphthalene pyrolysis.

Table I. Product distribution by thermal pyrolysis of methoxynaphthalene

Conversion	28.5 %
Selectivity of	
naphthol	17.3
methylnaphthol	7.1
methylnaphthalene	10.7
naphthalene	4.1
naphthaldehyde	1.9
methylmethoxynaphthalene	2.2
CO	14.4
CH_4	16.1
CO_2	0.35
H_2	1.4

390 C, 90 minutes, ρ_r water = 0

Selectivity: (mol product/mol ether consumed) x 100

Radicals may further react with molecular H_2 formed in reaction 4, thus accounting for the less-than-stoichiometric ratio of H_2 to CO found in the experiment, or recombine with each other (reaction 7) to form polycondensates.

According to the model proposed, the ratio of naphthol to CH_4 would be unity provided oxynaphthyl- and methylradicals are equally reactive in reactions 2 and 3. The experiment shows a ratio of 0.49/0.46, indeed close to unity. Also the ratio of CO to naphthalenes is predicted to be unity provided the naphthyl radical preferentially saturates itself by H abstraction, over recombination with other radicals. The experimental ratio of 0.41/0.42 is again in agreement with this model. Finally the ratio of naphthalene to naphthol indicates the rates of the propagation reactions 4, 5 relative to the initiation reaction 1. The experimental value of 0.42/0.49 shows that the methylene-naphthylether radical is formed mainly through reactions 1 and 3; the chain transfer reactions 4 and 5 are slow relative to initiation 1 and to reactions 6, 7 otherwise naphthalene and CO would have been formed in concentrations well over naphthol.

The formation of polycondensates likely results from recombinations of large radical fragments. The ratio of "polycondensed" oxygen to polycondensed naphthyl structures with a value close to 1 may suggest leading recombinations to be those of oxynaphthyl radicals with each other and with the parent ether. Recombinations are important because more than half of the converted parent is recovered as polycondensed matter.

Reaction Patterns in SC Water. The reaction pattern of methoxynaphthalene changes drastically in SC water. As a first result one observes an increase of conversion when water densities increase beyond the critical density (Figure 2). More dramatic is

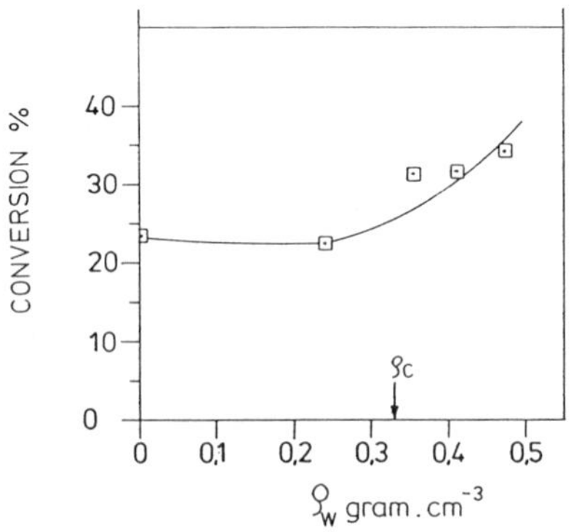

Figure 2. Conversion of methoxynaphthalene vs. pure water density.

the change in product distribution with the density. This is illustrated in Figures 3 and 4 as selectivity diagrams. Increase of the water density preferentially favors the formation of 2-naphthol while such products as (methyl)naphthalenes,methylmethoxynaphthalene and naphthaldehyde are gradually depressed. Figure 3 also shows the inverse effect of water density on naphthyl-polycondensation reactions; only 6 % of consumed naphthyl structure prevails as poly condensates at the highest water density, this figure is 56 % at zero density.

The methoxy group is converted into CO, CH_4 and methyl groups at subcritical densities, in addition to small amounts of CO_2 and aldehyd functional group (naphthaldehyde). Small concentrations of hydrogen were found over the entire density range in quantities approximately equimolar to CO_2.

At SC densities the methoxy group is preferentially converted into methanol, at the cost of mainly CO but also CH_4. Figure 4 illustrates clearly that approximately only 50 % of consumed methoxy groups are recovered while the remainder is incorporated in polycondensates. Quite clearly the water density has no effect on this selectivity pattern and this is in striking contrast to the polycondensation of naphthyl functional groups.

Organic oxygen, recovered in identified products, increases from 34 % at thermal pyrolysis conditions to over 100% at SC densities. (Figure 5) The latter means that products contain more organic oxygen than could be provided for by consumed methoxynaphthalene. This illustrates also the occurance of hydrolysis reactions at SC densities.

The reaction paths changing with the water density consequently are the result of a radical type pyrolysis transforming gradually to hydrolysis as leading mechanism. The product distribution at hydrolytic conditions is remarkably simple, in contrast to pyrolysis conditions, with an almost complete suppression of aromatic condensation. It follows therefore that SC hydrolysis provides a clean-cut mechanism for conversion of this ether.

Mechanism and Kinetics of SC Water Hydrolysis. Hydrolysis in liquid water is known to proceed through ionic intermediates; as such the rate is affected by the ionic strength of the solvent. On the other hand the observation that a reaction is affected by solvent ionic strength, e.g. as the result of dissolved inert salts, is general proof of an ionic mechanism. This test was also applied to SC water hydrolysis.

From Figure 6 it follows that the conversion rate of methoxy-naphthalene is positively affected by already small concentrations of NaCl dissolved in the SC water; the higher the water density the stronger the effects and the larger the contribution of hydrolysis to total conversion. Product selectivity is also markedly affected, as is illustrated with Figures 7 and 8 by addition of only 1.01 % wt NaCl to SC water.

Hydrolysis is known to be catalyzed by protons in solution. This again is also found to be true in SC water. Table II shows that addition of HCl in 68 ppm to SC water enhanced the rate constant by one order-of-magnitude; on the other hand addition of NaOH, hence OH anions, in a similar concentration did not measurably affect the rate.

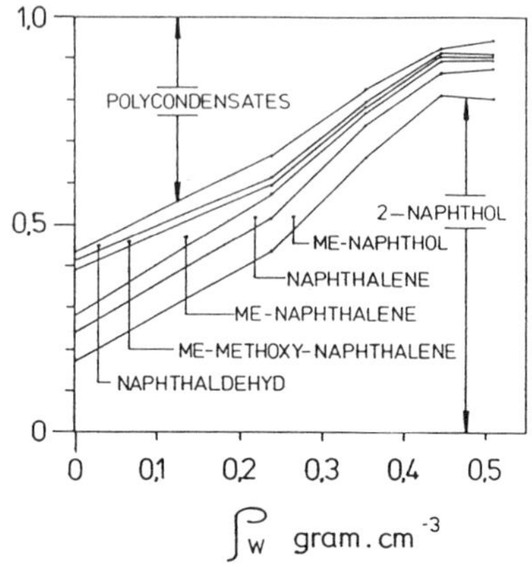

Figure 3. C-balance of consumed naphthyl groups (pure water experiments)

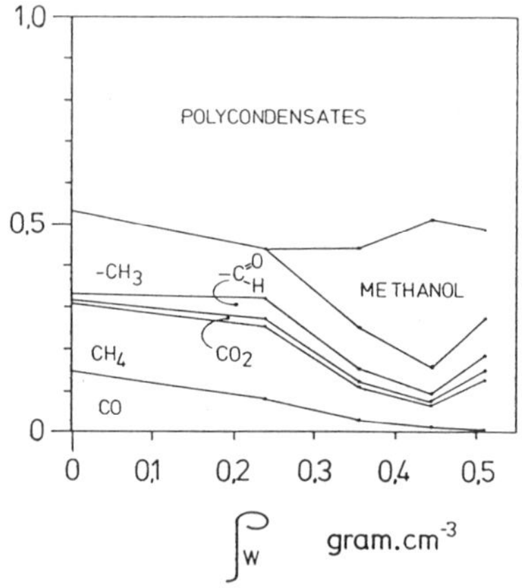

Figure 4. C-balance of consumed methoxy groups (pure water experiments)

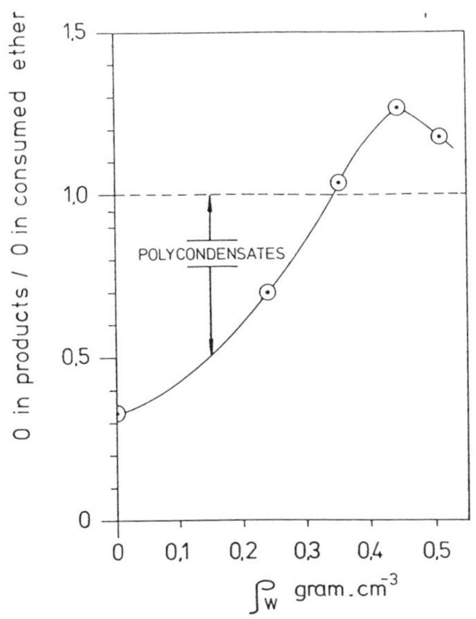

Figure 5. Balance of organic oxygen (pure water experiments)

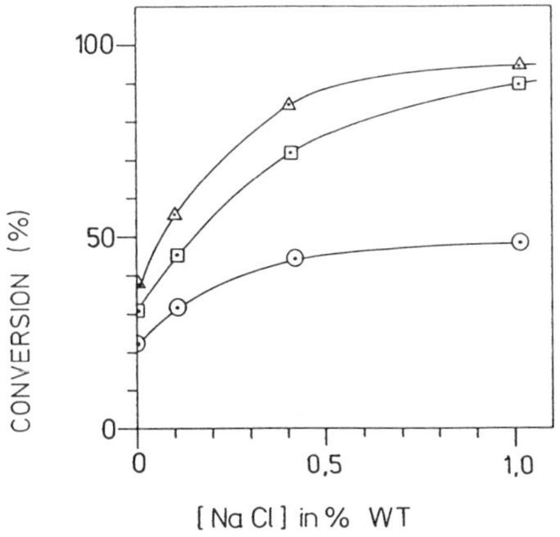

Figure 6. Conversion of methoxynaphthalene in SC aqueous NaCl

⊙ ρ_w = 0.25 gram/cm³

⊡ ρ_w = 0.35 gram/cm³

△ ρ_w = 0.45 gram/cm³

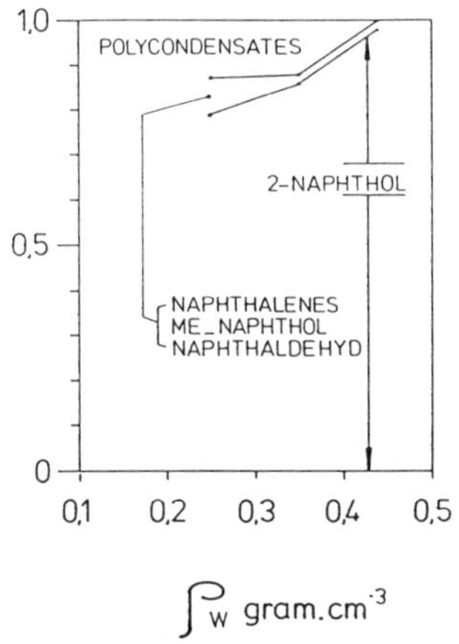

Figure 7. C-balance of consumed naphthyl groups in 1.01 % wt aqueous NaCl

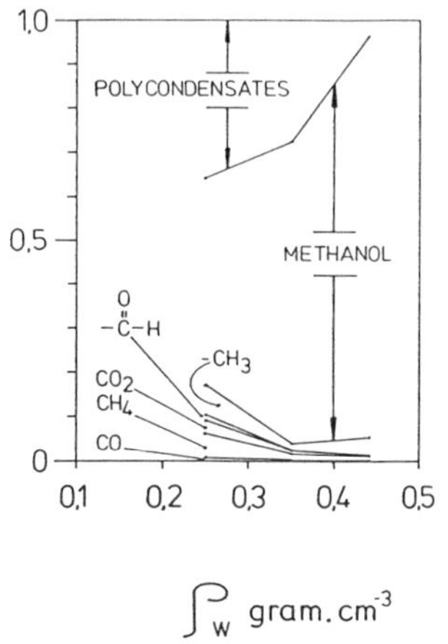

Figure 8. C-balance of consumed methoxy groups in 1.01 wt % aqueous NaCl

Table II. Effects of dissolved acid and alkali on methoxy naphthalene conversion in SC water

Additive	Conversion %	First-order k**, min^{-1}
None	38.3-42.0	5.67×10^{-3}
NaOH, 94 ppm*	40.8	"
HCl, 68 ppm*	91.7	50.2×10^{-3}

390 C, 90 minutes, $\rho_w = 0.45$ gram/cm³

* wt ppm added to water.

** from ether conversion after 90 minutes at reaction temperature; autoclave heating/cooling had negligible influence on conversion.

Conclusively the decomposition of methoxynaphthalene in SC water is a proton-catalyzed ionic reaction, all in agreement with proton-catalyzed hydrolysis. The mechanism proposed is illustrated in Figure 9. The rate-determining step is assumed to be the decomposition of the protonated ether, either in a unimolecular A-1 mechanism or in a bimolecular A-2 mechanism where water is involved in the formation of the transition state complex. The essential difference between both mechanisms is that A-2 regenerates a protium ion in a cyclic process by proton transfer, while A-1 requires a new ionization of water for every mole of ether converted. Which of both mechanisms dominates remains uncertain at this stage.

The kinetic rate expression for hydrolysis is derived by assuming steady-state for all reaction intermediates. Assuming further that the rate of hydrolysis is first-order in ether, the following equations are obtained:

for Mechanism A-1: $$r_1 = k_1[E] \qquad (1)$$

and for Mechanism A-2: $$r_2 = k_2[E] \qquad (2)$$

with $$k_1 = k_{10}\,\frac{[H_3O^+]}{[H_2O]}\cdot\rho_w \qquad (3)$$

and $$k_2 = k_{20}\,[H_3O^+].\rho_w \qquad (4)$$

The term $[H_3O^+]$ in Equations 3 and 4 follows as

$$[H_3O^+] = K_w^{0.5}/\gamma_\pm \qquad (5)$$

from the ionic dissociation constant of water, which is defined (7)

as $$K_w = [H_3O^+][OH^-]\,\gamma_+\gamma_- \qquad (6)$$

with $$[H_3O^+] = [OH^-] \qquad (7)$$

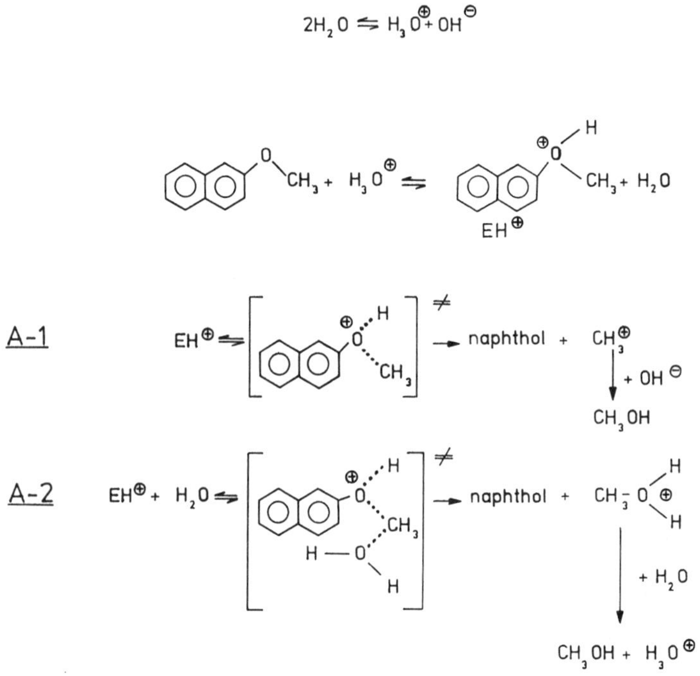

Figure 9. Hydrolysis mechanism of methoxynaphthalene in SC water

and $$\gamma_+ = \gamma_- \tag{8}$$

The latter represents the activity coefficients of H_3O^+ resp. OH^- ions in solution.

Multiplication of $[H_3O^+]$ by ρ_w is required for dimensional consistency; $[H_3O^+]$ in Equation 6 has the dimension of grammole/kg while all other concentrations terms are expressed in grammole/l. By substitution of Equations 5, 6, 7 and 8 in Equations 3 and 4 follows

$$\log k_1 = \log k_{10} + \log \rho_w + 0.5 \log K_w - \log \gamma_\pm - \log [H_2O] \tag{9}$$

and

$$\log k_2 = \log k_{20} + \log \rho_w + 0.5 \log K_w - \log \gamma_\pm \tag{10}$$

The activity coefficient $\gamma_\pm$ is determined by the ionic strength and the dielectric constant of the solvent, in first approximation according to the Debye-Hückel limiting law for dilute solutions,

$$\ln \gamma_i = - Z_i^2 \alpha (\mu)^{0.5} \tag{11}$$

where Z_i = charge of ion i

μ = ionic strength of the solution which is defined as $0.5 \Sigma C_i Z_i^2$ (12)

with C_i = concentration of ion i, grammole/l

The summation is taken over all ions present in the solution. The parameter α is defined as

$$\alpha = \frac{e^3 (2\pi N)^{0.5}}{\underline{k}^{1.5}} \cdot \frac{1}{(\varepsilon T)^{1.5}} \tag{13}$$

N, e and $\underline{k}$ are independent of solvent properties and temperatures; ε however is a function of both solvent density and temperature. By substitution of Equations 11, 12 and 13 in equations 9 and 10 it follows that

$$\varepsilon^{1.5} \log k_{1,2} = \varepsilon^{1.5} \log k^*_{1,2} + C^* (\mu)^{0.5} \tag{14}$$

with

$$\log k^*_1 = (\log k_{10} + \log \rho_w + 0.5 \log K_w - \log [H_2O]) \tag{15}$$

or

$$\log k^*_2 = (\log k_{20} + \log \rho_w + 0.5 \log K_w) \tag{16}$$

Equation 15 represents the A-1 mechanism and Equation 16 the A-2 mechanism.

Equation 14 correlates the rate constant $k_{1,2}$ for ether hydrolysis in SC water with solvent properties ε, μ and K_w and predicts a linear increase with the square-root of the ionic strength of the SC solution. This was tested with rate data obtained at water densities of resp. 0.25, 0.35 and 0.45 gram/cm³. The pure water values of ε and K_w at each density and a reaction temperature of 390 °C were obtained from the literature (7, 8) and substituted in Equation 14. These values are listed in Table III.

Table III. Values for ε and K_w of SC water at 390 C

density, gram/cm³	ε (8)	K_w (7)
0.25	3.59	8.91×10^{-18}
0.35	5.89	9.91×10^{-16}
0.45	8.20	3.34×10^{-14}

The ionic strength of the SC solution was varied by dissolving NaCl in concentrations nearly up to saturation (9). The ionic strength was calculated from the solution concentrations of H_3O^+, OH^-, Na^+ and Cl^- ions. The Na^+ and Cl^- concentrations were calculated from literature data on NaCl ionization in SC water at the prevailing densities (10); the H_3O^+ and OH^- concentration followed from K_w again at prevailing densities (7). The contributions of the latter two ions to total ionic strength is negligible as salt ion concentrations are two orders of magnitude larger.
Assuming that pyrolysis selectivity is not affected by SC water, rate constants of pyrolysis at various SC water densities were calculated with naphthalenes as pilot compounds and the "dry" experiment as the reference pyrolysis rate. The rate constant of hydrolysis $k_{1,2}$ subsequently followed as the difference of total ether conversion rate constant and pyrolysis rate constants.

It was found that, in agreement with Equation 14, a linear relation exists between the rate constant $k_{1,2}$ and $\mu^{0.5}$ at each density, and that $k^*_{1,2}$ is density dependent. The values of $k^*_{1,2}$ followed from the log $k_{1,2}$ vs. $\mu^{0.5}$ data by linear extrapolation to zero ionic strength at each single density (Table IV).

Table IV. Values for $k^*_{1,2}$ in relation to SC water density

density, gram/cm³	log $k^*_{1,2}$
0.25	- 3.15
0.35	- 2.73
0.45	- 2.65

By substitution of these values of $k^*_{1,2}$ in Equation 14 all rate data at the different NaCl concentrations and solvent densities could be correlated with $\mu^{0.5}$ by a single straight line. This is illustrated in Figure 10.

With the same data values for k_{10} and k_{20} were

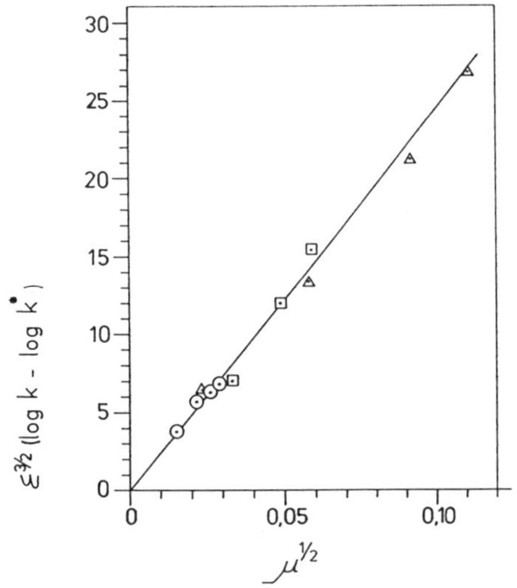

Figure 10. Correlation of Equation 14

⊙ ρ_w = 0.25 gram/cm³ ; NaCl: 0.106-1.012 % wt

⊡ ρ_w = 0.35 gram/cm³ ; NaCl: 0.106-1.013 % wt

△ ρ_w = 0.45 gram/cm³ ; NaCl: 0.0095-1.013 % wt

calculated from Equations 14, 15, 16 by substitution of the appropriate values for ρ_w, K_w, $[H_2O]$, C* (slope in Figure 10) and μ. A plot of these values, according to

$$\frac{d \ln k_{10,20}}{d P} = - \frac{\Delta V^{\neq}}{RT} + \Sigma \nu_i \beta \tag{17}$$

is illustrated in Figure 11. The slope of the lines is equal to the right-hand term of Equation 17. With the A-1 mechanism $\Sigma \nu_i$ equals zero; $\Delta V^{\neq}_{10}$ follows therefore directly from the slope value as 2377 cm^3/grammole. With the A-2 mechanism $\Sigma \nu_i$ equals minus 1 and the solvent compressibility β therefore enters the calculation. This results in values for $\Delta V^{\neq}$ from 3300 cm^3/grammole at 24 MPa down to 2900 cm^3/grammole at 31 MPa. The positive sign of $\Delta V^{\neq}$ indicates a decrease of the rate constant with the pressure; the magnitude of approximately 3000 cm^3/grammole is about two orders larger than values traditionally obtained in liquid systems. An explanation for the positive sign is not immediately available but it suggests that the transition is accompanied by expulsion of water molecules, which coordinate strongly in the supercritical range with solute molecules.

Rate constant of Pyrolysis

It is assumed that pyrolysis kinetics have a first-order ether dependence and that pyrolysis reactions are complementary to hydrolysis reactions. The first-order rate constant k_p follows now from

$$k_t = k_p + k_h \tag{18}$$

with k_t as the apparent first-order rate constant for total ether conversion and k_h as the first-order rate constant for hydrolysis. (Equations 1, 2). The values found are plotted vs $\mu^{0.5}$ in Figure 12. The scatter of the data is inherent to the calculation of k_p as the small difference of two relatively large numbers. Nevertheless the graph indicates k_p to be independent of solvent ionic strength and also of reaction pressure, an observation to be expected for pyrolysis with radical-type, non-ionic reaction intermediate.

Conclusions

The decomposition of methoxynaphthalene occurs by two parallel mechanisms; hydrolysis predominates at SC water density while thermal pyrolysis is dominant at zero and subcritical water densities. The hydrolysis is proven to be a proton catalyzed mechanism and is positively affected by dissolved NaCl, all in agreement with the secondary salt effect rate law. This rate law has traditionally been applied to liquid media but the current work has proven that the same rate law applies also for SC water.

The dielectric constant changes significantly within the SC range of densities; its effect on the rate constant has now also been proven to be in agreement with correlations proposed in the

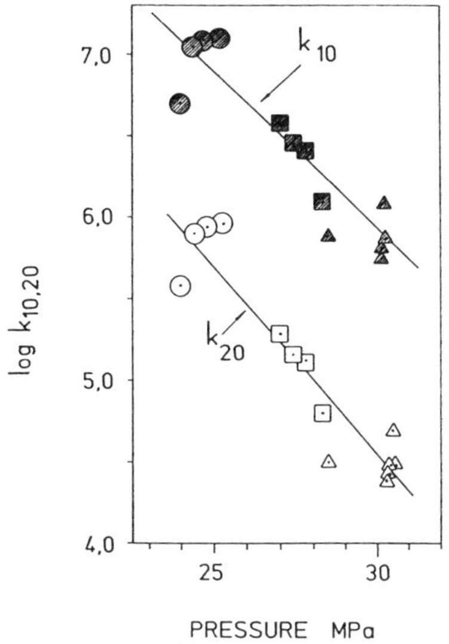

Figure 11. Correlation of Equation 17
Symbol legend: Figure 10

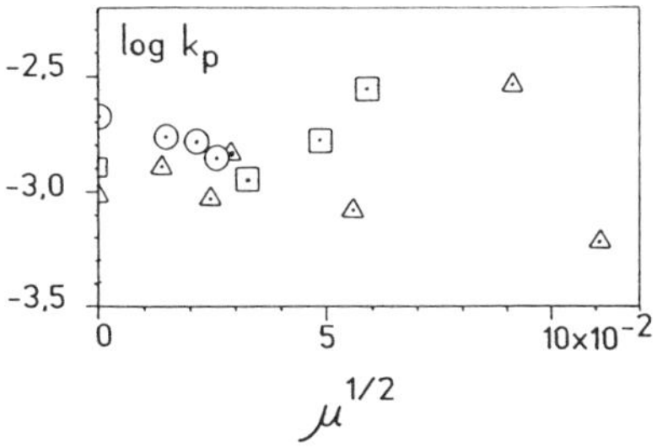

Figure 12. Correlation of pyrolysis rate constant with ionic strength and solvent density
Symbol legend: Figure 10

past for liquid solvents. Water, at SC densities, therefore behaves as an expanded liquid and supports hydrolysis reactions of components which do not react at subcritical temperatures.

Legend of Symbols

r	rate of hydrolysis, grammole/min
$k_{1,2}$	hydrolysis rate constant
$k^*_{1,2}$	hydrolysis rate constant at zero ionic strength (Equation 14)
γ_+	activity coefficient of H_3O^+ in solution

γ_- activity coefficient of OH^- in solution

Z_i charge of ion i

μ ionic strength of solution

C* proportionality constant in Equation 14; equals $Z_i^2 e^3 (2\pi N)^{0.5}/(\underline{k}T)^{1.5}$

C_i solution concentration of ion i, grammole/l

N Avogadro Number, 6.022×10^{23} grammole^{-1}

e charge of electron, 1.6022×10^{-19} C

$\underline{k}$ Boltzmann constant, 1.380×10^{-23} J/K

ε dielectric constant of solvent

T temperature of solution, K

[E] solution concentration of ether, grammole/l

$[H_2O]$ solution concentration of water, grammole/l

$[H_3O^+]$ molality of protium ion, grammole/kg

P static pressure of reaction system, Pa

$\Delta V^{\neq}$ activation volume, cm^3/grammole

R gas constant, 8.3134 Pa m^3/mol K

ν_i stoichiometric coefficient of reactant i involved in the formation of 1 grammole of transition complex

β isothermal compressibility of reaction mixture = $-\frac{1}{V}$ dV/dP

V reaction volume, l

ρ_w density of SC water, kg/l

Literature Cited

1. Penninger, J.M.L.; In Supercritical Fluid Technology; Penninger, J.M.L; Radosz, M.; McHugh, M.A.; Krukonis, V.J., Eds.; Elsevier Science Publ.: Amsterdam, 1985; p 309.
2. Lawson, J.R.; Klein, M.T.; Ind.Eng.Chem.Fundam. 1985, 24, 203.
3. Townsend, S.H.; Klein, M.T.; Fuel 1985, 64, 635.
4. Townsend, S.H.; Abraham, M.A.; Hupper, G.L.; Klein, M.T.; Paspek, S.C.; Ind.Eng.Chem.Res. 1988, 27, 143.
5. Antal, M.J.; Brittain, J., DeAlmeida, C.; Ramayya, S.; Roy, J.C.; In Supercritical Fluids; Squires, T.G.; Paulaitis, M.E.; Eds.; ACS Symp. Ser. 329, ACS Washington, DC, 1987
6. Penninger, J.M.L.; Fuel 1988, 67, 490.
7. Marshall, W.L.; Franck, E.; J.Phys.Chem.Ref.Data 1981, 10, 295.
8. Fogo, J.K.; Benson, S.W.; Copeland, C.S.; J.Chem.Physics 1954, 22, 209.
9. Baierlein, H.; Ph D Thesis Erlangen-Nürnberg 1983
10. Fogo, J.K.; Benson, S.W.; Copeland, C.S.; J.Chem. Physics 1954, 22, 212

RECEIVED May 23, 1989

Chapter 17

Fundamental Kinetics of Methanol Oxidation in Supercritical Water

Paul A. Webley and Jefferson W. Tester

Department of Chemical Engineering, Massachusetts Institute of Technology, Cambridge, MA 02139

The destruction of hazardous chemical wastes by oxidation in supercritical water is a promising new technology which has several advantages over conventional methods of toxic chemical waste disposal. Although the feasibility of the supercritical water oxidation process has been demonstrated, there is little kinetic information available on the underlying reaction mechanisms. We have recently determined the oxidation kinetics of several model compounds in supercritical water, and now report on our results of the oxidation of methanol, a common industrial solvent, in supercritical water. Global kinetic expressions are presented and our attempts to model the reaction using a free–radical mechanism with 56 elementary reactions are discussed. The inability of the elementary reaction model to represent oxidation in supercritical water is demonstrated and future model modifications are discussed.

In recent years, several innovative hazardous waste treatment processes have emerged that offer alternatives to existing waste treatment technologies. A particularly difficult problem is encountered when treating dilute aqueous wastes containing organic solvents, especially chlorinated hydrocarbons. A new technology involving oxidation in supercritical water via the MODAR process (1,2) has shown great promise as an effective treatment method.

In supercritical water oxidation (SCWO), organics, air and water are brought together in a mixture at 250 atm and temperatures above 400°C. Organic oxidation is initiated spontaneously at these conditions and the heat of combustion is released within the fluid resulting in a rise in temperature to 550–650°C. Organics are oxidized rapidly with conversions in excess of 99.99% at reactor residence times of less than 1 minute. Heteroatoms such as chlorine are oxidized to acids, which are precipitated out as salts by adding a base to the feed (3). For aqueous wastes containing 1 to 20 wt% organics, supercritical water oxidation is less costly than controlled incineration or activated carbon treatment and more efficient than wet oxidation at lower temperatures (200°C or less) and pressures (150 atm or less) (3). Oxidation of organics to carbon dioxide and molecular nitrogen is complete without the formation of noxious by–products such as NO_x (4). In addition, the supercritical water oxidation process is not restricted to aqueous organic wastes. Any pumpable stream including slurries of biomass or soil can be fed to the reactor.

0097–6156/89/0406–0259$06.00/0

Previous Work

The oxidation of several different organics in supercritical water has been demonstrated by several authors. Price (5) was able to eliminate 88–93% of the liquid TOC (total organic carbon), although the depressurized gas phase contained up to 11% carbon monoxide. Modell et al. (1) oxidized several toxic chlorinated hydrocarbons in supercritical water, destroying at least 99.99% of the organic chlorides and 99.97% of the TOC. Cunningham et al. (6) used this process to completely destroy a variety of biopharmaceutical and organic solvent wastes. Although these workers demonstrated the effectiveness of the process, their objective was to obtain complete conversion or total destruction and consequently their results contain no kinetic parameters or mechanistic details at well defined conditions of temperature, pressure and composition.

Kinetic data for oxidation in supercritical water were first reported by Wightman (7) for phenol and acetic acid. Single rate experiments were also conducted for ten other compounds, including ammonia. Previous work by Helling and Tester (8),(9) and Webley and Tester (10) determined the oxidation kinetics of carbon monoxide, ethanol and methane in supercritical water using an isothermal continuous flow reactor. Although Helling and Tester (9) also examined ammonia oxidation, the reactor operating temperature was limited to 550°C, at which ammonia conversions of less than 5% were observed. Reaction parameters could therefore not be established for ammonia.

Experimental

The apparatus used for the present study is an isothermal, isobaric plug–flow reactor consisting of 4.24 m of coiled Inconel 625 tubing. In all experiments, water is in excess of 99.9% by mole. The assumption of constant density is therefore an excellent one and facilitates data analysis. The reactants (oxygen and organic) are fed separately to the reactor in the form of aqueous solutions. These feed streams are pre–heated to the reactor temperature before mixing at the reactor inlet. After reaction, the products are cooled to room temperature, the mixture depressurized and the gas and liquid phases (mostly water) separated. Both gas and liquid flowrates are measured and their compositions are determined by gas chromatography. The experimental apparatus is that of Helling and Tester (8) with the exception of the sand bath used for the high temperature reactor control. A new sand bath, capable of extended operation at 700°C, has been installed.

For low conversions (less than 20%) of organic, the assumption of a differential reactor ($dc/ct \simeq \Delta c/\Delta t = ([C]_{in} - [C]_{out})/\tau$) is a good one and rates are calculated on this basis. For higher conversions (greater than 20%), the global stoichiometric oxidation equation is numerically integrated and the kinetic parameters are obtained by regression to conversion data. Rates are then calculated from the global rate expression.

Results

Fourteen oxidation runs were conducted to determine the oxidation kinetics of methanol in supercritical water. The temperature range covered was 450–550°C at a pressure of 243 atm (24.6 MPa). Reactor residence times ranged from 8.5 to 12.4 seconds. The oxidation run at 450°C resulted in such low conversions of methanol (<1%) that experimental errors were too large to draw any quantitative conclusions. The data from this run were not included in the data analysis but are shown in Table I. In addition to oxidation runs, one pyrolysis run was attempted. Unfortunately, the complete exclusion of oxygen from the reactor system is difficult and this run must be regarded as an oxidation run in the presence of very low concentrations of oxygen. The results of all the runs are shown in Table I, where conversion of methanol indicates loss of methanol, regardless of the reaction products.

The Arrhenius plot for an assumed first–order oxidation of methanol in supercritical water is shown in Figure 1 where error bars shown are at the 98% confidence limit. Also shown is the weighted least–squares regression line to the data, where the weights were taken as proportional to k_i^2/σ_i^2 (11). The variances, σ_i^2 for each value of ln(k) were estimated from a Monte Carlo error analysis using estimates of errors in the measured data. From the regression line, the activation energy was estimated to be 478.6±68.0 kJ/mol, where the stated error is at the 98% confidence limit. The first order rate expression for the oxidation of methanol in supercritical water is therefore:

$$-\frac{d[MeOH]}{dt} = 10^{29.4\pm5.1}\exp(-478.6\pm68.0/RT)[MeOH] \quad (1)$$

where [ ] indicates concentration in mol/cm^3 and the reaction rate has units of mol/cm^3.s. Since runs were conducted over ranges of methanol and oxygen concentrations, regression with respect to concentration is possible. A weighted least–squares regression of the data to the functional form:

$$-\frac{d[MeOH]}{dt} = A\exp(-E_a/RT)\,[MeOH]^a[O_2]^b \quad (2)$$

was performed. This regression gave an activation energy of 466.6±63 kJ/mol and a pre–exponential of $10^{28.4}$ (mol/cm^3)$^{1-a-b}$ s^{-1}. The overall weighted regression gave

$$-\frac{d[MeOH]}{dt} = 10^{28.4\pm6.0}\exp(466.6\pm63/RT)$$
$$\times[MeOH]^{0.73\pm0.3}[O_2]^{0.23\pm0.4} \quad (3)$$

where all errors are at the 98% confidence level.

A t–test on the order of the reaction with respect to oxygen at a 95% confidence level indicates that the order is not statistically different from zero. In addition, the order of the reaction with respect to methanol is not statistically different from one although the regression Equation 3 does provide a better fit to the data than Equation 2. We emphasize that Equation 3 is no more than a useful correlative equation for representation of methanol oxidation over our experimental operating range. The global parameters A and E_a in Equation 3 have very limited mechanistic or physical significance (if any) since each will include complex combinations of pre–exponentials and activation energies for the elementary reactions occurring in the oxidation process. Very few elementary reactions have pre–exponential factors exceeding 10^{18}, and activation energies are usually less than 450 kJ/mole. The high pre–exponential factor and fractional powers on the methanol and oxygen concentration are clear indicators that the reaction mechanism is a complex one.

Table I. Summary of Experimental Conditions and Results for Oxidation of Methanol in Supercritical Water

T (C)	[MeOH]/[O_2] feed	Residence time seconds	Conv. %	Reaction Rate 10^6mol/cm^{3}min	ln(k) 1/s
450	1:1.4	11.4	<1.0	0.22	−7.0
480	1:1.3	10.3	1.0	0.22	−6.9
500	1:1.2	9.7	4.2	0.88	−5.4
550	1:1.3	8.5	94.2	17.4	−1.1
530	1:1.7	8.9	20.7	2.9	−3.7
520	1:1.7	9.1	15.7	2.3	−4.0
550‡	1:0.1	8.4	9.3	1.3	−4.5
520	1:3.2	9.1	15.9	1.2	−4.0
520	1:0.5	8.9	14.7	5.2	−4.0
520	1:0.8	9.1	10.7	3.1	−4.4
520	1:1.0	8.8	9.6	2.0	−4.5
520	1:1.2	8.9	12.7	2.0	−4.2
520	1:0.9	8.9	7.1	1.1	−4.8
520	1:0.5	9.0	9.0	1.4	−4.6

‡ Pyrolysis run at very low O_2 concentration

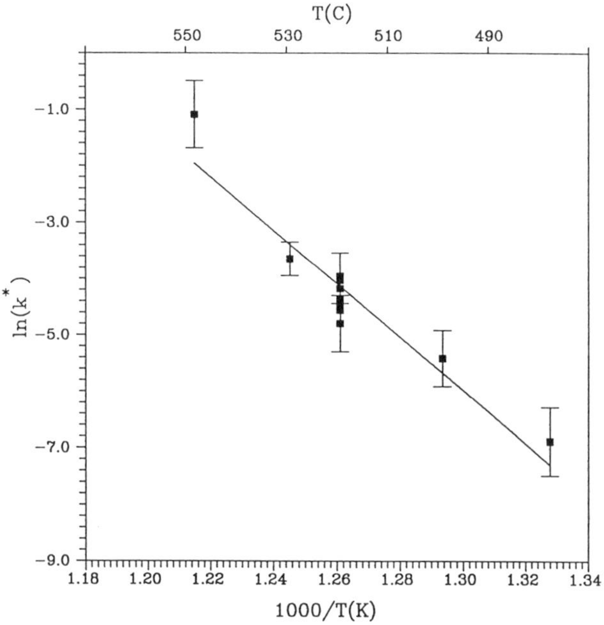

Figure 1. Arrhenius Plot for the First–Order Oxidation of Methanol in Supercritical Water.

Discussion

The oxidation of methanol in supercritical water in the temperature range 480°C to 550°C yielded both carbon monoxide and carbon dioxide for residence times of order 8.4 to 11.4 seconds. As in gas–phase oxidation of methanol (12), the formation of carbon monoxide is assumed to be an intermediate in the complete oxidation to carbon dioxide. In addition, hydrogen was detected in the reactor outlet. The ratio $[H_2]/[CO_2]$ in the outlet from the reactor varied from 0.24 to 0.46 at 520°C over the range of feed concentrations indicated in Table I. Prior work on the oxidation of carbon monoxide in supercritical water (8) showed that $[H_2]/[CO_2]$ in the reactor outlet ranged from 0.2 to 0.4 at 520°C, in good agreement with the present data. In that study, hydrogen was produced from the water–gas shift reaction. The hydrogen produced during methanol oxidation therefore appears to be a result of the water–gas shift reaction. The oxidation of methanol can be viewed as proceeding through two global steps. The first is direct oxidation to carbon monoxide and hydrogen:

$$CH_3OH + O_2 \longrightarrow CO + H_2O + H_2 \tag{4}$$

followed by direct oxidation of carbon monoxide to carbon dioxide or oxidation via the water–gas shift reaction;

$$CO + \frac{1}{2}O_2 \longrightarrow CO_2 \tag{5}$$

$$CO + H_2O \longrightarrow CO_2 + H_2 \tag{6}$$

We emphasize that Equations 4,5 and 6 are global representations of the oxidation of methanol to carbon monoxide and carbon dioxide and do not represent actual molecular pathways. Unlike methane oxidation in which the initial oxidation to carbon monoxide was found to be rate limiting (10), both the initial oxidation of methanol to carbon monoxide (Equation 4) and subsequent oxidation to carbon dioxide (Equations 5 and 6) are important. In addition, water competes with oxygen for oxidation of carbon monoxide to carbon dioxide.

As mentioned earlier, one pyrolysis run was conducted at 550°C. Unfortunately, some oxygen was present in the feed to the reactor, as evidenced from its detection in the product gas. The concentration of oxygen in the product gas was however at least an order of magnitude lower than in a comparable oxidation run. The low methanol conversion of 9.3% as compared to 94.2% for a comparable oxidation at 550°C demonstrates that pyrolysis rates are much lower than oxidation rates, despite the apparent independence of the rate on oxygen concentration. This behavior has been observed by several previous workers (12,13) in the gas phase oxidation of methanol and is discussed later. The pyrolysis run yielded relatively large quantities of hydrogen in the product gas from the reactor with a ratio $[H_2]/[CO_2]$ in the reactor outlet of 1.1. The water–gas shift reaction cannot alone be responsible for the large quantity of hydrogen produced during this run. For this reason, hydrogen has been added to the global equation 4. In addition, the elementary reaction model for methanol oxidation (presented later) indicates that hydrogen can be produced from the initial decomposition of and free radical attack on methanol. No methane or formaldehyde were detected in the product from the reactor. Although our gas chromatograph has a high sensitivity for methane (0.1% mole), the detection of formaldehyde is more difficult, due to possible adsorption onto the column packing. It is possible therefore that formaldehyde was present in small quantities in the products from the reactor. However, the good carbon mass balance closures (usually better than 3%) indicates that if present, formaldehyde would be in very low concentrations.

Mechanistic Considerations

The high concentration of water (greater than 99% by mole) and the relatively high densities (up to 0.1 g/cm^3) in the reaction system suggests liquid–like, ionic reactions typical of kinetics in room temperature water are possible in supercritical water. However, there is much evidence to indicate that in our operating range, supercritical water is a "gas–like" medium that is unable to stabilize and therefore support ions and hence the reaction mechanism is more likely a free–radical one. It is difficult to establish an unambiguous criteria for "liquid–like" or "gas–like" behavior in the near supercritical region. Bondarenko et al., (14) have adopted the momentum and character of the molecular rotational motion as an indication of one or the other state of aggregation. They assumed that all molecules in free internal rotation participate as a gas. This corresponds to the presence of P– and R–branches in the IR spectrum of the bands that are being investigated. From spectroscopic measurements, they concluded that above 500°C at 250 atm, the supercritical water phase was essentially gas–like. In addition, Franck and Roth (15) have measured both the IR spectrum and Raman spectrum of water in the supercritical region. These measurements indicated limited hydrogen bonding above 400°C with almost no bonding above 500°C and 250 atm.

The inability of supercritical water to solvate ions is demonstrated by the very low solubility of inorganic salts in our operating range (16). At 500°C and 250 atm, the dissociation constant of water is 10^{-23} and the dieletric constant is close to one, much different from its room temperature and pressure value, 80. The poor solvating and ionic properties of supercritical water therefore indicate that ionic reactions are unlikely to occur. Perhaps the best evidence in favor of a free–radical mechanism is the work of Antal et al., (17). They found that the dehydration of ethanol to ethylene in supercritical water with ion product greater than 10^{-14} proceeded via a heterolytic bond cleavage whereas dehydration in supercritical water for $K_w << 10^{-14}$ proceeded through a homolytic free–radical pathway.

The above considerations indicate that the oxidation of methanol in supercritical water *in our operating range* is probably a free–radical mechanism. It is of interest therefore to compare our results with gas–phase methanol oxidation studies. Major differences will highlight the role played by water in the oxidation process.

Comparison with Gas–Phase Methanol Oxidation Work

The oxidation of methanol in the gas phase has been investigated using both flow and static experiments. Bell and Tipper (13) investigated the slow combustion of methanol in a Pyrex vessel in the temperature range 430–470°C at pressures below atmospheric. The kinetics of the reaction were studied by means of pressure–time curves. The rate was found to be proportional to the square of the methanol concentration and independent of the oxygen concentration. An overall activation energy of 255 kJ/mol was measured. The products of the reaction were carbon monoxide and water with small amounts of formaldehyde, hydrogen peroxide, carbon dioxide and hydrogen. Although formaldehyde was not observed in our oxidation experiments, this may have been due to the inherent difficulties in detecting formaldehyde on the gas chromatograph, as previously mentioned. Bell and Tipper (13) attributed the formation of hydrogen to the reaction

$$H + CH_3OH = H_2 + CH_2OH \quad (7)$$

with the H atoms coming from decomposition of HO_2, CH_2OH or CHO radicals. In our oxidation studies, the hydrogen produced is chiefly attributed to the water–gas shift reaction, although the pyrolysis run indicates that hydrogen may be produced from the initial decomposition of methanol. It seems unlikely that the water–gas shift occurred to any appreciable extent in the work of Bell and Tipper (13) since addition of water to the reaction system did not result in increased formation of hydrogen. Experiments done in the presence of water showed an increase in rate of methanol oxidation by a factor of two for the addition of 25 Torr water to 100 Torr oxygen and 100 Torr methanol at 440°C. This observation has important implications for our oxidation studies in which water is the dominant component. Bell and Tipper attributed the increase in rate to a heterogeneous catalytic effect, assuming that the water adsorbed onto the surface of the vessel, covering adsorption sites for oxygen molecules thus reducing the rate of chain–termination and increasing the overall reaction rate.

It is not known to what extent catalysis occurs in our supercritical water oxidation apparatus. Previous investigations of a possible catalytic effect by Helling and Tester (8) indicated that no significant catalysis occurred. However, the surface to volume ratio was only increased by 52%. A catalytic packed bed reactor with a surface to volume ratio thirty times that of our present reactor is currently under construction to determine the extent of catalysis.

Aronowitz et al, (12) studied the kinetics of the oxidation of methanol in an adiabatic, turbulent flow reactor which allowed for chemical sampling along the length of the reactor. Experiments were performed at atmospheric pressure in the temperature range 950–1030 K. Major products were CO, CO_2 and H_2O with smaller amounts of H_2 and HCHO. In addition, trace quantities of hydrocarbons CH_4, C_2H_6 and C_2H_4 were detected at methanol rich conditions. In our pyrolysis run, methanol was in excess although oxygen was present. This very fuel–rich condition did not yield any measurable amounts of hydrocarbons, suggesting that the chain mechanism responsible for hydrocarbon formation in gas phase methanol oxidation is either absent or greatly suppressed in supercritical water.

Using semi–global modeling techniques, Aronowitz et al, (12) correlated the rate of methanol oxidation in fuel–lean conditions (equivalence ratios, φ less than 1) by the expression:

$$-d[CH_3OH]/dt = 10^{11.53}\exp(-229.3/RT)[CH_3OH]^{0.81} \quad (8)$$

while fuel–rich data ($\varphi > 1$) were represented by

$$-d[CH_3OH]/dt = 10^{31.5}\exp(-510.9/RT)[CH_3OH]^{1.55} \quad (9)$$

Application of Equation 8 to oxidation in supercritical water at 550°C gives a reaction rate an order of magnitude lower than that observed and a conversion of 9.4% compared to an observed conversion of 94%.

Significantly, Equations 8 and 9 show no oxygen dependence. Previous measurements of methanol pyrolysis by Aronowitz et al., (18) indicated that rates of pyrolysis are an order of magnitude lower than oxidation rates. This is in agreement with our oxidation studies in which pyrolysis at 550°C was an order of magnitude slower than oxidation at 550°C. It is therefore surprising to see no explicit dependence on oxygen concentration. Classically, oxygen dependence in hydrocarbon systems has been explained by the importance of the branching reaction:

$$H + O_2 = OH + O \qquad (10)$$

Therefore, the lack of oxygen dependence may imply that the above reaction is of minor importance in methanol oxidation. However, since oxidation rates are significantly faster than pyrolysis rates, there should be some effect of oxygen upon the methanol oxidation rate.

Elementary Reaction Modeling

In order to explain the data of Aronowitz et al (12) and previous shock–tube and flame data, Westbrook and Dryer (19) proposed a detailed kinetic mechanism involving 26 chemical species and 84 elementary reactions. Calculations using this mechanism were able to accurately reproduce experimental results over a temperature range of 1000–2180 K, for fuel–air equivalence ratios between 0.05 and 3.0 and for pressures between 1 and 5 atmospheres. We have adapted this model to conditions in supercritical water and have used only the first 56 reversible reactions, omitting methyl radical recombinations and subsequent ethane oxidation reactions. These reactions were omitted since reactants in our system are extremely dilute and therefore methyl radical recombination rates, dependent on the methyl radical concentration squared, would be very low. This omission was justified for our model by computing concentrations of all species in the reaction system with the full model and computing all reaction rates. In addition, no ethane was detected in our reaction system and hence its inclusion in the reaction scheme is not warranted. We have made four major modifications to the rate constants for the elementary reactions as reported by Westbrook and Dryer (19):
1) The more recent rate constants from the compilation of Tsang and Hampson (20) and Tsang (21) were used as a basis for low pressure, ideal gas reaction systems, typical of gas–phase combustion systems.
2) The rate constants for unimolecular dissociation reactions were adjusted to the high pressure of the supercritical water system by using high pressure rate constants, k_∞, where applicable. For dissociations still in the low pressure regime (bimolecular dissociations), rate constants were calculated assuming the strong collision assumption to be valid. Although this is rigorously not true, relatively high efficiencies have been reported for dissociation reactions in which water is the collision partner (22). This unusually large collision efficiency may be partly responsible for the high reaction rates observed in supercritical water. Where literature values for high pressure rate constants and strong collision fall–off curves were available, these were used. For reactions 61 and 89 (Table II), RRKM calculations were performed to determine the rate constants. The program "Fall–Off", available from the Quantum Chemistry Exchange Program (23) and modified by Shandross and Howard (24) was used for the RRKM calculations. Transition states were located at the centrifugal barrier on the minimum energy path from reactants to products. Bond vibration and rotation frequencies were estimated using the techniques of Benson (25).
3) The effect of pressure and system non–idealities on bimolecular reaction rates has been incorporated by using fugacity coefficients and compressibility

factors as described by Simmons and Mason (26). For the bimolecular reaction;

$$\nu_a A + \nu_b B \rightarrow AB\ddagger \rightarrow \text{products} \tag{11}$$

the ratio of the rate constant k to its low pressure value k_o, is given as:

$$\frac{k}{k_o} = \frac{(\varphi_a Z)^{\nu_a} (\varphi_b Z)^{\nu_b}}{(\varphi_{ab} Z)} \tag{12}$$

where Z is the system compressibility and φ_i is the fugacity coefficient of species i. Since our reaction system is very dilute in oxygen and organic (water is in excess of 99.8% by mole), Z is very close to the value for pure supercritical water. In using the above expression, we have assumed, as a first approximation, that the fugacity coefficients for all species (including transition states) except water are unity. It should be recalled that our operating range for methanol oxidation is well beyond the mixture critical point (which is close to the critical point of pure water) where large non–idealities occur. Above 500°C at 250 atm, the compressibility factor of water is close to unity and the system is "gas–like" with fugacity coefficients close to one. For oxidation closer to the critical point (374°C, 218 atm), we expect larger deviations from ideality and fugacity coefficients significantly different from unity. For stable species such as O_2, N_2, H_2 and CO_2, we have verified this assumption by computing fugacity coefficients using the equation of state of Christoforakos and Franck (27). The parameters in this equation were determined by fitting the equation to high–temperature, high–pressure PVT data for the binary systems O_2/H_2O (28), N_2/H_2O (29), CO_2/H_2O (30) and H_2/H_2O (31). Above 450°C, the fugacity coefficients for these species were all close to unity. Fugacity coefficients of radicals and transition states were assumed to equal one. The only correction made therefore was to include the compressibility factor and fugacity coefficient (when appropriate) of supercritical water as calculated from the steam tables (32). At most, this correction amounted to a factor of two and did not change the modeling results significantly.

4) The reverse rate constants for the elementary reactions used in the present work were calculated from the forward rate constants and the equilibrium constant by assuming microscopic reversibility. Standard states used in tabulations of thermodynamic data are invariably at 1 atm and the temperature of the system. Since concentration units were required for rate constant calculations, a conversion between K_p and K_c was necessary. Values of K_p were taken from the JANAF Thermochemical tables (1984). K_c was calculated from the expression:

$$K_c = K_p \left[\frac{P_{atm}}{RT}\right]^{-\Delta\nu} K_\Phi \tag{13}$$

where;

$$K_\Phi = \Pi\, [\, \varphi_i^{\nu_i}] \tag{14}$$

Once again, all fugacity coefficients except that of water were assumed equal to one.

The elementary reaction set used for the present analysis together with the modified forward rate constants used is shown in Table II. All fourteen methanol oxidation runs were simulated using this model. The stiff O.D.E.

Table II. Elementary Reaction Model for Methanol Oxidation in Supercritical Water

	Reaction	Rate (mol/cm³.s) log A	n	Ea/R,(K)	Ref.
(1,2)[a]	$CH_3OH+M \longrightarrow CH_3+OH+M$	16.0	0.0	45223	36
(3,4)	$CH_3OH+O_2 \longrightarrow CH_2OH+HO_2$	13.3	0.0	22600	21
(5,6)	$CH_3OH+OH \longrightarrow CH_2OH+H_2O$	12.7	0.0	664	38
(7,8)	$CH_3OH+O \longrightarrow CH_2OH+OH$	5.6	2.5	1550	21
(9,10)	$CH_3OH+H \longrightarrow CH_2OH+H_2$	7.3	2.1	2450	21
(11,12)	$CH_3OH+H \longrightarrow CH_3+H_2O$	12.7	0.0	2667	19
(13,14)	$CH_3OH+CH_3 \longrightarrow CH_2OH+CH_4$	1.5	3.2	3609	21
(15,16)	$CH_3OH+HO_2 \longrightarrow CH_2OH+H_2O_2$	11.0	0.0	6330	21
(17,18)[a]	$CH_2OH+M \longrightarrow CH_2O+H+M$	14.8	0.0	14914	37
(19,20)	$CH_2OH+O_2 \longrightarrow CH_2O+HO_2$	12.1	0.0	0.0	21
(21,22)[a]	$CH_4+M \longrightarrow CH_3+H+M$	13.7	0.6	53800	35
(23,24)	$CH_4+H \longrightarrow CH_3+H_2$	4.4	3.0	4406	20
(25,26)	$CH_4+OH \longrightarrow CH_3+H_2O$	5.3	2.4	1060	20
(27,28)	$CH_4+O \longrightarrow CH_3+OH$	9.0	1.5	4330	20
(29,30)	$CH_4+HO_2 \longrightarrow CH_3+H_2O_2$	11.3	0.0	9350	20
(31,32)	$CH_3+HO_2 \longrightarrow CH_3O+H_2$	13.3	0.0	0.0	20
(33,34)	$CH_3+OH \longrightarrow CH_2O+H_2$	12.6	0.0	0.0	19
(35,36)	$CH_3+O \longrightarrow CH_2O+H$	13.9	0.0	0.0	20
(37,38)	$CH_3+O_2 \longrightarrow CH_3O+O$	18.3	–1.6	14710	20
(39,40)	$CH_2O+CH_3 \longrightarrow CH_4+HCO$	3.7	2.8	2950	20
(41,42)	$CH_3+HCO \longrightarrow CH_4+CO$	14.1	0.0	0.0	20
(43,44)	$CH_3+HO_2 \longrightarrow CH_4+O_2$	12.6	0.0	0.0	20
(45,46)[a]	$CH_3O+M \longrightarrow CH_2O+H+M$	14.5	0.0	12990	37
(47,48)	$CH_3O+O_2 \longrightarrow CH_2O+HO_2$	10.8	0.0	1310	20
(49,50)[a]	$CH_2O+M \longrightarrow HCO+H+M$	13.8	0.6	45100	35
(51,52)	$CH_2O+OH \longrightarrow HCO+H_2O$	9.5	1.2	–225	20
(53,54)	$CH_2O+H \longrightarrow HCO+H_2$	8.3	1.8	1510	20
(55,56)	$CH_2O+O \longrightarrow HCO+OH$	13.3	0.0	1550	20
(57,58)	$CH_2O+HO_2 \longrightarrow HCO+H_2O_2$	12.3	0.0	5870	20
(59,60)	$HCO+OH \longrightarrow CO+H_2O$	13.5	0.0	0.0	20
(61,62)[b]	$HCO+M \longrightarrow H+CO+M$	16.0	–0.1	10228	20
(63,64)	$HCO+H \longrightarrow CO+H_2$	14.1	0.0	0.0	20
(65,66)	$HCO+O \longrightarrow CO+OH$	13.5	0.0	0.0	20
(67,68)	$HCO+HO_2 \longrightarrow CH_2O+O_2$	14.0	0.0	1510	19
(69,70)	$HCO+O_2 \longrightarrow CO+HO_2$	13.7	0.0	850	20
(71,72)[d]	$CO+OH \longrightarrow CO_2+H$	10.8	0.0	0.00091	20
(73,74)	$CO+HO_2 \longrightarrow CO_2+OH$	14.2	0.0	11900	20
(75,76)[c]	$CO+O+M \longrightarrow CO_2+M$	15.4	0.0	2184	39
(77,78)	$CO_2+O \longrightarrow CO+O_2$	13.2	0.0	26500	20
(79,80)	$H+O_2 \longrightarrow O+OH$	17.2	–0.9	8750	20
(81,82)	$H_2+O \longrightarrow H+OH$	4.0	2.8	2980	20
(83,84)	$H_2O+O \longrightarrow OH+OH$	9.7	1.3	8605	20
(85,86)	$H_2O+H \longrightarrow H_2+OH$	8.3	1.9	9265	20
(87,88)	$H_2O_2+OH \longrightarrow H_2O+HO_2$	12.2	0.0	160	20
(89,90)[b]	$H_2O+M \longrightarrow H+OH+M$	16.3	0.0	52900	20
(91,92)[c]	$H+O_2+M \longrightarrow HO_2+M$	18.8	–1.0	0.0	20
(93,94)	$HO_2+O \longrightarrow OH+O_2$	13.2	0.0	–200	20

Continued on next page.

Table II (cont.) Elementary Reaction Model for Methanol Oxidation in Supercritical Water

	Reaction	log A	Rate (mol/cm³.s) n	Ea/R,(K)	Ref.
(95,96)	$HO_2+H \longrightarrow OH+OH$	14.2	0.0	440	20
(97,98)	$HO_2+H \longrightarrow H_2+O_2$	13.8	0.0	1070	20
(99,100)	$HO_2+OH \longrightarrow H_2O+O_2$	16.2	–1.0	0.0	20
(101,102)	$H_2O_2+O_2 \longrightarrow HO_2+HO_2$	13.7	0.0	20000	20
(103,104)[b]	$H_2O_2+M \longrightarrow OH+OH+M$	17.9	0.0	22900	20
(105,106)	$H_2O_2+H \longrightarrow HO_2+H_2$	13.7	0.0	4000	20
(107,108)[c]	$O+H+M \longrightarrow OH+M$	18.7	–1.0	0.0	20
(109,110)[b]	$O_2+M \longrightarrow O+O+M$	18.3	–1.0	59380	20
(111,112)[b]	$H_2+M \longrightarrow H+H+M$	19.7	–1.4	52530	20

[a] Unimolecular Decomposition
[b] Bimolecular Reaction
[c] Termolecular Reaction
[d] Non–Arrhenius Rate form: $k = 10^{10.8}\exp(0.00091xT)$

routine LSODE (33) was used to integrate the 17 simultaneous rate equations and the Greens Function routine ODESSA (34) was used to obtain first order sensitivity coefficients.

In all cases the model predicted complete conversion of methanol to carbon monoxide and carbon dioxide. Concentration profiles for the major species are shown in Figure 2. We see that the oxidation of methanol is complete in less than 1 second. During this period the oxygen concentration drops rapidly and the concentration of carbon monoxide increases to a maximum. Subsequent oxidation of the carbon monoxide to carbon dioxide occurs at a much slower rate, evidenced by the gradual increase in CO_2 concentration and decrease in O_2 concentration. In addition, the maximum concentration of H_2 predicted by the model occurs at the same time as the maximum in carbon monoxide concentration. Thereafter, hydrogen oxidation to water occurs and the hydrogen concentration gradually decreases. The hydrogen predicted by the model is a consequence of the initial steps:

$$CH_3OH + H \longrightarrow CH_2OH + H_2 \quad (15)$$
$$CH_3OH + OH \longrightarrow CH_2OH + H_2O \quad (16)$$

followed by;

$$CH_2OH + M \longrightarrow CH_2O + H + M \quad (17)$$

where Equation 17 is the major route for the removal of CH_2OH in our system. Subsequent recombination of hydrogen atoms and reaction with water via:

$$H_2O + H \longrightarrow H_2 + OH \quad (18)$$

lead to hydrogen formation. Therefore, although hydrogen formation is predicted by the model, the source of the hydrogen is incorrect since we have attributed its formation to the water–gas shift reaction. This was confirmed by conducting simulations of previous carbon monoxide oxidation runs (8). Very little hydrogen was observed. The inability of current models to account quantitatively for hydrogen produced by the water–gas shift reaction is a major limitation.

In addition to production of carbon monoxide, carbon dioxide and hydrogen, the model predicts the formation of both methane and formaldehyde in small quantities. The concentrations predicted for methane and formaldehyde (both less than 1×10^{-9} mol/cm^3) are below the detectability limits of our analytical instruments and therefore their presence cannot be verified at present.

In order to determine the major routes of methanol removal, a sensitivity analysis was done. The abstraction reaction

$$CH_3OH + OH \longrightarrow CH_2OH + H_2O \quad (19)$$

provides the major route for removal of methanol. In addition, the methanol concentration is sensitive to the reactions:

$$H_2O + HO_2 \longrightarrow H_2O_2 + OH \quad (20)$$
$$HO_2 + OH \longrightarrow H_2O + O_2 \quad (21)$$

These reactions demonstrate the importance of the hydroxyl and hydroperoxy radicals in the mechanism. The HO_2 radical serves both to produce OH radicals through Equation 20 due to the high concentration of water, and to scavenge OH radicals through Equation 21.

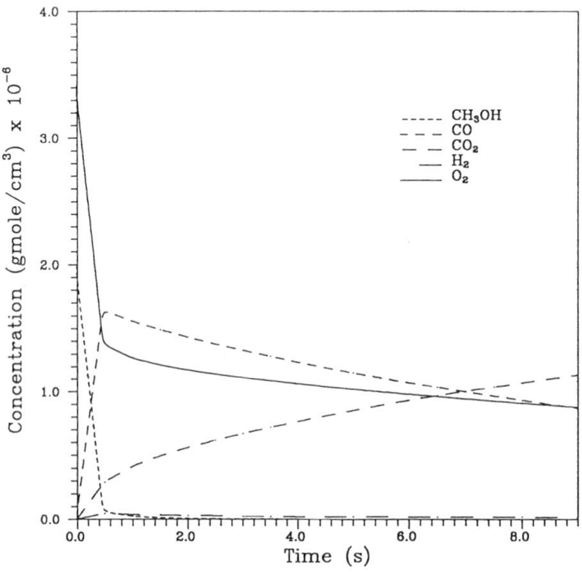

Figure 2. Concentration Profiles of Major Species. Methanol Oxidation Simulation. T=520°C.

Since a pyrolysis experiment was attempted, it was of interest to simulate a pyrolysis run. At 520°C, the mechanism predicted a conversion of 2.9%. At 550°C, the conversion increased to 13.8%. Although this is comparable to the pyrolysis run attempted at 550°C, (9% conversion), it must be recalled that we were unable to exclude oxygen completely from the reaction system and hence it was not a "true" pyrolysis run. In fact, simulation of our pyrolysis run with the initial concentrations of oxygen and methanol as estimated for that run resulted in a methanol conversion of 90%.

The inability of the current elementary reaction model to account even qualitatively for our experimental observations leads to speculation on the role played by water in the oxidation process. In addition to the possible formation of loosely bonded "cages" of water molecules around reactant molecules, the formation of solute–solvent complexes is likely. As the critical point is approached, radical recombinations (which proceed with virtually no activation energy) may become diffusion controlled, thereby lowering their rate over that predicted by gas–phase models. In addition, "caging" and the formation of solute–solvent complexes may lead to a change in activation energy for activated reactions. The activation energy will either be increased or decreased from its gas–phase value depending on the polarity of the transition state and the ability of supercritical water to stabilize the transition state molecules. Future modeling will attempt to include hydrogen–bonding and solvent–solute effects on reaction kinetics. In addition, future experimental work will examine the effect of pressure on oxidation kinetics in supercritical water, since the above–mentioned solvent effects are likely to change significantly with the density of the system.

Conclusions

The oxidation kinetics of methanol in supercritical water were measured over the temperature range 450°C to 550°C at 250 bar. The global oxidation process is highly activated, increasing from no conversion at 480°C to complete conversion at 550°C. In addition, the reaction kinetics are independent of oxygen concentration over the range of oxygen concentrations investigated, and are approximately first–order in methanol concentration. Although there is much evidence to suggest that the reaction mechanism is a free–radical one, a free–radical gas–phase elementary reaction mechanism modified for high pressure kinetic effects, is unable to predict the observed conversions. In addition, the model does not predict any hydrogen production from the water–gas shift reaction. It should be noted that the model represents a first–order approximation to the oxidation kinetics in supercritical water. Solvent effects, such as solute/solvent complexes and caging have not been included.

Acknowledgements

We acknowledge the partial support of NASA and the U.S.Department of Energy through the Los Alamos National Laboratory for this project.

Literature Cited

1. Modell,M.; Gaudet,G.G.; Simson,M.; Hong,G.T.;Bieman,K. Southwest Research Institute Eighth Annual EPA Research Symposium, Cincinnati, Ohio, March 9–10, 1982.
2. Staszak,C.N.; Malinowski,K.C.; Killilea,W.R. Environ.Prog. 1987, 6(2), 39–43.
3. Thomason,T.B.; Modell,M. Haz.Waste. 1984, 1, 453–467.

4. Timberlake,S.H.; Hong,G.T.;Simson,M.;Modell,M. SAE Technical Paper Series # 820872. 12th Int.Soc.Conf.Env.Sys. San Diego, CA July 19–21, 1982.
5. Price,C.M. S.M.Thesis, Massachusetts Institute of Technology, Cambridge, MA, 1981.
6. Cunningham,V.L.; Burk,P.L.; Johnston,J.B.; Hannah,R.E. Paper 50c, AIChE Summer National Meeting, Boston, MA, August 26, 1986.
7. Wightman,T.J. S.M.Thesis, University of California, Berkely, CA 1981
8. Helling,R.K.; Tester,J.W. Energy & Fuels, 1987, 1, 417–423.
9. Helling,R.K.; Tester,J.W. Environ.Sci.Tech., 1988, 22(11), 1319–1324.
10. Webley,P.A.; Tester,J.W. SAE Technical Paper Series #881039, 18th Int.Soc.Conf.Environ.Sys. San Francisco, CA 1988.
11. Héberger,K.,; Kemény,S.; Vodiczy,T. Int.J.Chem.Kinet. 1987, 19, 171–181.
12. Aronowitz,D.; Santoro,R.J.; Dryer,F.L.; Glassman,I. 17th Int. Symp. Comb. 1979, 633–644.
13. Bell,K.M.; Tipper,C.F.H. Proc.Roy.Soc.London. 1956, 238(A), 256–268.
14. Bondarenko,G.V.; Gorbatyi,Yu.E.; Edel'shtein,V.M. Dokl.Akad.Nauk.SSSR. 1974, 14(2), 30–33.
15. Franck,E.U.; Roth,K. Disc.Farad.Soc. 1967, 43, 108–114.
16. Martynova,O.I. Jones, 1976, 131–138.
17. Antal,M.J.; Brittain,A.; DeAlmeida,C.; Ramayya,S.; Roy,J.C. in Supercritical Fluids; Squires,T.G.; Paulaitis,M.E., Eds.; ACS Symposium Series No. 329; American Institute of Chemical Engineers: Washington, DC, 1988; 77–86.
18. Aronowitz,D.; Naegli,D.W.; Glassman,I. J.Phys.Chem. 1977, 81(25), 2555–2559.
19. Westbrook,C.K.; Dryer,F.L. Comb.Sci.Tech. 1979, 20, 125–140.
20. Tsang,W.; Hampson,R.F. J.Phys.Chem.Ref.Data. 1986, 15(3), 1087–1276.
21. Tsang,W. J.Phys.Chem.Ref.Data. 1987, 16(3), 471–508.
22. Troe,J. J.Phys.Chem. 1979, 83(1), 114–126.
23. Gilbert,R.G. Program FALLOFF, Quantum Chemistry Program Exchange, Section IX: Chemical Reactions, #460., 1988
24. Shandross,R., Howard,J. Modification of FALLOFF, Massachusetts Institute of Technology, Cambridge, MA 02139, 1989.
25. Benson,S.W. Thermochemical Kinetics; Wiley–Interscience: New York, 1976.
26. Simmons,G.M.; Mason,D.M. Chem.Eng.Sci. 1972, 27, 89–108.
27. Christoforakos,M.; Franck,E.U. Ber.Bunsenges.Phys.Chem. 1986, 90, 780–789.
28. Japas,M.L., Franck,E.U. Ber.Bunsenges.Phys.Chem. 1985, 89, 1268.
29. Japas,M.L., Franck,E.U. Ber.Bunsenges.Phys.Chem. 1985, 89, 793.
30. Todheide,K., Franck,E.U. Z.Phys.Chem.Neue.Folge. 1963, 37, 1963.
31. Seward,T.M., Franck,E.U. Ber.Bunsenges.Phys.Chem. 1981, 85, 2.
32. Keenan,J.H.; Keyes,F.G.; Hill,P.G.; Moore,J.G. Steam Tables. Thermodynamic Properties of Water Including Vapor, Liquid and Solid Phases; Wiley–Interscience, 1978.
33. Leis, J.R. Sc.D Thesis, Massachusetts Institute of Technology, Cambridge, MA 02139, 1986.
34. Leis,J.R.; Kramer,M.A. to be published in ACM Trans.Math.Software, 1988.
35. Larson,C.W.; Patrick,R.; Golden,D.M. Comb. & Flame. 1984, 58, 229–237.
36. Spindler,K.; Wagner,H.Gg. Ber.Bunsenges.Phys.Chem. 1982, 86, 2–13.

37. Greenhill,P.G.; O'Grady,B.V.; Gilbert,R.G. Aust.J.Chem. 1986, 39, 1929–1942.
38. Greenhill,P.G.; O'Grady,B.V. Aust.J.Chem. 1986, 39, 1775–1787.
39. Baulch,D.L.; Drysdale,D.D.; Horne,D.G.; Lloyd,A.C. Evaluated Kinetic Data for High Temperature Reactions, Vol.1: Homogeneous Gas Phase Reactions of the H_2–O_2 System, Butterworths, London, 1972.

RECEIVED May 1, 1989

Chapter 18

Thermodynamic Analysis of Corrosion of Iron Alloys in Supercritical Water

Shaoping Huang[1], Kirk Daehling[1], Thomas E. Carleson[1], Pat Taylor[1], Chien Wai[1], and Alan Propp[2]

[1]University of Idaho, Moscow, ID 83843
[2]EG&G, Idaho, Inc., Idaho Falls, ID 83415

A thermodynamic analysis was conducted for corrosion of iron alloys in supercritical water. A general method was used for calculation of chemical potentials at elevated conditions. The calculation procedure was used to develop a computer program for display of pH-potential diagrams (Pourbaix diagrams). A thermodynamic analysis of the iron/water system indicates that hematite (Fe_2O_3) is stable in water at its critical pressure and temperature. At the same conditions, the analysis indicates that the passivation effect of chromium is lost. For experimental evaluations of the predictions, see the next paper in the symposium proceedings.

High temperature and high pressure processing of materials often involves the use of supercritical fluids. Corrosion studies are quite essential for evaluation of the equipment in supercritical fluid operations. Previous electrochemical measurements for alloys in supercritical fluids are rare (1-3). The reported measurements (3) show that passivation of iron alloys is different at supercritical conditions compared to ambient conditions. The study of the electrochemistry of iron alloys can lead to control of corrosion of equipment utilizing the alloys. Thermodynamic analysis provides the information about stable species, i.e. corrosion products under given temperatures and pressures.

Thermodynamic property data of chemical species at high temperatures and high pressures are rare. Fortunately, extrapolation of thermodynamic properties into elevated conditions provides a way to conduct the thermodynamic analysis semi-quantitatively or qualitatively. A detailed review of the extrapolation methods is available (4).

The pH-potential diagram or Pourbaix diagram, the graphical presentation of a stable species within a pH-potential region, is a valuable tool for analyzing the electrochemical equilibria in aqueous solutions. With a diagram, the reaction products can be determined under given conditions. If a redox reaction product of a

0097-6156/89/0406-0276$06.00/0

metal is a soluble ionic species, its solid oxide will not be expected as a stable species, therefore passivation of this metal in such conditions is impossible. The diagrams for various metals and their oxides at room temperature are available (5). The equilibrium calculation and the graphical presentation has been developed into a FORTRAN IV (G level) 360/91 IBM computer program (6). The program is capable of calculating and plotting pH-potential diagrams for systems consisting of two elements, one metal and one nonmetal, in the presence of water at 25°C and at specific input activities of the species. The diagram is in the form of an array of symbols representing the fields of stable species.

For thermodynamic analysis of the corrosion of iron alloys in supercritical water, the above computer program was modified based upon standard thermodynamic property extrapolation methods.

THEORETICAL AND EMPIRICAL APPROACH

General Principles. For a given chemical reaction at equilibrium, a stoichiometric balance is:

$$n_I M(I) + n_{II} M(II) + n_O O_2 + n_H H^+ + n_W H_2O + n_L L = 0 \quad (1)$$

where: L represents the nonmetal species with a charge i; M(I) and M(II) represent species containing the given metal element M in different oxidation states, I and II; and a, b, ...h are the stoichiometric coefficients of the corresponding species. If nitrogen is assumed to be an inert gas, this equation describes most systems consisting of air and water. At constant temperature and pressure, the equilibrium condition is,

$$\Delta G = \Delta G° + RT \ln(a_I^{n_I}\ a_{II}^{n_{II}}\ f_O^{n_O}\ a_H^{n_H}\ a_W^{n_W}\ a_L^{n_L}) = 0 \quad (2)$$

where, R is the ideal gas constant, T is the absolute temperature, f_O is the fugacity (approximately equal to the partial pressure) of oxygen above the solution, a_i is the activity of species considered, and $\Delta G°$ is the standard Gibbs free energy change for the reaction, which is given by the following equation,

$$\Delta G° = n_I \mu°_I + n_{II} \mu°_{II} + n_O \mu°_O + n_H \mu°_H + n_W \mu°_W + n_L \mu°_L \quad (3)$$

where $\mu°_i$ is the standard chemical potential of corresponding species at the given temperature and pressure.

The method for extrapolation of chemical potentials to elevated temperatures and pressures is based on the Gibbs-Duhem equation. The temperature and pressure dependency of the chemical potential of a species i can be expressed as (7, pp.144),

$$\mu_i(T,P) = \mu_i°(T_o,P_o) - s_i°(T_o,P_o)\,(T-T_o) + v_i°(T_o,P_o)\,(P-P_o)$$

$$\int_{To}^{T} [\int_{To}^{T} (\partial s_i(T,P_o)/\partial T)_p \, dT$$

$$- \int_{Po}^{P} \alpha_{pi}(T,P)\, v_i(T,P)\, dP]_{Po}\, dT$$

$$+ \int_{Po}^{P} [\int_{To}^{T} \alpha_{pi}(T,P_o)\, v_i(T,P_o)\, dT$$

$$- \int_{Po}^{P} \kappa_{Ti}(T,P)\, v_i(T,P)\, dP]\, dP \qquad (4)$$

where: the superscript ° denotes the thermodynamic properties at reference state (e.g., at T_o and P_o); μ_i is the chemical potential; s_i is the entropy; v_i is the specific volume; α_{pi} is the coefficient of thermal expansion; κ_{Ti} is the isothermal compressibility; T is the temperature of the system; and P is the pressure.

In equation 4, $\mu_i°(T_o,P_o)$ characterizes the given species at the starting point, i.e. at the reference point. The partial molar entropy, $s_i°(T_o,P_o)$, is the slope of the line representing $\mu_i(T,P)$ versus T while P is a constant, and the partial molar volume, $v_i°(T_o,P_o)$, is the slope of the line representing $\mu_i(T,P)$ versus P while T is constant. By integration of the terms, $[\partial s_i(T,P_o)/\partial T]_p$ or $C_p(T,P_o)/T$, $\alpha_{pi}(T,P_o)\, v_i(T,P_o)$ and $\kappa_{Ti}(T,P)\, v_i(T,P)$, one can determine these partial molar properties.

From equation 2, 3 and 4, the reaction equilibrium status can be calculated. For the detailed procedure of calculation and determination of the predominant species in a pH-potential region, see references 5 and 6.

The first three terms of equation 4 can be easily calculated from the available data (8). The fifth term is zero. The temperature and pressure dependency of the coefficient of thermal expansion, α_{pi}, and the isothermal compressibility, κ_{Ti}, up to elevated conditions are seldom available. Usually, their contribution to the chemical potential value is small, except for water near its critical point. Therefore, it is assumed that the sixth and seventh terms of equation 4 for all species other than water are negligible. In other words, the volumes of solid species are regarded as constants and the volumes of species in aqueous solutions are considered to be the same as water. This assumption is adequate because the concentrations of the corrosion product in the solution are usually very low.

For the fourth term, the correlations for the entropies of species and for the specific volume of water are presented below.

Entropies of Species. The entropy of a solid species at temperature T can be obtained from the following equation.

$$s_i(T,P_o) = s_i°(T_o,P_o) + a\ln T + 10^{-3}bT - 5x10^{-7}cT^2 - 5x10^4 d/T^2 - B' \qquad (5)$$

This equation is derived from an empirical expression for heat capacity, where a, b, c, d and B' are constants which can be found in most of the handbooks of chemistry and physics (e.g. 8). For a dissolved species, however, very few experimental data are available for elevated temperatures and pressures.

For estimation of thermodynamic properties of dissolved species, one can use the Entropy Correspondence Principle (9), where the entropy of an ion at a given temperature is regarded as a function of the charge, the dielectric constant, mass, radius, and other variables. The function depends mainly upon the choice of the standard state, solvent, and temperature. The temperature dependency of entropy was derived based on the above principles and experimental data. By conducting the a square regression on Criss-Cobble's data (9), we obtained the following equation for calculating the entropies of species in aqueous solution.

$$s_i^*(T,P_o) = s_i^{\circ *}(T_o,P_o) + [A + B\, s_i^{\circ *}(T_o,P_o)](T-T_o) \qquad (6)$$

Where the superscript * denotes the entropy value relative to a standard state (the "absolute entropy") and A and B are empirical constants. The new defined entropy value is different from the conventional thermodynamic scale value where S_{H^+} is zero at all temperatures. The absolute entropy (S_i^*) and conventional entropy (S_i) of a species i are related by:

$$S_i(T) = S_i^*(T) - Z_i\, S_{H^+}^*(T) \qquad (7)$$

where Z_i is the charge of the ionic species, i.

The values are:

for simple cations, except H^+,

$$A = 0.132723 \text{ cal/molK}^2 \text{ and } B = -0.00164\text{K}^{-1};$$

for simple anions not containing oxygen,

$$A = -0.17473 \text{ cal/molK}^2 \text{ and } B = -0.00009\text{K}^{-1};$$

for oxygenated anions,

$$A = -0.38445 \text{ cal/molK}^2 \text{ and } B = 0.005852\text{K}^{-1};$$

for acid oxygenated anions, $XO_n(OH)_1^{m-}$,

$$A = -0.40004 \text{ cal/molK}^2 \text{ and } B = 0.011233\text{K}^{-1};$$

and for H^+,

$$s_{H^+}^*(T,P_o) = -33.5042 + 0.09462T \qquad (8)$$

where $s_{H^+}^*$ has the unit of cal/mol-K and T is in degrees Kelvin. The temperature range of the data for the above equations is from 298K to 573K. The deviation between the calculated entropies and Criss-Cobble's data is smaller than 2%.

Specific Volume of Water. The data for the specific volume of water are available over a wide range of temperatures and pressures. Normally high degree polynomials (usually at least eight degrees) are used as the state equations of both steam and liquid water (10, 11).

A new model is suggested here for the volume explicit state equation of water from ambient to supercritical conditions. The model, consisting of two parts, is shown below,

$$V = (1-\delta)V_l + \delta[1+\delta(V_g-1)] \quad (9)$$

where V is the specific volume of water, δ is a transition function, V_l is the state equation for liquid water, and V_g is the state equation for both steam and supercritical water. From ambient to critical conditions, the state equation for liquid water was found to be:

$$V_l = 1.0250-7.905x10^{-4}t+6.10x10^{-6}t^2+2.032x10^{-2}t/P \quad (10)$$

For temperatures up to 600 °C and pressures up to 330 atm, the state equation for steam and supercritical water, V_g was obtained as:

$$V_g = -5.450 + 7.604t/P \quad (11)$$

where t in °C, P is in atm, and V_l and V_g both have unit of liters/kilogram. The transition function δ changes at the boiling point from zero to one. This function was found to be

$$\delta = \frac{1}{1 + \exp[-100(t/t_b-1)]} \quad (12)$$

where t_b, the boiling point, is a function of pressure. Using the Antoine equation, t_b was determined as

$$t_b = \frac{4365.46}{255.291 - \ln P} - 12.3096 \quad (13)$$

A plot of the predicted volumes at 97, 194 and 291 atm versus temperature is shown in Figure 1. The experimental values are also plotted for comparison. For temperatures lower than critical, the error of the model (error = predicted/experimental - 1) is smaller than ±1.5%. For temperature larger than critical and lower than 600 °C, the error is not larger than ±10%.

RESULTS AND DISCUSSION

pH-Potential Diagram Program. The pH-potential diagram program was modified according to the above principles. The resulting program is capable of calculating the electrochemical equilibria and plotting Pourbaix diagrams at given temperatures and pressures. The input data include temperature, pressure, and the following information about each species: activity, standard chemical potential, standard entropy, molecular weight, density, phase state, molecular constituents, charge, and the thermodynamic constants - a, b, c, d, and B′ as mentioned above.

An additional routine was developed for the extrapolation of pH

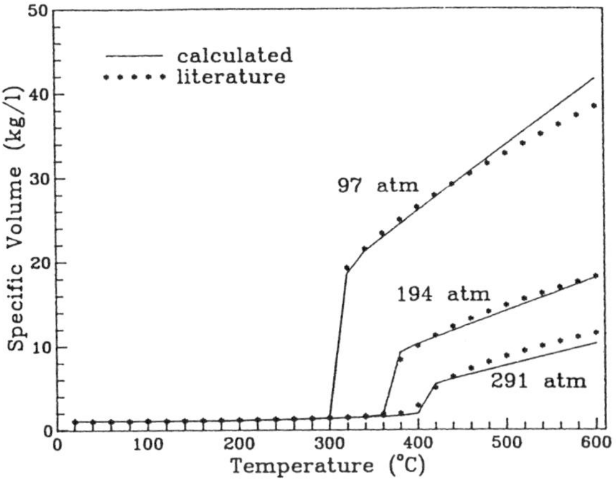

FIGURE 1 TEMPERATURE AND PRESSURE EFFECTS ON SPECIFIC VOLUME OF WATER

values from ambient conditions. Given the desired temperature, pressure and pH value of the solution, this routine provides the corresponding pH value and the neutral pH value at the specified condition, as well as the pH-potential equilibrium equations for the following two half redox reactions of water at given conditions.

$$H_2 = 2H^+ + 2e \qquad (14)$$

$$2H_2O = O_2 + 4H^+ + 4e \qquad (15)$$

This calculation relies on the prediction of the dissociation constant of water, K_w, at elevated temperatures. The prediction involves the dielectric constant of water and a power series consistent with non-electrostatic interactions in the absence of a dielectric medium (12). The dissociation constant of water was calculated based upon the assumption that the entropy of the dissociation of water is the sum of the contributions of the electrostatic and nonelectrostatic interactions. The deviation between the predicted and the experimental (12) values of pK_w is smaller than 0.01.

Diagrams for Iron and Chromium. Since chromium is a major component in iron alloys, the pH-potential diagrams for both iron and chromium were determined. Figure 2 and 3 are the diagrams for iron and chromium in water at its critical point. For comparison, Figure 4 and 5 show corresponding diagrams at ambient conditions, identical to those generated by Pourbaix et al. (5). The Y-axis values of potential are relative the standard hydrogen electrode (SHE). All species are at unit molarity. In the calculations, the fourth term of equation 4 was neglected, since the contribution of this term to the total value of chemical potential is only about 10% for anionic species containing oxygen, while for other species it is less significant. In the diagrams, a solid line indicates the border between two species. The vertical dashed line indicates the neutral pH value of water at the given temperature and pressure. The area between the two diagonal dashed lines indicates the stable region of water - the upper one where water is oxidized to oxygen and the lower one where water is reduced to hydroxide.

Passivation of iron under critical conditions is predicted. Hematite (Fe_2O_3) may still be the main corrosion product in the neutral water pH (pH 7.2) region, but the passivation potential range is narrower and shifts to negative potentials, compared with regions on the diagram for ambient conditions. For chromium, no solid chromium oxide stable species is predicted within the stable region of neutral water. This indicates chromium oxidation without any passivation oxide film formation.

Corrosion of Iron Alloys. Other recent work (3) presents experimentally measured values of the open circuit potentials of iron alloys in water at ambient and supercritical conditions. For iron and its alloys (1080 carbon steel, 304 stainless steel and 316 stainless steel), the open circuit potentials varied from -0.112 to +0.055 volt in water at its critical point, and varied from -0.138 to -0.060 volt in water at ambient conditions. These values were

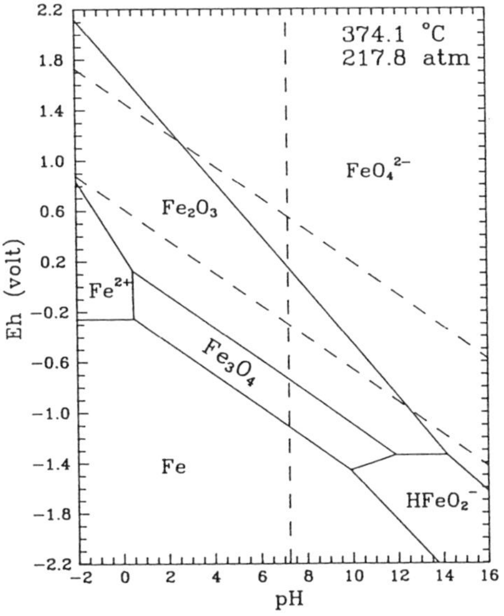

FIGURE 2 pH-POTENTIAL DIAGRAM FOR Fe-H_2O SYSTEM AT CRITICAL CONDITION

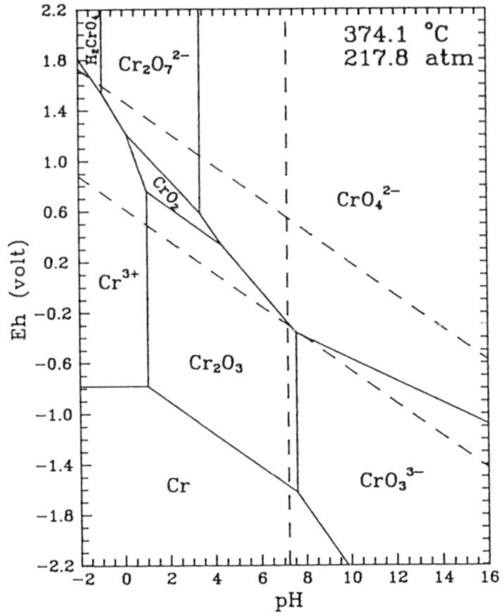

FIGURE 3 pH-POTENTIAL DIAGRAM FOR Cr-H_2O SYSTEM AT CRITICAL CONDITION

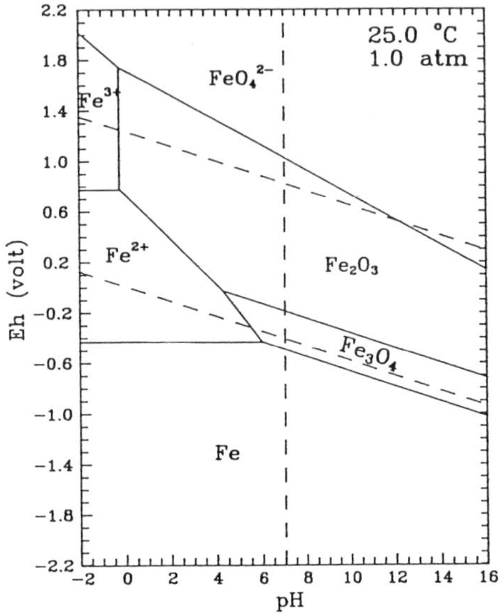

FIGURE 4 pH-POTENTIAL DIAGRAM FOR Fe-H_2O SYSTEM AT AMBIENT CONDITION

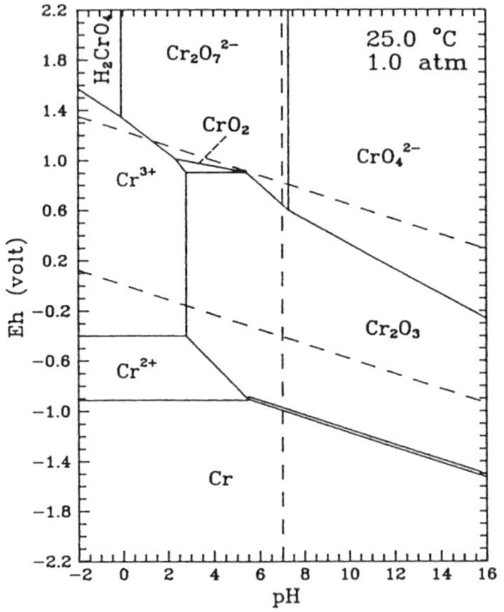

FIGURE 5 pH-POTENTIAL DIAGRAM FOR Cr-H_2O SYSTEM AT AMBIENT CONDITION

measured versus a silver-silver chloride/0.1M KCl reference electrode. Potentials versus a standard hydrogen electrode were calculated from those versus the silver-silver chloride electrode. The calculated potentials are 0.0 volt(SHE) and -0.1 volt(SHE) at critical and ambient conditions respectively (13). As mentioned, the neutral pH value is 7.0 at ambient and 7.2 at critical point. Locating the pH and open circuit potential values in the Pourbaix diagrams shown in Figures 2 to 5, Cr_2O_3, Fe_2O_3 (hematite) and Fe_3O_4 (magnetite) are the oxidation products at ambient conditions, while CrO_4^{2-} and Fe_2O_3 may be the corrosion products at critical conditions. It seems that chromium may join in the formation of the passivation film for iron alloys at ambient conditions, but at supercritical conditions, this passivation is lost due to the formation of a soluble oxidation product of chromium. This result is consistent with the phenomenon observed in the polarization analysis of pure iron and 304 stainless steel (3) where the exchange or corrosion current density of 304 stainless steel was found to be three times smaller than that of pure iron at ambient conditions, while they were comparable at supercritical conditions. Also, the measured open circuit potential falls inside the calculated stable region of water. In other words, the electrochemical measurements are consistent with the thermodynamic predictions.

CONCLUSIONS

A computer program was developed for electrochemical equilibria calculations and graphical pH-potential diagram presentation of a one-metal/one-nonmetal/water system. The program can be used for temperatures and pressures exceeding the supercritical point of water. The calculations show that hematite (Fe_2O_3) is the oxidation product of iron in supercritical water, and the oxidation product of chromium in supercritical water is an ionic species, CrO_4^{2-}. Passivation effect of chromium is lost in supercritical water.
A general extrapolation method of entropies and other related thermodynamic properties to elevated temperatures was developed from Criss-Cobble's data and models. The method was based upon the Entropy Correspondence Principle.

A simple volume explicit state equation of water was expressed as one equation for all phases: liquid, steam and supercritical water. This model used a transition function to switch equations at a phase change.

Thermodynamic predictions were consistent with experimentally measured corrosion rates and open circuit potentials. The results indicate enhanced corrosion of stainless alloys containing chromium may be expected in supercritical water. These corrosion rates appear comparable to those for mild steel or iron.

ACKNOWLEDGEMENT

This project is sponsored by EG&G Idaho Company under a research contract from the Bureau of Mines.

LITERATURE CITED

1. Flarsheim, W. M.; Tsou, Y.; Johnston, K. P.; Bard, A. J. J. of Phys. Chem. 1986, 90, 3857.
2. McDonald, A. C.; Fan, F. F.; Bard, A. J. J. of Phys. Chem. 1986, 90, 196.
3. Huang, S.; Daehling, K.; Carleson, T. E.; Taylor, P.; Wai, C.; Abdel-latif, M.; Propp, A. University of Idaho, "Electrochemical Measurements of Corrosion of Iron Alloys in Supercritical Water", next paper in this symposium.
4. Duby, P. In High Temperature Corrosion: March 2-6, 1981, San Diego, California; Rapp, R. A., Ed.; National Association of Corrosion Engineers: International Corrosion Conference Series, NACE-6, 1983; p. 353.
5. Pourbaix, M. Atlas of Electrochemical Equilibria in Aqueous Solutions; Pergamon Press: Oxford, Great Britain, 1966.
6. Verhulst, D.; Duby, P. The Thermodynamic Properties of Aqueous Inorganic Copper Systems; INCRA Monograph IV, The Metallurgy of Copper, 1985, p. 105.
7. Modell, M.; Reid, R. C. Thermodynamics and Its Applications 2nd edition, Prentice-Hall: New Jersey, 1983.
8. Weast, R. C.; Astle, M. J., Eds. CRC Handbook of Chemistry and Physics, 62nd edition CRC Press Inc., 1981.
9. Criss, C. M.; Cobble, J. W. J. Am. Chem. Soc. 1964, 71, 5385-5393.
10. Burham, C. W.; Holloway, J. R.; Davis, N. F. Am. J. Sci. 1969, 267(A), 70.
11. Pistorius, C. F. W. T.; Sharp, W. T. Am. J. Sci. 1960, 258, 757.
12. Helgeson, H. J. of Phys. Chem. 1967, 71, 3121.
13. Huang, S.; Daehling, K.; Carleson, T.E.; Propp, A. "Construction and Calibration of an Internal Silver/Silver Chloride Reference Electrode for Supercritical Fluid Studies", to be published.

RECEIVED May 2, 1989

Chapter 19

Electrochemical Measurements of Corrosion of Iron Alloys in Supercritical Water

Shaoping Huang[1], Kirk Daehling[1], Thomas E. Carleson[1], Masud Abdel-Latif[1], Pat Taylor[1], Chien Wai[1], and Alan Propp[2]

[1]University of Idaho, Moscow, ID 83843
[2]EG&G, Idaho, Inc., Idaho Falls, ID 83415

Electrochemical potentiostat measurements have been performed for the corrosion of iron, carbon steel, and stainless steel alloys in supercritical water. The open circuit potential, the exchange or corrosion current density, and the transfer coefficients were determined for pressures and temperatures from ambient to supercritical water conditions. Corrosion current densities increased exponentially with temperature up to the critical point and then decreased with temperature above the critical point. A semi-empirical model is proposed for describing this phenomenon. Although the current density of iron exceeded that of 304 stainless steel by a factor of three at ambient conditions, the two were comparable at supercritical water conditions. The transfer coefficients did not vary with temperature and pressure while the open circuit potential relative to a silver-silver chloride electrode exhibited complicated behavior.

When potassium iodide was dissolved in supercritical ethanol and then precipitated upon reducing the pressure, the reversible solvating power of supercritical fluids (SCF) was demonstrated (1). Since the last decade, this technology has been widely investigated (2,3). The electrochemical study of metals in supercritical water conditions is essential for an understanding of corrosion behavior. These studies are important for the evaluation of boiler tube failures in nuclear and conventional power plants, metallurgical processing, and geochemistry of the earth crust.

There are few published electrochemical studies of supercritical water (temperatures and pressures above 374 ^{o}C and 218 atm). Some of the data pertain to the power generation industry; for example, a 1975 report summarizes some industrial experience with corrosion of steam generator tubing in pressurized water reactors (4). Most of the other literature references concern

0097–6156/89/0406–0287$06.00/0

research results. A 1986 report presented, for temperatures up to 300 °C, the standard potential of the half reactions of copper(II) system in Na_2SO_4 and KCl solutions, as well as the diffusion coefficient of Cu(II) (5). W. M. Flarsheim et al. (6) conducted electrochemical measurements for $NaHSO_4$, KBr, KI, and hydroquinone solutions from 25 °C to 350 °C. J. B. Silver, et al. (7) studied the corrosion and the deposition of Magnetite on two kinds of steels (2.25Cr-1.0Mo and 9.0Cr-0.1Mo) in high temperature water under heat fluxes up to 860 kw/m^2. For high purity water, polarization curves were measured at room temperature and 290 °C by M. Hishida, et al. (8). They studied AISI 304 S.S., carbon steel (STS 42), Ni, Cr, and Co metals. They used a dynamic IR compensated potentiostat which suppressed the effect of solution resistance between the working and reference electrodes. The anodic polarization curves showed no active current peak except for carbon steel. Other electrochemical studies conducted near the critical point of water include those of Iding (9) and Carter (10).

Most of the data available in the literature are for subcritical conditions. Corrosion studies of iron alloys in supercritical water have not been reported. For supercritical fluid extraction and corrosion studies, a supercritical fluid reactor system for temperatures up to 530 °C and pressures up to 300 atm was constructed. This system was used to determine the electrochemical behavior of type 304 stainless steel (304 S.S.), 316 S.S., 1080 carbon steel (1080 C.S.), and pure iron in supercritical water.

EXPERIMENTAL

Construction of Apparatus. The schematic of the apparatus for supercritical corrosion studies is shown in Figure 1. The important components include: a type 396-89 Simplex Minipump which can accurately meter (between 46 and 460 ml/hr) a wide variety of solvents at pressures up to 6000 psi (about 400 atm); an EG&G Model 362 Scanning Potentiostat; the electrochemical cell; an IBM PC computer with interface hardware for electrochemical potential and current, temperature, and pressure measurement and control; and a 316 stainless steel reactor, which holds the supercritical fluid for the measurements. The alloy was selected for excellent corrosion resistance properties and relatively low cost when compared with other exotic alloys such as Hastelloy C.

The reference electrode is an internal silver-silver chloride type. It was based on a design of A. K. Agrawal, et al. (11). For supercritical water studies, the electrode body was made from 902 precision machinable ceramic (Cotronics Corporation Brooklyn, NY). A type No. 29 Sauereisen low expansion cement was used for connection of the electrode body and the silver-silver chloride wire. The concentration of the KCl solution in the electrode for this study is 0.1 mol/l. The iron alloy working electrode dimensions were 20mm x 5mm x 1.5mm. The body of the reactor served as the auxiliary electrode.

Measurements. Polarization curves were obtained by remote operation of the supercritical system by the computer and potentiostat. The effect of the IR drop in pure water was determined by comparing the

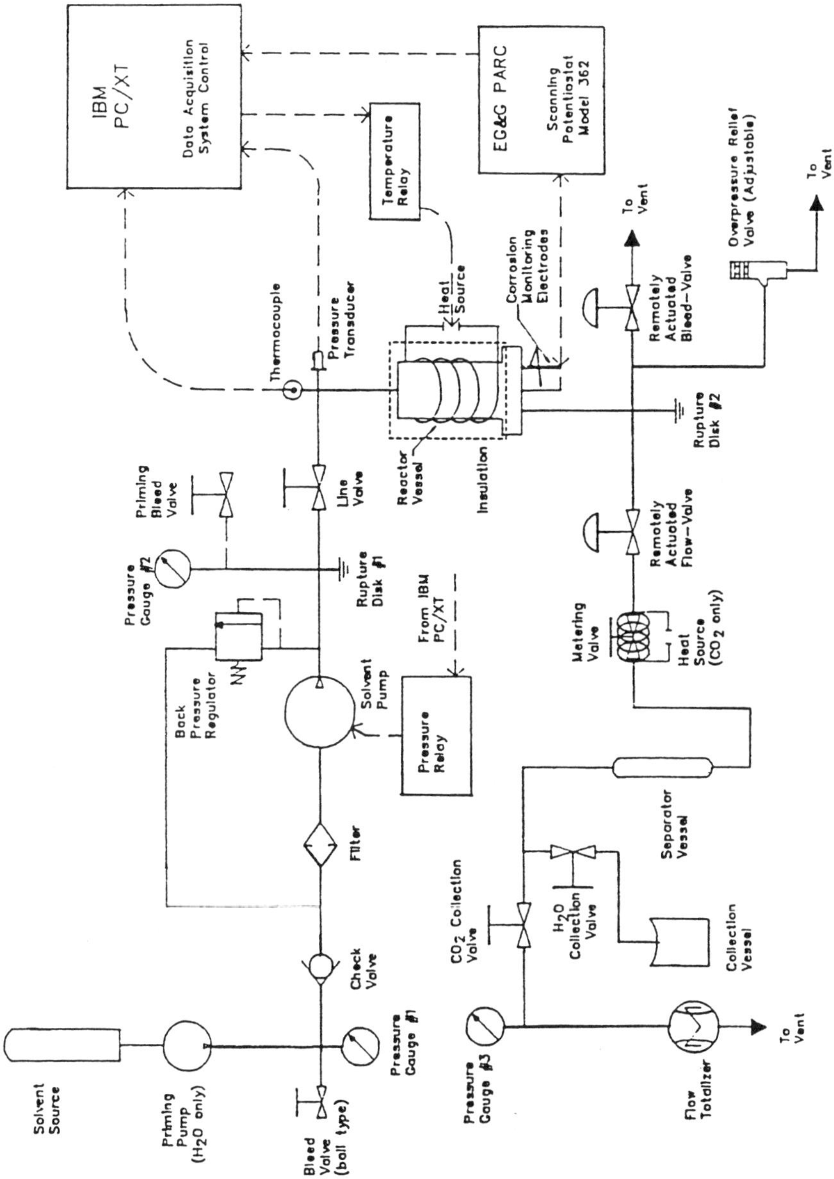

FIGURE 1 APPARATUS SCHEMATIC FOR CORROSION STUDY

results obtained with a 0.005 mol/l Na_2SO_4 solution to those of pure water at the same supercritical condition. The polarization curves are shown in Figure 2. The polarization behavior is similar except that the current density for pure water is less than that for the sodium sulfate solution and the curve for pure water has a larger potential range. Nevertheless, the effect of the IR drop should be quite small when the current density is close to zero. For potentials within +/- 0.2 mV of the open circuit potential, the current density is low and the two polarization curves are close to each other.

Although the potentiostat has an adjustable IR compensation circuit, this was not used. When it was used, the scan rate varied. Compensation did not seem to have much effect. Some researchers have developed their own IR compensation circuit for the high resistance solution (8), some added a supporting electrolyte (12,13), and some simply ignored it (11). Since the behavior of electrolytes under supercritical conditions is not well known, no supporting electrolyte was added to eliminate the IR drop in this experiment. Furthermore, the IR circuit was not used for this study either, since the desired electrochemical data, such as exchange current density, open circuit potential, and transfer coefficients, can be obtained from the polarization curves without IR compensation.

The polarization curves showed that a large scan rate affects the polarization curves. This phenomenon can be illustrated by the capacitance effect of the electric double layer on the working electrode surface. Based upon the model of M. Hishida, et al. (8), assume C_{dl} is the capacitance of electric double layer, R_f is the polarization resistance, and R_{sol} is the solution resistance. A characteristic relaxation time can be defined as:

$$\theta = \frac{C_{dl}\, R_f\, R_{sol}}{R_f + R_{sol}} \qquad (1)$$

If the time increment between each deviation of the applied potential is much larger than θ, i. e. if the scan rate is slow enough, the output current density of the polarization curves will not be affected. The experimental results showed this behavior for scan rates smaller than 10 mV/sec. The scan rate applied for this study was 5 mV/sec.

Most of the experimental work concerned ASAI 304 Stainless Steel. Other materials such as 316 S.S., pure iron, and 1080 C.S. were also studied. The water used in this study was glass distilled and nitrogen deaerated (oxygen concentration less than 0.5 ppm).

RESULTS AND DISCUSSION

Theoretical Background. Since the concentration of the oxygen in the studied water is low, a pure water/iron alloy system is assumed in the polarization data analysis. In this case, for a multi-step reaction,

$$M \longrightarrow M_{oxn} + ne \qquad (2)$$

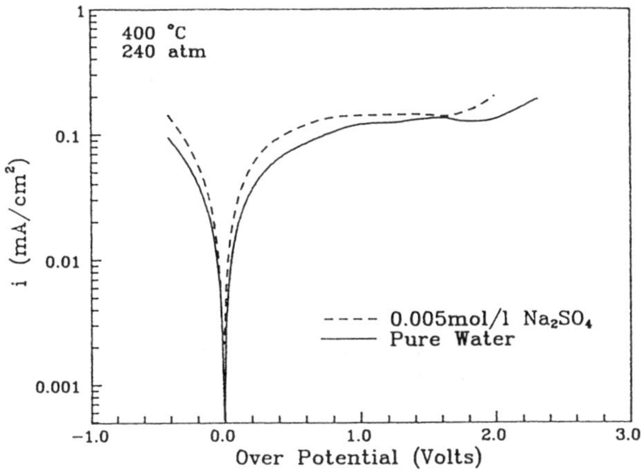

FIGURE 2 POLARIZATION CURVES FOR PURE WATER AND 0.005mol/l Na_2SO_4

the total current density is found to be (16),

$$i = i_o\{\exp[(\frac{n-\tau}{\epsilon} - r\beta)\frac{F\phi}{RT}] - \exp[-(\frac{\tau}{\epsilon} + r\beta)\frac{F\phi}{RT}]\} \tag{3}$$

where,

n is the total number of charges transferred;
τ is the reaction steps before the rate determining step;
ϵ is the stoichiometric number of the rate determining reaction;
β is the symmetry factor;
r is the number of electrons transferred in a rate determining step;
F is the Faraday constant;
R is the ideal gas constant;
T is the absolute temperature of the system;
ϕ is the over potential;
i_o is the exchange current density.

Usually, the terms

$$\alpha_a = \frac{n - \tau}{\epsilon} - r\beta \tag{4a}$$

and

$$\alpha_c = \frac{\tau}{\epsilon} + r\beta \tag{4b}$$

are called transfer coefficients.

From the output of the potentiostat, the total current density, i, can be obtained. The over potential, ϕ, can be calculated by the following equation,

$$\phi = E - (E_o + E_{IR}) \tag{5}$$

where,

E is the potential applied by the potentiostat,
E_o is the open circuit potential vs. the same reference electrode of the applied potential,
E_{IR} is the potential drop due to the resistance of the solution, IR drop.

Applying Ohm's law, one can express the IR drop as,

$$E_{IR} = i\, R_{sol} \tag{6}$$

where the R_{sol} is the resistance of the solution, assumed to be constant. Therefore, the over potential can be calculated based upon the data from the potentiostat.

Polarization Analysis. Polarization curves were obtained for pure iron, 1080 carbon steel, ASAI 316 type and 304 type stainless steel near the critical point of water. Passivation was indicated based upon the curves for 304 S.S. and 316 S.S.. The electrodic model discussed above was used to fit the polarization data from the potentiostat. Based on the discussion about the IR drop in the experimental section, the potential range used for fitting the model was selected as $E_o \pm 0.2$ volts. The false position method, also called DUD method (14), was used to perform the nonlinear regression for searching the parameters. R_{sol} was not significant within the selected potential range. This result verified the assumption that the IR drop does not have a significant effect on the polarization curves when the over potential is small.

Table I and Table II show the respective values of the exchange current densities and open circuit potentials for pure iron, 304 S.S. (15), 316 S.S., and 1080 C.S., near the critical point of water. The values at ambient conditions are also listed. Reproducibility of the exchange current density was found to be +/- 2 microamperes/square centimeter and +/- 25 millivolts for the open circuit potential for values near the critical point of water.

Although the exchange current density of pure iron is three times larger than those of stainless steel at ambient conditions, they become comparable at supercritical conditions. It seems that there exists a maximum value of the exchange current density around the critical temperature of water for each of the investigated materials. This may be due to the change in dielectric constant or conductivity as the critical point of water is passed. Polarization curves also show that the passivation voltage ranges for the two types of stainless steel near the critical point are much smaller than those at ambient conditions. This indicates that passivation may be lost around the critical point. A detailed study of 304 S.S. was conducted from ambient to supercritical conditions (temperature up to 530 °C and pressure up to 300 atm). The results are listed in Table III. A plot of i_o vs. temperature at about 230 atm is given in Figure 3. It shows that at a fixed pressure, the exchange current density increases exponentially with temperature up to the critical point, reaches a maximum, and then decreases.

Assuming that the pressure has a quadratic effect on the exchange current density, and the temperature has an exponential effect as observed, one can use the DUD method to conduct a least squares statistical search. The correlation for the exchange current density with temperature and pressure within the experimental range was found to be:

$$i_o = i^\circ \, (1 + a\, P_r + b\, P_r^2) \left[\sigma \exp\left(-\frac{\Delta G_a}{RT}\right) - (1-\sigma) \exp\left(-\frac{\Delta G_b}{R(T-T_m)}\right) \right] \qquad (7)$$

where,

i_o = exchange current density (mA/cm^2),

P_r = P/P_c, reduced pressure, ($0.9 \le P_r \le 1.4$, in this model), observed pressure divided by the critical value P_c,

T = temperature (K), (300K T 800K, in this model),

TABLE I EXCHANGE CURRENT DENSITY i_o (mA/cm^2)

MATERIAL	P(atm)	T(°C) 336.7	374.1	411.5
Pure Iron	196	0.00625	0.01803	0.01434
	218	0.01527	0.01535	0.01321
	240	0.01209	0.00927	
	ambient		0.00023	
1080 Carbon Steel	196	0.01882	0.01534	0.01014
	218	0.02225	0.01705	0.01270
	240	0.01631	0.01187	0.01426
	ambient		0.00072	
304 Stainless Steel	196	0.01248	0.01260	0.03307
	218	0.00919	0.01050	0.01129
	240		0.01010	0.01190
	ambient		0.00007	
316 Stainless Steel	196	0.00899	0.01035	0.00655
	218	0.01325	0.02607	0.00905
	240	0.00737	0.00955	0.01605
	ambient		0.00009	

TABLE II OPEN CIRCUIT POTENTIAL E_o (Volts)

MATERIAL	P(atm)	T(°C) 336.7	374.1	411.5
Pure Iron	196	-0.232	-0.054	-0.029
	218	-0.060	-0.025	-0.010
	240	-0.114	-0.178	
	ambient		-0.162	
1080 Carbon Steel	196	-0.250	-0.020	-0.052
	218	-0.240	-0.157	-0.122
	240	-0.260	-0.166	-0.167
	ambient		-0.170	
304 Stainless Steel	196	0.073	-0.066	-0.070
	218	0.103	-0.193	-0.377
	240		-0.106	-0.171
	ambient		-0.143	
316 Stainless Steel	196	0.073	-0.066	-0.070
	218	0.103	-0.193	-0.045
	240	0.127	-0.200	-0.200
	ambient		-0.219	

TABLE III i_o and E_o OF 304 STAINLESS STEEL IN WATER FROM AMBIENT TO SUPERCRITICAL CONDITIONS

Temperature (^oC)	Pressure (atm)	Exchange Current Density (mA/cm^2)	Open Circuit Potential (Volts)
25.3	237.6	0.00011	-0.062
99.7	235.4	0.00031	0.106
178.0	231.1	0.00102	0.123
334.7	222.4	0.0092	-0.270
370.0	189.7	0.0125	-0.257
374.1	204.9	0.0126	-0.230
377.8	228.9	0.0105	-0.221
393.5	246.3	0.0101	-0.106
413.0	218.0	0.0113	-0.377
420.9	246.3	0.0119	-0.171
452.2	233.3	0.0095	-0.258
503.1	235.4	0.0093	-0.354
534.5	234.5	0.0093	-0.284
374.1	261.6	0.0109	-0.257
424.8	261.6	0.0137	-0.197
479.6	261.6	0.0092	-0.458
503.1	261.6	0.0099	-0.488
534.5	261.6	0.0103	-0.311
374.1	285.6	0.0121	-0.248
397.4	285.6	0.0100	0.096
392.5	300.8	0.0123	0.103
420.9	300.8	0.0153	-0.011
471.8	300.8	0.0152	-0.320
526.6	327.0	0.0112	-0.300

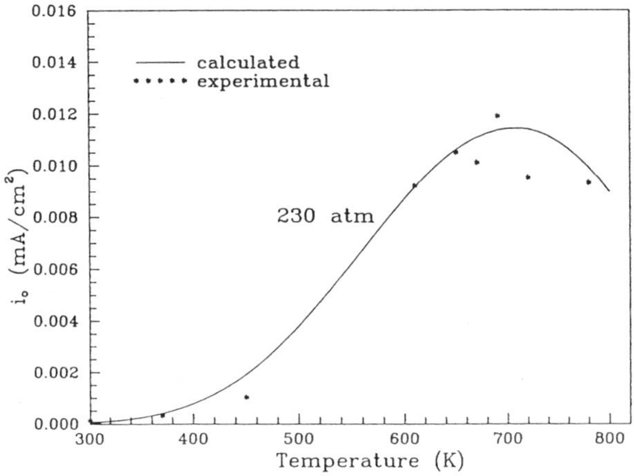

FIGURE 3 TEMPERATURE EFFECT ON EXCHANGE CURRENT DENSITY OF 304 S.S.

T_m = 273.15 K, parameter, also the melting point of water,
R = 1.9872 cal/molK, ideal gas constant,
i° = 24.39826 mA/cm^2, parameter,
a = -1.43275, parameter,
b = 0.634506, parameter,
σ = 0.540677, parameter,
ΔG_a = 6437.17 cal/mol, parameter,
ΔG_b = 4301.85 cal/mol, parameter.

Calculated exchange current densities at 230 atm are plotted with the experimental values in Figure 3. The deviation between the experimental result and the model value is smaller than ±0.002 mA/cm^2 under supercritical conditions.

Based upon the electrodics, the exchange current density is related to the activation energy (16, pp.1152). For an Arrhenius relation for the effect of temperature upon corrosion rate, ΔG_a and ΔG_b are analogous to activation energy and equal 6.44 kcal/mol and 4.30 kcal/mol, respectively. The literature indicates activation energies for mass transfer limited processes range between 1 and 3 kcal/mol and for reaction limited between 10 and 20 kcal/mol (17). Based upon this criteria, corrosion of the 304 S.S. in pure water in the experimental system may lie between the mass transfer and reaction rate limited cases.

The open circuit potential shows a complicated behavior. It is reported to change with time (18). This behavior may be due to the charge change in the electric double layer on the electrode. Besides the exchange current density and the open circuit potential, the electrodic model, equation 3, also provides the parameters for the transfer coefficients, which give the information about the corrosion mechanism. The following parameters were determined from the non-linear regression for the polarization curves (regression based upon 40 points per curve and 19 curves): n (the total number of charges transferred in the reaction) = 4, τ (the reaction steps before the rate determining reaction) = 2, r (the electrons transferred in the rate determining reaction) = 0, and ϵ (the stoichiometric number of rate determining reaction) = 4. Therefore, the transfer coefficients for the total reaction are

$$\alpha_a = 0.5 \text{ and } \alpha_c = 0.5 \tag{8}$$

For other reaction mechanisms, these coefficients vary from 0 to 2.5 (16, pp.1008). Since the number of electrons transferred in the rate determining reaction was found to be zero, the symmetry factor was not be able to be determined. In most cases this factor is assumed to be 0.5 (16).

CONCLUSIONS

A supercritical fluid operating system was developed for obtaining the polarization curves of materials. The cell IR drop was found to be negligible when the applied potential was near the open circuit potential.

The exchange current density, open circuit potential and transfer coefficients were calculated for different alloys around the supercritical point of water. Passivation of 304 or 316 stainless steel appears to be lost around the critical point. Extensive experiments were conducted on 304 S.S. in the supercritical region. A semi-empirical equation was proposed for the temperature and pressure effects on the exchange current density. This equation is valid from ambient to supercritical water conditions for temperatures up to 530 °C. Based upon the equation, an extremum point was found for the exchange current density of 304 S.S.. The calculated value is 433 °C, which agreed with the experimental results.

Based upon an Arrhenius relation for the corrosion rate, the activation energy for corrosion of 304 S.S. in supercritical pure water fell between values corresponding to a mass transfer limited and reaction rate limited process.

ACKNOWLEDGEMENT

This project is sponsored by EG&G Idaho Company under a research contract from the Bureau of Mines.

LITERATURE CITED

1. Hannay, J. B.; Hogarth, J. Proc. Roy. Soc. 1879, Sec. A, 29, 324.
2. McHugh, M.; Krukonis, V., Ed. Supercritical Fluid Extraction, Butterworths: New York, 1986.
3. Squires, T. G.; Paulaitis, M. E., Ed. Supercritical Fluids, Chemical and Engineering Principles and Applications, ACS Symposium Series No. 329; 1987.
4. Weeks, J. R. "Corrosion of Steam Generator Tubing in Operating Pressurized Water reactors", In Corrosion Problems in Energy Conversion and Generation; Electrochem. Sci., 1975
5. McDonald, A. C.; Fan, F. F.; Bard, A. J. J. of Phys. Chem. 1986, 90, 196.
6. Flarsheim, W. M.; Tsou, Y.; Johnston, K. P.; Bard, A. J. J. of Phys. Chem. 1986, 90, 3857.
7. Silver, P. J. B.; Tomlinson, L.; Hurdus, M. H.; Ashmore, C. B. NACE Corrosion 1985, 41, 157.
8. Hishida, M.; Takabayashi, J.; Kawakubo, T.; Yamashima, Y. NACE Corrosion 1985, 41, 570.
9. Iding, M. E.; Vermilyea, D. A. NACE Corrosion 1975, 31, 51.
10. Carter, J. P.; Cavino, B. S.; Driscoll, T. J.; Riley, W. D. NACE Corrosion 1984, 40, 205.
11. Agrawal, A.K.; Staehle, R. W. NACE Corrosion 1977, 33, 418.
12. Andresen, P. L.; Indig, M. E. NACE Corrosion 1982, 38, 531.
13. Fuller, G. A.; McDonald, D. D. NACE Corrosion 1984, 40, 474.
14. Ralston, M. L.; Jennrich, R. I. Technometrics 1979, 1, 7.
15. Daehling, K. Master Thesis, University of Idaho, Moscow, 1987.
16. Bockris, J. O'M.; Reddy, A. K. N. Modern Electrochemistry, Vol. 2, 1970; Plenum Press.
17. Levenspiel, O. Chemical Reaction Engineering; John Wiley and Sons: New York, 1972.

18. Syrett, B. C. "The Application of Electrochemical Techniques to the Study of Corrosion of Metallic Implant Materials", In Electrochemical Techniques for Corrosion, Baboian, R., Ed., NACE 1977.

RECEIVED May 2, 1989

Chapter 20

Phase and Reaction Equilibria Considerations in the Evaluation and Operation of Supercritical Fluid Reaction Processes

Said Saim, Daniel M. Ginosar, and Bala Subramaniam

Department of Chemical and Petroleum Engineering, University of Kansas, Lawrence, KS 66045

Equilibrium conversion and critical behavior plots can aid both in the rational selection of supercritical fluid (SCF) solvents as well as in the design of experiments that yield optimum operating conditions in the dense supercritical region. In this study, Gibbs' rigorous thermodynamic criteria are used to predict phase and reaction equilibria for the supercritical isomerization of 1-hexene over a Pt/γ-Al_2O_3 catalyst. Simulated equilibrium conversions agree well with experimental results obtained in a stirred autoclave reactor. Results from batch and continuous experiments attest to the potential of SCF reaction media to significantly reduce catalyst coking when operating under favorable conditions.

The unique solvent and transport properties of supercritical fluids (SCF's) have been exploited in chemical reaction schemes in a variety of ways. Some of the reported advantages of SCF reaction schemes over conventional processing include lower operating temperatures and improved yield for pyrolysis reactions, regeneration of catalysts by SCF extraction of the deactivating agents at milder operating temperatures, in situ reaction/ separation schemes for isomerization reactions, and improved product selectivity and rates in the case of free-radical polymerization and hydrocarbon oxidation reactions. Examples of each of these applications are summarized in recent review papers (1, 2). More recently, interesting rate enhancement effects on catalytic dehydrogenation of toluene near critical conditions are reported by Gabitto and coworkers (3). Extreme solvent effects on the rate constant of the unimolecular decomposition reaction of small amounts of α-chlorobenzyl methyl ether in an SCF solvent (1-1 difluoroethane) are reported by Johnston and Haynes (4). A detailed model for supercritical reaction-extraction of a solid component from a bed of porous particles is presented by Triday and Smith (5).

0097–6156/89/0406–0301$06.00/0

One of the first questions that arises in the development of SCF reaction schemes concerns the criteria for selecting an SCF medium for a given reaction. The selection depends upon several factors, viz., density, transport properties, inertness or reactivity of the chosen solvent, toxicity, and phase and reaction equilibria, to mention a few. The two latter factors, viz., phase and reaction equilibria, may be as important as the other selection factors and have, as yet, received little attention.

The relevance of phase and reaction equilibria considerations to the SCF solvent selection process may be better understood by considering the critical loci of mixtures of CO_2 with some hydrocarbons as shown in Figure 1 (6). The critical point of the reaction mixture varies widely with its composition. For practical purposes, the SCF operating temperature is confined to the vicinity of the critical point ($T = 1.0\text{-}1.2\ T_c$); i.e., the dense supercritical region. Clearly, the choice of the SCF solvent medium (either pure hydrocarbon or a mixture of hydrocarbon and CO_2) will determine the single-phase operating region in pressure-temperature space. The operating pressure and temperature will in turn determine the overall reaction equilibrium as well as reaction kinetics. However, most reported applications of SCF reactions fail to address the effect of SCF operating conditions upon reaction equilibrium. In this sense, the choice of SCF solvent in these studies seems to be more arbitrary than rational.

The objective of this paper is to demonstrate the importance of phase and reaction equilibria considerations in the rational development of SCF reaction schemes. Theoretical analysis of phase and reaction equilibria are presented for two relatively simple reactions, viz., the isomerizations of n-hexane and 1-hexene. Our simulated conversion and yield plots compare well with experimental results reported in the literature for n-hexane isomerization (6) and obtained by us for 1-hexene isomerization. Based on our analysis, the choice of an appropriate SCF reaction medium for each of these reactions is discussed. Properties such as viscosity, surface tension and polarity can affect transport and kinetic behavior and hence should also be considered for complete evaluation of SCF solvents. These rate effects are not considered in our equilibrium study.

Theoretical Analysis of Phase and Reaction Equilibria

For an isobaric and isothermal conversion process, it is desired to calculate the composition of the product mixture at equilibrium. The first step in the theoretical procedure is to iteratively solve the reaction equilibrium problem for composition of the equilibrium reaction mixture assuming that the reactant phase is initially critical or supercritical. Critical properties of the reaction mixture and fugacities are then estimated. Operating temperature and pressure are then re-evaluated such that the reaction mixture is constrained to the supercritical region. The iterative process is continued until satisfactory convergence is achieved.

The condition that the Gibbs energy of a system at a given temperature and pressure be a minimum at equilibrium is applied to determine the equilibrium composition of the reacting mixture (7).

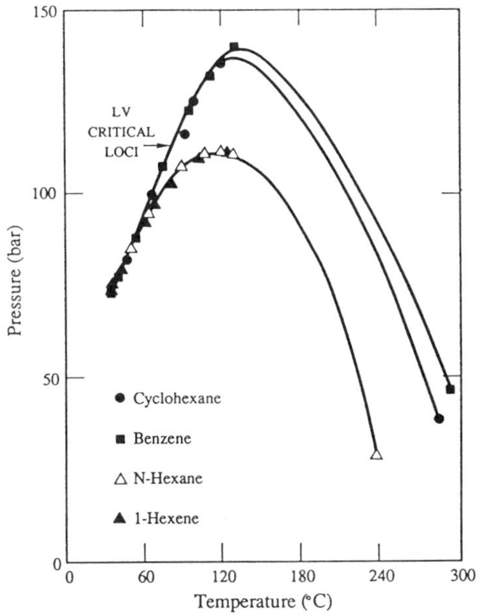

Figure 1. Critical loci of CO_2-hydrocarbon binary mixtures. (Reprinted from ref. 6. Copyright 1975 American Chemical Society.)

No a priori specification of the stoichiometric equations is required for this method. The only data needed are the standard Gibbs energies of formation for all the chemical species expected to be present at equilibrium along with physical property data.

The mathematical formulation leads to a constrained Gibbs energy minimization problem, subject to conservation of the total amounts of the individual chemical elements that make up the chemical species. This constraint is incorporated into the problem via the method of Lagrange multipliers. Details of the procedure are given elsewhere (8).

The phase behavior that is exhibited by a critical or supercritical mixture of several components is usually not simple. Street (9) reports six classes of phase behavior diagrams. In the simplest classes of systems (classes 1 and 2), the critical lines are continuous between the critical points of pure components. Study of reaction equilibrium at SCF conditions requires knowledge of critical properties of the reacting mixture at various levels of conversion. Three different approaches to evaluate critical properties are available, viz., empirical correlations, rigorous thermodynamics criteria and the theory of conformal solutions (10). The thermodynamic method is more general and reliable because it is consistent with the calculation of other thermodynamic properties of the reacting mixture (11).

The thermodynamic definition of the critical state was first enunciated by Gibbs (12). Mathematically, the criterion requires that for an n-component mixture, the following two Jacobians be equal to zero at the vapor-liquid critical point:

$$U = \begin{bmatrix} \frac{\partial^2 G}{\partial x_1^2} & \frac{\partial^2 G}{\partial x_1 \partial x_2} & \cdot & \cdot & \frac{\partial^2 G}{\partial x_1 \partial x_{n-1}} \\ \frac{\partial^2 G}{\partial x_2 \partial x_1} & \frac{\partial^2 G}{\partial x_2^2} & \cdot & \cdot & \frac{\partial^2 G}{\partial x_2 \partial x_{n-1}} \\ \vdots & \vdots & \vdots & \vdots & \vdots \\ \frac{\partial^2 G}{\partial x_{n-1} \partial x_1} & \frac{\partial^2 G}{\partial x_{n-1} \partial x_2} & \cdot & \cdot & \frac{\partial^2 G}{\partial x_{n-1}^2} \end{bmatrix}$$

$$M = \begin{bmatrix} \frac{\partial U}{\partial x_1} & \frac{\partial U}{\partial x_2} & \cdot & \cdot & \frac{\partial U}{\partial x_{n-1}} \\ \frac{\partial^2 G}{\partial x_2 \partial x_1} & \frac{\partial^2 G}{\partial x_2^2} & \cdot & \cdot & \frac{\partial^2 G}{\partial x_2 \partial x_{n-1}} \\ \vdots & \vdots & \vdots & \vdots & \vdots \\ \frac{\partial^2 G}{\partial x_{n-1} \partial x_1} & \frac{\partial^2 G}{\partial x_{n-1} \partial x_2} & \cdot & \cdot & \frac{\partial^2 G}{\partial x_{n-1}^2} \end{bmatrix}$$

where G = G (T, P, x) and the partial derivatives are obtained at constant P, T and $X_{k,k \neq i,n}$. The Gibbs free energy can be written in terms of P, T and x using an equation of state (EOS). However, since most equations of state are pressure explicit, it is necessary to express the partial derivatives of the Gibbs free energy in terms of the Helmholtz free energy as follows:

$$G = A + PV$$

where A = A (T, V, x), and is obtained through integration of the thermodynamic EOS

$$P = -\left(\partial A / \partial V\right)_{T,\ x}$$

The Peng-Robinson EOS was recently found to yield the best values of binary critical volumes for class 1 systems and to perform reliably for other critical properties as well (13). The expression of A in terms of T, V, and x along with the relationships between second and third derivatives of Gibbs and Helmholtz free energies are given by Peng and Robinson (11). Further algebraic manipulations lead to a system of two determinant equations. For any given mixture of known composition, the system to be solved is of the form:

$$U = U\ (T,\ V)$$
$$M = M\ (T,\ V)$$

Newton's numerical method was used to solve the above system of equations. Details of the solution procedure are given by Peng and Robinson (11).

Isomerization products and reactants have more or less identical physico-chemical properties. For example, critical properties of 1-hexene and its isomers are similar, especially among the isomers. For this reason, this critical property determination procedure was applied to only binary and ternary mixtures of 1-hexene and isomers (as a pseudo-component) in CO_2. Binary interaction parameters were taken from Vera and Orbey (14). Figures 2-5 show critical property loci of binary mixtures of 1-hexene and CO_2. Maximum deviation from reported experimental critical pressures in Figure 1 (6) is less than 2%.

Theoretical Results and Discussion

The theoretical analysis presented in the previous section was applied to two model SCF isomerization systems. In one scheme, the catalytic isomerization of n-hexane is carried out using supercritical CO_2 as solvent medium. CO_2 also acts as activator for the Lewis acid catalyst promoted with hydrogen bromide or hydrogen chloride. Reported advantages over conventional processing include better mass transfer characterisitics, better selectivity towards isomerization products, and easier separation of reactants, product isomers and catalyst (15).

The second example involves alumina-catalyzed isomerization of 1-hexene. In this case however, the reaction mixture itself is

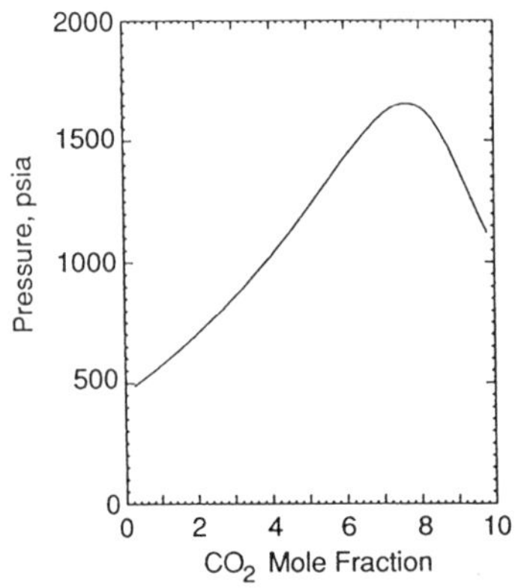

Figure 2. Critical pressure locus for CO_2 – 1-hexene mixtures.

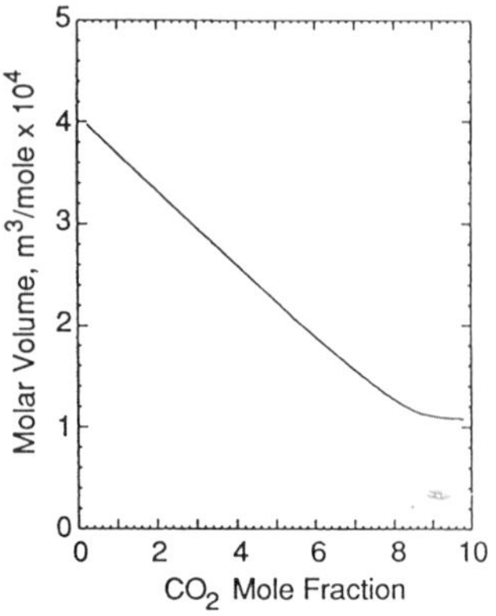

Figure 3. Critical volume locus for CO_2 – 1-hexene mixtures.

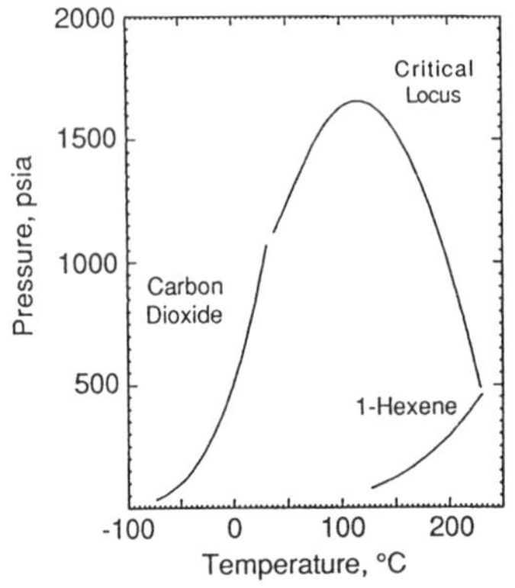

Figure 4. Temperature–pressure critical locus for CO_2 – 1-hexene mixtures.

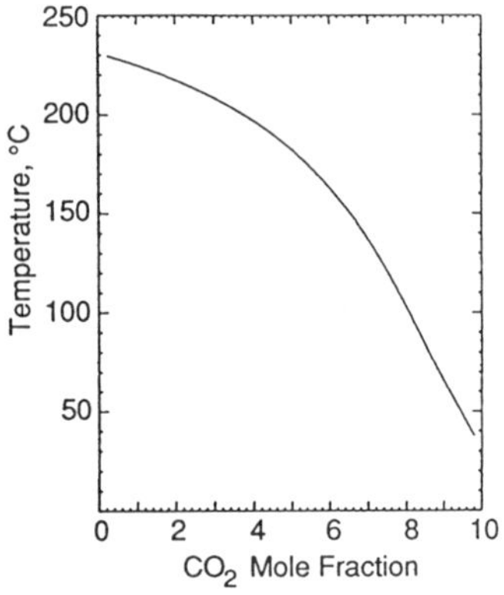

Figure 5. Critical temperature locus for CO_2 – 1-hexene mixtures.

employed as the SCF solvent medium (16). Reported advantages in this case include in situ catalyst regeneration at mild temperatures as compared to conventional processing and continuous control of coking, fouling or poisoning.

Isomerization of n-hexane. In the case of n-hexane isomerization in a background of CO_2, operating pressure has negligible effect on equilibrium conversion (8). Hence, there is no thermodynamic advantage for operating farther than necessary from the critical pressure of the reacting mixture. However, high pressures may be practically desirable for improving the solvent power of the SCF.

It was also found that equilibrium conversion decreases sharply with increasing temperature. In order to exploit the unique solvent properties of SCF's, operating temperatures are usually restricted to the 1.0-1.2 T_c range. A simple criterion for assessing potential SCF solvents is therefore to evaluate the effect of temperature on reaction equilibrium and on kinetics when operating in the 1.0-1.2 T_c range of the reaction mixture. For n-hexane isomerization in CO_2 studies, the operating temperatures employed by Kramer and Leder (15) are in the 1.1-1.3 T_c range (of the reaction mixtures); i.e., 80-150°C. This temperature range is both thermodynamically and kinetically favorable. Furthermore, these temperatures favor the formation of branched isomers and thus can lead to a high octane number product (8). Hence CO_2 may be considered as a good SCF solvent for n-hexane isomerization.

On the other hand, n-hexane by itself may not be as good an SCF reaction medium as CO_2. This is so because, as per our criterion , the operating temperatures should be from 234 to 336°C (i.e., 1.0-1.2 T_c) and at these temperatures, the overall conversion as well as selectivity to branched isomers are thermodynamically less favored (8).

It should be pointed out however that Kramer and Leder (15) used CO_2 as the SCF solvent because it also acts as a catalyst activator. Thus, factors other than temperature and pressure effects on reaction equilibrium can also dictate the choice of the SCF solvent.

Isomerization of 1-hexene. Figure 6 shows simulated equilibrium yields and conversions of 1-hexene versus temperature at the specified pressure. Cracking products were not considered in our analysis. Increase in temperature is seen to cause a slight decrease in equilibrium conversion and to have little effect on the isomer selectivities. Simulated equilibrium conversion at 250°C and 7,250 psia is 97.2%. This value compares with the experimental value of 40% obtained by Tiltsher et al. (16) in a catalytic flow reactor. Clearly , there is room for improving the experimentally reported conversion.

Higher temperatures will enhance both isomerization as well as cracking reaction rates. However, at higher operating temperatures, the ability of the solvent to extract the precursor compounds will suffer because of decrease in solvent density and hence in solvent power. The inability to extract coke precursor compounds will lead to eventual deactivation of the catalyst. Furthermore, the overall rate of reaction may be limited by

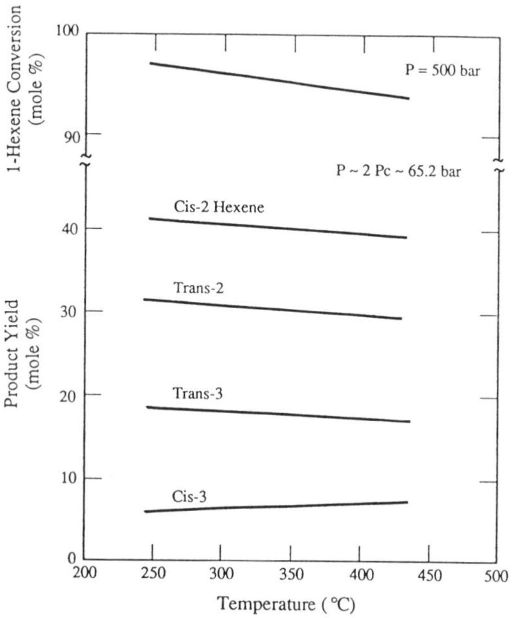

Figure 6. Variation of equilibrium conversion and yield of 1-hexene isomerization reactions with temperature. (Reprinted with permission from ref. 8. Copyright 1988 Pergamon.)

external mass transfer and internal pore diffusion resistances. Hence, there exists an optimum operating temperature for continuous maintenance of catalytic activity. These temperature effects are discussed in more detail elsewhere in this paper.

In order to study the effect of SCF operating conditions on reaction equilibrium and kinetics, an experimental facility has recently been completed and successfully tested as described in the following section. The heterogeneous catalytic isomerization of 1-hexene over Pt/γ-Al_2O_3 catalyst is chosen as the model reaction system; i.e., a reaction system that undergoes simultaneous deactivation by coking.

Experimental Unit

The experimental unit consists of an Autoclave Engineers' Supercritical Extraction Screening System (SCESS) equipped with a 300 cc packless autoclave reactor in place of the standard extractor vessel. The schematic of the experimental unit is shown in Figure 7. The unit is composed of three sections: the feed preparation section, the autoclave reactor section and the sampling section.

In the feed section, the mainly 1-hexene reactant (98.60% 1-hexene, 0.95% cis-3 hexene, 0.25% trans-2 hexene and 0.15% cis-2 hexene, supplied by Ethyl corporation) is introduced from an Instrumentation Specialties Company (ISCO) model 314 metering pump into a flowing stream of liquefied CO_2. The 1-hexene/CO_2 mixture is then fed to a high pressure positive displacement pump contained in the SCESS. In this pump, fluids can be pressurized up to 6,000 psig and total flow rate can be adjusted between 46 ml/hr and 460 ml/hr. Since feed to the pump must be in a liquid state, both the 1-hexene/CO_2 mixture and the pump head are sufficiently cooled by circulating chilled water at 5°C. The system pressure is controlled by means of an adjustable back pressure regulator.

The pressurized fluid from the pump is then introduced into the bottom of the reactor via a sparger tube. The reactor is heated by an external electrical furnace and reactor temperature is controlled by a proportional controller. The maximum operating pressure for the reactor depends upon the operating temperature. The reactor vessel is rated at 6,000 psig at ambient temperatures; the rating decreases to 5,000 psig at 480°C.

The reaction takes place on the catalyst housed in three stationary beds in the reactor. The catalyst used for the 1-hexene isomerization reaction is a commercial E-302 reforming catalyst, supplied by Engelhard corporation. The bifunctional catalyst is composed of 0.6 wt% Platinum supported on 1/16" right cylindrical gamma-alumina extrudates. To minimize external mass-transfer resistances and to achieve CSTR behavior, the fluid phase containing the reactants is kept mixed by an impeller powered by a 0.75 hp MagneDrive assembly that can provide stirring speeds up to 3,000 rpm. Unconverted reactant, product and the SCF medium exit via a port located at the top of the reactor.

The reactor effluent is then lead to a pressure reducing metering valve. To ensure single phase flow, the exit line leading to this metering valve is maintained at the same temperature as the

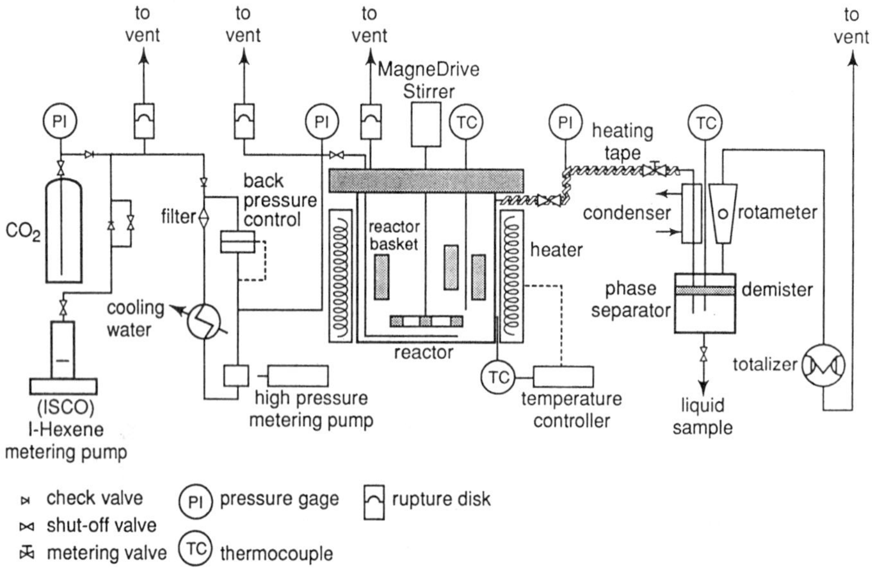

Figure 7. Schematic of experimental unit.

reactor. The metering valve serves to reduce the reactor effluent pressure to near atmospheric pressure and to control the total flow. The metering valve is kept sufficiently heated with a heating tape to overcome cooling by CO_2 expansion which may otherwise cause solidification of the reactor effluent stream. The expanded stream is then condensed by circulating chilled water at 5°C and the liquid mixture (1-hexene and its isomers) is separated from the gas in an externally cooled phase separator. The flow rate of the gas stream exiting the phase separator is monitored on a rotameter, and the total volume is measured in a flow totalizer. Liquid samples are taken from the phase separator and analyzed with a Hewlett-Packard 5890A gas chromatograph equipped with a thermal conductivity detector and an HP3393A computing integrator.

All piping, valves and fittings used in the system are rated to withstand up to 10,000 psig. To provide protection from over-pressure, three safety heads with rupture disks are employed. In the feed preparation section, a 1/4" rupture disk with a burst rating of 1,500 psig is used to protect the ISCO metering pump and the CO_2 cylinder from over-pressure. Following the high-pressure positive displacement metering pump, another 1/4" rupture disk with a burst rating pressure of 6,150 psig is installed. This rupture disk is used to protect the pump from over-pressure due to downstream blockage. The reactor is protected with a 1/8" rupture disk with a burst rating of 5,800 psig. The rupture disks are rated at approximately 20°C. All safety heads are connected to high-pressure hoses which vent to the building exhaust.

Experimental Considerations

Choice of the initial reactant mixture (1-hexene + CO_2) composition not only determines the critical properties of the reaction mixture but also dictates reactor conditions for dense supercritical operation. For example, an equimolar feed mixture of 1-hexene and CO_2 has a critical temperature of 183°C and a critical pressure of 1,235 psia. Hence operating the reactor at 230°C (1.1 T_c) and 2,600 psig (2.1 P_c) will ensure operation in the dense supercritical region (0.38 gm/cc). However, it is also necessary that this operating temperature be thermodynamically and kinetically favorable and within the operational limit of our experimental unit.

For 1-hexene isomerization, lower temperatures slightly favor equilibrium conversion. Therefore, decreasing reaction temperature through the addition of a low T_c solvent such as CO_2 is not thermodynamically unfavorable. However, the reaction may become kinetically limited. On the other hand, as seen from Figure 6, because temperature does not significantly affect equilibrium conversion, it may seem reasonable to employ 1-hexene itself as the solvent medium to allow higher operating temperatures. The critical temperatures of 1-hexene and its isomers are relatively high (231 - 248°C) thus favoring reaction kinetics. However, the overall rate of the catalytic isomerization of pure 1-hexene may become limited by external mass-transfer and internal pore-diffusion resistances at the higher operating temperatures (T = 1.1-1.2 T_c).

While external mass-transfer limitations can be minimized by providing adequate mixing, internal pore-diffusion limitations are not. Noting that catalyst coking rates increase with temperature, and maintenance of catalyst activity depends upon the ability of the dense supercritical reaction mixture to extract the coke-precursors from the catalyst surface before they undergo oligomerization, a mass-transfer limited reaction rate can lead to rapid catalyst deactivation by coking and eventual pore-plugging. A possible solution to such a problem would be to dilute the 1-hexene feed with a suitable amount of CO_2 so that the critical temperature of the reaction mixture is reduced to such levels as to provide reasonable isomerization reaction rates yet low coking rates at dense supercritical conditions of temperature and pressure.

It should be noted however that though the smaller size of CO_2 molecules (as compared to hexene molecules) may impart better transport properties to the SCF reaction mixture, oligomer extraction may actually be hindered by the presence of CO_2 due to (a) lower reaction temperature and hence lower vapor pressure for the oligomers, and (b) greater disparity between molecular structures of CO_2 and the oligomers.

Our research program is thus aimed at gaining a fundamental understanding of the behavior of heterogeneous catalytic reactions at SCF conditions based on phase and reaction equilibria considerations as well as kinetic and mass-transfer rate considerations. Such an understanding is essential to the rational design of SCF reaction schemes.

This paper presents results from initial experimental investigations of the isomerization of 1-hexene over a commercial Engelhard reforming catalyst. Both batch and continuous runs were performed. The objective of the batch run is to compare experimentally obtained equilibium conversion and product selectivities with those obtained through our theoretical procedure. The continuous runs were performed to compare reaction behavior at subcritical and supercritical conditions and to verify if continuous maintenance of catalyst activity is indeed possible at dense supercritical conditions.

Experimental Procedure and Results

The start-up procedure consists of first flushing the system with about three reactor volumes of CO_2 to purge the equipment of any oxygen that might lead to unwanted oxidation reactions. Reactor temperature is then raised to the operating value. After the desired temperature is reached, a predetermined amount of 1-hexene is pumped into the reactor. At the end of this step, CO_2 is pumped into the reactor until the desired pressure is reached. In case of a continuous run, constant reactant and product flow rates are then established by adjusting the flow control knobs on the pump and on the metering valve at the exit of the reactor. Liquid samples are collected every 15 or 30 minutes from the separator vessel. The non-condensed gas, consisting of essentially CO_2, passes through the rotameter and the flow totalizer. Total gas flow is noted at each sampling time.

The liquid samples are analyzed for 1-hexene and its four isomers, viz., cis-2, trans-2, cis-3 and trans-3 isomers, by gas chromatography. Distinct peak resolution was obtained for all isomers by using an HP 50 m, 0.32 mm bore, high performance fused silica capillary column coated with a 0.52 mm film of dimethylpolysiloxane.

Both batch and continuous runs were performed. Batch runs were performed mainly to check the validity of the predicted equilibrium conversions. A batch conversion of 90% was obtained after 11 hours of operation when 1-hexene isomerization was performed at a supercritical temperature of 265°C (1.07 T_c) and subcritical pressures varying between 250 and 150 psig (0.54-0.33 P_c). The liquid product collected in the phase separator bore no color and contained virtually no oligomers. Combined conversions to 2-hexenes and 3-hexenes, are about 69% and 21% respectively. These values compare well with the simulated equilibrium values of 72.4% and 24.3% respectively (see Figure 6).

In another batch run of an equimolar mixture of 1-hexene and CO_2 at a temperature of 350°C (1.37 T_c) and a pressure of 1,900 psig (1.55 P_c), up to 9% of the light brown product (based on chromatographic peak area %) consisted of higher molecular weight compounds. These compounds eluted in a narrow time band (about 1 minute) and their retention times are close to that of naphthalene (C_{10}). No higher molecular weight oligomeric structures were detected at these conditions.

The results of two continuous runs, one at subcritical and the other at supercritical conditions, are reported. In the subcritical run, a mixture of 30 mole% 1-hexene in CO_2 was reacted at T = 225°C (1.2 T_c) and P = 1,000 psig (0.5 P_c). Catalyst loading was 6.0 grams and the reactor residence time was approximately 7 minutes. The results of this run are shown in Figure 8. Catalyst deactivation is clearly seen to occur even after a prolonged time period of about 14 hours. The conversion of 1-hexene decreased from 51% to 32% in 14 hours giving a deactivation rate of roughly 1.4%/hour. This result confirms the inability of light subcritical fluids to maintain constant catalyst activity.

Figure 9 illustrates the results of a run at 230°C (1.1 T_c) and 2,600 psi (2.1 P_c) for an equimolar feed mixture of 1-hexene in CO_2. Catalyst loading was 1.0 gram and reactor residence time was about 1 hour. Whereas the product from the previous subcritical run was light clear, the product from this run bore a light brown color throughout the whole run, indicating the presence of higher molecular weight oligomer compounds than hexene. Because of the longer residence time used in this run, steady state was established only after about four hours. Larger transient changes in product yield and conversion were observed during the start up period. Thereafter no significant change in the product composition was observed for about seven hours as indicated by the conversion versus time plots (see Figure 9). Linear regression of conversion vs time data between four and 11 hours indicates that the slope of the regressed line is not statistically different from zero. This run implies that no deactivation of the catalyst occurs when operating at dense supercritical conditions.

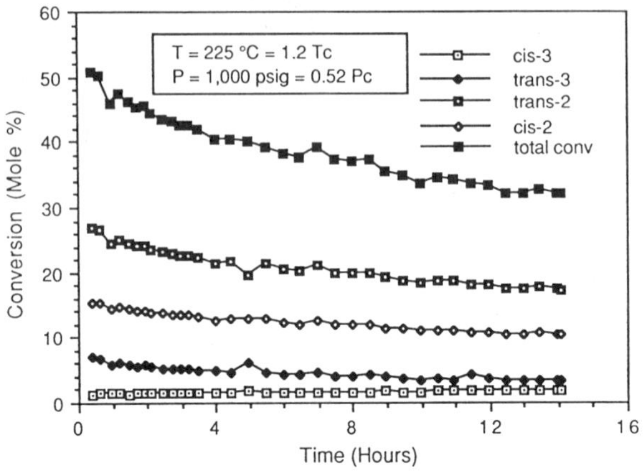

Figure 8. Isomerization of 1-hexene over Pt/γ-alumina catalyst in a CSTR at subcritical conditions.

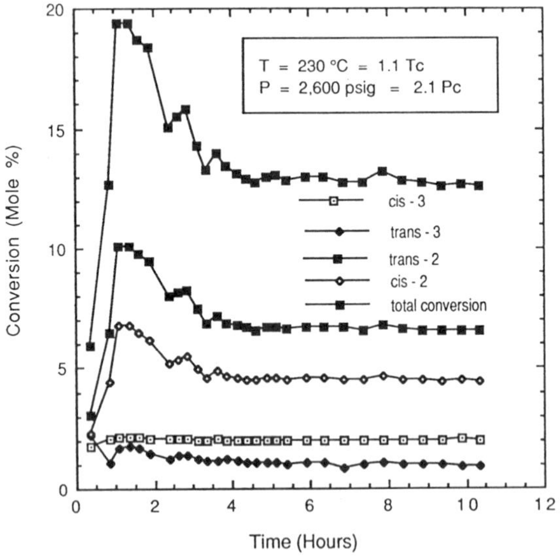

Figure 9. Isomerization of 1-hexene over Pt/γ-alumina catalyst in a CSTR at supercritical conditions.

The qualitative trends exhibited by our continuous runs agree well with those reported by Tiltscher and coworkers (16). Whereas Tiltscher et al. (16) employed a shell catalyst consisting of γ-Al_2O_3/Al-metal with very low catalytic activity, low catalytic active surface, and a thickness of the catalytic layer of less than 1 mm, this study used a high activity, high surface area industrial Pt/γ-Al_2O_3 catalyst. Notice however that the steady state conversion values are relatively low (20 and 40%) in both cases. Investigations aimed at improving the reaction rate while maintaining constant catalyst activity are under way. As discussed in the previous section, catalyst deactivation can occur even under dense supercritical conditions if the overall reaction rate is subject to mass-transfer limitations. To assist in the interpretation of rate data, theoretical criteria for detection of external mass-transfer and pore-diffusion intrusions are being developed. These studies will also provide insight into the type of coke precursors formed and thereby into the mechanism of coke formation.

Conclusions

Phase and reaction equilibria considerations are essential to the rational evaluation of SCF solvents as reaction media. For our study of the catalytic isomerization of 1-hexene, theoretical predictions of equilibrium conversion and product yield agree well with experimental results obtained in a stirred autoclave reactor. Since maintenance of catalyst activity depends upon the ability of the dense supercritical reaction mixture to extract the coke-precursors from the catalyst surface before they undergo progressive oligomerization to high molecular weight compounds, a mass-transfer limited reaction rate can lead to rapid catalyst deactivation by coking and eventual pore-plugging. We have shown that under favorable supercritical operating conditions, it is possible to significantly reduce catalytic coking of an industrial reforming catalyst.

Acknowledgements

This material is based upon work supported by the National Science Foundation under Grant No. CBT-8709276 and by the Kansas University General Research Fund (GRF 3509-20-0038).

Literature Cited

1. Subramaniam, B.; McHugh, M.A. Ind. Eng. Chem. Process Des. Dev. 1986, 25, 1.
2. Tiltscher, H.; Hofmann, H. Chem. Eng. Sci. 1987, 42, p. 959.
3. Gabitto, J.; Hu, S.; McCoy, B.J.; Smith, J.M. AIChE J. 1988, 34, p. 1225.
4. Johnston, K.P.; Haynes, C. AIChE J. 1987, 33, p. 2017.
5. Triday, J.; Smith, J.M. AIChE J. 1988, 34, p. 658.
6. Leder, F.; Irani, C.A. J. Chem. Eng. Data 1975, 20, p. 323.
7. Walas, S.M. Phase Equilibria in Chemical Engineering; Butterworth: Stoneham, 1985; p. 481.

8. Saim, S.; Subramaniam, B. Chem. Eng. Sci. 1988, 43, p. 1837.
9. Street, W.B. In Chemical Engineering at Supercritical Fluid Conditions; Paulaitis, M.E.; Penninger, J.M.L.; Gray, R.D., Jr.; Davidson, P., Eds.; Ann Arbor Science: Ann Arbor, MI, 1983, p. 3.
10. Spencer, C.F.; Daubert, T.E.; Danner, R.P. AIChE J. 1973, 19, p. 522.
11. Peng, D.; Robinson, D.B. AIChE J. 1977, 23, p. 137.
12. Gibbs, J.W. Collected Works; Yale U. Press: New Haven, Conn., 1928; Vol. 1, p. 55.
13. Palenchar, R.M.; Erickson, D.D.; Leland, T.W. In Equations of State: Theories and Applications; Chao, K.C.; Robinson, R.L., Eds.; American Chemical Society: Washington, D.C., 1986, p. 132.
14. Vera, J.H.; Orbey, H. J. Chem. Eng. Data 1984, 29, p. 269.
15. Kramer, G.M.; Leder, F. U. S. Patent 3 880 945, 1975.
16. Tiltscher, H.; Wolf, H.; Schelchshorn, J. Angew. Chem. Int. Ed. 1981, 20, p. 892.

RECEIVED May 1, 1989

Chapter 21

Kinetic Model for Supercritical Delignification of Wood

Lixiong Li[1] and Erdogan Kiran[2]

Department of Chemical Engineering, University of Maine, Orono, ME 04469

A reaction-diffusion model for delignification of red spruce by supercritical methylamine-nitrous oxide mixtures has been developed. The overall model has the form:

$$-d[X]/dt = k_1\{[X][S]^{1.5}\} - k_2\{[X]_0-[X]\}$$

where $[X]_0$ and [X] are the weight fraction of lignin in wood at time 0 and t, [S] is the weight fraction of methylamine in the extraction fluid, k_1 and k_2 are the rate constants for chemical reaction and diffusional resistance, respectively. The model has been shown to predict the experimental data over a wide range of temperatures (from 170 to 185°C), pressures (from 172 to 276 bars) and compositions (from 0 to 100% methylamine). The temperature and pressure dependence of the rate constants k_1 and k_2 have been determined and are discussed in terms of activation energies and volumes of activation.

Supercritical fluids display attractive solvent characteristics which can be manipulated by either the pressure or temperature. Using supercritical fluids as reaction media, simultaneous reaction and separation are also achievable. This methodology has recently been applied to the reactive separation of wood constituents, especially lignin, by supercritical fluids (1-4). Delignification processes using supercritical fluids are of potential industrial importance (5,6) and there is a need for the development of kinetic models which could permit a priori prediction of the rate of lignin removal. The present paper discusses such a model.

[1]Current address: Center for Energy Studies, Separations Research Program, The University of Texas, 10100 Burnett Road, Austin, TX 78758

[2]Address correspondence to this author.

0097-6156/89/0406-0317$06.00/0

Development of Kinetic Model

General Considerations. Any kinetic model that describes delignification of wood by supercritical fluids must take into account the factors related to the structure of wood, the heterogeneous nature of the reaction, and the factors associated with high pressure reaction kinetics.

Wood is a porous material composed of elongated hollow cells (fibers) which are held together by the middle lamella (an intercellular substance containing mostly lignin and pectic substances). The cell wall displays a layered structure and contains different amounts of the major wood components, cellulose, hemicelluloses and lignin. Even though middle lamella has the highest lignin concentration, the majority of lignin (about 70-80%) is located in the cell walls (7,8). Thus, for complete lignin removal, chemicals must penetrate and react through the cell wall. From a consideration of typical cell dimensions for a softwood tracheid (3 mm length, 0.03 mm width, 4 μm cell wall thickness), it can be calculated that the inner surface area of the cell wall is about 10,000 times larger than its cross sectional area and hence provides the primary surface for reaction and diffusion during delignification. Delignification can be treated as a fluid-solid reaction occurring at the cell wall interface. It may be envisioned to proceed through steps involving (1) diffusion of reactants from the bulk of the fluid phase to the cell wall interface; (2) diffusion of the fluid from the interface to the bulk of the cell wall; (3) chemical reaction between the reactants from the fluid phase and lignin in the cell wall; and (4) diffusion of products within the cell wall and finally out of the cell wall into the bulk of the fluid phase.

Realistic models of delignification should also consider structural changes, in particular changes in porosity of wood accompanying lignin removal. General kinetic models which consider such changes for solid reactions (9,10) require parameters such as the change in distribution of pore sizes with reaction which are difficult to access.

It turns out that often many complex heterogeneous reactions can be globally described by simpler models. For example, delignification reactions in conventional pulping such as kraft pulping (which involves NaOH and Na_2S as cooking chemicals) are usually treated by a simple relationship of the form:

$$-dL/dt = kL \qquad (1)$$

where L is the residual lignin content at time t (calculated on the original wood weight, wt %), and k is the apparent rate constant (11-13). This is a pseudo-first-order rate equation with respect to lignin under the assumption that alkali concentration in the cooking liquor does not change substantially. All the factors associated with any structural changes (such as porosity) have been combined in the apparent rate constant (k).

Clearly, such global relationships are of limited use for an understanding of the details of the actual reaction process, especially the diffusional effects which may change significantly as the reaction proceeds. In the treatment of diffusional factors in condensed phase reactions, the rate of reaction between reactants A and B can be explicitly expressed by a chemical reaction term and a diffusion term (14), such as:

$$-\partial C_A/\partial t = K_A - D_A \partial^2 C_A/\partial r^2 \qquad (2)$$

where C_A is the concentration of A which is a function of the distance from the origin of coordinates in the direction of diffusion (r) and time (t). K_A is the rate of change of C_A caused by the chemical reaction and D_A is the diffusion coefficient of A. If this rate equation were used to describe the conventional kraft delignification reaction, one could represent the lignin in wood as reactant A and cooking liquor in lumen (cell cavities) as reactant B. Since the diffusion of lignin in wood can be neglected, D_A would be zero. If the delignification reaction can be assumed to occur at the interface and follow a first-order kinetics with respect to lignin, that is $K_A = kC_A$, then, it is easy to see that Equation 2 reduces to Equation 1. However, there is a subtle difference between the two rate equations. In Equation 1, L is the residual lignin content based on initial wood weight (wt %) which is an averaged value of the lignin throughout the cell wall (bulk solid phase). In Equation 2, C_A is the lignin concentration at the interface where delignification reaction takes place. Because of the experimental difficulty of obtaining positional and time dependent variation of lignin concentration at the interface, Equation 2 can not be conveniently used for kinetic modeling of delignification. In the present model, our approach has been to take the simplicity of Equation 1 and combine it with the notion of including a term which would represent diffusional factors in a way similar to Equation 2, but involving more readily accessible quantities.

Before discussing the present model, it is instructive to examine also some aspects related to the pressure dependence of the reactions. As is well known, at low pressures, the rate constant k is considered to be a function of temperature only, and its temperature dependence is expressed by the Arrhenius equation:

$$\ln k = \ln A - E/RT \qquad (3)$$

or equivalently,

$$d\ln k/dT = E/RT^2 \qquad (4)$$

where A is the frequency or pre-exponential factor, E is the activation energy, R is the gas constant, and T is the absolute temperature.

When a reaction takes place under high pressures, however, the effect of pressure on the reaction rate may not be neglected. The description of the pressure dependence of reaction rates (15) is based on the assumption that the reaction goes through a transition state involving a volume change between reactants and the transition state complex. The pressure dependence of rate constants is expressed in a form similar to the Arrhenius relationship (16-19) which is given by:

$$[d\ln k/dP]_T = -\Delta V^{\neq}/RT \tag{5}$$

where $\Delta V^{\neq}$ is the "volume of activation" and represents the volume change accompanying the formation of the transition state, (the difference between the partial molar volume of the transition state and the sum of the partial molar volume of the reactants):

$$\Delta V^{\neq} = (\bar{V}_{\text{transition state}}) - \Sigma(\bar{V}_{\text{reactant}}) \tag{6}$$

A task in high pressure kinetics is the determination of either volumes of activation or empirical relations of k as a function of pressure for specific reactions (20-23). Many functional forms relating k and P have been used for specific systems (17,24). A common form is the second order polynomial relationship, i.e.:

$$\ln k = \alpha + \beta P + \gamma P^2 \tag{7}$$

where α, β, and γ are constants to be determined. From Equation 5 and Equation 7, one then obtains a linear relation of $\Delta V^{\neq}$ with pressure:

$$\Delta V^{\neq} = \Delta V_0^{\neq} - 2\gamma RTP \tag{8}$$

where $\Delta V_0^{\neq}$ represents the volume of activation at P = 0 and is independent of pressure.

Formulation of the Present Kinetic Model. The starting point of our model for supercritical delignification is the assumption that the reaction is heterogeneous and occurs primarily at the surface of the cell wall. After the reactive supercritical fluid reaches the interface (the surface of the cell wall), reactions with lignin are assumed to occur at numerous locations on the surface of the cell wall. As the delignification reaction proceeds, the reaction areas are envisioned to develop as pores into the cell wall, leading to longer diffusion paths which can result in greater "diffusion resistance" for the fluid to penetrate further into the cell wall to react with remaining lignin. The size, shape, and distribution of these pores would largely depend on the original distribution of lignin in the cell wall, and the selectivity of the delignification reaction, i.e., the extent of removal of components other than lignin from the cell wall. Unless the pore openings are excessively large, one would anticipate greater diffusion resistance as delignification proceeds. Longer path length will also present a diffusion resistance for removal of the solubilized lignin.

To include a diffusion resistance term in the rate equation, one can assume that the diffusional resistance would be proportional to the diffusion path which as discussed above would be related to the amount of lignin that has been removed from the cell wall matrix. We therefore propose the following form of the rate expression to describe rate of delignification:

$$-d[L]/dt = k_1[L]-k_2([L]_0-[L]) \quad (9)$$

where $[L]_0$ and $[L]$ are the lignin concentrations in wood at time 0 and t, respectively, k_1 is the rate constant for chemical reaction, and k_2 is the overall rate constant for the change in the diffusion resistance. The first term in this rate equation accounts for the chemical reaction which is assumed to be first-order irreversible. The second term accounts for the diffusion resistance which makes a negative contribution to the overall rate. The observed delignification rate is thus regarded as the net result of reaction and diffusion processes. The rate constant for chemical reaction, k_1, would be a function of temperature and pressure. The other constant (k_2 may be a function of parameters such as diffusivity and thickness of diffusion layer and encompasses the structural changes such as the porosity of wood cell wall which increases with increasing value of $([L]_0-[L])$. It should be noted that because of the completely different nature of k_1 and k_2, the [L] terms cannot be combined and Equation 9 cannot be reduced to a simple first order relationship. It should be further noted that the model considers the reaction and diffusion processes after the supercritical fluid reaches the cell wall surface. Thus, the implication of Equation 9 that at time zero there is no diffusional resistance should be interpreted to mean that no pores in the cell wall (as a result of lignin removal) have yet been generated and the fluid is simply reacting with the lignin distributed on the outer surface layer of the cell wall.

Since lignin content in wood is normally expressed on weight basis, lignin concentration can be more appropriately expressed as:

$$[L] = N_L/W_0 = (W_L/M_L)/W_0 = [X]/M_L \quad (10)$$

where N_L is the number of moles of lignin in wood at time t; W_0 is the weight of initial wood sample; W_L and M_L are the weight and average molecular weight of the residual lignin in wood sample at time t, respectively; and $[X](=W_L/W_0)$ is the lignin weight fraction at time t based on initial wood.

Since the molecular weight of native lignin in wood is not known (no method to isolate native lignin from wood without degradation has yet been found) and since during the course of degradation M_L is continuously changing, evaluation of [L] directly from Equation 10 is difficult. Therefore, an indirect approach is suggested to express [L]. Because lignin molecules are made from many phenylpropane units which are inter-connected by aryl-ether bonds (about two thirds of total bonds) and carbon-carbon bonds, a delignifying agent can attack many sites of the same lignin molecule according to the distribution of a particular type of bond in the molecule. It would appear that the concentration of the reactive sites available in lignin molecules is a more appropriate parameter than the concentration of lignin molecules as a whole. Thus a convenient way to express lignin concentration in wood would be:

$$[L]_{rs} = N_{rs}/W_0 \tag{11}$$

where N_{rs} represents the total number of relative sites available in the lignin molecule and $[L]_{rs}$ is the concentration of lignin reactive sites available in the wood sample. The total number of reactive sites available in lignin molecules at time t can be estimated by the total weight (W_L) of lignin at time t divided by the average molecular weight (M_S) of the chain segment between two consecutive reactive sites in a lignin molecule. Thus $[L]_{rs}$ can be written as:

$$[L]_{rs} = [X]/M_S \tag{12}$$

where [X] is W_L/W_0 as before. Since the average molecular weight of the chain segment between two consecutive reactive sites may be considered to be a constant for lignin in a given wood species, M_S will not be affected by the extent of delignification. Therefore, after substitution for lignin concentration in Equation 9, M_S can be eliminated from both sides of the equation. Hence one obtains:

$$-d[X]/dt = k_1[X]-k_2([X]_0-[X]) \tag{13}$$

where $[X]_0$ and [X] are the weight percent of lignin in the wood sample at time 0 and t, respectively, both expressed with respect to initial weight of wood.

Evaluation of the Kinetic Model

Experimental Data. The experimental system has been described previously (1,2). It involves a tubular reactor for extraction and a separator trap for precipitation. In a typical run, about 3.5 grams of red spruce sawdust (16 mesh) was packed into the reactor (19 cc) and delignification was carried out for different time periods in the temperature range from 170 to 185°C and the pressure range from 172 to 276 bar. Methylamine (99.8% purity from Airco) and nitrous oxide (99.9% purity from Linde) at selected compositions were continuously pumped into the system at a constant flow rate (1 g/min). Delignification conditions were all above the critical temperature and the pressure of the fluids used. [The critical properties of methylamine, nitrous oxide and their binary mixtures have been reported elsewhere (25)]. The reaction mixture was then released to the separator (35 cc) where the dissolved lignin was precipitated at atmospheric pressure. After a desired period of delignification time, the residue remaining in the reactor and the precipitate collected in the separator were taken out and analyzed. Lignin contents were determined as the acid-insoluble lignin (Klason lignin) using a modified procedure (26).

Evaluation of the Kinetic Parameters. The analytical solution for [X] in Equation 13 is given by:

$$[X] = ([X]_0 + C_2/C_1)e^{-C_1 t} - C_2/C_1 \tag{14}$$

where:

$$C_1 = k_1 + k_2 \tag{15}$$

$$C_2 = -k_2[X]_o \tag{16}$$

In this study, the lignin content [X] in wood has been replaced by Klason lignin contents of the residues. It should be noted that Klason lignin only accounts for the acid-insoluble portion of the lignin in the wood residue and therefore represents the lower limit of the lignin that actually exist in the wood residue. The amount of soluble lignins that may remain in the residue has not been further investigated in this study.

Figure 1 compares the experimental data (Klason lignin contents) for delignification of red spruce by methylamine at 185°C and 276 bar with the predictions from the reaction-diffusion model (Equation 13) and from n^{th} order homogeneous reaction models. For the reaction-diffusion model, the rate constants k_1 and k_2 have been calculated to be 0.5163 and 0.2012 [1/hr], respectively. The figure demonstrates that the reaction-diffusion model proposed in this work is capable of predicting the experimental data over a much wider range. The predictions from the first-order (n=1) homogeneous reaction model grossly deviates from the experimental data points. The second-order reaction model shows substantial improvement over the first-order but still underpredicts the extent of delignification. Whereas the third-order reaction model over-predicts the extent of delignification. All four models seem to agree and converge at long extraction times. There is some deviation for the reaction-diffusion model in the prediction of lignin content at low extraction times (i.e., half hour data point). This may however be a consequence of the delay period in the start-up of extraction which introduces some uncertainty in total extraction time when extraction times are short. There is a time period over which the system pressure is increased to the extraction conditions.

Effect of Temperature on Reaction Rate. The Klason lignin contents of red spruce residue extracted by methylamine at four different temperatures along with the predicted curves from the reaction-diffusion model are shown in Figure 2.

The calculated (least square fit) values of k_1 and k_2 according to Equation 13 are given in Table I. The standard deviations expressed as wt% Klason lignin refer to deviation between lignin predicted (using the indicated values of k_1 and k_2 and Equation 13) and the experimental data.

The table shows that the rate constant k_1 increases with temperature as would be expected from the Arrhenius relationship (Equation 3). However, the constant for diffusion (k_2) is seen to decrease with increasing temperature under the experimental conditions according to:

$$\ln k_2 = \ln A^* + E^*/RT \tag{17}$$

where A^* is a constant and E^* is the activation energy for diffu-

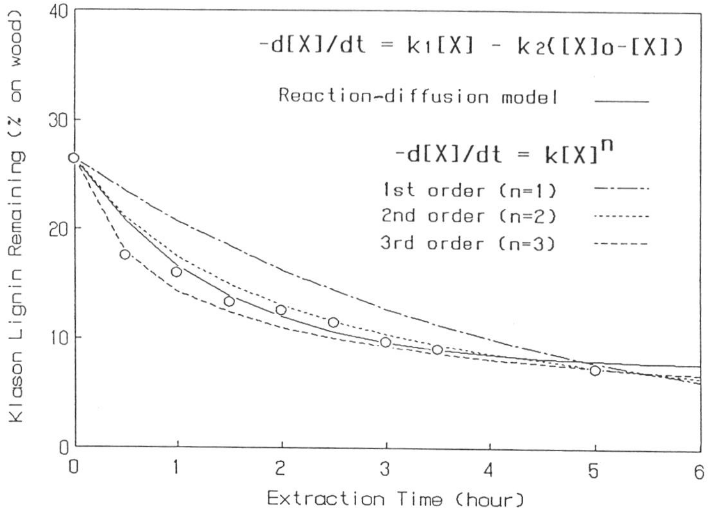

Figure 1. Comparison of the extent of delignification predicted by the homogeneous (first-order, second-order and third-order) models and by the reaction-diffusion kinetic model. The open circles are the measured Klason lignin contents in the residues obtained from methylamine extraction of red spruce at 276 bar, 185°C and 1 g/min solvent flow rate.

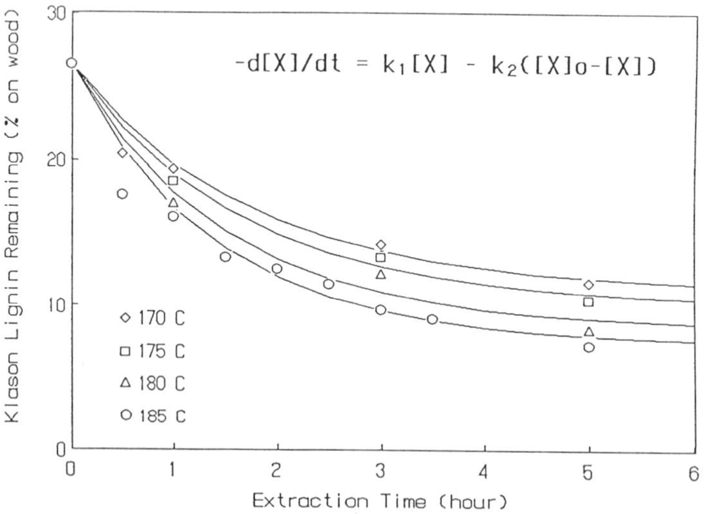

Figure 2. Comparison of the extent of delignification predicted by the reaction-diffusion model with the measured Klason lignin contents in the residues obtained from supercritical methylamine extraction of red spruce at 276 bar, 1 g/min solvent flow rate, and four different temperatures (170, 175, 180, and 185°C).

Table I. The Rate Constants for Delignification of Red Spruce by Supercritical Methylamine at 276 bar and Four Different Temperatures

	k_1 [1/hour]	k_2 [1/hour]	Standard Deviation [wt% Klason Lignin]
170°C	0.3375	0.2447	0.3703
175°C	0.3769	0.2322	0.5028
180°C	0.4541	0.2155	0.9483
185°C	0.5163	0.2012	0.5921

sion process. The decrease in k_2 with temperature indicates that the second term in Equation 13 becomes smaller (a decrease in the diffusion resistance term) with temperature and hence lead to an increase in the overall rate of delignification. Factors that contribute to this type of temperature dependence of the diffusion resistance term can be various. One of the factors that would influence k_2 is the viscosity of the solvent. A decrease in the solvent viscosity may facilitate diffusion. Even though there is no data in the literature on the viscosity of methylamine and methylamine-nitrous oxide mixtures at the specific extraction conditions, in most cases the viscosity of fluids is found to be inversely proportional to temperature (27). Another factor that may affect diffusion process is the density of the solvent. For example, self-diffusion coefficients of supercritical ethylene increases drastically with decreasing density of ethylene (28). Since both the fluid viscosity and density would become more favorable for easier diffusion with increasing temperature, the diffusional resistance to delignification is expected to decrease with increasing temperature. If the level of lignin removal ($[X]_0$-[X]) is maintained constant, then k_2 should decrease with increasing temperature, which is in accord with the observed temperature dependence of this quantity.

By a linear regression of Equation 3 using the data given in Table I, activation energy and frequency factors have been calculated and are given in Table II. (The standard deviation in this

Table II. The Activation Energy and Frequency Factor for Delignification Reactions in Supercritical Methylamine at 276 bar

	lnA	E/R [K]	A [1/hour]	E [KJ/mol]	Corr. coeff.	Standard deviation
$\ln k_1$	12.30	5935	2.197×10^5	49.32	0.9946	0.02398
$\ln k_2$	-7.455	2682	5.78×10^{-4}	22.29	0.9971	0.00795

table refers to the deviation between lnk calculated using the E and A values indicated in the table and the values based on Table I).

As shown, the activation energies are calculated to be 49 KJ/mol for the reaction term (k_1) and 22 KJ/mol for the diffusion resistance term (k_2), respectively. These values are in the lower range of reported activation energy values (40 - 150 KJ/mol) for kraft and high pressure delignification processes (11,13,29-31). In conventional pulping, three regions of delignification (which are often described by psuedo-first-order kinetic models) are observed (7,13). These regions, referred to as the initial, bulk, and residual delignification stages, show different activation energies, the low values are often associated with the initial delignification stage.

It should be noted that the temperature dependence of k_1 and k_2 given in Table II is based on data obtained at 276 bar. To estimate the rate constants at other pressure levels, one need to correlate the rate constants with pressure which is discussed in the following section.

Effect of Pressure on Reaction Rate. The variation of Klason lignin contents of red spruce residue with extraction pressure is shown in Figure 3. Based on these data points, the rate constants at different pressures have been estimated and are given in Table III.

TABLE III. The Rate Constants for Delignification of Red Spruce by Supercritical Methylamine at 185°C and Four Different Pressures

	k_1 [1/hour]	k_2 [1/hour]	Standard deviation [wt% Klason lignin]
172 bar	0.3375	0.1560	0.5167
207 bar	0.4691	0.1994	0.8235
241 bar	0.5432	0.2270	0.6972
276 bar	0.5163	0.2012	0.5921

They have been fitted with the parabolic relationship given by Equation 7 resulting in:

$$\ln k_1 = -4.057 + 0.02720P - 0.00005388P^2 \tag{18}$$

$$\ln k_2 = -5.968 + 0.03705P - 0.00007688P^2 \tag{19}$$

where k_1 and k_2 are in units of [1/hour] and P in [bar]. According to Equation 5, the volume of activation terms are expressed as:

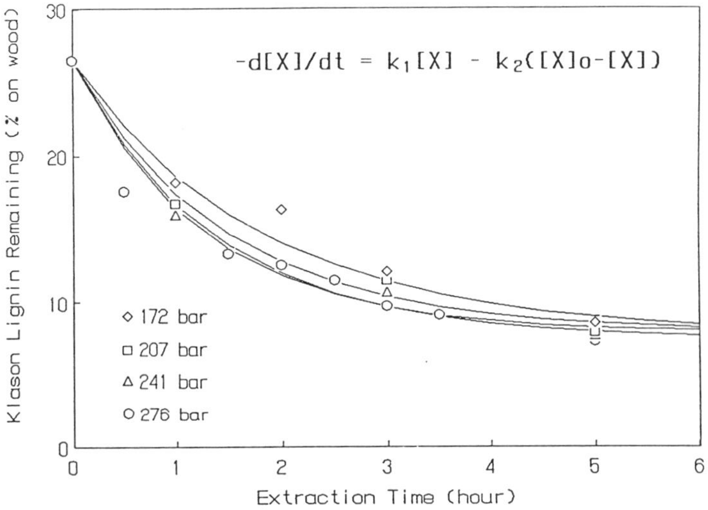

Figure 3. Comparison of the extent of delignification predicted by the reaction-diffusion model with the measured Klason lignin contents in the residues from supercritical methylamine extraction of red spruce at 185°C, 1 g/min solvent flow rate, and four different pressures (172, 207, 241, and 276 bar).

$$\Delta V_1^{\neq} = -1036 + 4.105P \qquad (20)$$

$$\Delta V_2^{\neq} = -1411 + 5.857P \qquad (21)$$

where $\Delta V_1^{\neq}$ and $\Delta V_2^{\neq}$ are in [cm^3/mol] and P is in [bar]. $\Delta V_1^{\neq}$ and $\Delta V_2^{\neq}$ show a sign change (and an accompanying decrease in k_1 and k_2) above 252 and 241 bar, respectively. This reversal in sign is not unusual since volumes of both the reactants and the transition state change with pressure and since there is no a priori reason for their having similar compressibilities. The plots of lnk versus pressure are in general curved towards the pressure axis (23).

Effect of Solvent Concentration on Reaction Rate. In the kinetic modeling of chemical (such as kraft) pulping, the effect of cooking liquor on delignification rate is sometimes considered. For example, the alkali concentration [OH^-] can be included in the rate equation (12,13,32). Since the extraction experiments in this study have been conducted under constant solvent flow (1 g/min) and the solvent ratios in binary fluid extractions have been maintained at a constant, one can combine the solvent concentration factor into an effective rate constant (k_{eff}). Therefore, Equation 13 can be rewritten as:

$$-d[X]/dt = k_{1eff}[X] - k_{2eff}([X]_o-[X]) \qquad (22)$$

with

$$k_{1eff} = k_1[S]^n \qquad (23)$$

and

$$k_{2eff} = k_2[S]^m \qquad (24)$$

where k_{1eff} and k_{2eff} are the effective rate constants for reaction and diffusion processes, respectively, [S] is the concentration of methylamine expressed as weight fraction in the binary extraction fluid, and n and m are some constants to be determined. If [S]=1 (corresponding to pure methylamine solvent), Equation 22 reduces back to Equation 13. If it is a binary solvent (methylamine-nitrous oxide), k_{1eff} and k_{2eff} are expected to have different values for k_1 and k_2, respectively.

Following similar procedure used for determination of rate constants (k_1 and k_2), k_{1eff} and k_{2eff} for methylamine-nitrous oxide extraction at 60 wt% of methylamine ([S] = 0.6), 185°C, 276 bar and 1 g/min solvent flow rate have been found to be 0.2399 [1/hour] and 0.1997 [1/hour], respectively. Substituting these values along with the values of k_1 and k_2 into Equation 23 and Equation 24, one solves n as 1.500 and m as 0.01465. A physical interpretation of n and m is that nitrous oxide acts as an inert which blocks the active reaction sites on the surface of the inner cell wall, hence reducing the observed rate constant. On the other hand, the diffusion process (transporting dissolved fractions into bulk phase) should not be affected by an inert as long as the inert is compatible with the active reagent and the dissolved materials. This explanation is supported by a near zero value of m, indicating that nitrous oxide has indeed no significant effect on the diffusion process.

Hence the final expression of the kinetic model for delignification of red spruce by supercritical methylamine and methylamine-nitrous mixture is expressed as:

$$-d[X]/dt = k_1[X][S]^{1.5}-k_2([X]_o-[X]) \quad (25)$$

where m has been set to zero for simplicity.

Figure 4 shows the delignification curves predicted with Equation 25 at five compositions of methylamine-nitrous oxide mixture, along with the experimental data points. The standard deviation of the Klason lignin content of red spruce residue from methylamine (60 wt%)-nitrous oxide mixture extraction is 0.2103% as compared to 0.5921% from pure methylamine extraction. There is only one experimental data point for each of the other three compositions (20 wt%, 40 wt%, and 80 wt%). The data point from methylamine (80 wt%)-nitrous oxide extraction appears to be noticeably different from the predicted delignification curve. But based on the experimental data adjacent to this solvent composition--100 wt% methylamine (8 points) and 60 wt% methylamine (5 points), the predicted curve for 80 wt% methylamine is considered more reliable than the single data point.

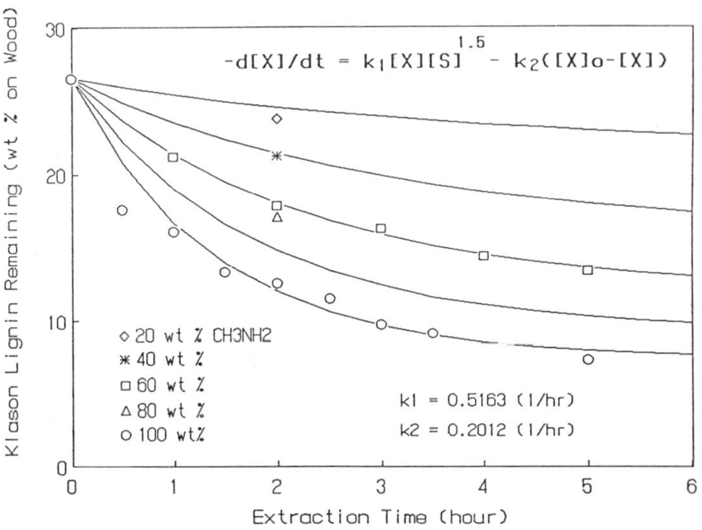

Figure 4. Comparison of the extent of delignification predicted by the reaction-diffusion model with the measured Klason lignin contents in the residues from supercritical methylamine extraction of red spruce at 272 bar, 185°C, 1 g/min solvent flow rate, and five different solvent compositions (20, 40, 60, 80, and 100 wt% of methylamine).

CONCLUSIONS

This study has shown that a reaction-diffusion model can effectively describe the supercritical delignification of wood, accounting for temperature, pressure, and solvent effects over a wide range. The residual lignin contents predicted by the model are in good agreement with the measured Klason lignin contents of the residues from supercritical methylamine-nitrous oxide delignification of red spruce.

ACKNOWLEDGMENT

This research has in part been supported by the National Science Foundation (Grant No. CBT-84168875).

LITERATURE CITED

1. Li, L.; Kiran, E. Ind. Eng. Chem. Res. 1988, 27, 1301-1312.
2. Li, L.; Kiran, E. Tappi J. 1989 (in press).
3. Li, L.; Kiran, E. Paper presented at the 195th ACS National Meeting, Toronto, Canada, 1988.
4. Li, L. Ph.D. Thesis, University of Maine, Orono, Maine, 1988.
5. Kiran, E. Tappi J. 1987, 70(11), 23-24.
6. Kiran, E. American Papermaker 1987 (November), 26-28.
7. Sjostrom, E. Wood Chemistry--Fundamentals and Applications; Academic Press: New York, 1981; Chapter 4.
8. Fengel, D.; Wegener, G. Wood-Chemistry, Ultrastructure, Reactions; Walter de Gruyter: London, 1983; Chapter 8.
9. Szekely, J.; Evans, J.W.; Sohn, H.Y. Gas-Solid Reactions; Academic Press: New York, 1976.
10. Doraiswamy, L.K.; Sharma, M.M. Heterogeneous Reactions: Analysis, Examples, and Reactor Design, Vol. 1: Gas-Solid and Solid-Solid Reactions; John Wiley & Sons: New York,1984.
11. Kleinert, T.N. Tappi 1966, 49(2), 53-57.
12. Yan, J.F. Tappi 1980, 63(11), 154.
13. Kondo, R.; Sarkanen, K.V. Holzforschung 1984, 38(1), 31-36.
14. Eyring, H.; Lin, S.H.; Lin, S.M. Basic Chemical Kinetics; John Wiley & Sons: New York, 1980; Chapter 9.
15. Evans, M.G.; Polanyi, M. Trans. Faraday Soc. 1935, 31, 875-895.
16. Koch, E. Non-Isothermal Reaction Analysis; Academic Press: New York, 1977; Chapter 2.
17. van Eldik, R. Inorganic High Pressure Chemistry--Kinetics and Mechanisms; Elsevier: New York, 1986.
18. Whalley, E. Trans. Faraday Soc. 1958, 55, 798-808.
19. Isaacs, N.S. Liquid Phase High Pressure Chemistry, John Wiley & Sons: New York, 1981; Chapter 4.
20. Lohmuller, R.; Macdonald, D.D.; Mackinnon, M.; Hyne, J.B. Can. J. Chem. 1978 56, 1739-1745.
21. Plamer, D.A.; Kelm, H. Coord. Chem. Rev. 1981, 36, 89-153.
22. Asano, T.; Okada, T. J. Phys. Chem. 1984, 88, 238-243.
23. Tiltscher, H.; Hofmann, H. Chem. Eng. Sci. 1987, 42(5), 959-977.

24. Eckert, C.A. Ann. Rev. Phys. Chem. 1972, 239-264.
25. Li, L.; Kiran, E. J. Chem. Eng. Data 1988, 33, 342-344.
26. Effland, M.J. Tappi, 1977, 60(10), 143-144.
27. Reid, R.C.; Prausnitz, J.M.; Poling, B.E. The Properties of Gases and Liquids, 4th ed.; McGraw-Hill: New York, 1987.
28. Jonas, J.; Lamb, D.M. In Supercritical Fluids; Squires, T.G.; Paulaitis, M.E., Eds; ACS Sumposium Series 329, 1987.
29. Olm, L.; Tistad, G.; Svensk Papperstidn. 1979, 82, 458-464.
30. Beer, R.; Peter, S. in Supercritical Fluid Technology, Penninger, J.M.L.; Radosz, M.; McHugh, M.A.; Krukonis, V.J., Eds; Elsevier: New York, 1985.
31. Norden, S.; Teder, A. Tappi 1979, 62(7), 49-51.
32. Kleppe, P.J., Tappi 1970, 53(1), 35-47.

RECEIVED May 1, 1989

RATE PROCESSES: CRYSTALLIZATION, HEAT, AND MASS TRANSFER

Chapter 22

Gas Antisolvent Recrystallization: New Process To Recrystallize Compounds Insoluble in Supercritical Fluids

P. M. Gallagher[1], M. P. Coffey[1], V. J. Krukonis[1], and N. Klasutis[2]

[1]Phasex Corporation, 360 Merrimack Street, Lawrence, MA 01843
[2]Air Force Armaments Laboratory, Eglin Air Force Base, FL 32542–5435

The recrystallization of nitroguanidine from N-methyl pyrrolidone and N,N-dimethyl formamide using supercritical fluids and gases near their vapor pressures as anti-solvents was investigated. The nitroguanidine used for the recrystallization study consisted of high aspect ratio needles, 5 x 100 microns; because of its low bulk density the as-produced nitroguanidine is not satisfactorily incorporated into explosives formulations at high solids loading. Depending upon the specific combinations of parameters, the particle size and size distribution could be varied over a wide range, e.g., spherical particles of 100 microns diameter (the desired shape and size), low aspect ratio crystals, unusual star-shaped clusters, loose spherical agglomerates, or monodisperse particles of one micron or less. The results presented for nitroguanidine define a quite general recrystallization concept for processing difficult-to-comminute solids.

The particle size and size distribution of solid materials produced in industrial processes are frequently not those desired for subsequent use of these materials, and as a result, comminution and recrystallization operations are carried out on a large scale in the chemicals, pharmaceuticals, dyes, polymers, and explosives industries in order to take a material from one size or size distribution and change it into another. The processes used to accomplish the particle size changes are as diverse as the materials they are practiced on. As examples of methods for particle size redistribution, there are simple crushing and grinding (which for some compounds are carried out at cryogenic temperatures) ball milling (with or without milling aids), air micronization (also called jet impingement or fluid energy milling), sublimation, and recrystallization from solution. The latter technique, recrystallization, can be carried out by several processes, e.g., thermal methods, using the temperature dependence of solubility, and anti-solvent methods using a non-solvent to decrease the solubility

0097–6156/89/0406–0334$06.25/0

of a solid which is dissolved in a solvent. Additionally, chemical reactions of soluble materials to produce an insoluble (and desired) product is practiced on an industrial scale in the formation of dyes and other products, the reactant concentration and product precipitation controlled to give the desired particle size distribution.

There are practical problems associated with many of the above processes; for example, there are many materials that are difficult to process by grinding or solution techniques for one reason or another. Certain dyes, chemical intermediates, biological and pharmaceutical compounds which are "waxy" or "soft", certain specialty polymers, and explosives are a few categories of difficult to process materials.

Supercritical fluids have been shown to be excellent recrystallization agents for a variety of materials, and their use in recrystallizing an explosive, nitroguanidine (NQ), is presented. A new process, GAS (gas anti-solvent) Recrystallization, is described.

Background

Supercritical fluids have been used as solvents for a wide variety of extractive applications[1]. Some specific examples are coffee and tea decaffeination[2,3], hops and essential oils extraction[4], fractionation of polymers[5,6], purification of reactive monomers[7], extraction of cholesterol from animal fats[8] and eggs[9], and others[1,4]. Some of the processes are in production, and others are in the research or advanced development stage. Gases such as carbon dioxide, the light hydrocarbons, e.g., ethane, ethylene and propane, and the chlorofluorocarbons have been used to advantage in specific instances. These processes are based on the pressure-dependent solvent properties of gases and liquids above their critical points first demonstrated in 1879 by Hannay and Hogarth[10]. Todd and Elgin first described the use of supercritical fluids for separating materials in 1955[11]; one or more materials in a mixture can be dissolved in a supercritical fluid at high pressure, and the material(s) can be recovered when the pressure is reduced. The separation of solutions of liquids using supercritical fluids was reported in 1959 by Elgin and Weinstock[12].

Recently, supercritical fluids have been investigated in an important "non-extractive" application, viz., Supercritical Fluid Nucleation[13,14,15,16,17,18,19]. In this process a solid is dissolved in a supercritical fluid at some temperature and pressure conditions, and the solution is expanded to some lower pressure level which causes the solid to precipitate. This concept has been demonstrated for a wide variety of materials including polymers[14], dyes[14] and steroids[19]. By varying the process parameters that influence supersaturation and nucleation rates particles can be obtained which are quite different in size and morphology from the parent material. Supercritical Fluid Nucleation can be an attractive recrystallizing method for many solids, especially for some difficult-to-comminute or -recrystallize materials such as pharmaceuticals used in dermal salves, injectable solutions, and ophthalmological preparations which require ultra-fine and uniform particles. As an example of the fine particle size formation capabilities of Supercritical Fluid Nucleation, Figures 1a

(a)

(b)

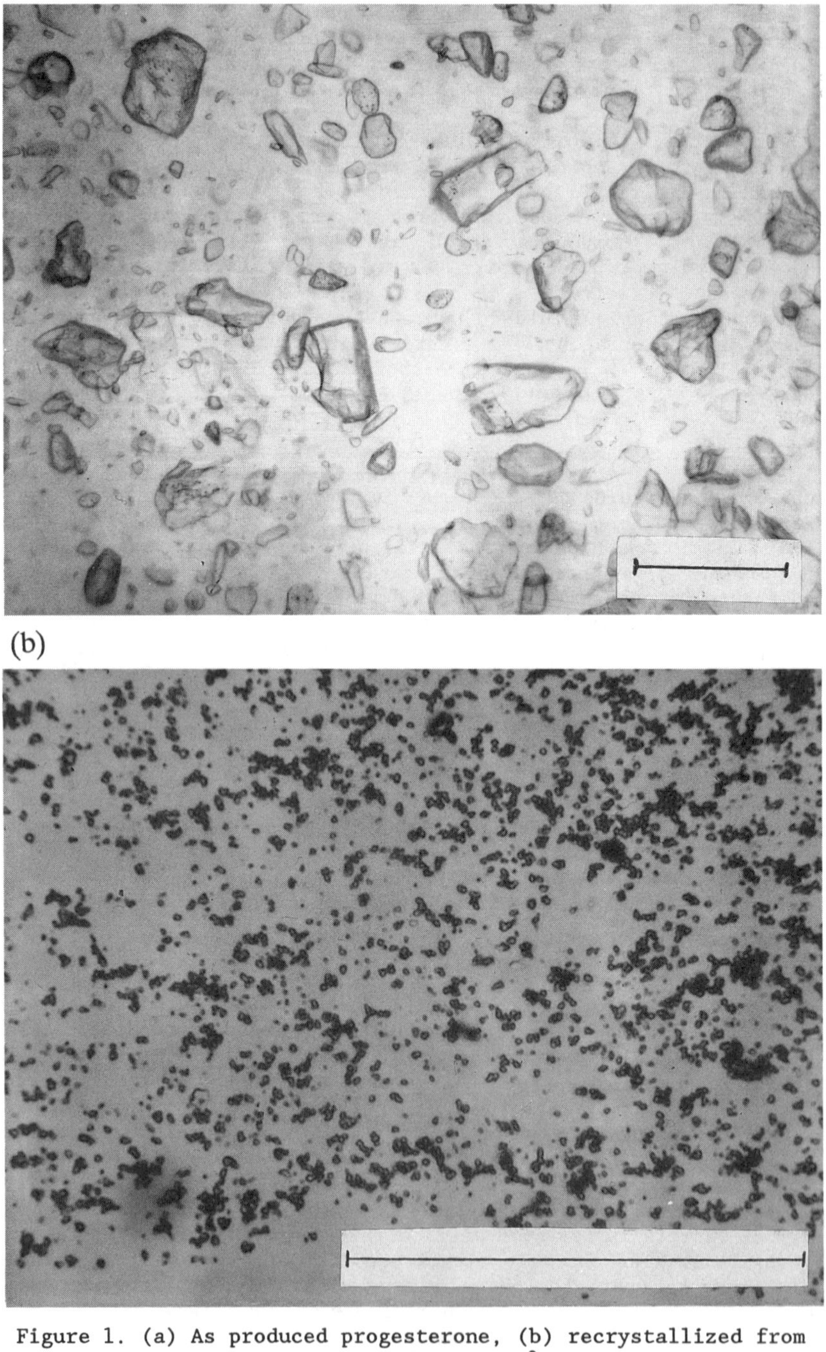

Figure 1. (a) As produced progesterone, (b) recrystallized from supercritical carbon dioxide, 4000 psi, 55°C expanded to ambient.

and 1b show parent (as-received) progesterone and progesterone recrystallized from supercritical carbon dioxide; the recrystallized progesterone exhibits a uniform 3 micron particle size.

One obvious requirement for the process of Supercritical Fluid Nucleation is that the solid material to be recrystallized dissolve in some supercritical fluid to an "appreciable" extent. However, not all solids dissolve in the simple supercritical fluids mentioned earlier, and, therefore, it is sometimes difficult to satisfy the solubility requirement. The use of supercritical fluids and, additionally, gases near their vapor pressures, as anti-solvents is an attractive alternative recrystallization technique for processing particularly difficult-to-comminute or -recrystallize solids that are insoluble in supercritical fluids. This method, termed GAS (gas anti-solvent) Recrystallization, exploits the ability of gases to dissolve in organic liquids and to lower the solvent power of the liquid for the compounds in solution, thus causing the solids to precipitate. Although the media for recrystallization in the respective processes are different, the principles of precipitation by GAS Recrystallization are basically the same as those in operation in Supercritical Fluid Nucleation. Furthermore, there are essentially no limitations for employing GAS Recrystallization provided that certain solubility relationships exist among the solid material, the liquid (solvent), and the gas (anti-solvent). If an organic liquid (which is a solvent for the solid to be recrystallized) and a gas are at least partially miscible, introduction of the gas into the liquid solution will result in expansion of the liquid phase, reducing its solvent power for the solid, and the solid will precipitate. Depending upon parameters which influence the supersaturation ratio and the rate of particle growth and which are discussed subsequently, the crystal size, size distribution, and shape can be varied over wide ranges.

Experimental Methodology

GAS Recrystallization of an explosive, nitroguanidine (NQ), was studied. NQ is not soluble in carbon dioxide, the light hydrocarbons, and the chlorofluorocarbons, but it is soluble in several organic liquids, e.g., N-methylpyrrolidone (NMP) and N,N-dimethyl formamide (DMF). It was expected that several gases would be found soluble in the two liquids, and thus, that GAS Recrystallization could be used to recrystallize NQ. Several gases were screened for their expansion characteristics with DMF and NMP using a standard 5000 psi Jerguson gauge (Jerguson Valve and Gage Company, now a division of Clark-Reliance, Strongsville, OH). Carbon dioxide, chlorodifluoromethane (CFC-22), and dichlorodifluoromethane (CFC-12) were found to be miscible with both liquids over wide temperature-pressure ranges.

In an interesting aside here, ethylene and ethane were found to be essentially insoluble in the two liquids even at pressure levels as high as 5000 psi. For example, ethane did not expand the liquids at all, from which it can be inferred that the solubility was nil, and ethylene expanded the liquids less than five percent. For comparison, liquid hexane was also found to be insoluble in DMF and NMP, and, thus, the insolubility of ethane is, perhaps, not unexpected considering the similarity in the solubility parameters of the two hydrocarbons, viz., about 7 $(cal/cc)^{1/2}$ for liquid hexane and about 6 for 5000 psi ethane at 20°C[20], and because the solubility parameter of DMF is so high, viz., 12.1 at 20°C[21]. On the other hand, the solubility parameter of carbon dioxide is also about 7 at 5000 psi 20°C[20]. In many systems hexane and carbon dioxide have been shown to be quite similar in their solubilizing characteristics[22,23,24], but with DMF and NMP carbon dioxide and hexane are very dissimilar in their behavior.

Figure 2 shows the room temperature expansion behavior of carbon dioxide-DMF. At low pressure levels the expansion follows Henry's Law, i.e., a straight line, $p=Hx$ path. At about 400 psi the expansion increases markedly, and as the pressure approaches the vapor pressure of carbon dioxide, carbon dioxide and DMF become miscible. The expansion behavior shown in Figure 2 is obtained by observing the increase in the volume of a known amount of the liquid that has been charged to a volume-calibrated Jerguson gauge as the carbon dioxide is introduced and mixed with the liquid at increasing pressure levels.

Higher temperature levels were not tested for the carbon dioxide-DMF system, but as an example of the general effect of temperature on expansion of liquids by carbon dioxide, Figure 3 shows the expansion of cyclohexanone, a liquid used in other GAS Recrystallization studies[25]; as is seen, higher temperature levels require higher pressure levels to reach similar expansion levels. There is a potential for amine-carbon dioxide reactions to occur in the nitroguanidine system which is the reason that carbon dioxide was not evaluated in depth; instead chlorodifluoromethane and dichlorodifluoromethane were tested as anti-solvents for GAS Recrystallization of NQ. Both chlorofluorocarbons expand DMF and NMP essentially identically. Since dichlorodifluoromethane is substantially more deleterious to the environment[26], and since the research was directed to the ultimate development of an industrial process for recrystallizing nitroguanidine, chlorodifluoromethane was chosen as the preferred gas anti-solvent. The expansion behavior of chlorodifluoromethane and NMP at two temperature levels is shown in Figure 4. As the figure shows (and as did also Figures 2 and 3), the gas need not be supercritical to expand a liquid solvent; hence, the "G" in the acronym GAS denotes the more general use of a "gas" as an anti-solvent.

Initial recrystallization tests were carried out in the Jerguson gauge to determine the parameters, viz., initial NQ concentration, rate of expansion, final pressure level, etc., that combine to yield particles of a certain kind. It is difficult, if not impossible, to predict *a priori* the particle size, size distribution, shape, and morphology that will result from specific combinations of experimental parameters, and, thus, it is informative to present some simple nucleation and particle growth theory insofar as the theory guided the course of the experimental investigation in its early stages.

Principles Underlying GAS Recrystallization

The equilibrium in an assembly of dissolved molecules which can

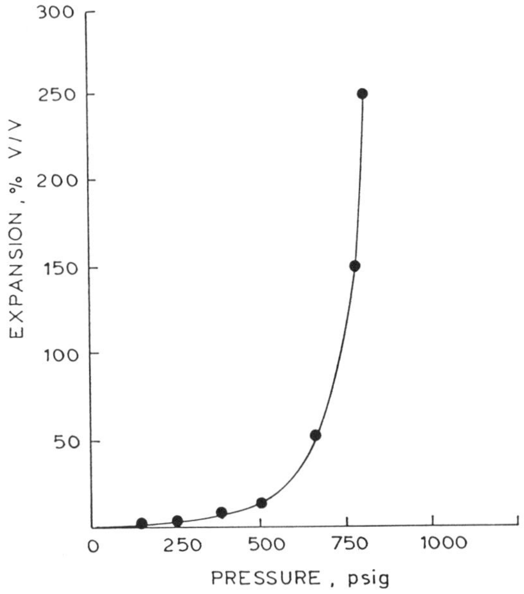

Figure 2. Volumetric expansion of DMF by carbon dioxide.

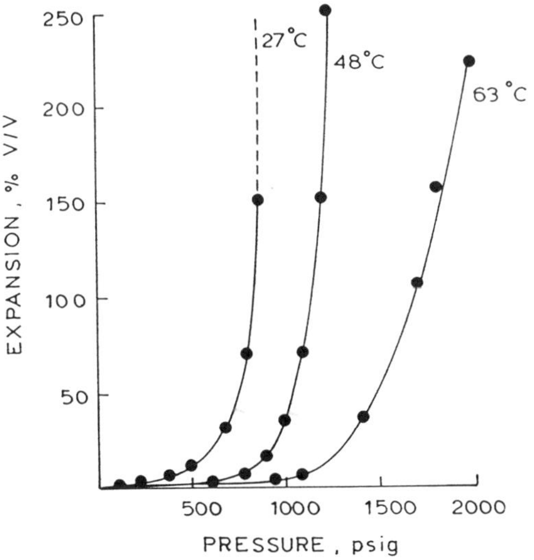

Figure 3. Volumetric expansion of cyclohexanone by carbon dioxide.

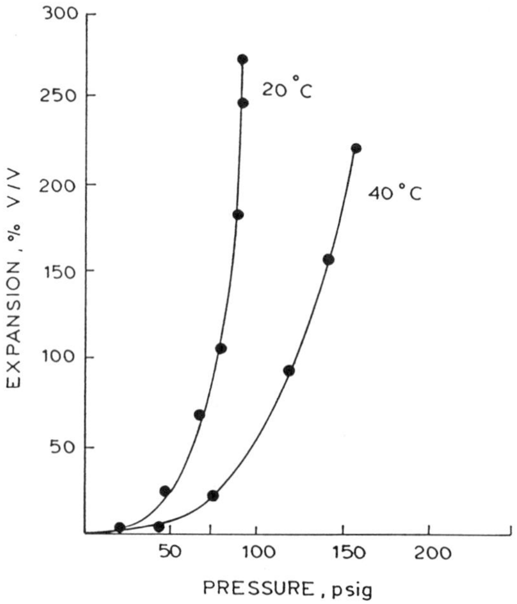

Figure 4. Volumetric expansion of NMP by chlorodifluoromethane.

combine to form a critical nucleus beyond which size favorable fluctuations will cause the nuclei to grow has been described by Gibbs[27]. (The term "critical" here does not relate to the critical point of a gas or liquid; it has been used by many authors to describe the specific nucleus size that can subsequently grow to form particles.) Gibbs presented the conditions of critical nuclei formation based upon free energy considerations[27], and Adamson[28] gives an excellent treatment of Gibbs' mathematical description. The formation of an embryo, i.e., an assembly of molecules of a size smaller than the critical nucleus, requires that an interface between two phases form. Thus, the free energy of the system will have to increase initially until an embryo reaches some critical diameter. Once the embryo is of sufficient size, i.e., of the critical diameter described by Gibbs (and, incidentally, Gibbs used the term "globules" to describe assemblies of molecules), there are two competing modes of lowering the particle free energy: the nucleus can grow indefinitely, or it can shrink and disappear. Critical nuclei will grow at the expense of subcritical embryos so that the final particle size will be dependent upon some initial critical nuclei concentration. Adamson also presents a readable development of the rate equations describing nucleation first derived by Becker and Doring[29]. Using the free energy considerations of Gibbs, Becker and Doring derived an equation for the rate of formation of nuclei of this critical size, viz.,

$$\text{rate} = Ze^{-\Delta G_{max}/RT} \qquad (1)$$

where

- Z is the collision frequency, calculable from classical kinetic theory and
- ΔG_{max} is the Gibbs free energy expression which derives from nucleus surface energy considerations and is given below.

The free energy relation is

$$\Delta G_{max} = B/[RT\ln S]^2 \qquad (2)$$

where

- B is a constant derivable from physical properties of the system,
- S is the supersaturation ratio. (For purposes of illustration, in a vapor-liquid system S is the ratio of actual component pressure to the normal vapor pressure; in a system involving a solute dissolved in a solvent, S is the ratio of actual concentration to saturation concentration.)

From Equations 1 and 2 it is seen that the nucleation rate is very strongly influenced by the supersaturation ratio. As an example of the rapid increase in the nucleation rate as supersaturation is increased, Adamson shows for the condensation of water vapor that the number of nuclei formed (per second per cc for water vapor at 0°C) is only 10^{-8} at a supersaturation ratio of 3.5 whereas it is 10^{3} at a ratio of 4.5. The ratio at which the nucleation rate changes from imperceptible (e.g., 10^{-8}) to very large (e.g., 10^{3}) is termed the critical supersaturation ratio. The rate of nucleation at that point is termed catastrophic, and the phenomenon is termed catastrophic nucleation.

Equations 1 and 2 are strictly applicable only to the homogeneous condensation of a vapor to liquid droplets. In a two or three component system such as anti-solvent recrystallization, for example, or for recrystallization from a melt, other factors such as viscosity of the medium, the mode of nucleation, i.e., whether homogeneous or heterogeneous, etc., come into play, and the equations are modified by these other factors. Equations 1 and 2 are, nevertheless, of pedagogical value in predicting some general effects of these other systems.

The final particle size that is achieved in any particular recrystallization situation is related to both the rate of formation of critical nuclei described above and by the rate of growth of these nuclei. The rate of molecule transport at the interfacial boundary is difficult to describe in quantitative terms, but the overall rate of growth of the nuclei, once they have been formed during the catastrophic period, and the effect of concentration on growth can be given by the mass transfer relation

$$\text{flux of material to surface} = kA\Delta C \qquad (3)$$

where

k is a mass transfer coefficient

A is the surface area of the particle at any particular instant

ΔC is the concentration driving force at any particular instant; ΔC is given by $C\text{-}C_{eq}$ (or in words, the concentration of the component in the gas minus its equilibrium concentration.)

In an anti-solvent recrystallization process, then, particle size and particle size distribution is determined by the interaction between the nucleation rate and the growth rate of crystals, on one hand, and by the rate of creation of supersaturation, on the other hand; all three are influenced by the manner of addition of the anti-solvent. Figure 5 is a qualitative picture of simultaneous events that occur when an anti-solvent is added to a solution of a solute that is to be recrystallized. The three zones shown in Figure 5, designated I, II, and III, denote three areas of supersaturation. Zone I is for a supersaturation less than 1, i.e., for actual solute concentrations less than saturation. No growth of particles will occur in this zone (and in fact if there are any particles that are "somehow" present, they will dissolve). In Zone II, the supersaturation is less than the critical value discussed earlier, but "some" nucleation can occur; particles that are present in this

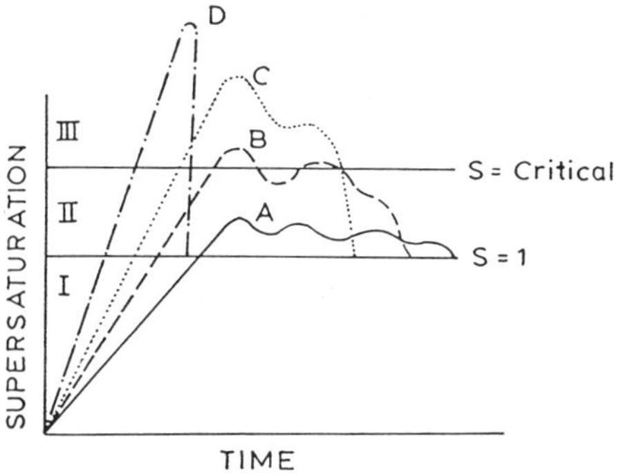

Figure 5. Variation of supersaturation with rate of addition of anti-solvent.

zone can grow by the diffusional mechanisms of Equation 3. Zone III represents very high supersaturation levels, and in this regime catastrophic nucleation can occur.

The four paths, A, B, C, and D, shown in Figure 5 represent the events for four different rates of addition of anti-solvent, and they depict the simultaneous creation of supersaturation (by anti-solvent addition) and the consumption of supersaturation (by nucleation and growth). The rate of addition as indicated by Curve A is low, and only very little nucleation occurs when the supersaturation exceeds S=1. The Path A rate of addition is so low that the rate of nucleation and the diffusional growth of those (few) nuclei formed maintain the supersaturation below the critical value until the concentration of solute in solution eventually falls to saturation, S=1. Curve B represents a higher rate of addition, high enough to exceed the critical supersaturation level. The solution is rather more quickly depleted of solute by the higher rate of nucleation and the higher overall growth rate (and several nucleation-supersaturation decreases are denoted by the relative maxima and minima in the curve). The addition via Curve C is a variant of Curve B, viz., a still higher rate of anti-solvent addition. Finally, the rate of addition depicted by Curve D is so high that almost all the solute in solution is consumed by the formation of nuclei, and the solute in solution is depleted almost solely by that mechanism.

These nucleation and growth events are depicted differently on a solution expansion-time plot in Figure 6. The starting point on the vertical axis is that amount of expansion that results in the onset of (visible) nucleation. In the GAS Recrystallization tests that were carried out in the Jerguson gauge, it was found convenient to compare results using as a starting point the first appearance of particles visible to the unaided eye. The appearance of particles, manifested as a haze, was termed the onset of nucleation, and the gas pressure which resulted in sufficient expansion to cause this nucleation to occur was termed the threshold pressure (THP). (The particles at this point are, of course, much larger than the "true" critical nuclei described by Gibbs; the Gibbs nuclei are assemblies of, perhaps, a few thousand molecules and, thus, they are invisible to the unaided eye.) Threshold pressure is a function of solution concentration, the higher the concentration, the lower the pressure required to initiate nucleation.

Figure 6 summarizes the various experimental expansion paths that were investigated during the studies; the expected (and experienced) results are noted on the expansion paths. Starting at THP, (for strict accuracy, at the amount of expansion at THP), if the subsequent expansion is essentially nil as shown by the Curve A expansion path, relatively few nuclei are formed as related earlier; they can grow to be large because there remains in solution a relatively large amount of solute after some nuclei are formed. Curves B and C depict the faster rates of addition of gas anti-solvent shown as Curves B and C in Figure 5; both rates of addition are similar in their effects on size and size distribution, i.e., both will result in a wide distribution, the specific final results a function of the particular rate of addition and concentration. Curve D represents an extremely high rate of expansion by rapid gas injection and pressure rise; essentially monodisperse and very small

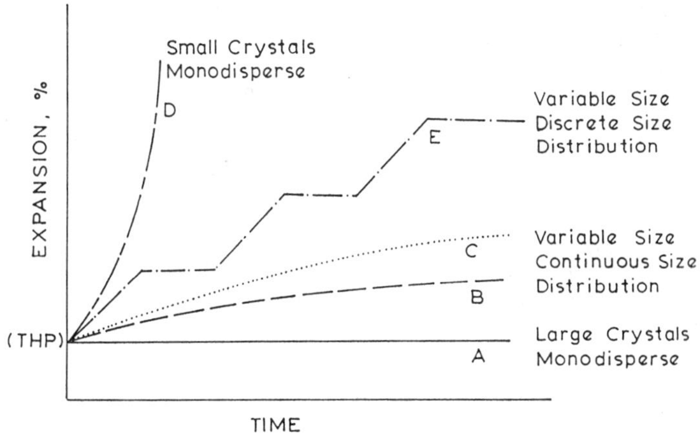

Figure 6. Expansion paths for anti-solvent addition.

particles will result. Another path, Curve E, which was not shown in Figure 5, is included in Figure 6 for completeness; a discrete particle size distribution is obtained by alternate gas injection and a hold at constant expansion for some period of time, the gas injection-hold time sequence continued until the solution is depleted in solute.

The various expansion paths shown pictorially in Figures 5 and 6 were investigated for their effects on particle size and size distribution, and generally the particle size and distribution results could be correlated to the experimental procedure. Occasionally, however, some inexplicable particle sizes and shapes were obtained. As stated earlier the initial tests were carried out in a Jerguson gauge. A recrystallization test is carried out in the following manner: an amount of NQ solution is charged to the cavity of the Jerguson gauge. Gas is admitted into the solution through a cotton filter located at the bottom of the cavity. The buoyant force propels the small bubbles of gas upward causing intimate mixing of gas and liquid resulting in dissolution of gas into the liquid and causing the liquid to expand. Particles of NQ form at "some" level of expansion as described earlier, and they settle to the bottom of the cavity. Depending upon the specific expansion path, a test is completed in a period of a few seconds or up to one hour. After the particles have settled, the solute-depleted solution is drained from the cavity through the filter, and the particles are trapped on the cotton. Fresh gas is then introduced until liquefaction pressure is reached; the solvent adhering to the particles is dissolved in the liquefied gas, and the solution is drained. The dried particles are sampled via a long retriever and are examined by optical microscopy. Depending upon particle size and morphology and depending upon the characteristics desired to be accented, particles were examined under magnification of 60 to 500X using either transmitted or reflected light.

Results and Discussion

Figure 7 is a photomicrograph of the nitroguanidine used for GAS Recrystallization studies. (The scale marker designates 100 microns on this and subsequent figures.) The primary particles are seen to be high aspect ratio needles, about 100 microns long by about 5 microns in cross section. With a very rapid expansion path, i.e., addition of gas within a period of a few seconds, the NQ particles that were formed were very small and regular, of the order of a few microns in size. (For most facile presentation conditions are given in the figure captions.) The ultra-fine particles that were formed were essentially the same for carbon dioxide injection or for chlorodifluoromethane or dichlorodifluoromethane injection into NQ solutions over a wide NQ concentration range from 1 to 10% (w/w). For example, Figures 8 and 9 show the material recrystallized by rapid injection of carbon dioxide and chlorodifluoromethane, respectively, into NQ-DMF solution; NQ-NMP solutions behave the same with the rapid expansion path.

Larger crystals of NQ and a wider distribution of particle sizes were produced with gas addition at an intermediate rate, e.g., by the addition via Curves B and E of Figure 6. Figure 10 shows the

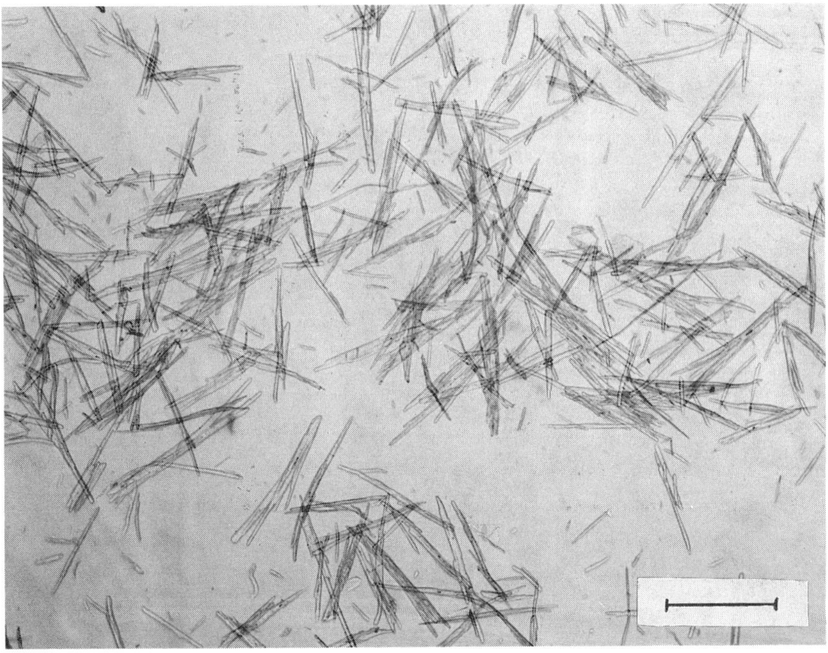

Figure 7. As received nitroguanidine.

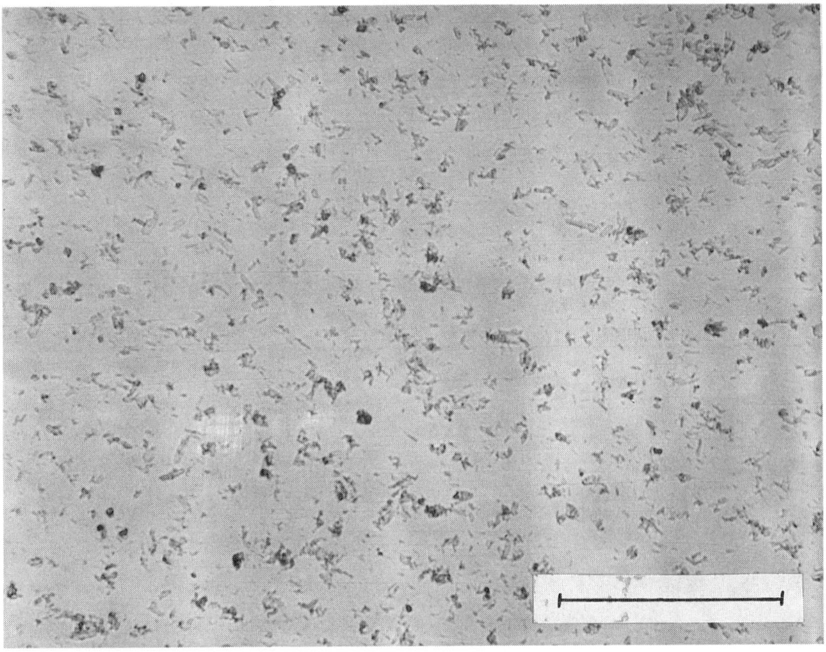

Figure 8. GAS Recrystallized NQ - small particles formed from 12% NQ-DMF, rapid (5 sec) injection of CO_2 to 750 psi, 20°C.

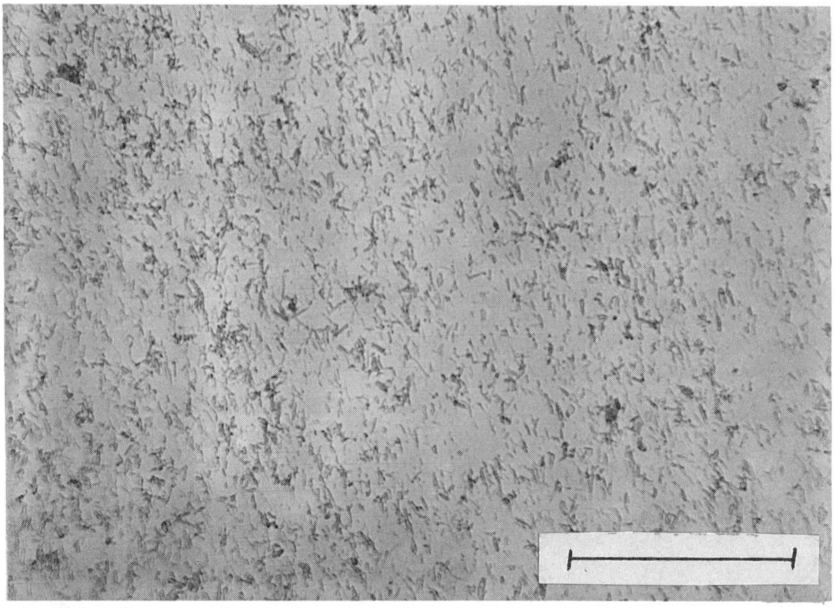

Figure 9. GAS Recrystallized NQ - small particles formed from 10% NQ-DMF, rapid (5 sec) injection of chlorodifluoromethane to 75 psi, 20°C.

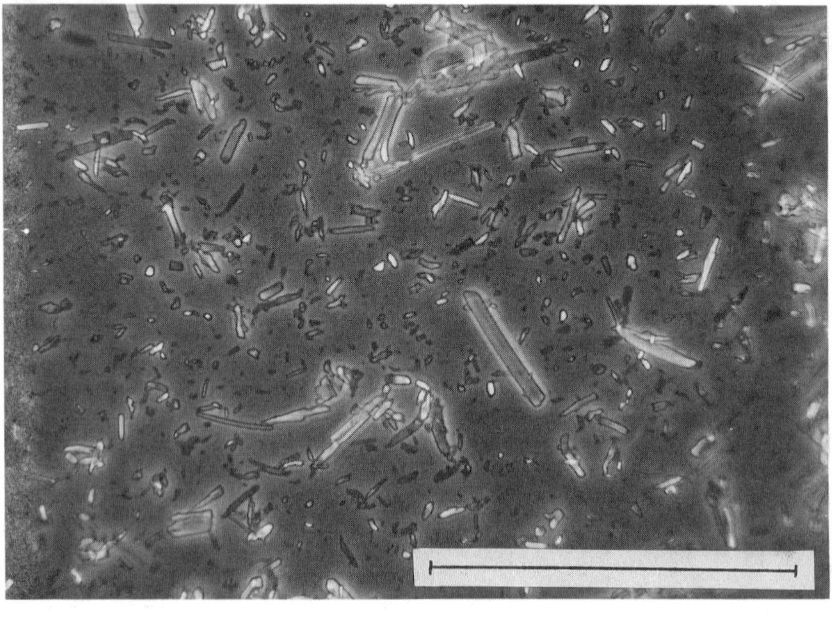

Figure 10. GAS Recrystallized NQ - continuously varying particle size formed from 10% NQ-DMF by continuous addition of chlorodifluoromethane to 180 psi, 20°C.

particles formed by continuous addition of anti-solvent, and a polydisperse particle size distribution is evident. A different type of distribution results with discrete or step-wise expansion of the solution; Figure 11 shows particles formed when gas was added in three steps, the first expansion to THP with a short hold, the next to an "intermediate" level of about 75% expansion, and finally a rapid rise to 150%. A trimodel distribution resulted. Interestingly, there is a propensity for elongated growth of NQ crystals in many of the tests, and even the very small particles shown in Figures 8 and 9 exhibit a slight aspect ratio.

Occasionally some "different shaped" NQ "particles" were obtained during the GAS Recrystallization tests. Figure 12 shows particles which resemble snowballs; the snowballs are quite friable and are actually loose agglomerates of primary particles. The primary needle-shaped particles that comprise the snowballs (and which are obtained by lightly smearing the snowballs on the microscopic slide) are shown in Figure 13. Another unusual structure was seen only at a certain concentration level of NQ in DMF. Figure 14 shows these particles, termed "starbursts", obtained during a series of tests investigating NQ concentration effects. The starbursts are, like the snowballs, also agglomerates of primary particles, but the starbursts are more coherent than the snowballs. Additionally, it is seen that the primary crystals comprising the starbursts are larger in cross-section than the needles comprising the snowballs of Figure 12. Concerning the concentration effects on starbursts, over the range of 1 to 10% initial concentration of NQ (in DMF), the starbursts formed at about 3%, but not at higher or lower NQ concentrations, nor were they formed in NMP solution.

If the expansion level is very low and maintained at THP, i.e., if expansion is carried out via Curve A addition shown in Figures 5 and 6, larger particles would be expected to form. Figure 15 shows such large particles formed by very slow addition to the threshold pressure with expansion subsequently maintained at this value. Large, dense, regular particles, i.e., spherical or cubical in shape, are desired for the explosive formulations.

Closing Remarks

GAS Recrystallization was directed to a specific explosive, nitroguanidine, but the process is quite general in its capabilities of recrystallizing virtually any solid material provided that the solid is soluble in some organic liquid and that some gas is soluble in the liquid sufficiently to expand it appreciably. The research is currently at the feasibility stage, and this paper reports some of the initial results. It is premature, therefore, to extend the results to, say, an economic evaluation of the process at some large production level. On the other hand, it is not premature to discuss some of the potential advantages of GAS Recrystallization. Gases that are soluble in liquids can be admixed "almost instantaneously", i.e., complete expansion can be made to occur literally within a few seconds, and, thus, extremely high supersaturation levels and nucleation rates can be attained resulting in the formation of extremely small particles not readily achievable by other processes. After the particles have been filtered, the solvent and anti-solvent

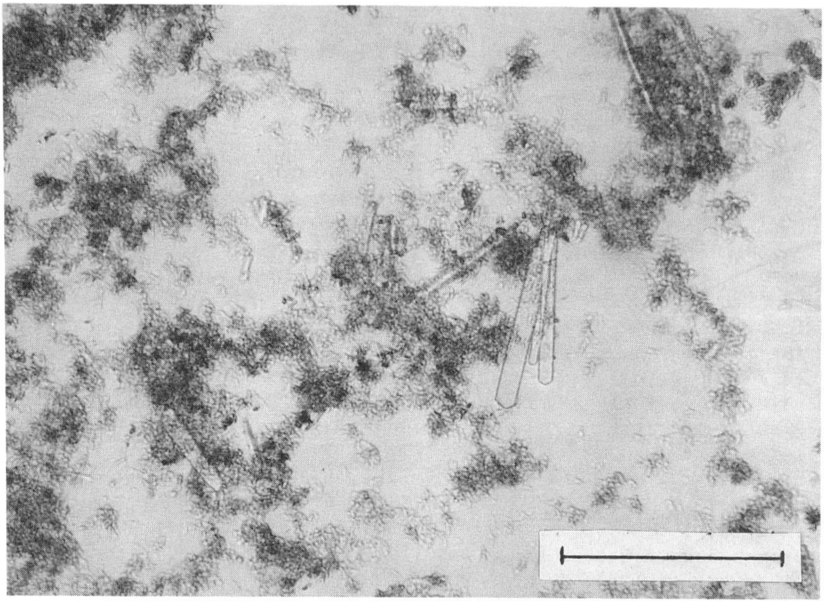

Figure 11. GAS Recrystallized NQ - discrete particle size distribution formed from 5% NQ-NMP, by stepwise addition of chlorodifluoromethane.

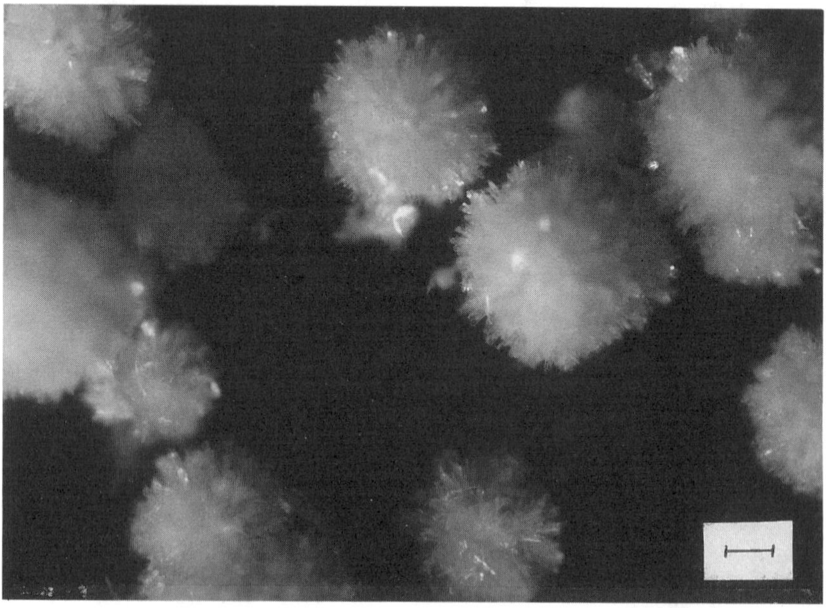

Figure 12. GAS Recrystallized NQ - snowballs formed from 8% NQ-DMF, moderate rate (5 min) injection of chlorodifluoromethane to 180 psi, 35°C.

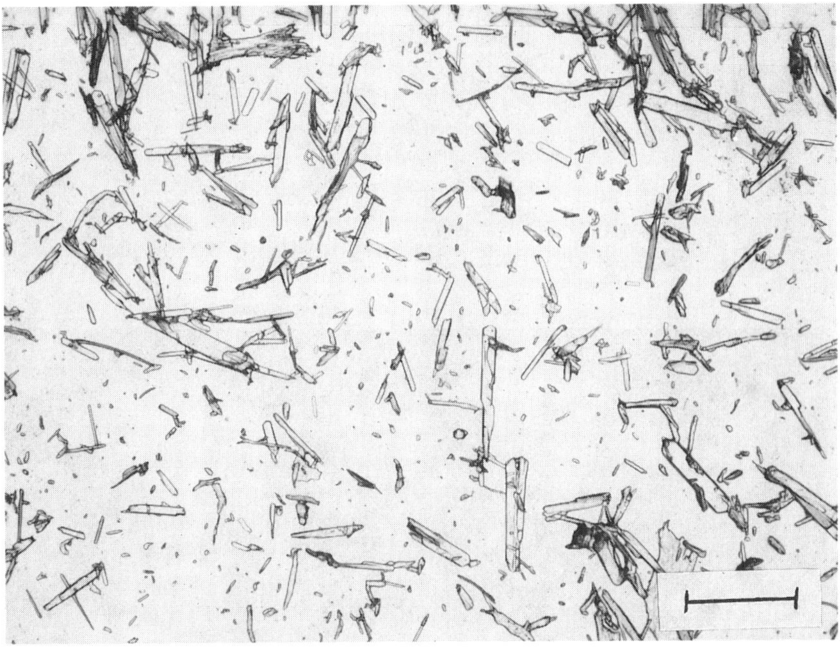

Figure 13. Primary particles comprising the snowballs of Figure 12.

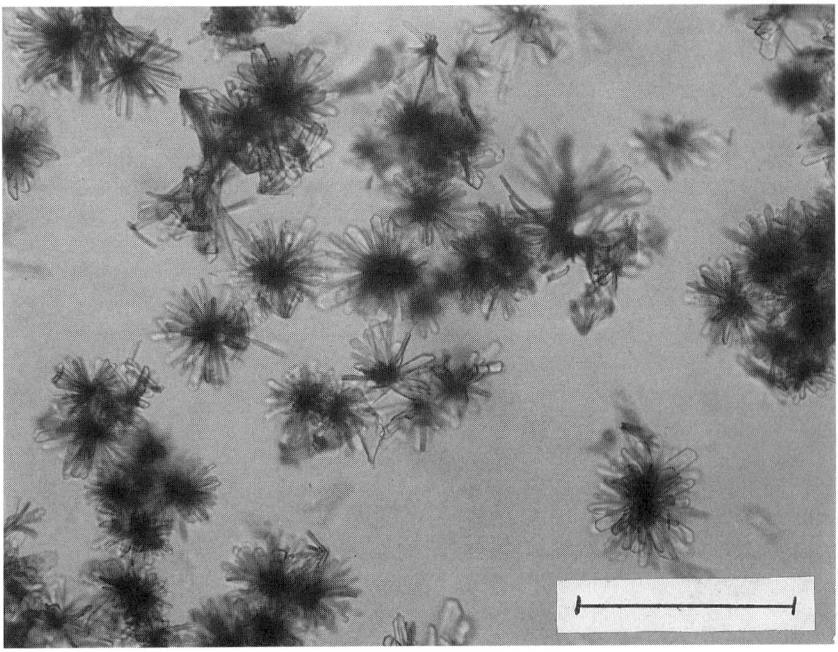

Figure 14. GAS Recrystallized NQ - starbursts formed from 3% NQ-DMF, 2.5 min injection of chlorodifluoromethane to 110 psi, 30°C.

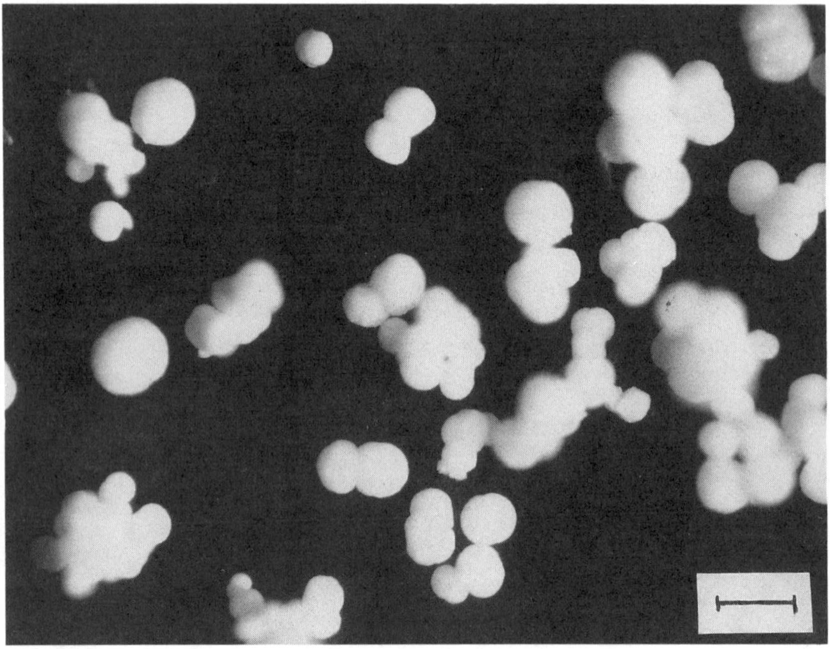

Figure 15. GAS Recrystallized NQ - spheres formed from 5% NQ-NMP very slow chloro-difluoromethane injection to 80 psi (the threshold pressure) and hold time for 30 min, 22°C.

solution can be separated by simple pressure decrease, and with the use of the chlorofluorocarbons, for example, operating pressure levels can be below 100 psi. Because non-polar (or only very slightly polar) gases are used in GAS Recrystallization, the energy requirements during pressure decrease and separation can be small relative to distillation of two polar liquids.

Although it was not discussed in the experimental section, the solid to be recrystallized cannot be simultaneously "too" soluble in the gas. For example, naphthalene cannot be readily recrystallized from liquid solution via GAS Recrystallization. Specifically, with the system naphthalene-toluene-carbon dioxide, only very narrow ranges of pressure and temperature can be used. As the carbon dioxide pressure is raised to promote expansion of the toluene, and thus a decrease in its dissolving power for naphthalene, the carbon dioxide itself becomes an increasingly good solvent for the naphthalene, and the naphthalene does not precipitate under the competing forces of a decreased solvating power of an expanded toluene solvent and the increasing solvating power of carbon dioxide (Krukonis, V.J. Unpublished data). On the other hand, if the solid material is soluble in the gas, Supercritical Fluid Nucleation exhibits the potential for recrystallizing the solid directly. GAS Recrystallization is an effective process for recrystallizing a wide variety of solid materials, but as for any new process that is to be evaluated, it should be subjected to a careful case-by-case evaluation for its economics relative to other recrystallization processes before it is carried to advanced development.

Acknowledgment

The funding for this work was provided by the Air Force Armament Laboratory (AFATL), Eglin AFB, FL 32542-5000, Contract F08635-87-C-0346 and is gratefully acknowledged. The complete results of the study are reported in Reference 30.

Literature Citations

1. McHugh, M. A., Krukonis, V. J. Supercritical Fluid Extraction: Principles and Practice, Butterworth Publishing, Boston, 1986; Chapter 10, Appendix.
2. Zosel, K. U.S. Patent 3,806,619, 1974.
3. Vitzthum, O., and Hubert, P. U.S. Patent 4,167,589, 1979.
4. Stahl, E., Quirin K. W. , Gerard, D. Dense Gases for Extraction and Refining; Springer Verlag, Berlin 1987, Ch. IV.
5. Yilgor, I., McGrath, J. E., Krukonis, V. J. Polym. Bull. 1984, 12, 499.
6. Elsbernd, C. S., Mohanty, D. K., McGrath, J. E., Gallagher, P. M., Krukonis, V. J. 194th ACS Mtg., New Orleans, September 1987.
7. Krukonis, V. J. Polymer News 1985, 11, 7.
8. Krukonis, V. J., Coffey, M. P., Bradley, R. L., M. Korycka-Dahl, M., Kroll-Conner, P. L. Center for Dairy Research Conference, Milkfat: Trends and Utilization, Madison, April 1988.
9. Best, D. Prepared Foods 1988, March, 124.

10. Hannay, J. B., Hogarth, J. Proc. Roy. Soc. (London) 1879, 29, 324.
11. Todd, D. B., Elgin, J. C. AIChE Jour. 1955, 1, 20.
12. Elgin, J. C., Weinstock, J. J. J. Chem. Eng. Data 1959, 4, 3.
13. Worthy, W. C&E News 1981, Aug 3, 16.
14. Krukonis, V. J. Ann. AIChE Mtg. San Francisco, November 1984.
15. Paulaitis, M. E., Krukonis, V. J., Kurnik, R. T., Reid, R. C. Chem. Eng. Rev. 1983, 1, 179.
16. Larsen, K. A., King, M. L. Biotech. Prog. 1986, 2, 73.
17. Petersen, R. C., Matson, D. W., Smith, R. D. J. Am. Chem. Soc. 1986, 108, 2100.
18. Krukonis, V. J. 1st Int. Symp. Supercrit. Fluids Nice, October 1988.
19. Coffey, M. P., Krukonis, V. J. Supercritical Fluid Nucleation - An Improved Ultra-Fine Particle Formation Process, Phasex Corp. Final Report to National Science Foundation 1988, Contr. ISI-8660823.
20. King, J. W. 188th ACS Mtg., Philadelphia, August 1984.
21. Rodriguez, F. Principles of Polymer Systems 2nd ed. 1982, Hemisphere Publishing, New York, p 25.
22. Hyatt, J. A. JOC 1984, 49, 5097.
23. Allada, S. R., IEC Prod. Res. Dev. 1984, 23, 344.
24. Dandge, D. K., Heller, J. P., Wilson, K. V. IEC Prod. Res. Dev. 1985, 24, 162.
25. Krukonis, V. J., Coffey, M. P., Gallagher, P. M. Exploratory Development on a New Process to Produce Improved RDX Crystals: Supercritical Fluid Anti-Solvent Recrystallization, Phasex Corp. Final Rept. to U.S. Army Ballistics Research Laboratory, 1988, Contr. DAAA15-86-C-007.
26. MacKerron, C. B., Hunter, D., Johnson, E., Ushio, S. Chem. Eng. 1988, Jan 18, 22.
27. Gibbs, J. W. The Collected Works of J. W. Gibbs, Vol. I, Thermodynamics, Yale University Press, New Haven, 1957, p 322.
28. Adamson, A. W. Physical Chemistry of Surfaces, Interscience Publishers, Easton, PA, 1963, pp 288-298.
29. Becker, R. and Doring, W. Ann. Physik 1935, 24, 719.
30. Krukonis, V.J. and Gallagher, P.M. Exploratory Development on a New Nitroguanidine Recrystallization Process: GAS Anti-Solvent Recrystallization, Phasex Final Rept. to U.S. Air Force Armament Laboratory, 1988, Contr. F08635-87-C-0346.

RECEIVED May 23, 1989

Chapter 23

Solids Formation After the Expansion of Supercritical Mixtures

Rahoma S. Mohamed, Duane S. Halverson, Pablo G. Debenedetti, and Robert K. Prud'homme

Department of Chemical Engineering, Princeton University, Princeton, NJ 08544–5263

The continuous formation of solids following the rapid expansion of a highly compressible supercritical mixture is an interesting alternative to conventional crystallization methods. Phase separation is caused by a mechanical perturbation which propagates rapidly and hence gives rise to essentially uniform conditions within the supersaturated medium. In this study, a systematic investigation of the influence of pre- and post-expansion conditions upon the crystallinity and particle size of Naphthalene and Lovastatin powders obtained by the rapid expansion of supercritical mixtures was performed. Naphthalene particles were produced continuously by expanding carbon dioxide-rich mixtures into a crystallizer whose temperature, pressure and composition were independently controlled. The mixture was expanded through a laser-drilled nozzle within which fluid residence times were less than 10^{-5} seconds. Naphthalene powders were found to be crystalline in all cases studied. Particle size of the solid product was consistently found to be a sensitive function of pre- and post-expansion temperature and of pre-expansion Naphthalene concentration, suggesting the possibility of an accurate control of product characteristics through small changes in process variables. The Lovastatin product formed upon the expansion of Lovastatin/CO_2 supercritical mixtures consisted of aggregates of sub-micron crystal units. Mean particle sizes in the range of 126-255 nm were measured for these units. The average particle size was found to be relatively insensitive to changes in solute concentration, or in pre- and post-expansion temperature for the conditions investigated in this study.

The solubility of solids in supercritical fluids is a very sensitive function of temperature and pressure. Unlike liquids, supercritical fluids are highly compressible and minor temperature or pressure changes lead to large changes in density and, therefore, solvent power (1). The expansion of supercritical solutions, therefore, results in a substantial solubility decrease. The solubility of Naphthalene in carbon dioxide at 45° C, for example, decreases by about two orders of magnitude upon reducing the pressure from 127 to 62 bars (2). If this

0097–6156/89/0406–0355$07.00/0

pressure reduction is rapid, it can give rise to very large supersaturation ratios. Here, the supersaturation ratios are generated not by changing the fluid's temperature (which inevitably gives rise to non-uniformities within the fluid due to the low thermal diffusivities of liquids and supercritical fluids) but by changing the fluid's pressure. The speed at which pressure perturbations propagate within the fluid allows precipitation from an essentially uniform medium and clearly distinguishes this process (precipitation following the rapid expansion of a supercritical solution) from conventional crystallization, where temperature gradients give rise to a wide distribution of nucleation rates within the solvent. The uniform conditions within the nucleating environment would thus be expected to result in the production of particles that are more even in morphology and narrower in size distribution than those produced by conventional crystallization techniques.

The striking effects on particle size and morphology which result upon the rapid expansion of supercritical solutions were already recognized by Hannay and Hogarth, who described the precipitated solids as "snow" in a gas and a "frost" on glass (3-4). The implications of this observation, however, were not realized until recently when Krukonis (4) reported on his pioneering investigation of the precipitation of aluminum isopropoxide, dodecanolactam, polypropylene, ß-estradiol, ferrocene, Navy blue dye and soy bean lecithin (a heat-labile phospholipid with desirable surfactant properties in emulsification applications, where its desirable morphology is as small particles) following the rapid expansion of supercritical solutions. Supercritical carbon dioxide was used with all the solids tested except for the polypropylene experiments, where propylene was employed instead. The extraction temperature and pressure were kept at 328 K and 345 bar (413 K and 241 bar for the polypropylene experiments), respectively. Optical and scanning electron microscopy analysis of the precipitated solids upon the expansion of the supercritical mixtures to atmospheric conditions revealed significant changes (with respect to the starting material) in the morphology and particle size distribution of all the solids tested. ß-estradiol precipitated as 1 μm (and smaller) particles, where the original solid loaded in the extractor consisted of a highly non-uniform distribution of crystalline particles that ranged in size from a few to 100 μm. The lecithin precipitation was accompanied by an order of magnitude size reduction, while the ferrocene and navy blue dye underwent a size decrease of approximately one to two orders of magnitude. Dodecanolactam exhibited a moderate decrease in particle size and, in addition, a highly specific shape in the form of needles that were about 1μm in diameter and 30 μm long. Polypropylene, on the other hand, exhibited a remarkable change in morphology, from large spherical lumps to fiber-like particles, while undergoing a modest size reduction.

Larson and King (5) investigated the extraction and precipitation of pharmaceutical products. Lovastatin, an anti-cholesterol drug, was extracted and precipitated using a mixture of carbon dioxide and 3% methanol. The addition of methanol enhanced the solubility by an order of magnitude at 379 bar and 313.2 K. Photomicrographs of the precipitated Lovastatin revealed a fivefold reduction in particle size with respect to starting material with complete retention of crystallinity, as confirmed by X-ray diffraction patterns.

Smith and coworkers (6-10) have recently conducted a series of investigations on the extraction and precipitation of a variety of organic polymers and inorganic materials including ceramic and preceramic powders. Silica powders were precipitated by rapid expansion following their extraction with supercritical water at 580-590 bar and pre-expansion temperatures higher than 723 K. The particle size and agglomeration characteristics of these powders were found to be sensitive to the concentration of silica before the expansion as well as to the presence of dilute salt solutions (KI, NaCl) in the supercritical mixture. Amorphous, 1 μm thick SiO_2 films, and a GeO_2 product whose morphology was dramatically altered from 5 μm porous agglomerates to ca. 1 μm spheres upon the reduction of the pre-expansion temperature by merely 30 K, were obtained. The organic polymers used included polystyrene, polypropylene, polyphenyl sulfone and polymethyl methacrylate, with pentane and propane as the supercritical solvents. Of particular interest in these studies was the observation that product morphology was extremely sensitive to pre-expansion temperature: polystyrene, for example, precipitated as fibers that were 1 μm in diameter and 100-1000 μm long at a pre-expansion temperature close to its melting point (ca. 443 K) and as 20 μm spheres at higher temperatures. Polypropylene exhibited a similar behavior. Polycarbosilane powders and films were also produced with morphology being determined primarily by the pre-expansion temperature and solute concentration. Furthermore, experiments with the expansion chamber under vacuum (1-1000 Pa) suggested that the morphology of the precipitated solids was affected by the post-expansion pressure. The effect of higher than atmospheric post-expansion pressures was not investigated.

The attractiveness of supercritical nucleation as a method for the production of uniform particles of controllable morphology and size distribution has been repeatedly pointed out by previous investigators (4-10). Crystalline and amorphous powders with particles that are uniform and with narrow size distributions are important for a variety of applications. In the case of pharmaceutical organic products, there is need for novel means of manufacturing drugs in the form of finely divided powders with a narrow particle size distribution, since conventional techniques of achieving the desired subdivision invariably result in wide particle size distributions, reduced activity of the drug, or a loss in storage stability due to an alteration of the desired crystalline morphology (11). In addition, the effectiveness of many injected drugs used in emergency situations is dependent to a large extent on the ability of the particles to dissolve rapidly, with oversized crystals having unacceptably slow dissolution rates.

Monodisperse polymer spheres in the size range 2-10 μm are important in a variety of applications, and are hence of interest in the context of supercritical nucleation. 8 μm polymer spheres can potentially be used as vehicles for drugs that can be selectively delivered to the liver (with larger particles trapped in capillaries and smaller particles flowing continuously through the circulatory system instead of being filtered). spheres, in addition, are widely used in chromatography packings. The current emphasis in high performance liquid chromatography is to use 3-5 μm spheres to achieve rapid mass transfer and high resolution. Uniform particles are required or else the chromatography bed cannot be uniformly packed and dispersion occurs with a loss of resolution. The ability to produce uniform particles from a wide range of polymeric materials would open up novel possibilities for packing media in chromatographic separations.

Precipitation from supercritical fluids is of interest not only in relation to the production of uniform particles. The thermodynamics of dilute mixtures in the vicinity of the solvent's critical point (more specifically, the phenomenon known as retrograde solubility, whereby solubility decreases with temperature near the solvent's critical point) has been cleverly exploited by Chimowitz and coworkers (12-13) and later by Johnston et al. (14). These researchers implemented an elegant process based on retrograde solubility for the separation of physical solid mixtures which gives rise to high purity materials.

The technological potential of particle formation using supercritical fluids calls for further experimental studies, designed to investigate in a methodical way the influence of operating conditions upon the characteristics of solid products. This work is part of an ongoing study aimed at identifying the thermodynamic and kinetic variables that influence and possibly control the morphology and particle size distribution of powders produced by the rapid expansion of supercritical mixtures. An important aspect of this work is the systematic exploration of the influence of post- and pre-expansion conditions (temperature, pressure and composition) upon product characteristics: the former have not been studied to date.

We begin by examining the well-characterized CO_2 /Naphthalene system, which has been thoroughly studied thermodynamically (2,15). Exploratory particle formation experiments have also been performed on this mixture (5). This allows a much better understanding of the underlying fundamentals through systematic changes in experimental conditions about reproducible base cases (16). Next, we examine the crystallization of a more complex compound, Lovastatin, an anti-cholestorol drug (Figure 1). Experiments are organized according to a two-level factorial design (17) which allows the identification of primary effects and interactions of perturbed variables on the morphology and particle size distribution of the precipitation products. The Lovastatin study is representative of the application of supercritical nucleation towards the production of crystalline pharmaceutical powders with small particle size.

Experimental

The experimental apparatus, shown in Figure 2, consists of two main units, extraction and precipitation. It has the unique feature of allowing the independent control of all relevant process variables, namely pre- and post-expansion temperature, pressure and composition. The solvent, carbon dioxide in this case (MG, "Bone Dry"), is delivered by a variable stroke dual-piston pump, A (Milton Roy Minipump), at constant rates of up to 460 ml/hr of liquid CO_2 and against pressures as high as 414 bars. The high pressure liquid then enters a preheater, E(0.32 O.D. X 51 cm stainless steel tubing) which is immersed in a water bath, D, whose temperature is controlled to within ±0.01° C by a heater-circulator, F (Braun Thermomix 1460). The fluid being heated to the extraction temperature is introduced into the extraction column, C (2.54 O.D. X 30 cm stainless steel column, Autoclave Engineers P/N CNLX16012) which is packed with alternate layers of glass wool and solute. The temperature at the exit of the preheater, in the extraction column and elsewhere in the apparatus is measured by type J thermocouples and displayed with a precision of ±0.1° C (Omega Digicator, model 412B). The extraction pressure is monitored by a digital gauge (model 901B Heise Gauge, 0-690 bar, ±0.05% accuracy). The saturated stream emerging from the extraction column can be diluted with fresh solvent delivered by the second piston of pump A: this allows the independent control of the supercritical mixture's composition prior to expansion. This composition can be determined using a High Performance Liquid Chromatography (HPLC) system which is similar to that employed by Larson and King (5). The fluid stream is diverted to the HPLC system through a Valco six-port injection valve with a 5 μL sampling loop. The contents of the loop are subsequently injected and swept into a Beckman Reversed Phase C_{18} Ultrasphere column with the acetonitrile/water mobile phase. The eluting streams are monitored by a Micrometrics dual variable wavelength UV detector (model 788) and the peak areas are computed using a Hewlett-Packard 3390A integrator. The peak areas were reproducible to within ± 5%.

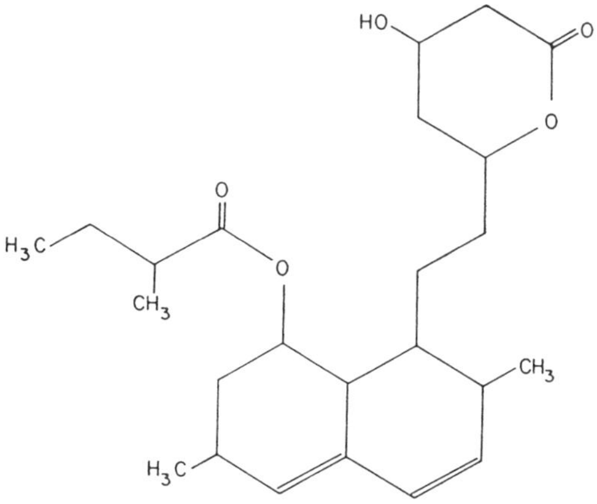

Figure 1. Chemical structure of Lovastatin.

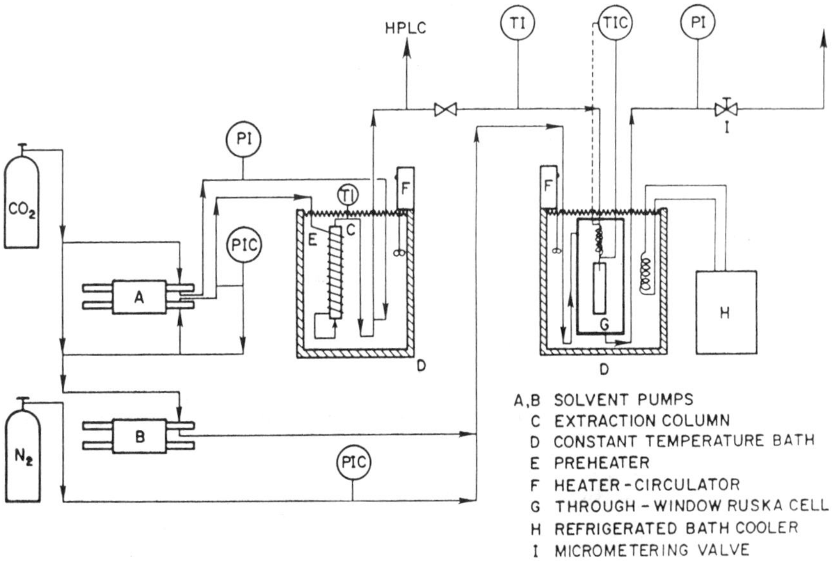

Figure 2. Experimental apparatus.

The supercritical mixture exiting the extraction unit enters the laser-drilled expansion nozzle, Figure 3 (made by Advanced Laser Systems, Waltham, MA). The nozzle is fabricated by laser-welding a stainless steel disc, 4.76 mm in diameter and 0.25 mm thick onto a 6.35 mm O.D. stainless steel tube. The disc is subsequently laser-drilled at its center to the specified nozzle diameter. The pre-expansion temperature is adjusted through a variable resistance heating cable (Watlow cable heater P/N 62H65A3X) controlled by an indicator controller (Watlow, model 808). The nozzle is designed for rapid expansion with typical residence times inside the laser-drilled orifice of 10^{-5} seconds. The expanded fluid then enters a high pressure crystallizer, G (Ruska Cell model 2329-800) which allows visual observation through a glass see-through window. The crystallizer is immersed in a second glass-walled bath, D, where silicone oil is used instead of water to protect the Ruska cell from corroding. The bath temperature is controlled to within ±0.01° C by another heater-circulator, F, for heating duty and by an immersion cooler, H (Neslab model IBC-4), together with circulator F, for cooling duty . The temperature inside the crystallizer, G, is also measured and recorded.

A second variable stroke pump, B, is used to control the pressure and composition of the nucleating medium inside the crystallizer. This is accomplished by filling the crystallizer through pump B, while maintaining a steady flow which is vented through a micrometering valve, I (Autoclave Engineers 30VRRM). The pressure inside the crystallizer is measured by a model 901A Heise Digital Gauge (0-414 bar, ±0.1% accuracy). An inert substance such as nitrogen (Figure 2) can also be continuously delivered into the crystallizer to lower the partial pressure of the supercritical fluid in the nucleating medium. The crystallizer is equipped with a glass tube (25 mm I.D. x 75 mm) for the collection of the precipitated particles. The expansion nozzle was positioned at the center of the glass tube and approximately 50 mm from its bottom. In this study, the collected particles were subsequently analyzed through optical (including polarized light) microscopy and X-ray diffraction. Lovastatin particles were further analyzed using Scanning Electron Microscopy. Particle size distributions of the Lovastatin samples were obtained using COULTER Model N4SD sub-micron particle analyzer which calculates the particle size distribution from the measurement of the sample diffusion coefficient by photon correlation spectroscopy.

All experiments in this study were performed at solvent flow rates between 0.1 and 0.5 standard liters per minute (SLPM) for the CO_2/Naphthalene system and between 0.2 and 1.5 SLPM for the CO_2/Lovastatin system. Increasing the CO_2 flow rate from 0.1 to 1.5 SLPM had no effect on the measured solubility of Lovastatin indicating the attainment of equilibrium at these conditions. The solubilities measured during the Lovastatin crystallization experiments were found to be in close agreement with the independent equilibrium measurements. Investigators who used similar extraction columns and solute loadings and an identical flow technique have reported equilibrium (i.e., saturation) conditions at the extractor's outlet and negligible effect on the measured solubilities upon increasing the solvent flow rate up to 2.17 SLPM (18), 0.5 SLPM (19), and 1.45 SLPM (12). In all of the above cases, agreement with Tsekhanskaya's Naphthalene/CO_2 data (2)was used as a criterion to test the accuracy of equilibrium data. The low solvent flow rates used in our study, therefore, guarantee the attainment of saturated conditions prior to expansion.

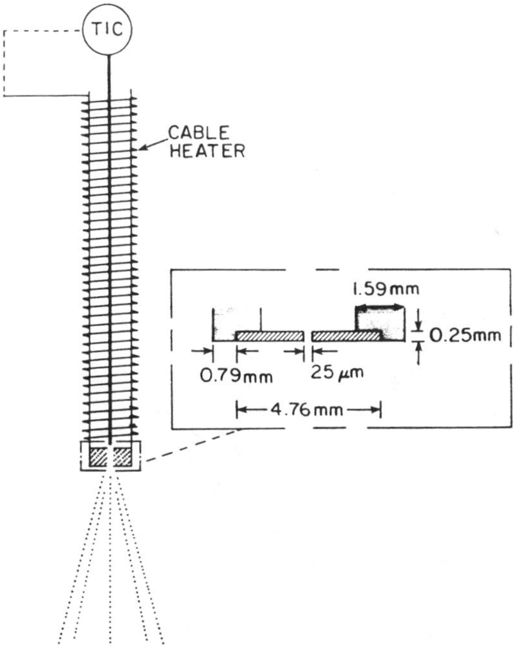

Figure 3. Expansion nozzle.

Results and Discussion

The nucleation and growth of dissolved solutes following the expansion of supercritical mixtures is a complex process in which not only the solute, but also the solvent, undergoes dramatic changes (depending upon pre- and post-expansion temperatures, the solvent can be caused to expand to a dilute gas, condense to a liquid, or solidify). Given these complexities, there is, at present, no quantitative model which can be used to predict particle size and morphology for a given set of operating conditions. The development of such a model must be based upon a clear understanding of the individual effects of each pre- and post-expansion condition on the product's characteristics. The present work is a part of an ongoing project in which several model systems are being studied in experiments specifically designed so as to investigate the effects of each process variable upon the crystallinity, morphology and size of the solid product. Our long-term goal is the development of a quantitative description of the supercritical nucleation process based on a sound understanding of the relevant physics.

In what follows we describe the main trends which we have so far found to be important and reproducible for both the CO_2/Naphthalene and CO_2/Lovastatin systems. A short, partial discussion of our Naphthalene experiments has already been published elsewhere (16).

Carbon Dioxide/Naphthalene System

Virgin Naphthalene particles (Fisher Scientific, N7-500) were charged into the column C (Figure 2) and extracted with supercritical carbon dioxide at 221.4 bar and 45° C. The mixture was subsequently expanded through the nozzle shown in Figure 3 to 18.7 bar and 45° C after being heated to 135° C. The precipitated particles following the expansion ranged in size from 30 to 135 μm (Figure 4a).

In the experiments that followed, the pre- and post-expansion temperature, pressure and composition were systematically perturbed around their values in this base case and the results were compared to identify the effects of each variable on the morphology and size of the precipitated Naphthalene particles. The pre-expansion temperature, while varied, was kept high enough in all experiments so that the jet of expanded fluid downstream of the nozzle was not visible, i.e. no condensation of carbon dioxide occurred upon expansion. The base case post-expansion pressure corresponds roughly to the solubility minimum of Naphthalene in CO_2 at 45° C (2). All of the experiments performed in this study are listed in Table I (Run 1 is the above mentioned base case).

Effect of Temperature. A change in the pre-expansion temperature of 25-35° C about the base case value of 135° C resulted in a pronounced effect on the particle size. Increasing the temperature to 170° C produced larger particles, while a decrease in temperature to 110° C resulted in the production of smaller particles (Runs 1,2 and 3; Figure 4). Similar experiments, performed at higher extraction temperatures and pressures (higher solute concentrations), however, revealed little effect of the pre-expansion temperature on the particle size (Runs 4,5,6 and 7, Figure 5). Extraction conditions in this study were always outside of the retrograde region (20-21) and hence an increase in temperature always resulted in an increase in Naphthalene concentration in the supercritical fluid.

The post-expansion temperature (in contrast to the pre-expansion temperature), was found to be important both at high and low Naphthalene concentrations. Crystallizer temperatures from 8° C to -8° C resulted in the production of Naphthalene crystals with similar size distributions that ranged

Table I. Experimental Conditions and Results

Run	Extraction		Precipitation		Pre-expansion	y^a	Approximate
No.	Pressure (bar)	Temperature (°C)	Pressure (bar)	Temperature (°C)	Temperature (°C)	x10[4]	size[c] (μm)
			Pre-Expansion Temperature (moderate Concentration)				
2	221.2	45	18.1	45	170	260	38-225
1[b]	221.4	45	18.2	45	135	260	30-135
3	221.2	45	17.9	45	110	260	8-75
			Pre-Expansion Temperature (high Concentration)				
6	362.3	55	18.0	45	170	573	4-51
4	362.0	55	17.9	45	135	573	4-38
5	361.9	55	17.7	45	135	573	4-39
7	362.0	55	17.8	45	110	573	2-35
			Post-Expansion Temperature				
1	221.4	45	18.2	45	135	260	30-135
8	221.7	45	18.3	7	135	260	6-29
18	148.4	45	18.0	45	135	192	15-83
9	148.9	45	18.1	8	110	192	2-15
10	148.6	45	18.0	-8	110	192	4-19
4	362.0	55	17.9	45	135	573	4-38
11	361.8	55	18.2	7.5	135	573	2-16
			Naphthalene Concentration				
12	221.5	55	17.9	45	135	444	6-32
1	221.4	45	18.2	45	135	260	30-135
4	362.0	55	17.9	45	135	573	4-38
15	362.2	35	17.9	45	135	188	14-78
			Post-Expansion Pressure				
16	221.2	45	70.0	45	135	260	15-113
17	221.7	45	44.6	45	135	260	15-105
1	221.4	45	18.2	45	135	260	30-135
			Pre-Expansion Pressure				
15	362.2	35	17.9	45	135	188	14-78
13	221.4	35	18.1	45	135	174	38-188
14	221.3	35	17.9	45	135	174	23-165
18	148.4	45	18.0	45	135	192	15-83
			Reproducibility				
4	362.0	55	17.9	45	135	573	4-38
5	361.9	55	17.7	45	135	573	4-39
13	221.4	35	18.1	45	135	174	38-188
14	221.3	35	17.9	45	135	174	23-165

a mole fraction of naphthalene (2).
b base case.
c from optical microscopy, using 500x magnification.

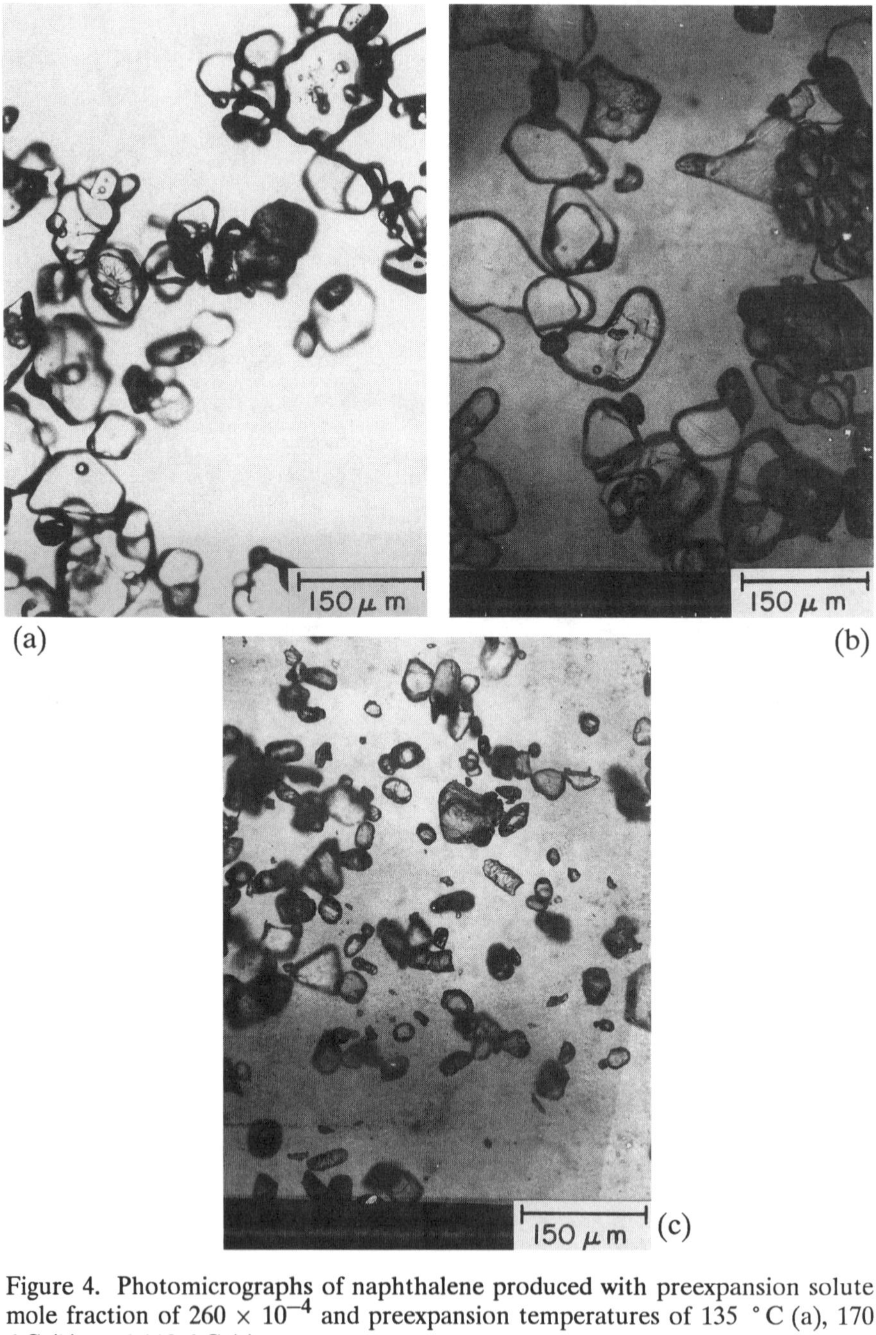

Figure 4. Photomicrographs of naphthalene produced with preexpansion solute mole fraction of 260×10^{-4} and preexpansion temperatures of 135 °C (a), 170 °C (b), and 110 °C (c).

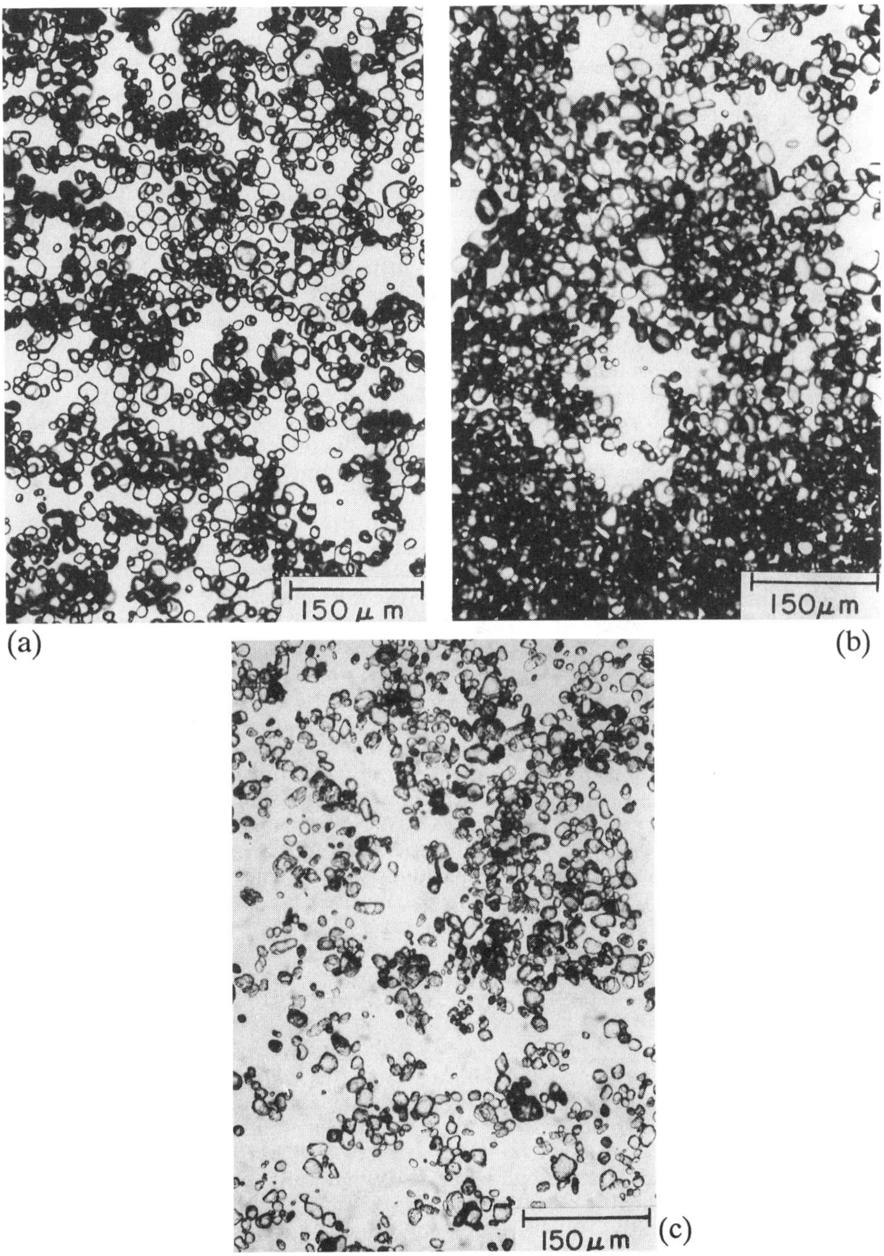

Figure 5. Photomicrographs of naphthalene produced with preexpansion solute mole fraction of 537×10^{-4} and preexpansion temperatures of 135 °C (a), 170 °C (b), and 110 °C (c).

from approximately 2 to 29 μm (Runs 8,9 at 7 and 8° C respectively: Figure 6a; Run 11 at 7.5° C: Figure 6b; Run 10 at -8° C: Figure 6c).

This large reduction in particle size with respect to the base case upon lowering the post-expansion temperature is very interesting, since it suggests a decoupling of nucleation and growth processes. This separation, which is very difficult to attain in conventional crystallization processes, would allow a very sensitive control of the size of the precipitated product.

Effect of Concentration. The concentration of solutes in the supercritical fluid is easily altered by a change in either the extraction temperature or pressure. Increasing the mole fraction of Naphthalene before expansion was found to result in a decrease in the size of the precipitated particles (see Runs 1,12,4,15; Figure 7). An increase in the Naphthalene concentration leads to higher supersaturation ratios in the fluid upon expansion. According to the classical theory of nucleation (9,22,23), this leads to higher nucleation rates. Since, according to this theory, the particle volume is inversely proportional to the nucleation rate, our results appear to be consistent with simple theoretical predictions. Smith and coworkers (9) , who reported an increase in particle size with concentration of both silica and germanium oxide in supercritical water (contrary to our findings), speculated that the production of large particles at the higher concentrations was due to the coalescence of the increased number of nuclei in the expanding jet. Condensed vapor drops produced during decompression in studies of homogeneous nucleation in cloud chambers (9, 24) revealed a decrease in particle size upon increase in solute concentration which is in agreement with the results of our experiments. It is not clear whether the difference between our apparatus and that of Smith and coworkers (6-10), particularly the expansion nozzle (the one used in this study is 20 times shorter and has an L/D ratio 10 times smaller), and/or the different systems being studied are responsible for these conflicting trends.

Effect of Pressure. Increasing the post-expansion pressure (i.e., the pressure inside the Ruska cell) leads to a decrease in supersaturation and a concurrent increase in the solute's partial pressure within the nucleating fluid. In the present study, a moderate decrease in particle size resulted upon post-expansion pressure increase (Runs 1,16 and 17). The smallness of this effect suggests a rough balance between lower supersaturation (which leads to lower nucleation rates and hence larger particle sizes) and higher solute partial pressure [which leads to higher nucleation rates and hence smaller particle sizes (9, 23)].

Results obtained upon changing the pre-expansion pressure (at constant pre-expansion temperature and solute concentration, as well as post-expansion temperature and pressure) proved inconclusive (Runs 18,13,14,15).

Crystallinity of Naphthalene Particles. The degree of crystallinity of the resulting powders is important both from the standpoint of producing pharmaceutical products, and since it sheds some light on the mode of particle growth in the post-expansion jet. X-ray diffraction patterns obtained for the virgin Naphthalene and particles produced by the rapid expansion of supercritical mixtures (Runs 1 and 8) are presented in Figure 8. The patterns are superimposable indicating identical structure and retention of crystallinity of the produced particles. Furthermore, optical examination of the particles in polarized light excluded the possibility of contamination with non-crystalline or isometric phases. All other experiments produced similar results.

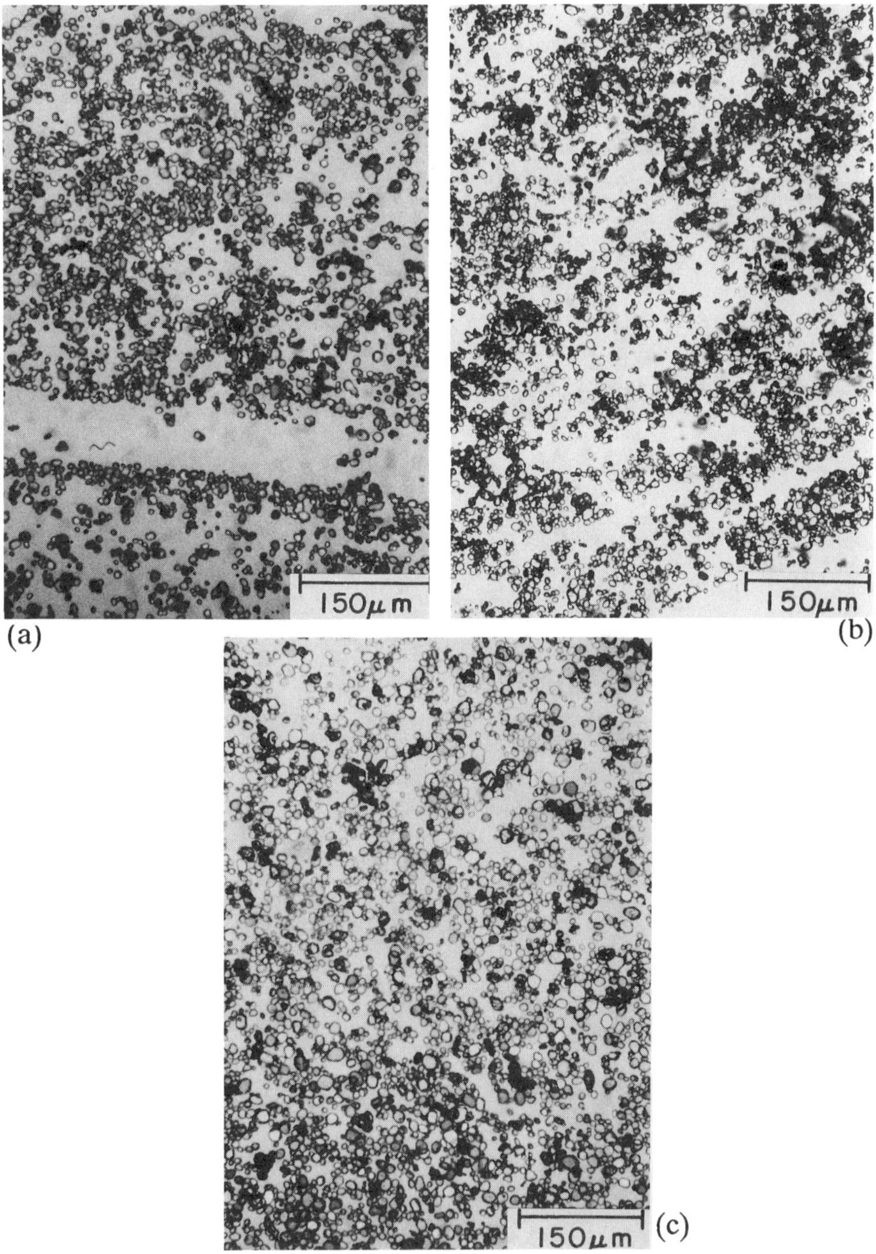

Figure 6. Photomicrographs of naphthalene powders produced upon cooling the crystallizer to 8 °C (a), 7.5 °C (b), and –8 °C (c).

(a)

(b)

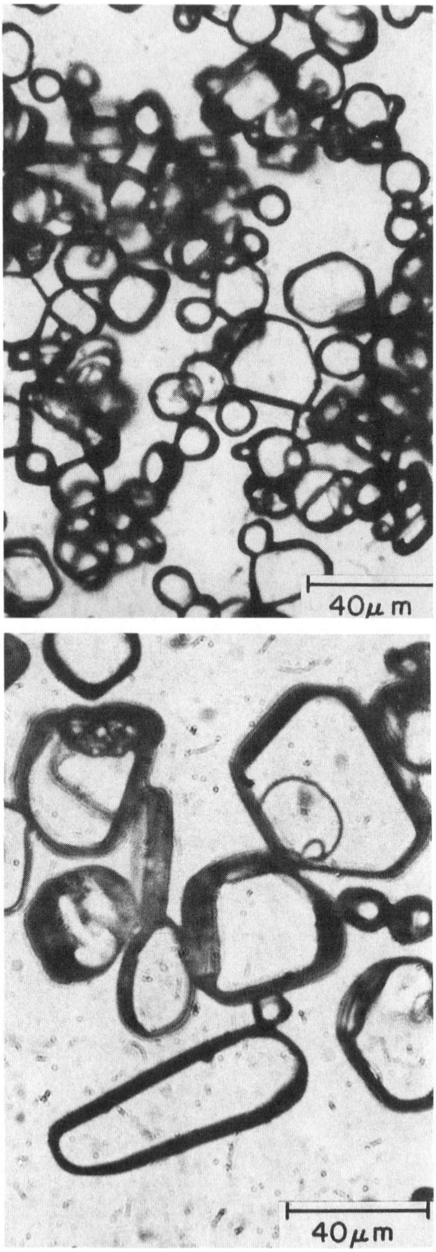

Figure 7. Photomicrographs of powders obtained with preexpansion naphthalene mole fractions of 573×10^{-4} (a) and 188×10^{-4} (b).

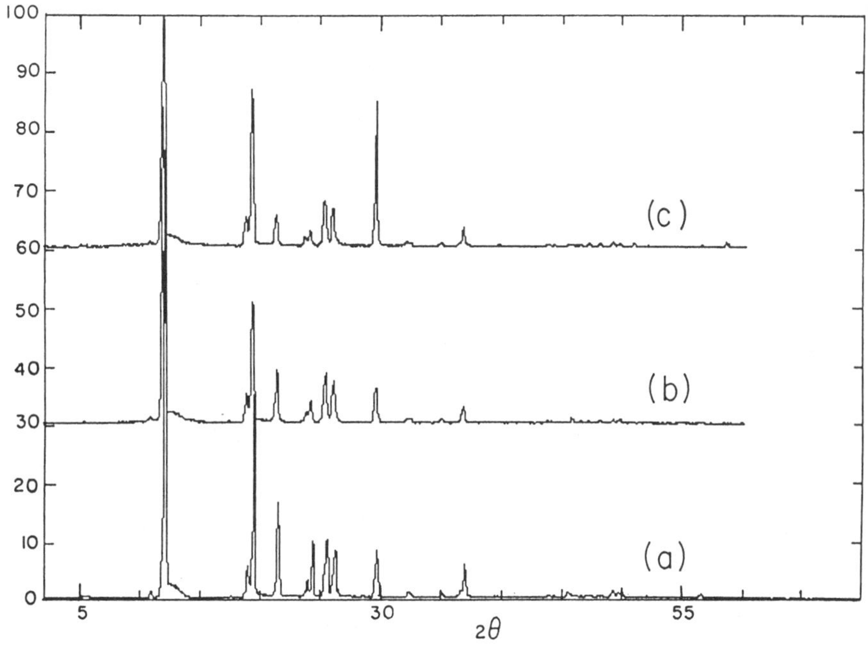

Figure 8. X-ray diffraction patterns from original naphthalene (a) and from powders produced in run 1 (b) and run 8 (c). Vertical axis represents the intensity in arbitrary units. All other experiments produced similar results.

Carbon Dioxide/Lovastatin System

In order to investigate the application of supercritical nucleation to the production of pharmaceutical drugs, we studied the extraction and crystallization of Lovastatin (an anti-cholesterol drug; Figure 1) in CO_2. There is interest in producing sub-micron Lovastatin crystals which, with conventional processing techniques, can only be obtained after milling (11). Larson and King (5) have previously explored the crystallization of Lovastatin using a mixed solvent system containing 3-5% methanol in carbon dioxide. The addition of methanol resulted in an enhancement in the solubility of Lovastatin in the supercritical phase of almost an order of magnitude. The methanol, however, had a low vapor pressure (relative to carbon dioxide) and thus condensed following the expansion. The precipitated particles dissolved in the condensed methanol and subsequently recrystallized. Since we are interested in crystallization from a supercritical (as opposed to liquid) phase, this recrystallization must be avoided. This can be done by simply increasing the pre- or post-expansion temperature or lowering the pressure so that methanol remains in the vapor phase. Since methanol condensation occurred in our apparatus at temperatures as high as 70° C, increasing the temperature was an unacceptable solution in light of the results obtained from the Naphthalene crystallization study, which revealed major effects of the pre- and post-expansion temperatures on the size of the precipitated particles. Lowering the precipitation pressure, while conceivable, was not implemented in this study, in which the lowest possible pressure inside the Ruska cell was atmospheric. It is hence important that any co-solvent used for solubility enhancement be volatile enough so as not to condense following the expansion. We have found that the solubility of Lovastatin in pure carbon dioxide at 55° C and 75° C, though low, was sufficient for crystallization studies. Accordingly, we report here on experiments with no co-solvent added to the carbon dioxide. Work is currently in progress on the selection of a suitable co-solvent for this system which combines solubility enhancement and adequate volatility.

The equilibrium solubility of Lovastatin in carbon dioxide at 55° C and 75° C and for pressures up to 400 bar was obtained using the HPLC apparatus and reported in Table II and Figure 9. The data exhibit both the abrupt change in solubility above the solvent's critical point, as well as the retrograde behavior, both of which characterize supercritical extraction processes. The solubilities were reproducible to within ± 5%.

The three process variables which were found to have significant effects on the precipitated Naphthalene crystals, namely solute concentration and pre- and post-expansion temperatures were further examined in the CO_2/Lovastatin system. Eight experiments were selected according to a two-level factorial design (17; Figure 10) and identified in order to study the primary and secondary effects of the three process variables (as well as their interaction) on the morphology and particle size distribution of the precipitation products. The experiments were all conducted at extraction and precipitation pressures of 379 and 2 bars, respectively. A 25 μm diameter orifice was used throughout.

An SEM photomicrograph of the Lovastatin charged into the extractor (as supplied by Merck, Sharp & Dohme Research Laboratories) is presented in Figure 11a. These particles were produced by conventional crystallization and subsequent grinding. The Lovastatin particles produced following the rapid expansion of supercritical mixtures at the conditions given in Figure 10 formed aggregates whose elementary units appear to range in size from 0.1 to 1 μm (Figure 11b, c, and d). Aggregation is likely to have occurred upon impact of the precipitating particles on the collection tube inside the crystallizer. Particle size distributions obtained following the sonication of each sample in heptane for 3.5

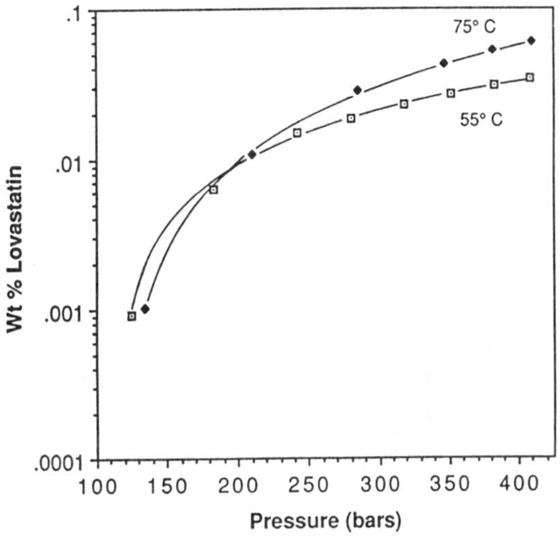

Figure 9. Equilibrium solubility of Lovastatin in CO_2.

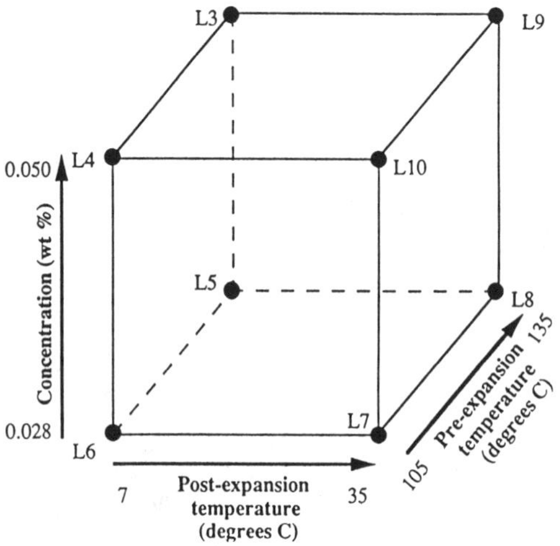

Figure 10. Two-level factorial design of the CO_2/Lovastatin experiments.

(a)

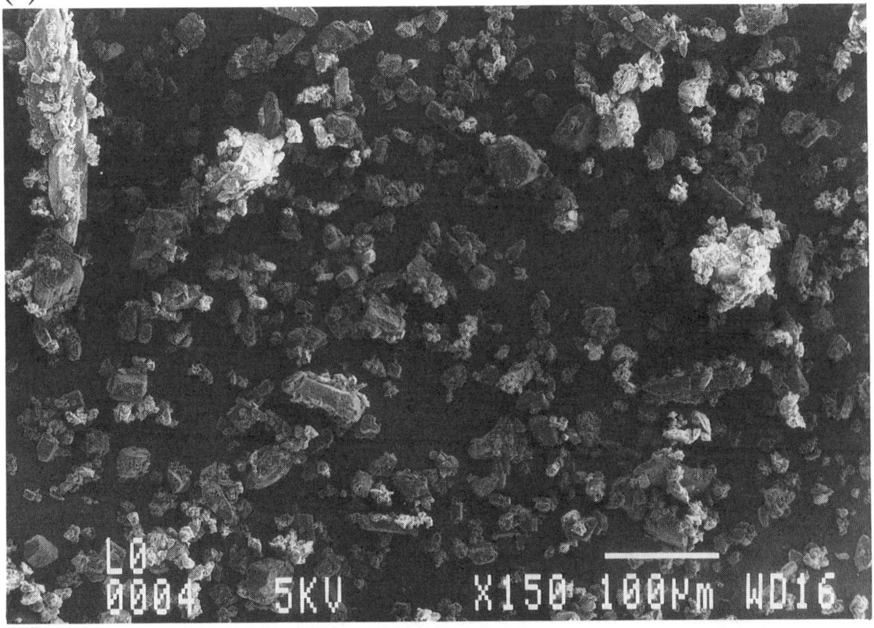

(b)

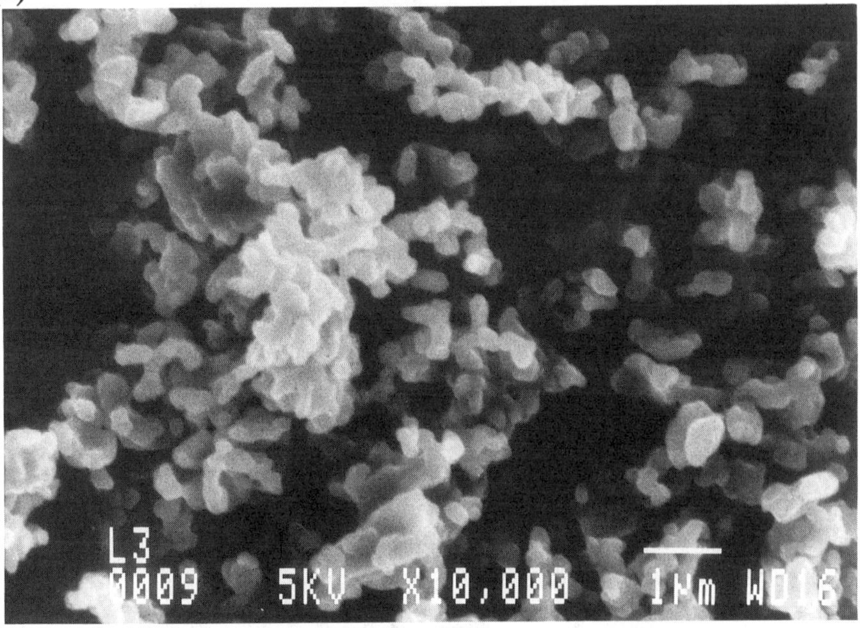

Figure 11. SEM photomicrographs for Lovastatin charged into extractor (a) and samples formed in experiment L3 (b). *(Continued on next page.)*

(c)

(d)

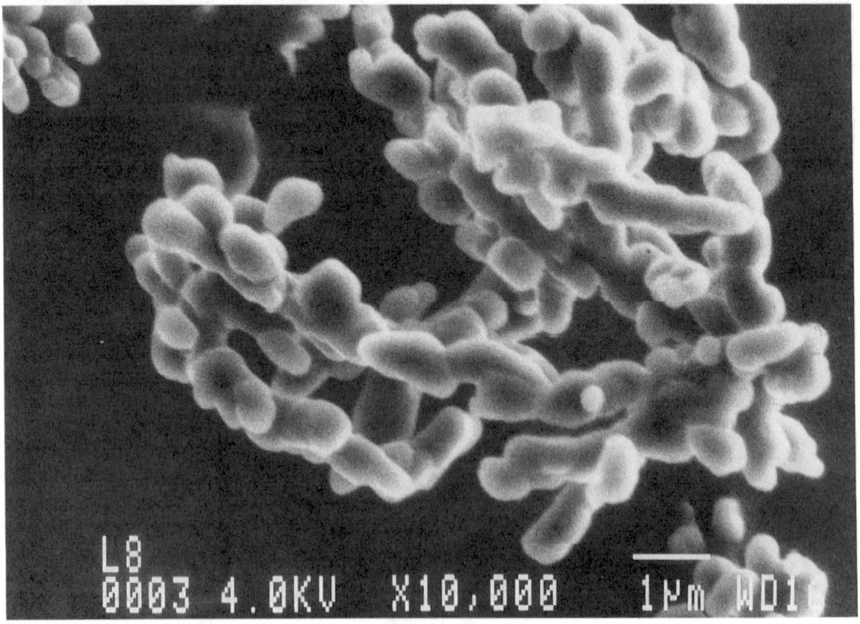

Figure 11. Continued. SEM photomicrographs for samples formed in experiments L7 (c) and L8 (d).

Table II. Solubility of Lovastatin in Supercritical Carbon Dioxide

Pressure (bar)		Solubility (wt%)
	at 55° C	
124.5		0.0009
182.5		0.0063
242.5		0.0151
280.2		0.0187
318.4		0.0230
351.6		0.0273
381.9		0.0310
408.6		0.0341
	at 75° C	
134.3		0.0010
209.9		0.0107
285.1		0.0284
346.7		0.0419
381.1		0.0526
409.2		0.0600

minutes in order to break-up the aggregates revealed mean particle sizes that ranged from 126 to 255 nm with standard deviations ranging from 42 to 67 nm (Figure 12). The particle sizes obtained are in agreement with the size of the elementary units shown in the SEM photomicrographs. The effect of sonication on the measured particle size, i.e. the degree of aggregate break-up, is shown in Figure 13 for experiment L4. The measured particle size was found to decrease with increase in sonication time up to 3 minutes beyond which no further decrease in size was observed. It was, hence, assumed that a sonication time of 3.5 minutes was sufficient to break up the aggregates and was used for analysis of all samples thereafter. Particle size analyses for experiments L5, L6 and L10 were not performed as the samples collected in both experiments were found to be insufficient for analysis with COULTER model N4SD. X-ray diffraction patterns obtained for the precipitated Lovastatin particles indicated the complete retention of the crystalline structure following the expansion.

In contrast to Naphthalene crystallization, Lovastatin particles were found to be rather insensitive to changes in process conditions. The important result, however, is that a crystalline product with loosely aggregated sub-micron size units was obtained in all cases, although there was evidence of some fusion between the crystallite units in the high post-expansion temperature runs (L7,L8,L9,L10).

Conclusions

In conventional crystallization processes, supersaturation (and hence nucleation) are caused by a thermal perturbation. Because of the inherently low thermal diffusivities of liquids, this approach is always accompanied by the existence of temperature non-uniformities within the supersaturated liquid. This, in turn, gives rise to rather wide product size distributions. In the rapid expansion of a highly compressible supercritical mixture, on the other hand,

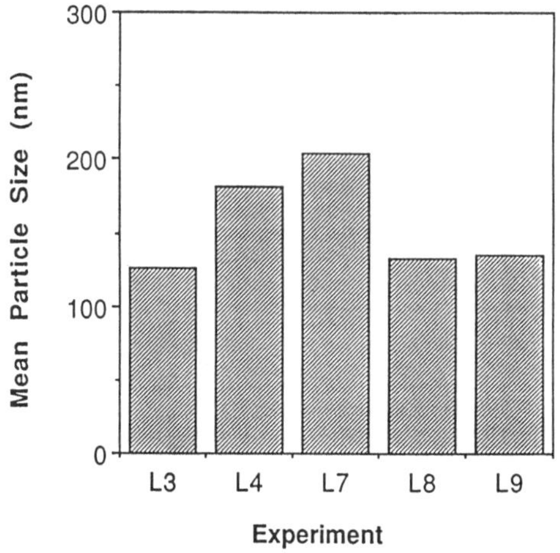

Figure 12. Histogram of mean particle size.

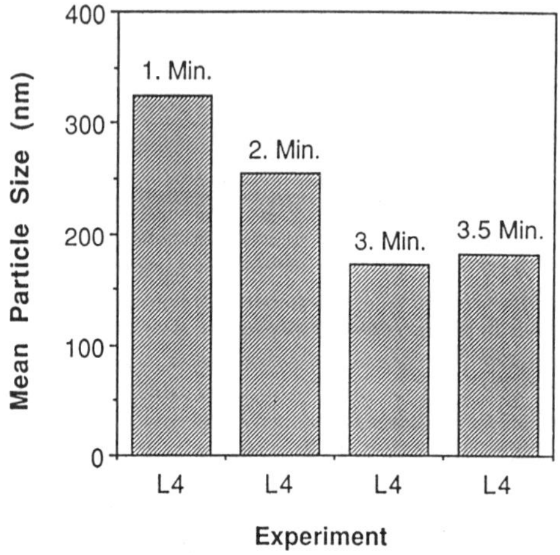

Figure 13. Effect of sonication time on measured mean particle size for samples formed in experiment L4.

supersaturation is caused by a mechanical perturbation. The speed with which this perturbation propagates gives rise to much more uniform conditions within the supersaturated fluid and hence to a narrower particle size distribution.

We have performed a systematic investigation of the influence of pre and post-expansion conditions upon the particle size and crystallinity of powders obtained via the continuous expansion of mixtures of Naphthalene in supercritical carbon dioxide. Particle size was found to be a sensitive function of pre-expansion temperature (at moderate solute concentration), post-expansion temperature, and solute concentration in the supercritical mixture. The fact that Naphthalene crystals as small as 2 μm could be easily produced by lowering the post-expansion temperature suggests the feasibility of separating nucleation and growth, and of a sensitive control over product characteristics through small changes in easily manipulated process variables.

The crystallization of Lovastatin (an anti-cholesterol drug) following the rapid expansion of CO_2-rich solutions was studied via a two-level factorial experimental design. In this approach, those process conditions which were found to be important in the Naphthalene study were systematically varied in order to assess their influence upon Lovastatin crystallization, as well as their mutual interactions. The Lovastatin powders were found to be crystalline. Aggregated powders consisting of sub-micron units were obtained in all cases. Lovastatin crystallization was found to be relatively insensitive to changes in process conditions within the range of experimental variables studied here. Judging from the crystallinity of the powders, the increased architectural complexity of the Lovastatin molecule (vis-a-vis Naphthalene) does not impose constraints upon the rapidity with which the supercritical mixture can be expanded without producing amorphous solids.

The reproducibility of our observations, as well as the fact that the powders were found to be crystalline in all cases studied herein, are encouraging results. It is now necessary to develop a quantitative model of the supercritical nucleation process, and to investigate experimentally the precipitation of molecules whose architecture is sufficiently complex to limit the rapidity with which the supercritical mixture can be expanded. Such investigation will further our understanding of supercritical nucleation, thereby (it is hoped) paving the way for the practical application of this promising technology to the continuous, efficient production of solids with controllable morphology and narrow size distribution.

Acknowledgements

The authors gratefully acknowledge the financial support of the National Science Foundation (Grant No. CBT-8657010), of Merck, Sharp and Dohme Research Laboratories, and of the donors of the Petroleum Research Fund, administered by the American Chemical Society. The authors also wish to express their appreciation to Karen Larson of Merck, Sharp and Dohme Research Laboratories for her valuable suggestions throughout the course of this study and for her relentless efforts with the Lovastatin particle size analysis.

Literature Cited

1. Kumar, S.K.; Johnston, K. P. Journal of Supercritical Fluids 1988, 1(1), 15.
2 Tsekhanskaya, Yu. V.; Iomtev , M. B. ; Mushkina, E. V. Russ. J. Phys. Chem. 1964, 38(9), 1173 .
3. Hannay, J. B. ; Hogarth, J. Proc. Roy. Soc. London 1879, A29, 324 .
4. Krukonis, V., "Supercritical Fluid Nucleation of Difficult to Comminute Solids," presented at the AIChE Annual Meeting, San Fransisco, November (1984).

5. Larson, K. A. ; King, M. L. Biotech. Prog. 1986, 2(2),73 .
6. Petersen, R. C.; Matson , D. W. ; Smith, R. D. J. Am. Chem. Soc. 1986, 108(8), 2100
7. Matson, D. W.; R. C. Petersen, R. C. ; Smith, R. D. Adv. Ceramic Mater. 1986, 1(3), 242 .
8. Matson, D. W.; R. C. Petersen, R. C. ; Smith, R. D. Mater. Lett. 1986, 4, 429 .
9. Matson, D. W. ; Fulton, J. L. ; Petersen, R. C. ; Smith, R. D. Ind. Eng. Chem. Res. 1987, 26, 2298 .
10. Matson, D. W.; R. C. Petersen, R. C. ; Smith, R. D. Advances in Ceramics 1987, 21, 109 .
11. Rosas, C., private communication, 1987.
12. Pennisi, K. J. ; Chimowitz, E. H. J. Chem. Eng. Data 1986, 31, 285 .
13. Chimowitz, E. H. ; Pennisi, K. J. AIChE J. 1986, 32(10), 1665.
14. Johnston, K. P. ; Barry, S. E. ; Read, N. K. ; Holcomb, T. R. Ind. Eng. Chem. Res. 1987, 26, 2372 .
15. McHugh, M. A. ; Yogan, T. J. J. Chem. Eng. Data 1984, 28, 210.
16. Mohamed, R. S. ; Debenedetti, P. G. ; Prud'homme, R. K. AIChE J. 1989, 35(2), 325.
17. Box, G. E. P. ; Hunter, W. G. ; Hunter, J. S., Statistics for Experimenters; John Wiley & Sons, New York, 1978, p. 306.
18. Kurnik, R. T. ; Holla , S. J. ; Reid, R. C. J. Chem. Eng. Data. 1981, 26, 47
19. Chang, H. ; Morrell, D. G. J. Chem. Eng. Data 1985, 30, 74 .
20. Debenedetti, P. G. ; Kumar, S. AIChE J. 1988, 34(4), 645 .
21. MacKay, M. E. ; Paulaitis, M. E. Ind. Eng. Chem. Fund. 1979, 18(2), 149 .
22. McCabe, W. L. ; Smith, J. C. Unit Operations of Chemical Engineering, McGraw-Hill, New York, 1967, p 785 .
23. Turner, J. R. ; Kodas, T. T. ; Friedlander, S. K. J. Chem. Phys. 1988, 88(1), 457.
24. Wegener, P. P. Nonequilibrium Flows; Wegener, P. P., Ed.; Marcel Dekker: New York, 1969; pp. 163-243.

RECEIVED May 1, 1989

Chapter 24

Solid–Fluid Mass Transfer in a Packed Bed Under Supercritical Conditions

G.-B. Lim[1], G. D. Holder[1], and Y. T. Shah[2]

[1]Department of Chemical and Petroleum Engineering, University of Pittsburgh, Pittsburgh, PA 15261
[2]College of Engineering and Applied Science, University of Tulsa, Tulsa, OK 74104–3189

Experimental gas-solid mass transfer data are presented for the well defined supercritical CO_2-naphthalene system at 10-200 atm and 35°C. These data are compared with low pressure gas-solid and liquid-solid systems. It has been found that both natural and forced convection are important under these conditions and that mass transfer rates at near-critical conditions are higher than at lower or higher pressure.

The use of supercritical fluids in separation processes has received considerable attention in the past several years and the fundamentals of supercritical fluid (SCF) extraction and potential applications have been described in a recent review article (1). It is generally known that supercritical conditions enhance the dissolution of solid particles. In comparison with liquid solvents, supercritical fluids have a high diffusivity, a low density and a low viscosity, thus allowing rapid extraction and phase separation. Little information is available in the literature however, on mass transfer coefficients between supercritical fluids and solids.

Experimental Apparatus and Procedure

A schematic diagram of the experimental apparatus used in this study is shown in Figure 1. Liquid carbon dioxide is pumped into the system via a high-pressure Milton-Roy liquid pump. Pressure is controlled by using a back pressure regulator and pressure fluctuation is dampened with an on-line surge tank. The system consists of a preheater which allows the solvent to reach the desired temperature and a cylindrical extraction vessel which is 104 cm^3 in volume, 14.8 cm in length and 3.45 cm in diameter. The extraction vessel is packed with a single section naphthalene pellets which have been made from pure melted naphthalene using a die. The height of the active packing in the bed can be changed by using inert packing at the bottom and the top of the bed. Because the surface area of the pellets decreases during an extraction, the average area and particle diameter is used assuming that uniform extraction of the particles takes place. For a section two layers thick, this should be valid. Only a small portion (5 to 20%) of the naphthalene was extracted, and the

0097–6156/89/0406–0379$06.00/0

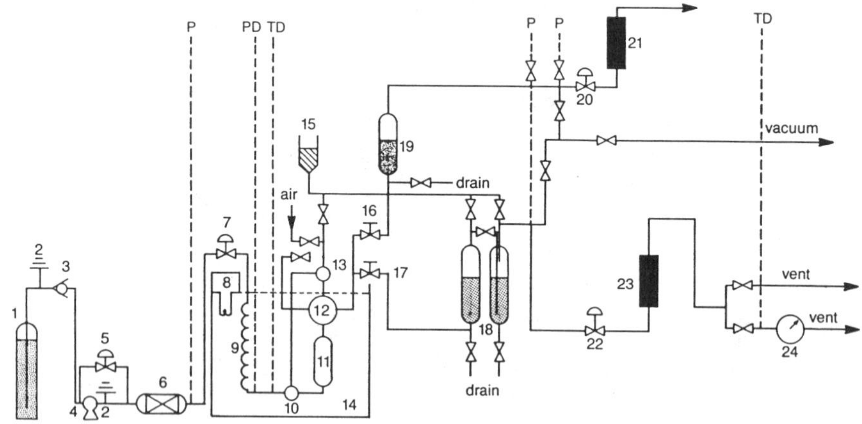

1. Liquid CO_2 cylinder
2. Relief valve
3. Check valve
4. Liquid pump
5. Back pressure regulator
6. Surge Tank
7. Pressure regulator
8. Temperature controller
9. Preheater
10. Three-way valve
11. Extractor
12. Four-way valve
13. Three-way valve
14. Water bath
15. Solvent tank
16. Metering valve
17. Metering valve
18. Sample tanks
19. Solid trap
20. Back pressure regulator
21. Rotameter
22. Back pressure regulator
23. Rotameter
24. Wet test meter

P: Pressure gauge
PD: Pressure transmitter to Data logger
TD: Thermocouple to Data logger

Figure 1. Schematic diagram of the experimental apparatus

pellets maintained their integrity and did not break up. The inert packing material consisted of glass beads with a diameter similar to that of the pellets. An advantage in using the inert pellets is to reduce end effects in the packed bed being used as the extractor. Pressure at the inlet of extractor is measured using a pressure transducer. The temperature of the extractor is also measured at the inlet using a thermocouple (type K).

The fluid mixture coming out of the extractor is depressurized to atmospheric pressure by passing it through a heated metering valve and a back pressure regulator. The instantaneous flow rate of the gas leaving the extractor is measured using a rotameter and the total amount of gas flow is measured with a calibrated wet-test meter.

The mass of precipitated solid is found as described below. With this value and the total amount of gas flow through the wet-test meter, the mole fraction of solids in the supercritical fluid can be readily determined. The temperature and pressure in the wet-test meter are also measured.

The sample collectors are high pressure bombs which contain glass beads with toluene which dissolves the extract (naphthalene) from the carbon dioxide. These vessels are operated at 30 to 35 atm where the solubility of the solid in the carbon dioxide is near a minimum. The second vessel is redundant and is used to guarantee that all of the extract is collected and to reduce entrainment losses. No naphthalene was found in this second vessel during current experiments. To determine the amount of extract collected, the amount of toluene (with dissolved extract) is weighed. A sample of the toluene-extract solution is then injected into a gas chromatograph to determine what portion of the solution is extract. Finally, the bypass, from valve 12 to 16, is designed to insure steady-state flow through the extraction vessel 11 before samples are taken.

The whole apparatus is rated for a pressure of 330 atm. All measured temperatures and pressures are recorded on a data logger at regular time intervals. The SCF-solid mass transfer coefficients were measured as a function of flow rate and pressure (or density). The ranges of system parameters examined in this study are shown in Table I.

TABLE I. Ranges of System Parameters in Experimental Study

System:	Naphthalene-Supercritical CO_2
Pellet Characteristics:	
Material:	Naphthalene
Shape:	Cylindrical
Size:	
Length (cm) :	0.45
Diameter (cm) :	0.45
Height of Bed (cm):	0.9 (2 layers)
Temperature of Bed (^oC):	35
Pressure (atm):	10-200
Flow Rates (STD liter/min at 0^oC and 1 atm):	0.9-33.5
Reynolds Number:	$2 < Re < 70$
Schmidt Number:	$2 < Sc < 11$
Grashof Number:	$78 < Gr < 3.25 \times 10^7$

Correlations for Diffusion Coefficient, Viscosity and Density of CO_2-Naphthalene Mixture

The development of mass transfer models require knowledge of three properties -- the diffusion coefficient of the solute, the viscosity of the SCF, and the density of the SCF phase. These properties can be used to correlate mass transfer coefficients. At 35°C and pressures lower than the critical pressure (72.83 atm for CO_2) we use the diffusivity interpolated from literature diffusivity data (2,3). However, a linear relationship between log Dv and ρ at constant temperature has been presented by several researchers (4,5) who correlated diffusivities in supercritical fluids. For pressures higher than the critical, we determined an analytical relationship using the diffusivity data obtained for the CO_2-naphthalene system by Iomtev and Tsekhanskaya (6), at 35°C.

$$D_v = 3.3531 \times 10^{-4} \times 10^{-0.6860\rho} \ [\mathrm{cm^2/sec}] \quad (1)$$

The above equation (1) was used for calculating diffusivities at pressures higher than critical pressure in this study. The viscosity of the system is approximated as that of pure CO_2, using experimental data of Stephan and Lucas (7). In addition, the density of the CO_2 -naphthalene mixture is needed at different mixture compositions. These values are obtained from the modified Peng-Robinson equation of state (8,9). The standard form of P-R EOS (9) can be written as

$$P = \frac{RT}{v-b} - \frac{a}{v(v+b) + b(v-b)} \quad (2)$$

By considering both temperature and density effects on the binary interaction parameter, its standard form can be modified as:

$$k_{ij} = \alpha_{ij} + \beta_{ij}\,\rho = \alpha_{ij} + \beta_{ij}/v \quad (3)$$

Then the modified P-R EOS becomes

$$P = \frac{RT}{v-b} - \frac{a' - c - d/v}{v(v+b) + b(v-b)} \quad (4)$$

where

$$a' = \sum_i\sum_j y_i y_j \sqrt{a_i a_j} \quad (5)$$

$$b = \sum_i y_i b_i \quad (6)$$

$$c = \sum_i\sum_j y_i y_j \sqrt{a_i a_j}\,\alpha_{ij} \quad (7)$$

$$d = \sum_i\sum_j y_i y_j \sqrt{a_i a_j}\,\beta_{ij} \quad (8)$$

$$\alpha_{ii} = \beta_{ii} = 0 \quad (9)$$

Then α_{ij} and β_{ij} were optimized to minimize the percent absolute average relative deviation (% AARD) between the calculated solubility and the experimental solubilty measured here, and the experimental solubilities at 35°C reported by Tsekhanskaya et al. (11) and by McHugh and Paulaitis (12). The equilibrium solubilities of naphthalene in CO_2 used to calculate the mass transfer coefficients are given in Table II. The optimized values of α_{ij}, β_{ij}, and %AARD are 0.0402, 6.5384 and 9.23 respectively. Prediction of the solubility with these two optimized parameters is given in Figure 2 with data of Tsekhanskaya et al. (11), McHugh and Paulaitis (12) and our experimental solubility data (below the critical pressure).

Table II. Equilibrium Solubility of Naphthalene in CO_2 at 35°C

P(atm)	$y^* \times 10^4$ [a]	P(atm)	$y^* \times 10^4$ [b]
10	0.5374	79.4	59.0
50	0.9651	90	90.0
60	1.6719	100	102.6
72.83	5.3583	120	125.6
		150	147.0
		200	170.0

a: This work (experimental).
b: Tsekhanskaya et al. (11), McHugh and Paulaitis (12) data.

Cell Model

Since the present experimental study was carried out with two layers of particles and for small Reynolds numbers, the effects of axial dispersion and natural convection on the experimental results would be important. In the present study, a cell model is used to analyze the data. For a small number of mixers, Kramers and Alberda (13) derived a relationship

$$n-1 = \frac{Lu}{2E_a} = \frac{L}{d_p}\frac{Pe,a}{2} \qquad (10)$$

Although approximate, equation 10 is used in this study for determining the number of perfect mixers to be used in the cell model below.

Mass Transfer Coefficients from the Cell Model

The cell model is a generalization of a class of models such as the completely mixed tanks-in-series model and the back-flow mixed tanks-in-series model. The common characteristic of this model is that the basic mixing unit is a completely mixed or stirred tank. This model has been employed extensively from the early days of chemical engineering to the present (14).

Even though no experimental data on axial dispersion have been published for supercritical fluids, we can approximate its effect as described below. For supercritical systems, the value of the Schmidt number, around 10, is intermediate to the values for gases (Sc ≈ 1.0) and liquids (Sc ≈ 1000). By comparing the order of magnitude of Schmidt

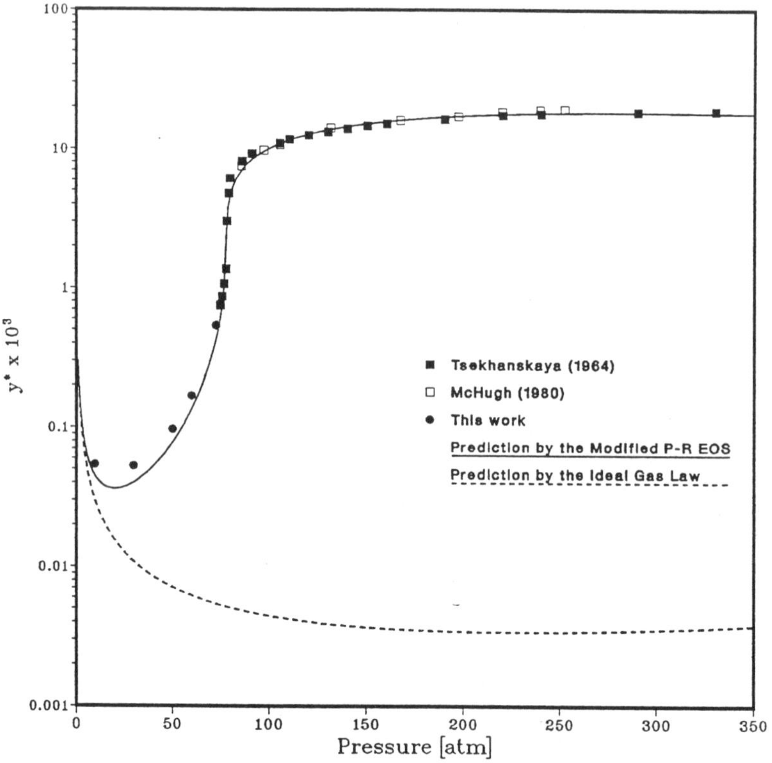

Figure 2. Solubilities of naphthalene in carbon dioxide at 35°C

numbers of gases, supercritical fluids and liquids, we assume that the value of Pe,a for SCFs is close to the value of Pe,a for gases and is approximately equal to 2.0 when Re is greater than 1.0.

Since complete mixing is assumed in a cell, the mole fraction of a solute in an out-going stream from the nth cell is y_n. If the bed is viewed as a series of n perfect mixing cells each having surface area A_T/n and constant mass transfer coefficient k_y, then for the steady-state mass transfer the material balance gives

$$k_y = \left(\frac{V_T}{A_T/n}\right) \left[\left(\frac{y^*}{y^*-y_n}\right)^{1/n} - 1\right] \tag{11}$$

In the present study, all the experiments were carried out with two layers of particles. As mentioned above, we can assume that the value of Pe,a for SCF is approximately equal to 2.0 when Re is greater than 1.0. Then, the number of perfect mixers in a packed bed can be determined by equation 10. With two layers of particles, the number of cells is three; that is, n = 3 in equation 11. If a different model is used the mass transfer coefficients will change slightly.

Effect of Natural and Forced Convection

After mass transfer coefficients under supercritical conditions are determined, they need to be correlated as a function of the significant independent variables. The mass transfer between a fluid and a packed bed of solid can be described by correlations of the following form.

$$Sh = f\ (Re, Sc, Gr) \tag{12}$$

where Sh, Re, Sc and Gr are, respectively, the Sherwood number, the Reynolds number, the Schmidt number and the Grashof number. Recently, Debenedetti and Reid (4) pointed out that, in the case of supercritical fluids, buoyant effects had to be considered because supercritical fluids showed extremely small kinematic viscosities as a result of their high densities and low viscosities. The effect of buoyant forces is more than two orders of magnitude higher in supercritical fluids than in normal liquids at the same Reynolds number.

When natural convection is controlling the effect of Reynolds number is unimportant, and the general expression reduces to

$$Sh = f\ (Sc, Gr) \tag{13}$$

For large Schmidt number (usually a liquid-solid system) Karabelas et al. (15) proposed the following relationship for natural convection.

$$Sh = a\ (Sc\ Gr)^b \tag{14}$$

If forced convection is controlling equation 13 will not correlate the data. In the forced convection regime, the Grashof number is unimportant and the general expression is reduced to,

$$Sh = h(Re,\ Sc) \tag{15}$$

For a wide range of Reynolds numbers, the following relationship was found to be effective,

$$Sh = cRe^{d}Sc^{1/3} \tag{16}$$

When both natural and forced convection are important, equations 13 and 15 cannot correlate the data successfully. Mandelbaum and Böhm (16) suggested the following correlation for combined natural and forced convection.

$$\frac{Sh}{(Sc.Gr)^{1/4}} = e \left[\frac{Re}{Gr^{1/2}}\right]^{f} \tag{17}$$

Results and Discussion

Mass Transfer Correlation. The development of a correlation for the mass transfer coefficient is based upon the mass transfer coefficients which were obtained through the use of the cell model proposed by Kramers and Alberda (13). Buoyant effects become important under supercritical conditions, because of the small kinematic viscosities which are a consequence of the high densities and low viscosities. Consequently, it is necessary to consider both forced and natural convection when attempting to correlate mass transfer coefficients under supercritical conditions.

The domain of the mixed convection regime depends on the fluid, the flow configuration and the flow pattern (17). It is usually defined by a region $a < Gr/Re^{n} < b$, where a and b are the lower and upper bounds of the domain, respectively and Gr/Re^{n} is the buoyanct force parameter, n being a constant that varies with the flow configuration. Conversely, when the buoyant force is the dominant mode of transport, Re^{n}/Gr or some power of it, becomes the important parameter for mixed convection.

A common approach to mixed convection is to compute the Sherwood number as the sum of the powers of pure natural and forced convection (18):

$$Sh^{3} = Sh_{n}^{3} + Sh_{f}^{3} \tag{18}$$

In this case one has to provide a reliable correlation for both Sherwood numbers (for natural and forced convection). This approach was applied to the present system in a single step, that is fitting the entire data set with equation 18 using equations 14 and 16. Then equation 18 became a four parameter (a,b,c,d) correlation, achieving its best fit with about 16% AARD. However, it was observed that for each pressure, better correlations could be obtained by resorting to correlations for pure natural convection around the critical pressure and for forced convection away from critical pressure. Despite being more accurate, this latter approach would introduce the problem of providing a supplementary chart showing the region of validity for each correlation in terms of relevant dimensionless numbers. In addition, the existence of many regions is cumbersome.

Because of the latter considerations another approach was undertaken. Mandelbaum and Böhm (16) characterized mass transfer in the mixed region by two lumped expressions, $Sh/(Sc.Gr)^{1/4}$ and $Re/Gr^{1/2}$. This analysis is also in line with the theory for strongly buoyancy-driven transport (19), where the parameter $Re/Gr^{1/2}$ characterizes the relative magnitude of the forced convection effect on the transport mechanisms. In the present case, we found that this kind of relationship is the best for correlating the experimental results over the entire pressure range with a single expression. This correlation shown in Figure 3 has 15.3% AARD:

$$Sh/(Sc.Gr)^{1/4} = 1.692\ [Re/Gr^{1/2}]^{0.356} \qquad (19)$$

However, additional data could be useful in assessing the range of validity for the proposed relationship.

Discussion. The data and correlations at pressures of 10 atm to 200 atm are compared to literature values at low pressure (1 atm and 25°C) (20-22). Figure 4 shows that above the critical pressure mass transfer coefficients are less dependent on Re than below the critical pressure. At low pressure, the density gradient across the boundary layer is much smaller. Therefore, the effect of natural convection on mass transfer is very slight. Near the critical pressure, however, the effect of natural convection on mass transfer rates becomes important at low Reynolds number due to very large density gradients across the boundary layer.

In the high pressure region, the 100 atm data shows less dependency on Re than do data at 120 atm to 200 atm. As the pressure increases, the supercritical carbon dioxide becomes more dense and the effect of dissolved naphthalene has a less significant effect on the gas density. Therefore, the mass transfer coefficient decreases gradually with increasing pressure in this high pressure region. These effects on mass tranfser rate can be explained by the density difference between CO_2-naphthalene mixtures and pure CO_2 at different pressures (Figure 5). Figure 6 shows that the mass transfer coefficient increases at the same Reynolds number as the pressure increases from 10 atm to near the critical pressure. However, it increases dramatically near the critical point, has its maximum value, then it decreases gradually as the pressure increases beyond the critical pressure. These phenomena result from the fact that the effect of natural convection is strongest near the critical point. This can be also explained by the fact that the Grashof number (Gr), or Rayleigh number (Ra = Sc x Gr), has its maximum value near the critical point. These conclusions are supported by Figure 7 which shows that the mass transfer coefficients are much higher near critical conditions than at low pressures. Increasing the pressure from the critical causes a slow decrease in the mass transfer coefficient. Decreasing the pressure from the critical results in a rapid decrease. This is directly related to the density gradient (Figure 5) and consequently to the Grashof number which follows the same trends as the mass transfer coefficient as pressure is changed.

Several factors can affect this enhanced mass transfer. First, as Debenedetti and Reid (4) pointed out, the very low kinematic viscosities in conjunction with very high buoyant forces serve to enhance natural convection at the same Reynolds number. This is accentuated by large density differences that can occur as naphthalene dissolves in the CO_2. It is possible to have very large, negative partial molar volumes (i.e., -2000 cc/mole) for a solute at conditions near the critical point (Eckert et al., (23)) which causes the fluid density to depend strongly on composition. At 35°C and 100 atm, naphthalene's partial molar volume at infinite dilution is approximately -300 cc/mol. This can cause a significantly higher fluid

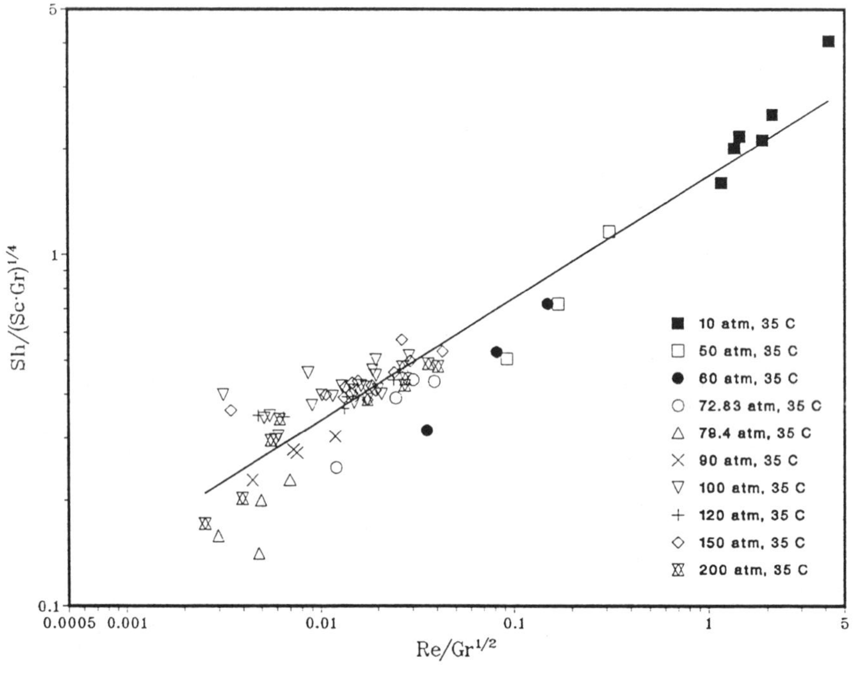

Figure 3. Correlation for combined natural and forced convection in the CO_2-naphthalene system at 35°C

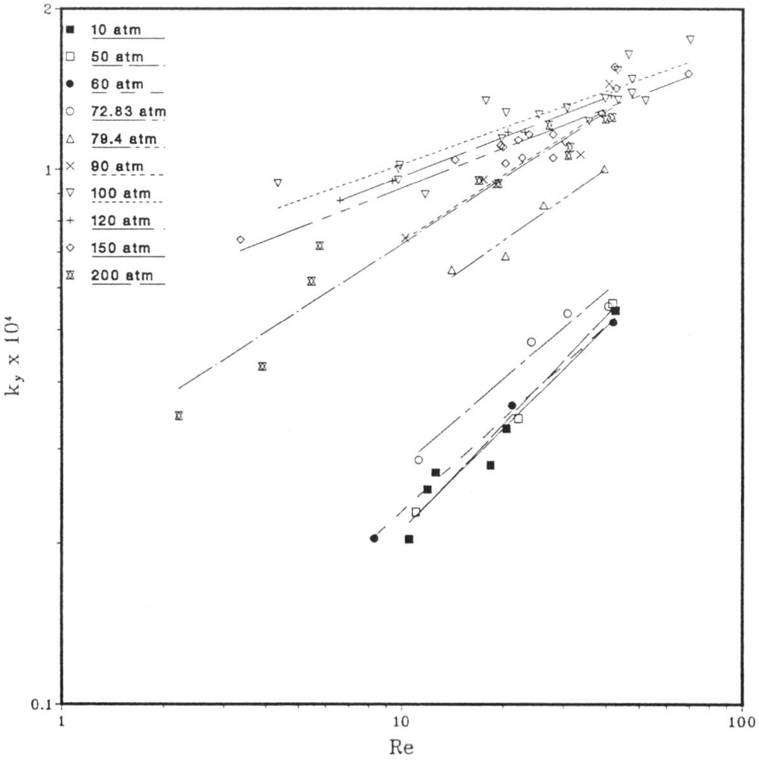

Figure 4. Correlation of k_y versus Reynolds number at 35°C

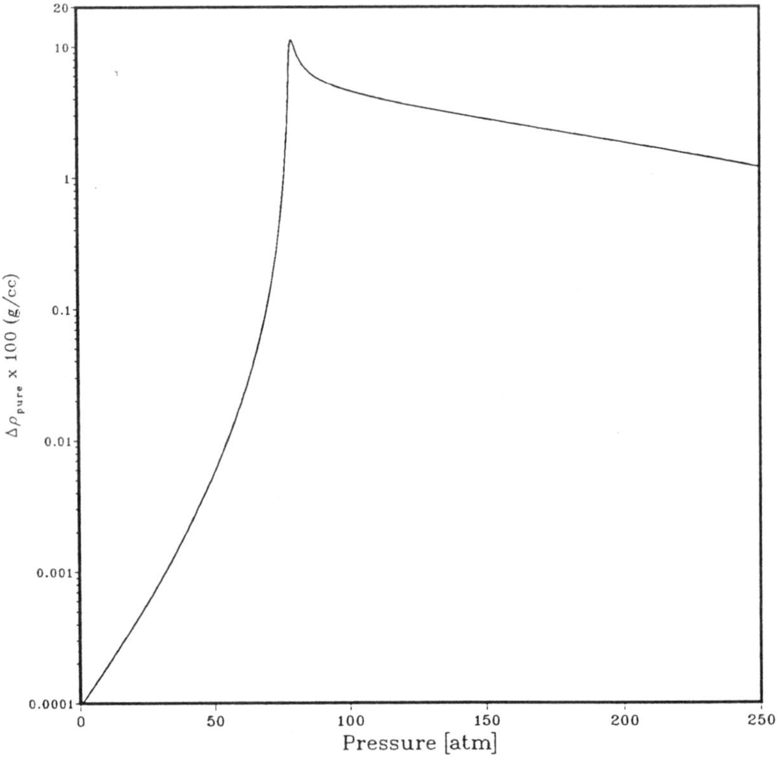

Figure 5. Effect of pressure on the density difference between equilibrium mixtures of naphthalene in CO_2 and pure CO_2 at 35°C ($\Delta\rho_{pure} = \rho^* - \rho_{CO_2}$)

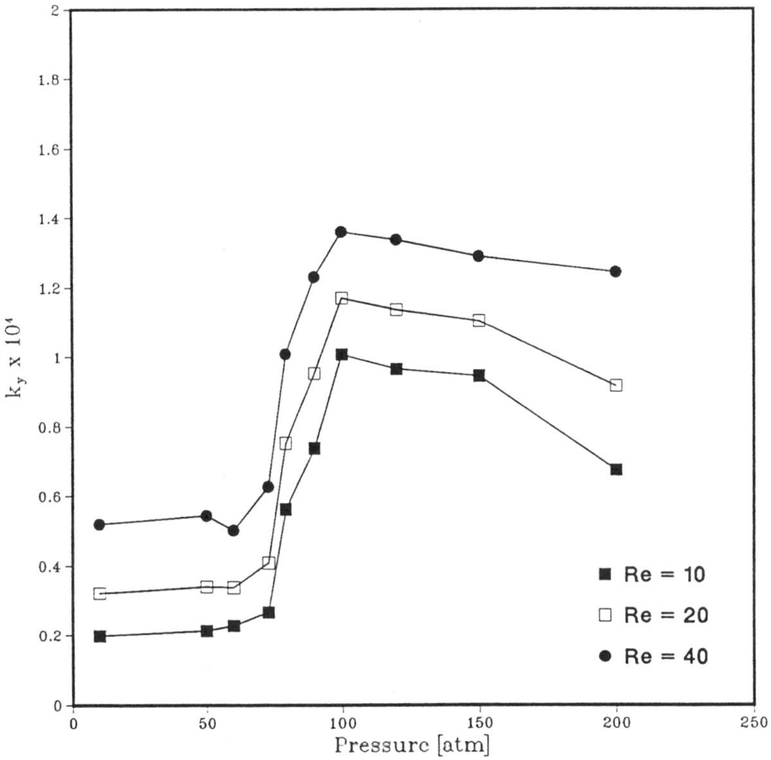

Figure 6. Effects of experimental pressure and Reynolds number on the mass transfer coefficient at 35°C

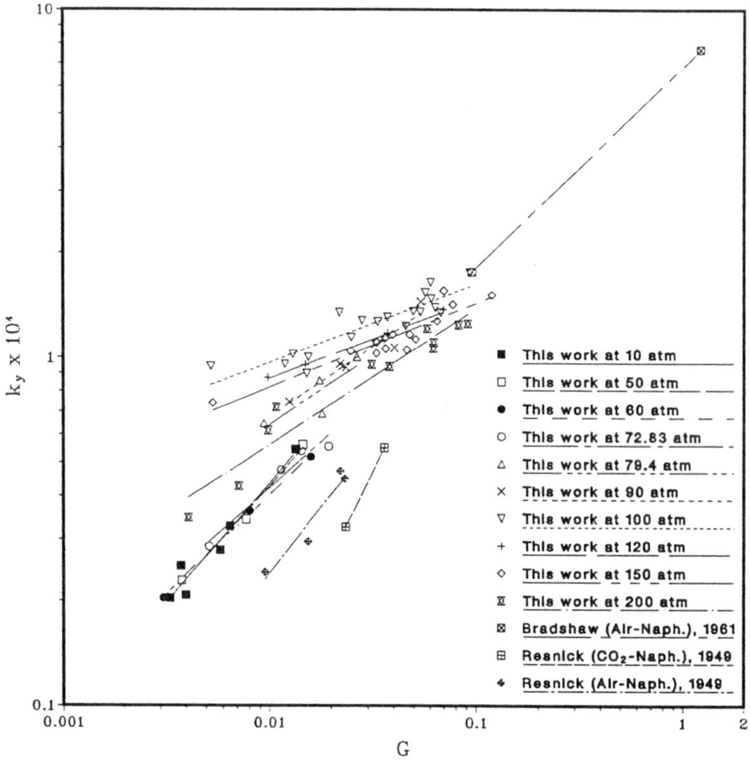

Figure 7. Comparison between supercritical mass transfer coefficients (10-200 atm and 35°C) and subcritical mass transfer coefficients (1 atm and 25°C)

density at the solid surface as compared to the bulk. Additionally, near critical conditions, the compressibility of the fluid begins to diverge and molecular level density fluctuations can occur and lead to additional convective contributions. The partial molar volume effects decrease rapidly as the pressure increases beyond the critical point and at 200 atm this effect is relatively minor. A decreasing density gradient explains why natural convection becomes less important.

Conclusions

At 35°C, the gas-solid mass transfer coefficient increases dramatically near the critical point, has its maximum value near 100 atm, and then decreases gradually as pressure increases. The mass transfer rate under supercritical conditions is much higher than at standard conditions (1 atm and 25°C) for liquid-solid and gas-solid systems, due to strong natural convection effects. Both natural and forced convection are important for supercritical mass transfer.

The results have implications for the sizing and control of process units including reactors and separation columns since high mass transfer rates will allow smaller reactors.

Acknowledgment

Authors gratefully acknowledge financial support from the U.S. Department of Energy under contract number DE-FG22-84PC71257. Dr. Gus Dassori's considerable efforts in this work are sincerely acknowledged.

Legend of Symbols

a	Peng-Robinson parameter [atm cm^6/gmole^2]
A_T	Total surface area of pellets in extractor [cm^2]
a'	Peng-Robinson mixture parameter defined in equation (5) [atm cm^6/gmole^2]
b	Peng-Robinson parameter [cm^3/gmole]
c	Peng-Robinson mixture parameter defined in equation (7) [atm cm^6/gmole^2]
d	Peng-Robinson mixture parameter defined in equation (8) [atm cm^9/gmole^3]
d_p	Average of diameter of sphere possessing the same surface area as a piece of packing [cm]
D_v	Molecular diffusivity [cm^2/sec]
E_a	Axial dispersion coefficient [cm^2/sec]
g	Gravitational acceleration [cm^2/sec]
G	Superficial mass velocity [g/cm^{2}sec]
Gr	Grashof number = $d_p^3 g \rho \Delta\rho / \mu^2$
k_{ij}	Peng-Robinson binary interaction parameter
k_y	Mass transfer coefficient [gmole/cm^{2}sec mole-fraction]
L	Total height of bed [cm]
M_{av}	Average molecular weight [g/gmole]
n	Number of perfect mixers
P	Total pressure [atm]
Pe,a	Axial Peclet number = ud_p/E_a
R	Gas constant = 82.0567 [atm cm^3/gmole°K]
Re	Reynolds number = $\rho d_p u_s/\mu$

Sc	Schmidt number = $\mu/\rho D_v$
Sh	Sherwood number = $k_y M_{av} d_p/\rho D_v$
T	Temperature [K]
u	Interstitial velocity [cm/sec]
u_s	Superficial velocity [cm/sec]
v	Molar volume [cm^3/gmole]
V_T	Total molal flow rate [gmole/sec]
y*	Equilibrium mole fraction of the solute (naphthalene)
y_n	Mole fraction of the solute (naphthalene) in stream outgoing from the nth cell

Greek Letters

α_{ij}	Adjustable parameter defined in equation (3)
β_{ij}	Adjustable parameter defined in equation (3) [cm^3/gmole]
μ	Viscosity [g/cm sec]
ρ	Average density over the bed length [g/cm^3]
ρ^*	Density at solid-solvent interface [g/cm^3]
ρ_{co_2}	Pure CO_2 density [g/cm^3]
$\Delta\rho$	Average density difference = $\rho^* - \rho$ [g/cm^3]
$\Delta\rho$ pure	Density difference between equilibrium mixtures of naphthalene in CO_2 and pure CO_2 [g/cm^3]

Literature Cited

1. Paulaitis, M.E.; Krukonis, V.J.; Kurnik, R.T.; Reid, R.C. Rev. Chem. Eng. 1983, 1, 179.
2. Morozov, V.S.; Vinkler, E.G. Russ. J. Phys. Chem. 1975, 49, 1404.
3. Vinkler, E.G.; Morozov, V.S. Russ. J. Phys. Chem. 1975, 49, 1405.
4. Debenedetti, P.G.; Reid, R.C. AIChE J. 1986, 32, 2034.
5. Feist, R.; Schneider, G.M. J. Sep. Sci. Tech. 1982, 26, 261.
6. Iomtev, M.B.; Tsekhanskaya, Y.V. Russ. J. Phys. Chem. 1964, 38, 485.
7. Stephan, K.; Lucas, K. Viscosity of Dense Gas; Plenum Press: New York, 1979; 75.
8. Mohamed, R.S.; Holder, G.D. Fluid Phase Equilibria 1987, 32, 295.
9. Lim, G.-B.; Mohamed, R.S.; Holder, G.D.; Bendale, P.G., Errata, Fluid Phase Equilibria 1988, 43, 359.
10. Peng, D.Y.; Robinson, D.B. Ind. Eng.Chem. Fund. 1976, 15, 59.
11. Tsekhanskaya, Y.V.; Iomtev, M.B.; Mushkina, E.V. Russ. J. Phys. Chem. 1964, 38, 1173.
12. McHugh, M.; Paulaitis, M. J. Chem. Eng. Data 1980, 25, 326.
13. Kramers, H.; Alberda, G. Chem. Eng. Sci. 1953, 2, 173.
14. Levenspiel, O., Chemical Reaction Enginering; John Wiley and Sons: New York, 1962, 290.
15. Karabelas, A.J.; Wegner, T.H.; Hanratty, T.J. Chem. Eng. Sci. 1971, 26, 1581.
16. Mandelbaum, J.A.; Bőhm, V. Chem. Eng. Sci. 1973, 28, 569.
17. Chen, T.S.; Armaly, B.F.; Aung, W. Mixed Convection in Laminar Flow Boundary-Flow; Natual Convection, Fundamentals and Applications: Kakac, S.; Aung, W.; Viskanta, R., Ed.; Hemisphere Publishing Co.: New York, 1985; p. 699-725.
18. Churchill, S.W. AIChE J. 1977, 23, 10.
19. Gebhart, B.; Jaluria, Y.; Maharjan, R.L.; Sammakia, B. Buoyancy-Induced Flows and Transport; Hemisphere Publishing Co.: New York, 1988.

20. McCune, L.K.; Wilhelm, R.H. Ind. and Eng. Chem. 1949, 41, 1124.
21. Petrovic, L.J.; Thodos, G. Ind. Chem. Fund., 1968, 7, 274.
22. Resnick, W.; White, R.T. Chem. Eng. Progress, 1949, 45, 377.
23. Eckert, C.A.; Ziger, D.H.; Johnston, K.P.; Kim, S. J. Phys. Chem. 1986, 90, 2738.

RECEIVED May 2, 1989

Chapter 25

Two-Phase Heat Transfer in the Vicinity of a Lower Consolute Point

Michael C. Jones

Center for Chemical Engineering, National Institute of Standards and Technology, Boulder, CO 80303

Experimental measurements of heat transfer coefficients are reported for three binary mixtures near their lower consolute points. Two of these, respectively n-pentane and n-decane in solution with supercritical CO_2, involve vapor--liquid equilibrium whereas the third, triethylamine--water, involves liquid--liquid equilibrium. Anomalously high heat transfer coefficients were found for the supercritical mixtures at compositions which condense on heating (retrograde condensation). These are attributed to thermocapillarity, and criteria are proposed for the values of dimensionless parameters which would indicate its presence. For cases considered, numerical values support this explanation. Additional support is provided by photographic evidence.

Design and scale-up of supercritical solvent extraction and processing plants wi.l require a knowledge and understanding of transport processes in supercritical fluid mixtures. We have begun a program to measure transfer coefficients of mixtures of organic liquids and carbon dioxide. In this work we report on heat transfer coefficients. Our emphasis has been on the important aspect of the phase diagram of binary mixtures characterized by the coexistence of vapor and liquid phases above the critical pressures of the two components. This two-phase region may be utilized to effect a preliminary separation of solute and solvent. At constant pressure, it is a region which typically possesses both upper and lower critical solution points in the neighborhood of which differences between the phases become attenuated: volume and enthalpy changes across the phase boundary, and interfacial tensions, decrease to zero as the critical point is approached. Phase behavior of several hydrocarbon--CO_2 systems is discussed by Schneider (1). It is also known by theory and experiment that the mutual diffusion coefficients decrease to zero as the mixture critical point is approached -- see Cussler (2). Therefore it is of interest to discover which, if any, of these unique properties are reflected in the transport coefficients. Our work is focussed on the lower critical solution point or lower consolute point. Such a point is indicated in Figure 1. which is the isobaric phase diagram for the mixture n-pentane--CO_2.

As is well known, on the solvent side of the lower consolute point is a region in which a condensate can be formed at constant pressure through the process of heating -- so-called retrograde condensation. On the solute side, heating should produce normal evaporation.

Previously, Durst and Stephan (3) observed some enhancement in heat transfer coefficients in natural convection in mixtures of n-heptane -- methane as the two-phase region was entered. We reported heat transfer results earlier (4) for free convection and cross flow about a heated horizontal cylinder in a supercritical n-decane--CO_2 mixture. These also showed enhancement in the two-phase region relative to single phase. At that time, analysis was hindered by the unavailability of estimates or measurements of some crucial mixture properties. Happily, this situation has been remedied somewhat and we can now draw some conclusions.

In this paper, we now report measurements of heat transfer coefficients for three systems at a variety of compositions near their lower consolute points. The first two, n-pentane--CO_2 and n-decane--CO_2 are supercritical. The third is a liquid--liquid mixture, triethylamine (TEA)--H_2O, at atmospheric pressure. It seems to be quite analogous and exhibits similar behavior. All measurements were made using an electrically heated, horizontal copper cylinder in free convection. An attempt to interpret the results is given based on a scale analysis. This leads us to the conclusion that no attempt at modeling the observed condensation behavior will be possible without taking into account the possibility of interfacial tension-driven flows. However, other factors, which have so far eluded definition, appear to be involved.

Phase diagrams

For the hydrocarbon--CO_2 systems studied here, at pressures above the critical pressure (7.383 MPa) and above the critical temperature (304.21 K) of CO_2, the isobaric x,T coexistence plots of liquid and vapor phases form simple closed loops. The minimum occurs at the lower consolute point or the Lower Critical Solution Temperature (LCST). Since pressure is usually uniform in the vicinity of a heat transfer surface, such diagrams serve to display the equilibrium states possible in a heat transfer experiment.

To illustrate, Figure 1 shows the x,T coexistence plot for n-pentane--CO_2 at 8.96 MPa. The calculated curve was obtained by Rainwater and Lynch (National Institute of Standards and Technology, personal communication, 1988) using the Leung-Griffiths (5) model as adapted, for example, in Rainwater and Moldover (6), with parameters

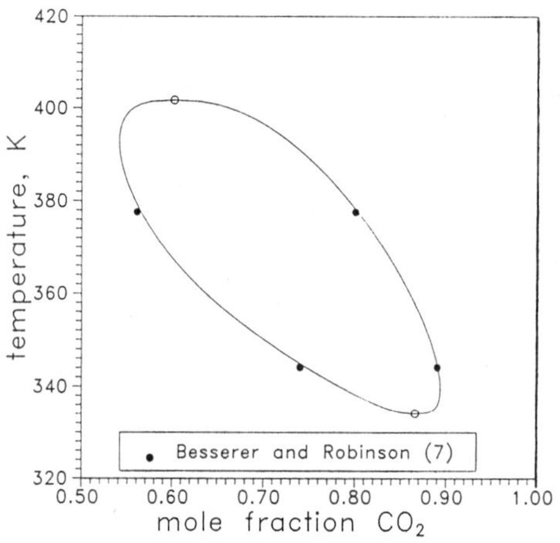

Figure 1. Isobaric phase coexistence plot for n-pentane--CO_2 at 8.96 MPa. Solid curve calculated -- see text. Upper and lower consolute points indicated by open circles.

derived from a global fit to the experimental data of Besserer and Robinson (7). The fit predicts the LCST to be 334.1 K at a CO_2 mole fraction of 0.8658. This method is not available for the n-decane--CO_2 system, but in this case fits were obtained to the data of Reamer and Sage (8) and Nagarajan and Robinson (9) for single pressures using the Peng-Robinson (10) equation of state. However, determinations of LCST compositions were not possible to better than a few mole percent here due to the failure of flash calculations to converge in the neighborhood of a critical point. At 10.4 MPa, we estimate the LCST to be 325 K at a CO_2 mole fraction of 0.93 ± 0.02 ; at 12.2 MPa, we find 335 K at a mole fraction 0.91 ± 0.02. For the TEA--water system, we relied on the experimental data, for example (11). The LCST at atmospheric pressure is 18.2 °C and TEA mole fraction 0.09.

In our heat transfer experiments, bulk fluid conditions were chosen to be just below the phase coexistence curve on either side of the lower consolute point. Thus, on heating at constant pressure, either evaporation of a liquid or condensation (retrograde) of a vapor took place once a small excess of test section surface temperature over bulk fluid temperature occurred. To be specific, the retrograde condensation region of the vapor--liquid phase coexistence curve of Figure 1 is the region of positive slope to the right of the LCST.

Experimental

A schematic of the experimental arrangement is shown in Figure 2. In the experiments, we measure steady state surface temperatures of a heated test section at selected power levels while bulk-fluid temperature and pressure are held constant. We express the results as conventional heat transfer coefficients using an average of the surface temperatures from up to four thermocouples. We find that under most conditions the point-to-point variation of surface temperature is less than 4 % of the overall temperature difference.

The experimental arrangement is similar to that reported in (4), but several changes were made to improve control, to improve visibility of the test section, and to reduce overall system volume.

To ensure repeatable heat transfer results we find it important to maintain good control of the bulk fluid temperature and pressure during a run. We achieve this by circulating the fluid through the centrifugal pump C and a large capacity temperature-controlled oil-bath heat exchanger H. Further, as power input to the test section increases, we compensate with the bath circulator/temperature controller BC/TC which circulates water through channels bored in the experimental vessel V and controls the exit water temperature. Mid-run adjustments to the pressure if necessary can be made by adjusting the volume of the 2-liter free-piston accumulator A. In this manner, bulk-fluid temperatures can be held constant to better than 1 K and pressures to .01 MPa over the entire range of power levels to the test section. The bulk fluid temperature is indicated by two platinum resistance thermometers suspended in the fluid in V and the system pressure by a calibrated precision Bourdon tube gauge.

The pressure vessel itself is bored from a block of aluminum alloy 6061-T6. It has a capacity of approximately 2 liters and is

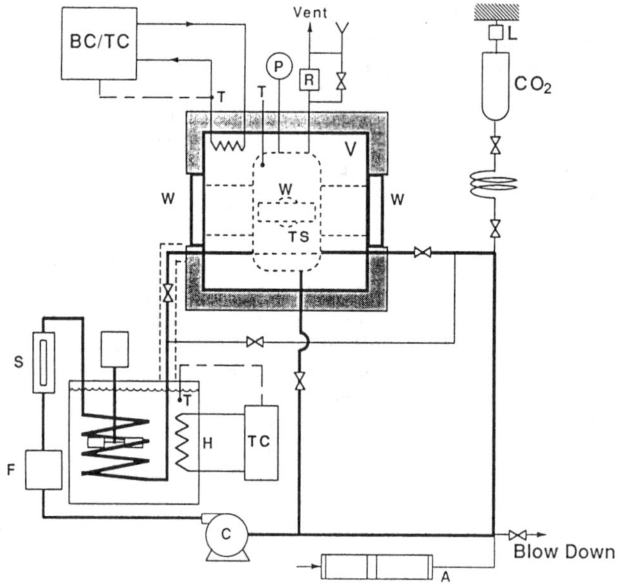

Figure 2. Schematic diagram of experimental system.

fitted with two sapphire windows of 41 mm diameter mounted on opposite walls and a third of 25 mm diameter mounted at right angles. With this arrangement the test section mounted inside can be viewed and photographed. For all results reported here, the test section is a machined copper cylinder of 11 mm diameter, 38 mm length, mounted horizontally between flush, filled-Teflon end pieces. It is drilled just under the surface to accept type K thermocouple junctions of 0.13 mm diameter wire and is bored out on axis to accommodate a 25-ohm ceramic resistor as heat source. The test section is protected from the turbulence of the circulating fluid by leading the latter through a wire screen inside V. Visual observations confirm that the fluid in the vicinity remains undisturbed until free convection from the test section begins.

Mixtures are made up by first pouring a measured volume of the hydrocarbon solute into the system. Then, the system is closed and pressurized as CO_2 is weighed in from a gas cylinder suspended from a load cell L capable of weighing to 0.01 kg. The resulting mole fraction is known to $\pm$ 0.0005.

All temperature measurements and conversions and power measurements are made automatically by a six-digit digital voltmeter, channel scanner, and microcomputer communicating via an IEEE 488 bus.

For the liquid--liquid system experiments at atmospheric pressure, we mounted the same copper test cylinder in a simple 2-liter water-jacketed beaker. This was filled with the test mixture and maintained at constant temperature by a water bath circulator. The instrumentation was identical to the supercritical system. Since this was a batch system, the experiments were of limited duration and the liquids were remixed after each test at a given power level.

Heat Transfer Results

The heat transfer coefficients h are shown in Figures 3, 4, 5 and 6 plotted against $\Delta T = T_W - T_B$. For small ΔT, accuracy is limited by thermocouple calibration error, which we estimate to be within $\pm$ 0.2 K. Thus, estimated errors in h range from $\pm$ 2 % at ΔT = 30 K, to $\pm$ 4% at 10 K, and up to $\pm$ 20 % at 1 K. Repeatability is within 4 % when bulk fluid temperature and pressure are held to the tolerances quoted in the previous section.

In Figure 3, heat transfer coefficients are shown for n-pentane--CO_2 at 8.9 MPa and bulk fluid CO_2 mole fractions of: 0.830 on the liquid side of the LCST, 0.865 precisely at the LCST, and 0.876 on the vapor side. For comparison, we also show results for pure carbon dioxide at the same bulk temperature and pressure. A similar set of results is shown for n-decane--CO_2 for each of two pressures in Figures 4 and 5. In Figure 4, at 10.4 MPa, the LCST of 325 K occurs at a CO_2 mole fraction x of 0.93 $\pm$ 0.02 according to our Peng-Robinson fit of the phase equilibrium data. Thus, only the results for x = 0.973 are clearly on the vapor side of the LCST and only those for x = 0.867 are on the liquid side. In Figure 5, for 12.2 MPa, the LCST has shifted slightly to x = 0.91 $\pm$ 0.02 and T = 335 K. Therefore, we expect the data for x = 0.940 to now be on the vapor side.

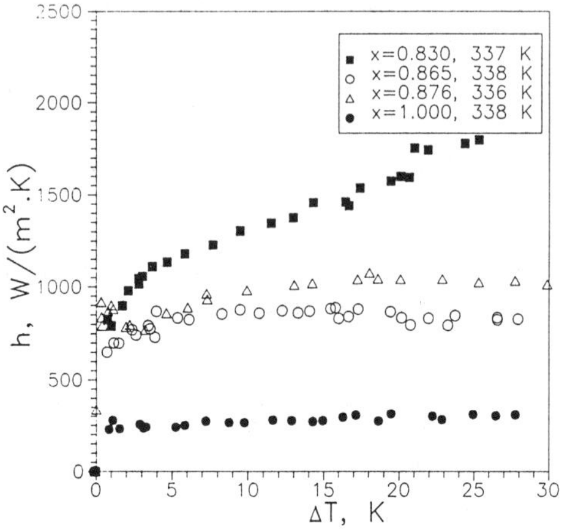

Figure 3. Heat transfer coefficients for n-pentane--CO_2 mixtures at 9.0 MPa.

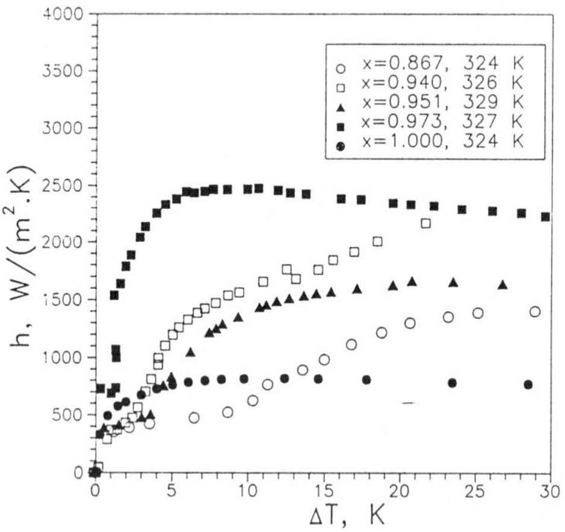

Figure 4. Heat transfer coefficients for n-decane--CO_2 mixtures at 10.4 MPa.

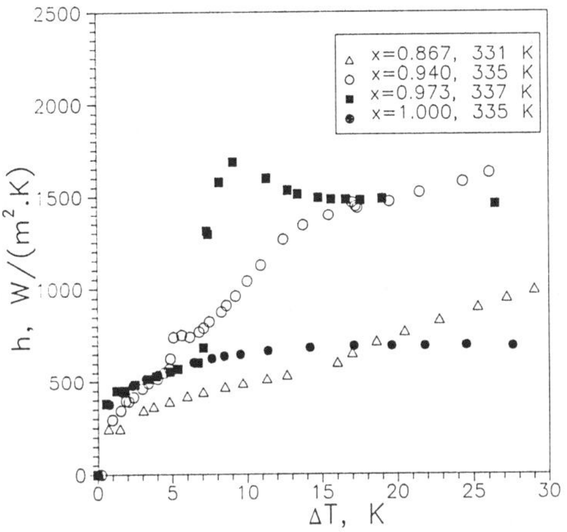

Figure 5. Heat transfer coefficients for n-decane--CO_2 mixtures at 12.2 MPa.

The most striking feature of these results is the steep rise in the heat transfer coefficient for the n-decane--CO_2 results at x = 0.973 at both pressures. This rise coincided in both cases with the onset of the condensation process as we observed droplets or streamers falling from the test section. In all but the n-decane-CO_2 mixture of x = 0.867 we saw some form of dense material falling from the test section, but it was not always clear whether this was an immiscible phase; the appearance was often more like that of density schlieren. We offer no explanation for the fact that, in the n-pentane--CO_2 results, the liquid side heat transfer coefficients are greater than for the vapor side in contrast to the n-decane--CO_2 system, but we stress that the vapor side condensation results for the former are still "enhanced" in the sense described below.

The results for the TEA--water mixtures at atmospheric pressure are shown in Figure 6. These are for TEA mole fractions of x = 0.05 and 0.59. The LCST is 18.2 ^oC at x = 0.09. We also obtained a very similar data set at the latter mole fraction, but we omitted it for clarity. For contrast and comparison, a data set for pure water is shown. These mixture results again show a sharp rise in heat transfer coefficient as "condensate" first appeared. In fact, the appearance was remarkably similar to the n-decane--CO_2 results for x = 0.973 discussed above, but the visibility of the phase separation was enhanced by the presence of a fine emulsion at the phase interface and the absence of strong refractive index gradients characteristic of the supercritical systems. This permitted the structure of the interface to be seen more clearly. In Figure 7 we show photographs that typify the appearance of the two phases. In all cases observed here, both in supercritical vapor--liquid and in liquid--liquid systems, the dense phase appears to wet the cylinder surface regardless of composition.

Interpretation - Scale Analysis

It is of importance to establish whether conventional relationships and correlations could satisfactorily be applied to supercritical mixtures. Whereas heat transfer coefficients in single-phase situations are approximately in accord with established correlations, a few simple calculations show that this is not the case in the two-phase region. Considering first evaporation, while it is well known that boiling heat transfer is highly variable, depending as much on the nature and preparation of the surface as on the system itself, nevertheless, we expected some correspondence between our data and empirical correlations. In fact we find for both supercritical systems that, on the liquid side of the critical point, observed heat transfer coefficients are two orders of magnitude below the predictions of the Rohsenow (12) correlation for a temperature difference of 10 K. Evidently, an understanding of this process in terms of conventional boiling is not possible and we postpone further discussion of the topic to future work.

Classical laminar film condensation, on the other hand, is more predictable and we anticipated greater success in treating our experimental data in which it occurred. An upper bound is set by the Nusselt expression (13) for a pure fluid; in binary mixtures, additional heat transfer resistance may be expected due to the

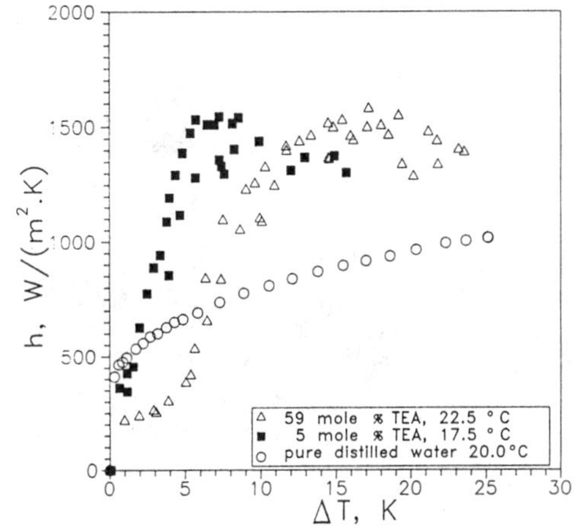

Figure 6. Heat transfer coefficients for TEA--water mixtures at atmospheric pressure.

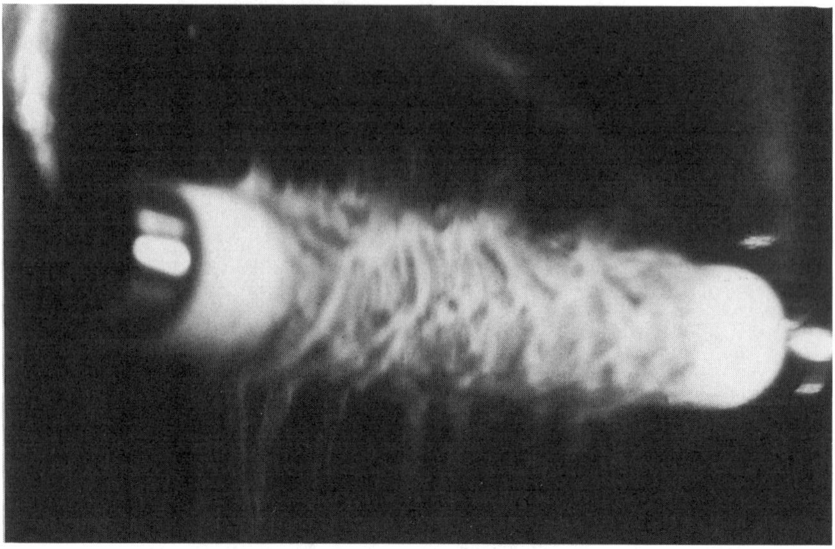

a

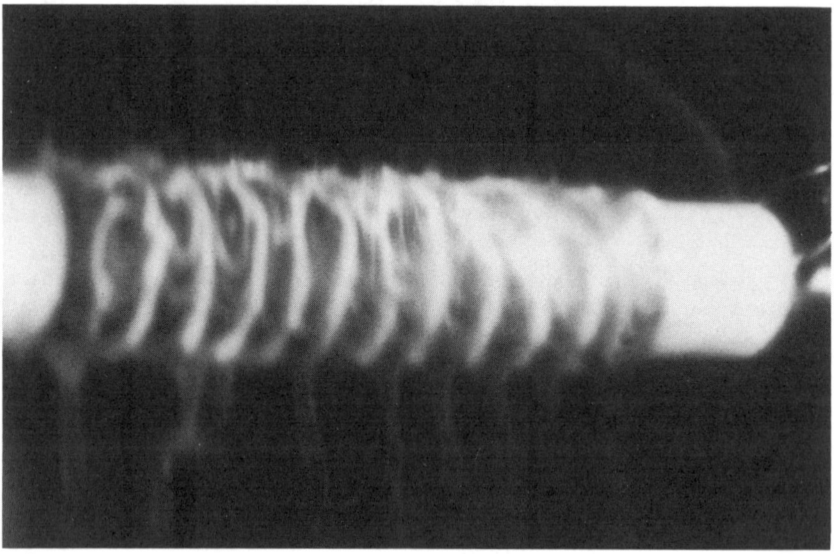

b

Figure 7. Photographs of phase formation on the heated horizontal cylinder in a solution of 9 mole % TEA in water. ΔT = 10 K. Photograph a shows the structure normally seen, while in photograph b the structure has become organized by the gravity vector without change in operating conditions.

accumulation of the more volatile component adjacent to the surface. The Nusselt expression for condensation on a horizontal cylinder of diameter D is

$$Nu = \frac{hD}{k} = 0.73 \left[\frac{\rho_L(\rho_L-\rho_V)g\Delta HD^3}{\mu k(T_W-T_B)}\right]^{1/4} \tag{1}$$

The quantity in square brackets is the film Rayleigh number, Ra_{film}, and will occur again below. Nu values calculated for the three two-phase systems studied here are listed in Table I for $(T_W-T_B) = 10$ K and are compared with the experimental values. We see that the experimental values for the supercritical systems are far in excess of this supposed upper bound. Thus, any explanation must not only account for a strongly diminished vapor resistance but must also predict a reduction in the liquid film resistance.

Table I Nusselt Numbers for Condensation

	System 1. $nC5-CO_2$ 8.96 MPa	System 2. $nC10-CO_2$ 10.4 MPa	System 3. TEA-H_2O 0.1 MPa
$Nu = 0.73\ Ra_{film}^{1/4}$	73	128	37
Nu_{exp}	130	287	26

The hypothesis advanced here is that the observed enhancement is due to thermocapillarity arising from the temperature dependence of the interfacial tension. Although we know of no theory for such a Marangoni effect in binary condensation *per se*, theory does exist for both mass transfer (14, 15) and heat transfer (16, 17) across an interface in the absence of bulk flow. We note that the sign of the derivative of the interfacial tension with respect to temperature is positive near a lower consolute point and that this is in the correct direction to sustain disturbances in condensation rate. Thus, in retrograde condensation, provided a critical temperature gradient normal to the interface is exceeded, a local increase in condensation flux toward the vapor liquid interface will result in its cooling. The consequent drop in tension will give rise to surface flow away from the disturbance which, thus, reinforces the initial flux disturbance. We postulate the presence of cellular convection on both sides of the interface promoting heat transfer in the liquid film and mixing in the vapor which tends to disperse the more volatile component.

A stability analysis will be required to confirm this postulate, but in the meantime we can establish by scale analysis the feasibility of such a mechanism in a qualitative, order-of-magnitude manner. This can be done in the following way. We suppose that steady film flow of condensing liquid is established by a balance of viscous and buoyancy forces. We then seek a criterion by analysis of the governing conservation equations and boundary conditions that will indicate the presence of thermocapillary forces of the same order of magnitude as the buoyancy forces. We discover two candidate criteria in what follows. The method we use has been outlined by Bejan (18).

For a vertical condensing liquid film, scale analysis of the 2-dimensional constant-property conservation equations of mass, momentum and energy, with appropriate boundary conditions, shows that $Ra_{film}^{1/4}$ is a measure of the slenderness ratio L/d of the film (18). Here L is taken as the streamwise length of the condensation film and d an average film thickness. This interpretation is valid in the limit where the momentum equation is a balance of viscous and gravitational forces. Under such conditions, a velocity scale is

$$U_r \sim \frac{g\Delta\rho d^2}{\rho\nu} \tag{2}$$

Under the same assumptions, the scale of shear stresses in the liquid film may be obtained by integration of the momentum equation to give

$$\tau_{yx}^{L} \sim g\Delta\rho d \tag{3}$$

On the other hand, if a liquid film experiences interfacial tension gradients, then the balance of tangential stresses at the interface gives rise to the key relationship

$$\tau_{yx}^{V} + \frac{\partial\sigma}{\partial x} = \tau_{yx}^{L} \tag{4}$$

and the scale of interfacial shear stress induced by tension gradients is $\partial\sigma/\partial x$. Therefore, we may take the following as a criterion for these stresses to be of significance in the buoyancy driven film flow:

$$\partial\sigma/\partial x \sim g\Delta\rho d \tag{5}$$

or

$$\frac{d\sigma}{dT}\frac{\partial T}{\partial x} \sim \frac{d\sigma}{dT}\frac{\Delta T}{L} \sim g\Delta\rho d \tag{6}$$

Introducing the slenderness ratio to eliminate d, we get the final form of our first criterion

$$\frac{Ra_{film}^{1/4}}{Bo} \sim 1 \tag{7}$$

where Bo is the modified Bond number $g\Delta\rho L^2/((d\sigma/dT)\Delta T)$. We call this Criterion A below.

We may propose an alternative criterion based on the observation by Ostrach (19) that the existence of a surface tension gradient on the free surface of a liquid layer implies its own reference velocity

$$U'_r = \frac{1}{\mu}\left(\frac{d\sigma}{dT}\right)\frac{d}{L}\,\Delta T \tag{8}$$

Dimensional analysis of the energy equation then gives a criterion for when convective terms become comparable in magnitude to conductive terms favoring the formation of thermal boundary layers.

$$Ma\left(\frac{d}{L}\right)^2 = \frac{U'_r d}{\alpha}\left(\frac{d}{L}\right)^2 \sim 1 \tag{9}$$

This defines the Marangoni number Ma. The expression for the slenderness ratio of our gravity-driven film is again used in Equation 9. We call this Criterion B. Having now introduced the reference velocity U'_r, we note that Criterion A as expressed by Equation 7. is the ratio U'_r/U_r.

In Table II, we list the various dimensionless groups defined above and the two criteria for enhancement of the heat transfer coefficient by interfacial tension-driven flow. Calculated values are given for the three mixtures for which we presented experimental data. All values pertain to a temperature difference ΔT of 10 K.

Table II Characteristic Dimensionless Parameters for Condensation

	Parameter	System 1. $nC5\text{-}CO_2$ 8.96 MPa	System 2. $nC10\text{-}CO_2$ 10.4 MPa	System 3. $TEA\text{-}H_2O$ 0.1 MPa
	$Ra_{film}^{1/4} \sim L/d$	100	175	51
A	$Ra_{film}^{1/4}/Bo$	0.07	0.21	0.44
	Ma	5551	4410	1708
B	$Ma(d/L)^2$	0.55	0.14	0.65
	Cr	$1.6(10^{-6})$	$5(10^{-4})$	$5(10^{-4})$

In the final row of the table, we have listed values for the Crispation number, Cr = $\mu\alpha/\sigma d$. This number arises in stability analyses of interfacial transport when deflection of the interface under normal stresses is permitted (15, 16). Its importance was demonstrated by Scriven and Sternling who showed that the effect is always destabilizing; a nondeflecting interface has Cr = 0 or infinite tension. Since we deal with vanishingly small interfacial tensions in the neighborhood of a critical solution point we were curious to see whether Cr was unusually high in our experiments. This would, perhaps, lead to qualitatively different behavior than for normal liquid films. In fact, the values shown are quite small; the small value of σ is offset by μ and α and we arrive at values of Cr not greatly different from normal liquids in air.

In preparing Tables I and II, we used the following sources for data on physical properties. For TEA - water mixtures we obtained interfacial tension and density data for coexisting phases from Derdulla and Rusanow (20). The same quantities for n-decane--CO_2 mixtures were obtained from Nagarajan and Robinson (9) while interfacial tensions for n-pentane--CO_2 were obtained from calculations by Rainwater and Lynch (National Institute of Standards and Technology, personal communication, 1988) based on the application of two-scale-factor universality described in Moldover and Rainwater (21). Thermal properties C_p and ΔH for the supercritical mixtures were calculated from the Peng-Robinson equation of state using the computer program EXCST described by Ely (22) with an extension for ΔH described by Jones and Giarratano (23). ΔH for the TEA--water system was estimated from the excess enthalpies of Chand et al. (24). Transport properties for the supercritical mixtures were also calculated by EXCST using the extended corresponding states model, while for TEA-- water mixtures, mole averages of pure component values were used.

According to Criterion A, we should have the best chance of seeing interfacial tension gradient effects in systems 2 and 3 where the parameter of the criterion is indeed ~ 1. However, even though this parameter is the largest for system 3, the experimental results do not display an enhancement over the heat transfer rate predicted by the Nusselt theory, Equation 1., despite the characteristic steep rise as the second phase is formed. On the other hand, system 1, with the smallest value for A, shows about 80% enhancement, and system 2, 120 % enhancement. Thus, success with A is mixed. The parameter for Criterion B is largest again for system 3. It is puzzling that for system 2, which shows the greatest heat transfer enhancement, Criterion B parameter is the smallest while system 3, having the largest values for both A and B parameters, shows no enhancement over the Nusselt theory. Influences other than buoyancy and thermocapillarity must certainly be operating.

We have evidence, for example, that coalescence of the nucleating dispersed phase on the heat-transfer surface may be a rate-limiting step in the phase change for liquid--liquid mixtures. We performed experiments with the system 2-butanol--water and found no rise in heat transfer coefficient above that of pure water. Visual observations showed that, while an emulsion was formed close to the heat transfer surface, no coalescence took place and no film was formed on the cylinder surface. The role of coalescence needs further investigation.

As evidence that interfacial tension effects do indeed exist, we show in figure 7 representative photographs of the test section in the TEA--water system. The structure of the condensate layer is made visible by the presence of a fine emulsion and absence of strong refractive index gradients more typical in the supercritical systems. In 7a, a fairly chaotic arrangement of cells is seen while in 7b, the cells appear to have become organized under the influence of gravity. We have recently obtained photographs for a supercritical system showing similar structures. However, it is more difficult to obtain photographs of publishable quality in the supercritical environment and we defer these to future publication.

A final note concerns what effect the precipitous decline of mutual diffusion coefficients might have as the critical point is approached. We saw no effect in our experimental results which might be attributable to this decline. Its effect must be either, masked by other mechanisms, or, as Cussler has suggested (University of Minnesota, personal communication, 1988), it may simply result in a steep concentration gradient over a very small distance in most engineering experiments where, as is here the case, a relatively large temperature spread is used and the resulting increase in mass transfer resistance is small.

Conclusions

Measurements of heat transfer rates in mixtures of hydrocarbons and supercritical CO_2 show enhancement when the two-phase region of the phase diagram is reached. Attempts to interpret the enhancement in terms of established theory and correlations fail, but evidence is presented to support the hypothesis that, in condensation, interfacial tension gradients give rise to flows which are strong enough to influence the results. The supercritical results are qualitatively similar to those obtained for the liquid--liquid system TEA--H_2O in which photographic evidence strongly suggests such flows. Paradoxically, this liquid--liquid system does not show the same heat transfer enhancement over that predicted by classical condensation theory.

Acknowledgments

We gratefully acknowledge the contributions to the experimental program of Patricia Giarratano and Laurel Powers of the NIST Chemical Engineering Science Division and Jo Hornback of the Colorado School of Mines. We also thank Jim Ely and Jim Rainwater of the NIST Thermophysics Division for their interest and continued help with fluid properties and phase behavior.

Legend of Symbols

Bo = Bond number defined in text
C_p = heat capacity of liquid
d = liquid film thickness
D = cylinder diameter

g = acceleration due to gravity
h = heat transfer coefficient
ΔH = heat of condensation
k = thermal conductivity
L = Reference dimension
Ma = Marangoni number defined in text
Ra_{film} = liquid film Rayleigh number defined in text
T = temperature
u = velocity in y direction
U_r U'_r = reference velocities defined in text
v = velocity in x direction
x = coordinate parallel to the wall
y = coordinate perpendicular to the wall

Greek symbols

α = thermal diffusivity
ν = kinematic viscosity
ρ = density; $\Delta\rho$ = density difference between phases.
σ = interfacial tension
τ = shear stress

Subscripts and subscripts

$_B$ = bulk fluid value
$_W$ = wall value
$_V$ = vapor phase
$_L$ = liquid phase

Literature Cited

1. Schneider, G. M.; In Extraction with Supercritical Gases; Schneider, G. M.; Stahl, E.; Wilke, G., Eds.; Verlag Chemie: Weinheim, 1980; pp 45-81.
2. Cussler, E. L. AIChE Journal 1980, 26, 43.
3. Durst, M.; Stephan, K. In Natural Convection; Kakac, S.; Aung, W.; Viskanta, R., Eds.; Hemisphere: New York, 1985; pp 827-841.
4. Jones, M. C.; Giarratano, P. J.; Powers, L. A.; Proc. Nat. Heat Trans. Conf., Pittsburgh, 1987; AIChE Symposium Series No. 257; vol 83, p 115.
5. Leung, S. S.; Griffiths, R. B. Phys. Rev. 1973, A8, 2670.
6. Rainwater, J. C.; Moldover, M. R. In Chemical Engineering at Supercritical Conditions; Paulaitis, M. E. et al. Eds.; Ann Arbor Science: Ann Arbor, 1983; pp 199-220.
7. Besserer, G. J.; Robinson, D. B. J. Chem. Eng. Data 1973, 18, 416.
8. Reamer, H. H.; Sage, B. H. J. Chem. Eng. Data 1963, 8, 509.
9. Nagarajan, N.; Robinson, R. L. J. Chem. Eng. Data 1986, 31, 168.
10. Peng, D. Y.; Robinson, D. B. Ind. Eng. Fundam. 1976, 15, 59.
11. Hales, B. J.; Bertrand, G. L.; Hepler, L. G. J. Phys. Chem. 1966, 70, 3970.
12. Rohsenow, W. M. Trans. ASME 1952, 74, 969.

13. Sparrow, E. M.; Gregg, J. L. J. Heat Transfer 1959, 81, 13.
14. Sternling, C. V.; Scriven, L. E. AIChE Journal 1959, 6, 514.
15. Scriven, L. E.; Sternling, C. V. J. Fluid Mech. 1964, 19, 321.
16. Zeren, R. W.; Reynolds, W. C. J. Fluid Mech. 1972, 53, 305.
17. Smith, K. A. J. Fluid Mech. 1966, 24, 401.
18. Bejan, A. Convection Heat Transfer; John Wiley: New York, 1984.
19. Ostrach, S. Ann. Rev. Fluid Mech. 1982, 14, 313
20. Derdulla, H. J.; Rusanow, A.I. Z. Phys. Chem. (Leipzig) 1970, 245, 375.
21. Moldover, M. R.; Rainwater, J. C. J. Chem. Phys. 1988, 88, 7772.
22. Ely, J. F. Proc. 63rd. Gas Processors Annual Convention, 1984, p 9.
23. Jones, M.C.; Giarratano, P.J. AIChE Journal 1988, 34, 2059.
24. Chand, A.; McQuillan, A. R.; Fenby, D. V. Fluid Phase Equil. 1979, 2, 263.

RECEIVED May 1, 1989

FOOD, PHARMACEUTICAL, AND ENVIRONMENTAL APPLICATIONS

Chapter 26

Extraction and Isolation of Chemotherapeutic Pyrrolizidine Alkaloids from Plant Substrates

Novel Process Using Supercritical Fluids

Steven T. Schaeffer[1], Leon H. Zalkow[2], and Amyn S. Teja

School of Chemical Engineering, Georgia Institute of Technology, Atlanta, GA 30332–0100

Pyrrolizidine alkaloids have long been known to be antitumor active and, more recently, have become of interest as anti-cancer agents. They occur naturally in several plant species, but are often difficult to extract and isolate from the plant material without degradation or the use of toxic solvents. The extraction of a model pyrrolizidine alkaloid, monocrotaline, from the seeds of Crotalaria spectabilis was investigated in this work.

The crushed seeds of Crotalaria spectabilis were first contacted with supercritical carbon dioxide and, as expected, the oils comprising the bulk of the seed material were preferentially extracted. The addition of ethanol and water as co-solvents in the fluid phase led to the appearance of monocrotaline in the extract. Monocrotaline contents as high as 24% of the total extract could be obtained with carbon dioxide-ethanol mixtures.

In order to increase the extract purity further, two additional processes were developed. The temperature-solubility cross-over point was utilized to obtain extracts containing as much as 50% monocrotaline. A second novel process incorporating ion exchange resins was also studied and yielded extracts containing 94 to 100% monocrotaline. Because this technique depends only on the basic character common to the pyrrolizidine alkaloids, it is expected to be equally effective in the extraction and isolation of the other members of this class and, in fact, could be extended to the other classes of basic alkaloids.

[1]Current address: E. I. du Pont de Nemours and Company, P.O. Box 2042, Wilmington, NC 28402

[2]Current address: School of Chemistry, Georgia Institute of Technology, Atlanta, GA 30332

0097–6156/89/0406–0416$06.00/0

Supercritical fluid extraction processes are particularly appropriate for the separation and isolation of biochemicals where thermal decomposition, chemical modification, and physiologically-active solvents are undesirable. Examples of these bioseparations include the extraction of oils from seeds using carbon dioxide (1), of nicotine from tobacco using carbon dioxide-water mixtures (2), and of caffeine from coffee beans again using carbon dioxide-water mixtures (3).

In the present investigation, supercritical carbon dioxide and carbon dioxide + co-solvent mixtures were used to extract and isolate a model pyrrolizidine alkaloid from its parent plant. Pyrrolizidine alkaloids have been used in herbal medicine to combat tumors as long ago as the fourth century A.D. (4) and to treat cancer since the tenth century A.D. (5). More recently, they have received increasing attention as chemotherapeutic drugs. Processes for their separation, however, are specific to each alkaloid, and either lead to chemical modification of the alkaloid or require the use of solvents which must then be completely removed from the extract.

The purpose of this study was to develop a general supercritical fluid based process for the separation and purification of pyrrolizidine alkaloids such as monocrotaline ($C_{16}H_{23}NO_6$, MW=325.3). Monocrotaline was selected as a model because of its role in the development of semisynthetic pyrrolizidine alkaloids (6) and because it occurs in several species of Crotalaria. The seeds of Crotalaria spectabilis served as the source of monocrotaline in this study.

Experimental

A dual-feed single-pass supercritical fluid extraction apparatus was constructed as shown in Figure 1. Pressurized carbon dioxide was filtered, liquified in an ice bath C, and pumped to the system pressure in an Eldex dual-channel metering pump E. The co-solvent was filtered and fed from a graduated cylinder D into the other head of the pump. The solvents were mixed in a Kenics static mixer I immersed in the constant temperature bath F. The homogeneous mixture then passed through an equilibrium cell packed with alternating layers of solute and glass beads. Equilibrium between the solute and the fluid was achieved by allowing sufficient fluid residence time in the equilibrium cell. This was confirmed experimentally by monitoring the composition of the supercritical mixture at the exit of the cell. An in-line temperature-controlled Mettler-Parr fluid densiometer L was used to determine the fluid density. The mixture was then depressurized in a heated micrometering valve O and the condensables (solute and a portion of the co-solvent) were collected in a collection vessel Q. Condensation was ensured by immersing the collection vessel in an ice bath. The gas mixture then passed through a bank of rotameters R for flow visualization, through a gas sampling valve S for composition analysis, and finally through a wet test meter V for flow totalization. A gas chromatograph U with a flame ionization detector was used for analysis of the carbon dioxide + ethanol stream. The tubing outside the bath was heated to the bath temperature to prevent deposition of material in the lines.

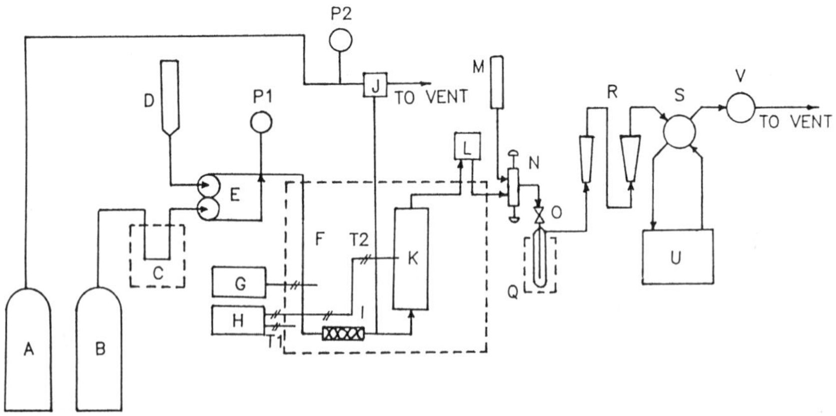

Figure 1: Schematic diagram of the dual-feed single-pass supercritical extraction apparatus.

Liquid solvent contained in M was used to flush the micro-metering valve of any deposited material.

The system pressure was maintained using a back-pressure regulator J pressurized with high pressure nitrogen from a tank A. The pressure was monitored by two calibrated Heise gauges P1 and P2. The bath temperature was controlled by a controller G and was monitored using two calibrated thermocouples T1 and T2. The amount of extract in the collection vessel was determined gravimetrically. The amount of monocrotaline in the extract was determined using a proton NMR technique developed by Molyneaux et al. (7).

Purity and Preparation of Materials. The carbon dioxide used was Coleman instrument grade with a purity of >99.9%. Pure ethanol (>99.9%) was obtained by reactive distillation of HPLC grade ethanol with magnesium turnings catalyzed with iodine. Water was obtained by double distillation. Ethanol used for valve and collection vessel flushing was 95 wt.% ethanol - 5 wt.% water. All solvents and co-solvents were dried to confirm the absence of solids. Monocrotaline (>99%) was obtained from the seeds of Crotalaria spectabilis using the method described by Gelbaum et al. (6).

The seeds of Crotalaria spectabilis were obtained in Clarke County, Georgia in November, 1984. These seeds were analyzed and found to contain 1.9 wt.% monocrotaline and 2.5 wt.% hexane-extractable lipid material. The seeds (5mm indiameter) were milled to 1 mm to break the hard outer coat and to expose the inner seed material containing the monocrotaline. The seed fragments were sieved to remove the 850+ micron fraction (predominantly outer coat fragments) and the 850- micron fraction was packed in the equilibrium cell.

Pure Component Studies

Before investigating the extraction of monocrotaline from Crotalaria spectabilis, the solubility of pure monocrotaline was measured. This was done to determine the magnitude of the solubility, to evaluate the effect of co-solvents, and to confirm the integrity of the extracted monocrotaline.

Carbon Dioxide - Monocrotaline System. The solubility of pure monocrotaline was measured at three temperatures (308.15, 318.15, and 328.15K) at pressures ranging from 8.86 MPa to 27.41 MPa. The solubilities ranged form $6\text{x}10^{-6}$ to $4.4\text{x}10^{-5}$ mole fraction (Figure 2). Long extraction times were used to reduce the experimental error to $1\text{x}10^{-6}$ mole fraction. These low solubilities are typical of large polar biomolecules.

Carbon Dioxide - Ethanol - Monocrotaline System. Since monocrotaline is soluble in ethanol, ethanol was selected as a co-solvent. The carbon dioxide - ethanol phase diagram of Panagiotopoulos and Reid (8) was used to ensure that no liquid phases were present at any time in the equilibrium cell. This was also confirmed by visual observation.

The solubility of monocrotaline in carbon dioxide + 5 mol% and 10 mol% ethanol was measured at three temperatures and pressures ranging from 10.34

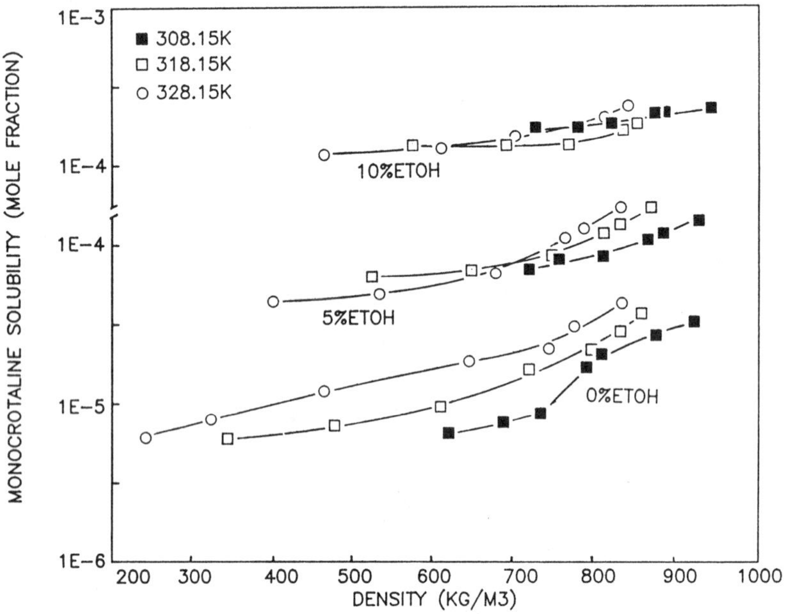

Figure 2: Solubility of monocrotaline in carbon dioxide, 95 mole% carbon dioxide + 5 mole% ethanol and 90 mole% carbon dioxide + 10 mole% ethanol.

to 27.41 MPa. The solubility was as high as 2.21×10^{-4} mole fraction (Figure 2). The largest increase due to the addition of ethanol was 25 fold indicating that ethanol is an effective co-solvent for monocrotaline. Also, it was found that the temperature-solubility cross-over point $(dy/dT)_P = 0$ advanced nearly linearly with ethanol concentration.

Complex Substrate Studies

Due to the relative success of the pure component solubility studies, the same series of experiments were carried out using the complex seed material. Three systems were investigated to evaluate the ability of supercritical fluids to extract monocrotaline from the seeds of Crotalaria spectabilis. Pure carbon dioxide was studied with the expectation that the oils would be preferentially extracted. Ethanol was added as a co-solvent to increase the solubility of monocrotaline. Also, due to its success in the extraction of caffeine and nicotine, water was used as a co-solvent.

Carbon Dioxide - Crotalaria Spectabilis System. The fluid phase concentration was found to be time-dependent (Figure 3). At the start of the extraction, the concentration was constant indicating that it was equal to the equilibrium concentration. After approximately one mass percent of the initial mass of the bed had been removed, however, the exit concentration began to decrease.

This result may be explained by a combination of intraparticle diffusion and bulk mass transfer processes. As material is extracted from the exposed areas of the seed, the solvent must travel further through the pores to reach the solute. Also, as the entrance portion of the bed becomes depleted of soluble components, the effective bed length decreases until the residence time is insufficient to achieve equilibrium. Similar effects were observed in seed oil extraction by Fattori (1) and Taniguchi et al. (9).

The solubility of the seed material in carbon dioxide was defined as the intial concentration prior to observation of these depletion effects. The solubility was found to range from 0.016 wt.% to 0.6 wt.% (Figure 4). Analysis of the extracts revealed that the monocrotaline content was very low (<0.05 wt.%), and that the extracts were predominantly lipid material as expected.

Carbon Dioxide - Ethanol - Crotalaria Spectabilis System. The solubility of the seed material increased by as much as 20 fold in the presence of ethanol (Figure 5). The presence of monocrotaline in the extracts became detectable at 2 - 4 mol% ethanol and increased markedly as ethanol concentration increased. The solubility of monocrotaline when extracted from the seed material was from 50% to 93% less than the comparable solubility (same ethanol concentration, temperature, and pressure) of monocrotaline. This is in contrast to the observations of Dobbs (10) and Kurnik et al. (11) who found that the solubility of a component is generally enhanced in the presence of other components. As the extraction proceeded and the seed bed became depleted of soluble components, the overall exit fluid phase concentration again decreased. However, the

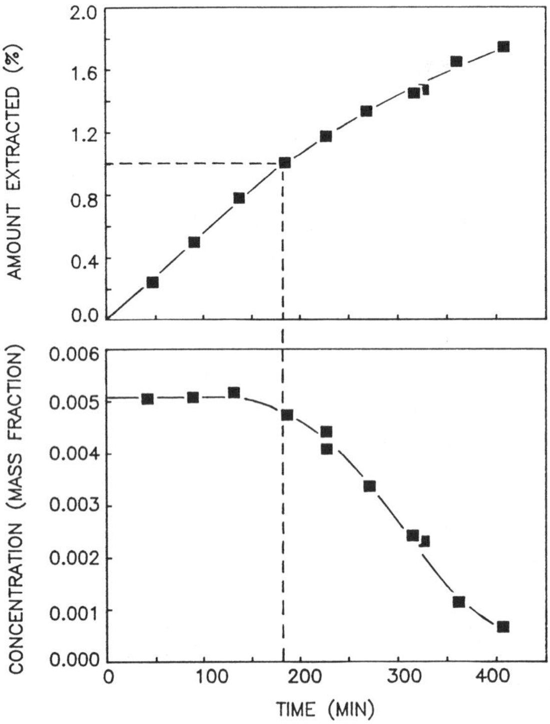

Figure 3: Amount extracted and concentration of Crotalaria spectabilis in carbon dioxide as a function of time of extraction at 308.15 K, 27.41 MPa.

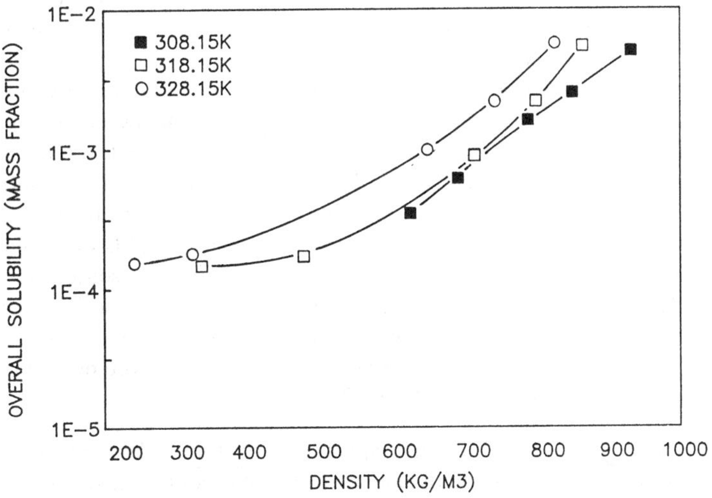

Figure 4: Overall solubility of Crotalaria spectabilis extract in carbon dioxide. No monocrotaline was detected in the extract.

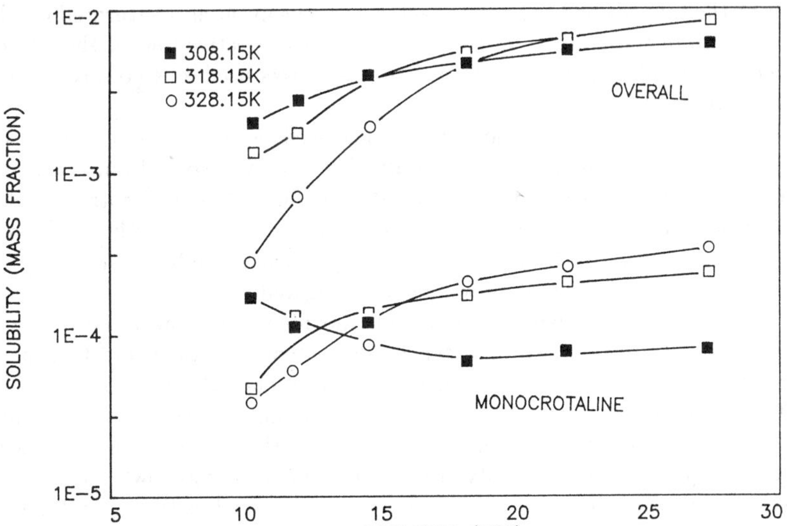

Figure 5: Overall and monocrotaline solubility in 95 mole% carbon dioxide + 5 mole% ethanol.

monocrotaline concentration increased by as much as 2 fold in this case (Figure 6). This result would indicate that the decrease in monocrotaline concentration was caused by the presence of the lipids in the fluid phase. The net result was that the selectivity of monocrotaline (or purity of monocrotaline in the extract) increased dramatically with seed bed depletion.

It was also observed that ethanol appeared at the outlet later and at lower concentrations than expected (Figure 7). The ethanol concentration then increased beyond the inlet concentration and finally settled at the inlet value. Since no bulk liquid phases were observed, it is believed that this was due to intraparticle retention-release phenomena. A possible explanation could be absorption of ethanol into lipid microdroplets and eventual release as the lipids were solubilized.

Overall solubilities as high as 1.08 wt.% and monocrotaline solubilities as high as 0.07 wt.% were observed. Selectivity toward monocrotaline increased with increasing ethanol concentration, with increasing temperature and decreasing pressure (Figure 8). Selectivities were as high as 24 wt.% before depletion effects became apparent, and as high as 40 wt.% after depletion.

Carbon Dioxide - Water - Crotalaria Spectabilis System. Water has been used as a co-solvent in the extraction of caffeine from coffee and nicotine from tobacco. To extract caffeine from coffee beans, Zosel (3) recommended the pre-saturation of the carbon dioxide with water before passing the fluid through the coffee bean bed. In the case of nicotine from tobacco, Hubert and Vitzthum (2) first soaked the tobacco with up to 25 wt.% water and subsequently passed carbon dioxide through the tobacco bed. The water acted as a co-solvent as it saturated the fluid phase. Both of these processes have proven highly selective toward the alkaloids present in the plant material. Therefore, water was used as a co-solvent in the present study.

In the first set of experiments, the crushed seeds were saturated with 25 wt.% water and extracted with pure carbon dioxide. The wet seeds formed a gummy material as the seed oils and water mixed. The fluid phase concentration of the seed components at all conditions was very low compared with the pure carbon dioxide extractions. Since these results are apparently due to considerable mass transfer limitations, the system was not investigated further.

In the second set of experiments carbon dioxide was pre-saturated with water by passing it through a bed of water at the system conditions and then passed through the equilibrium cell. Since the solubility of water in carbon dioxide is quite low (<0.8 mol%, see 12), the overall solubility of the seed components in the fluid phase was only slightly higher than the solubility in pure carbon dioxide (Figure 9). The solubility of monocrotaline was quite low ($<5\times10^{-5}$ mole fraction), and the selectivity toward monocrotaline was less than 2 wt.%. Based on these results, it was decided that a water co-solvent process was not effective in this situation, and, possibly, in other cases where significant amounts of oils are present.

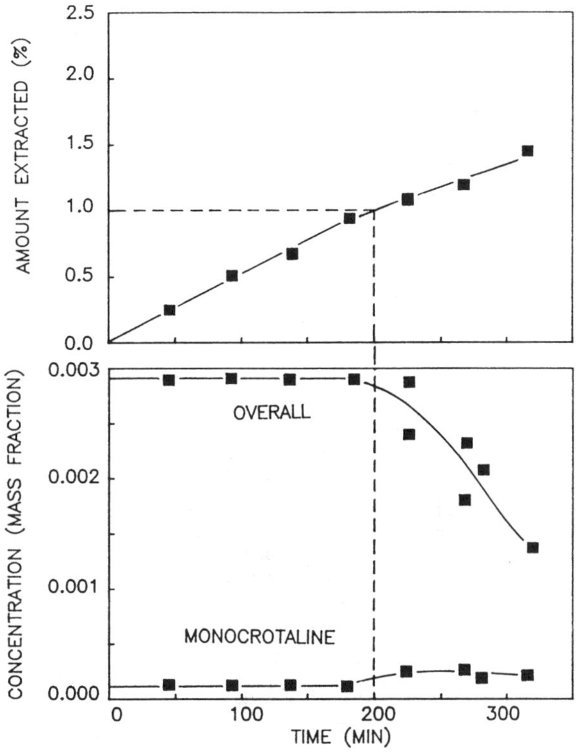

Figure 6: History of an extraction of Crotalaria spectabilis with 95 mole% carbon dioxide + 5 mole% ethanol at 318.15 K, 11.81 MPa.

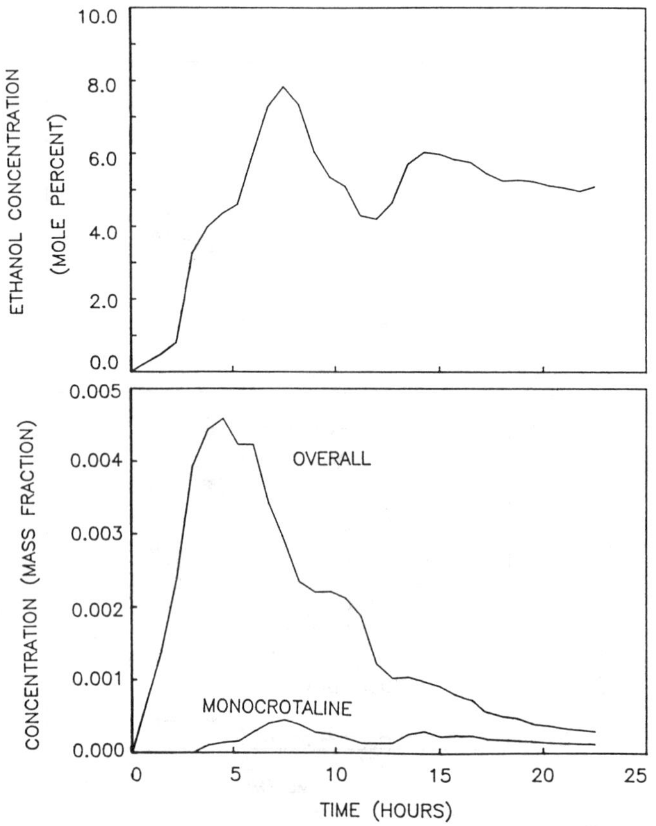

Figure 7: History of an extraction of Crotalaria spectabilis with 95 mole% carbon dioxide + 5 mole% ethanol at 318.15 K, 18.46 MPa. Note that the fluid residence time in the solute bed is 50 minutes.

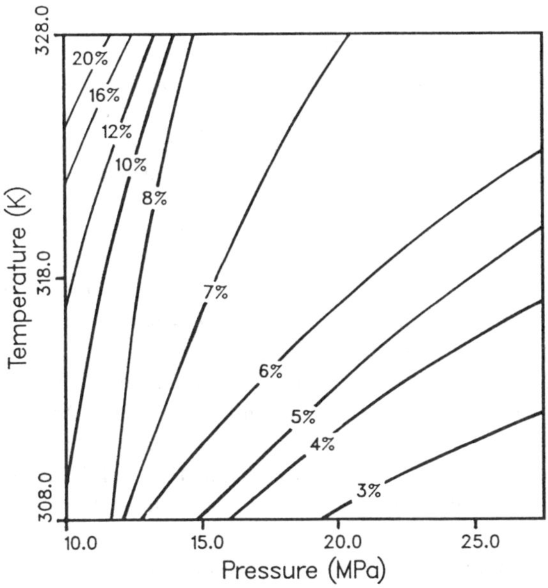

Figure 8: Monocrotaline selectivity over the range of experimental conditions for 92 mole% carbon dioxide + 8 mole% ethanol. Monocrotaline selectivity is defined as the monocrotaline solubility/overall solubility by mass.

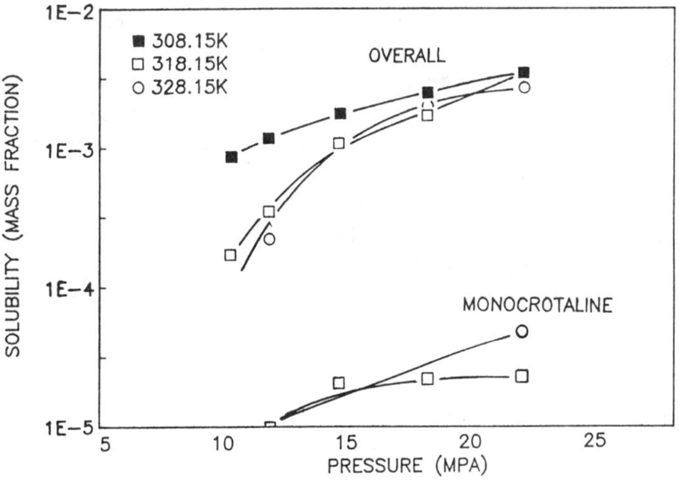

Figure 9: Overall and monocrotaline solubility in water-saturated carbon dioxide.

Isolation Studies

The objective of any separation process is to obtain high purity product. Monocrotaline selectivities as high as 40 wt.% were obtained in the above studies. Clearly, significant post-processing would be required to obtain pure monocrotaline from the extract. Two alternative supercritical fluid-based processes were therefore considered in order to obtain pure monocrotaline.

Isolation Process in the Cross-Over Region. Chimowitz and Pennisi (13) developed a process for the separation and isolation of components from mixtures by operating in the multicomponent temperature-solubility cross-over region. The cross-over point of a pure component $(dy/dT)_P = 0$ represents the pressure at which the dependence of solubility on temperature reverses itself. At lower pressures, the solubility is principally dependent on solvent density – raising temperature decreases density and thus solubility decreases. At higher pressures, solubility is principally dependent on solute sublimation pressure – raising the temperature increases sublimation pressure and thus solubility increases. The cross-over point is therefore unique for each solute-solvent system. When there are two solutes, cross-over points occur at different pressures. At an intermediate pressure, the temperature can therefore be manipulated to deposit either component.

A cross-over point was observed in our study by considering the soluble lipid seed material as a single psuedocomponent (Figure 10). A second independently temperature-controlled bath was inserted after the first to investigate cross-over region purification. In the first study, the seeds were extracted at 10 mol% ethanol, 328.15K and 14.77 MPa and passed through the second bath at 308.15K. Pure monocrotaline should have deposited, and the lipid material should have passed through unaffected had it been a single component. While the calculated amount of monocrotaline was deposited, a portion of the lipid material also fell out of solution due to the fact that the lipid material contains many components with a range of cross-over points. Fractionation of the lipid materials thus occurs as a result of the temperature change. The deposited material contained 13.6 wt.% monocrotaline, with the rest being lipid materials.

The converse of this process, namely extraction at 10 mol% ethanol, 308.15K, 14.77 MPa followed by an increase in the temperature, should result in only the lipid material being deposited. By operating the second bath at 363.15 K no monocrotaline was deposited as expected. The fraction of monocrotaline in the supercritical phase could thus be increased to 49 wt% in this fashion.

Ion Exchange Adsorption Process. Supercritical fluid extracts of Crotalaria spectabilis contain monocrotaline, a basic alkaloid, and non-polar lipid material. In the separation of caffeine from coffee, Zosel (3) recommended separation of the caffeine from the fluid phase by either adsorption onto activated carbon or absorption into liquid water. Activated carbon adsorption would be undesirable in the present case because the lipids would also be adsorbed and because desorption from activated carbon is quite difficult. Liquid water would absorb

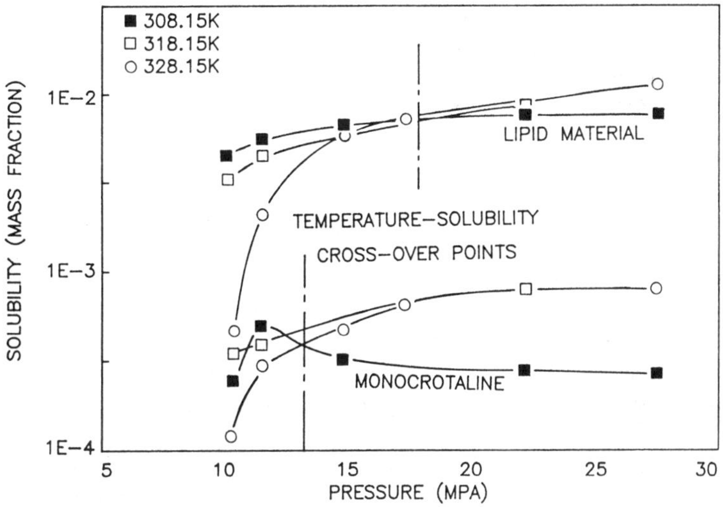

Figure 10: Lipid material and monocrotaline solubility in 92 mole% carbon dioxide + 8 mole% ethanol. The overall extract without the monocrotaline is assumed to consist of lipid material.

both monocrotaline and ethanol and result in a decrease in the ethanol concentration in the supercritical phase. This, in turn, would cause precipitation of all components from the fluid phase.

Since the conventional monocrotaline isolation process described by Gelbaum et al. (6) uses cation exchange resin, this same resin was used as a second stage adsorbent to remove monocrotaline from the fluid. After passing the extract fluid through the resin bed, entrapped materials were removed by flushing the resin with ethanol. The adsorbate was released and removed by flushing with 1N NH_4OH.

The effect of temperature, pressure, and co-solvent on the effectiveness of this process were investigated. For all cases, the adsorbate was found to be greater than 94 wt.% monocrotaline and averaged 98 wt.% (Table 1). The NMR spectra of the adsorbate indicated that no lipids were present and that the structure of the monocrotaline was intact. The unadsorbed and entrapped fractions were analyzed and found to contain no monocrotaline. Thus, the resin was both highly selective toward monocrotaline and highly efficient in its adsorption regardless of temperature, pressure, or co-solvent concentration.

Discussion

Our results indicate that single-stage supercritical extraction processes are limited in their ability to separate trace components from complex mixtures even with the use of co-solvents. However, two-stage processes such as the ones developed here can significantly improve the degree of separation. Utilization of the multicomponent temperature-solubility cross-over region is a promising isolation technique. However, when the substrate contains a wide variety of components, multiple separations may be necessary to complete the isolation and the yields may be low.

The use of ion exchange resins combined with supercritical extraction results in an attractive isolation technique. Since the pyrrolizidine alkaloids are generally basic, the process described here could be used for the isolation of other members of this class, and for other basic alkaloids. In designing a process using ion exchange resins, it is important that the co-solvent not deactivate the resin. In our example, water and ethanol are acidic in carbon dioxide and, therefore, do not deactivate the acidic resin. An industrial process based on this concept could be quite efficient and inexpensive since pressure reduction and subsequent solvent recompression are unnecessary (Figure 11).

There are several logical extensions of this work. Anion exchange resins could be used to remove acidic components, such as fatty acids from supercritical fluids. Also, since many biological materials contain more than one component with the same ionic character, several resin beds can be placed in series in order of increasing strength to adsorb components according to ionic strength.

Table 1. Results of Cation Exchange Resin Adsorption

Conditions	Experiments				
	1	2	3	4	5
Co-solvent	EtOH	EtOH	EtOH	EtOH	H2O
Average Concentration (mol%)	4.0	5.3	3.8	4.3	0.7
Temperature (K)	308.15	308.15	328.15	328.15	328.15
Pressure (MPa)	10.34	22.15	10.34	22.15	22.15
Mass of Solvent Used (g)	93.6	89.3	93.0	60.6	67.6
Solute Unadsorbed					
Total Mass (mg)	154.5	455.0	20.3	306.9	171.3
Mass Monocrotaline (mg)	0	0	0	0	0
Solute Entrapped in Resin					
Total Mass (mg)	30.4	39.9	2.4	29.8	20.7
Mass Monocrotaline (mg)	0	0	0	0	0
Solute Adsorbed on Resin					
Total Mass (mg)	9.3	8.9	3.0	12.7	3.3
Mass Monocrotaline (mg)	9.0	8.9	3.0	12.3	3.1
Monocrotaline Purity (%)	97	100	100	100	94

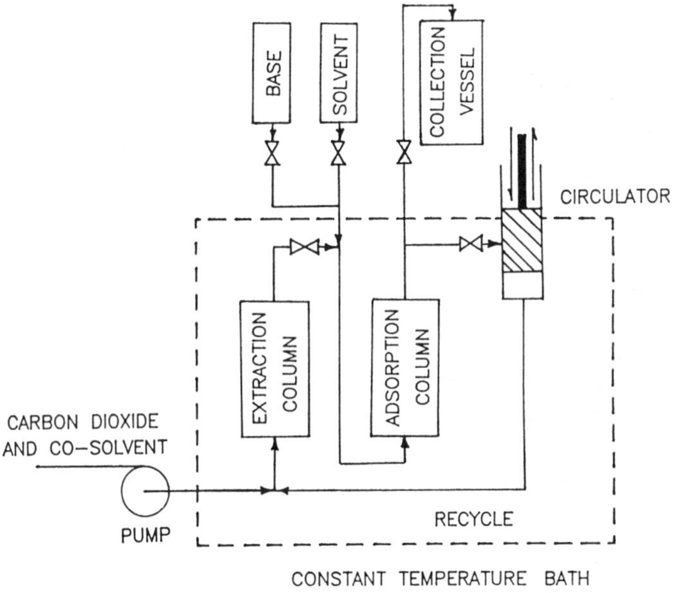

Figure 11: Potential industrial process for supercritical fluid isolation of basic alkaloids from complex substrates.

Conclusions

The major result of this work was the development of a new process which combines supercritical fluid extraction with ion exchange to isolate pyrrolizidine alkaloids from plant materials. Other findings include the following:

1. Time-dependence. It was shown that the concentration of the solute in the fluid phase depends on the previous history of the complex substrate.

2. Solubility Depression. In contrast to studies reported in the literature, it was found that the presence of more volatile components can significantly depress the solubility of non-volatile components in the supercritical phase (relative to their solubilities as pure components).

3. Co-solvent Retention-Release. Our results indicate that co-solvents can be retained in complex substrates, possibly by absorption into intraparticle liquids.

4. Cross-Over Region Purification. The multicomponent cross-over region can be used to isolate components from complex mixtures. However, clear-cut separations are difficult to achieve and multiple stages may be necessary.

5. Ion Exchange Resins. Ion exchange resins may be used to selectively remove ionic components from supercritical fluids in much the same manner as from liquids.

Literature Cited

1. Fattori, M. Ph.D. Dissertation, University of British Columbia, 1985.
2. Hubert, P.; Vitzthum, O. G. Angew Chem. Int. Ed. Engl. 1978 17, 710-715.
3. Zosel, K.; U. S. Patent 4,260,639, 1981.
4. Hunger, F. W. T. The Herbal of Psuedo. Apuleius, 1935.
5. Cockayne, O. Lech Book of Bold Leechdoms, Wortcunnings, and Starcraft of Early England, London, England, Vol. 2, 1864-1866.
6. Gelbaum, L. T.; Gordon, M. M.; Miles, M.; Zalkow, L. H. J. Org. Chem. 1982 47, 2501-2504.
7. Molyneaux, R. J.; Johnson, A. E.; Roitman, J. N.; Benson, M. E. J. Agric. Food Chem. 1979 27, 494-499.
8. Panagiotopoulos, A. Z.; Reid, R. C. ACS Symposium Series 300 1985 571-582.
9. Taniguchi, M.; Kamihira, M.; Tsuji, T.; Kobayashi, T. World Congress III of Chem. Eng. Tokyo, 1986 Paper 10b-105, 1040-1043.
10. Dobbs, J. M.; Johnston, K. P. Ind. Eng. Chem. Res. 1987 26, 1476-1481.
11. Kurnik, R. T.; Holla, S. J.; Reid, R. C. J. Chem. Eng. Data 1981 26, 47-51.
12. Wiebe, R.; Gaddy, V. L. J. Am. Chem. Soc. 1941 63, 475-477.
13. Chimowitz, E. H.; Pennisi, K. J. AIChE J. 1986, 32, 1665-1676.

RECEIVED May 1, 1989

Chapter 27

Supercritical Fluid Carbon Dioxide Extraction in the Synthesis of Trieicosapentaenoylglycerol from Fish Oil

W. B. Nilsson, V. F. Stout, E. J. Gauglitz, Jr., F. M. Teeny, and J. K. Hudson

U.S. Department of Commerce, National Oceanic and Atmospheric Administration, National Marine Fisheries Service, Northwest Fishery Center, Utilization Research Division, 2725 Montlake Boulevard East, Seattle, WA 98112

Supercritical fluid carbon dioxide (SC-CO_2) fractionation of fish oil ethyl esters (EE) was employed to prepare EE of two omega-3 fatty acids, all _cis_-5,8,11,14,17-eicosapentaenoic acid (EPA) and all _cis_-4,7,10,13,16,19-docosahexaenoic acid (DHA) in 90% purity and to separate the synthetic triacylglycerols (TG), trieicosapentaenoylglycerol (tri-EPA), and tridocosahexaenoylglycerol (tri-DHA) in $\geq$92% purity from other reaction mixture components. In the synthesis, glycerine reacted with EE and sodium glyceroxide catalyst to form TG. The desired TG (50-60 wt%) was accompanied by fatty acids (2-5%), mono- and diglycerides (10-15%), unidentified byproducts (5-10%), and unreacted EE (15-25%). The TG products were isolated using supercritical fluid fractionation employing pressure programming at 60 °C with pure CO_2 and CO_2 plus 4 wt% ethanol. Analyses of the isolated EE and TG indicated that reactant EE are incorporated into the TG with little or no selectivity regardless of chain length or degree of unsaturation.

Omega-3 (ω3) fatty acids have been the subject of many recent reports due to their reputed medicinal properties in the treatment of cardiovascular and autoimmune/inflammatory diseases (1-4). Most of these studies have suggested that consumption of fish oils, the main source of ω3 fatty acids in the human diet, has significant implications for long-term health and the prevention of numerous diseases. However, before ω3-containing substances can be recommended in nutritional supplements or as pharmaceutical agents, the physiological properties of individual ω3 fatty acids must be defined. Two ω3 fatty acids that have drawn widespread attention are all _cis_-5,8,11,14,17-eicosapentaenoic acid (EPA or 20:5ω3) and all _cis_-4,7,10,13 16,19-docosahexaenoic acid (DHA or 22:6ω3). The designation 20:5ω3 connotes a straight-chain, 20-carbon fatty acid with 5 methylene-interrupted double bonds beginning at the third carbon counting from the terminal methyl group.

The overall goal of the fish oil research program in our laboratory, which dates back to the mid-1950's, has been to develop various methodologies to produce materials for elucidating the nutritional and medicinal properties of ω3-containing substances not only in humans, but also in terrestrial (5) and aquatic animals (6). The diversity of fatty acids present in fish oils, and the inherent instability of highly unsaturated fish oils have hampered these efforts. In some early work, molecular distillation combined with forced urea complexing of the hydrolyzed native triacylglycerols, here referred to as triglycerides, effected isolation of nearly 90% DHA (7). The fate of other ω3 fatty acids was not determined, but the method may well be less effective for isolating shorter chain ω3 fatty acids. At best, the process required two molecular distillations at elevated temperatures (up to 180 °C) and one multi-stage urea fractionation, resulting in some degradation of the highly polyunsaturated products of interest.

Recently, supercritical fluid carbon dioxide extraction (SFE) has shown promise in applications requiring the separation of complex mixtures of high boiling unstable substances at significantly lower temperatures than used in processes such as fractional distillation. One of many examples is the purification of the monomer diacetone acrylamide (8). Supercritical fluid CO_2 (SC-CO_2) has the further advantage of avoiding the use of flammable organic solvents. Carbon dioxide is readily available, and does not leave toxic residues, making its use attractive in food and pharmaceutical applications. Augmenting the work of Eisenbach (9), two reports from this laboratory have shown that EPA and DHA in purities above 90% can be obtained from fish oil ethyl esters using SFE at temperatures as low as 80 °C (10-11). In addition, a recent report describes a countercurrent continuous process for producing large quantities of both fatty acid esters of similar purity (12).

Up to now, most physiological studies on individual ω3 fatty acids have used methyl or ethyl esters directly. Concerns have been raised over the suitability of this form of lipid for several reasons. Hydrolysis leads to highly toxic methanol or less toxic ethanol. Beyond the question of the liberated alcohol, the monoesters might themselves be toxic because of different intrinsic properties vis-à-vis the triglycerides. The monoester, because it is not the usual form or is much less viscous, may either be resistant to absorption or may interfere with other processes in the gastrointestinal tract. In fact, a recent report claims that when EPA-containing materials were fed to fasting human subjects, free fatty acids, arginine salts, and triglycerides were metabolized more rapidly and completely than ethyl esters (13).

The obvious alternative to monoesters, excluding the irritating free fatty acid, is the triglyceride composed of a single fatty acid, for example, EPA. Such triglycerides have reportedly been isolated in minor amounts from some fish (14), but synthesis from the pure ester or fatty acid is the only practical source. Although saturated fatty acids or their esters react readily to form triglycerides, highly unsaturated fatty acids are much more resistant. The method of Lehman and Gauglitz (15) requires free fatty acids and elevated temperatures. From our SFE work, substantial quantities of purified (ca. 90%) esters of EPA and DHA were available. We wished to incorporate them directly into triglycerides without the extra step of hydrolysis of the esters to the acids, especially since the acids are even more readily autoxidized than the highly reactive ω3 esters (16). The generation of ester reactants and the purification of the synthetic triglycerides are the main subjects of this paper.

Experimental

SFE Apparatus. Concentrates of the ethyl esters of EPA and DHA were obtained by fractionation of menhaden oil esters which previously had been urea fractionated (17). The apparatus is shown schematically in Figure 1. Unless otherwise noted, identification of vendors has been made elsewhere (10). The heart of the system consists of a pair of 10,000 psi double-ended, diaphragm-type compressors installed in parallel. Process pressure was controlled using a back pressure regulator. Compressed CO_2 was pumped through 1/4" OD high pressure 304 stainless steel (SS) tubing into a one-foot 1/4" ID SS pipe which sometimes was used as a preheater. The extraction vessel was a six-foot long, 3.5" OD x 1.25" ID 304 SS pipe (Temco, Inc., Tulsa, OK). (The use of trade-names in this publication does not imply endorsement by the National Marine Fisheries Service.) The column was packed with 0.16" Propak which is a protruded 316 SS distillation packing material (Scientific Development Co., State College, PA). The ester feedstock was loaded into the column through a port located 1 foot from the bottom. In addition to an internal thermocouple probe at the top of the column, several thermocouple ports were drilled at 7" intervals along the side of the column. The thermocouple probes, the tips of which lie at the center of the column, provided simultaneous measurement and control of the process temperature at several positions along the length of the column. Each probe was wired to an on/off temperature controller (Syscon Int. Inc., South Bend, IN) which controlled the process temperature at the probe position by supplying power to a silicon rubber heater wrapped around the column. As shown in Figure 1, there were 7 individually controlled heaters. Thus a thermal gradient could be introduced along the length of the column. In all ester fractionations, a gradient was used in which the temperature increased from the bottom to the top of the column.

SC-CO_2 Extraction Methodology. In a typical batch cycle, the column was loaded with ca. 100 g of esters. The system was then purged with CO_2 and the column heaters allowed to establish the desired temperature gradient. Once the column was pressurized, CO_2 passing through the esters at the bottom of the column dissolved a portion of the charge into the fluid phase. The ester-laden fluid was forced upwards through the increasing temperature gradient. Process temperatures and pressures were selected such that ester solubility decreased with increasing temperature, that is, underwent retrograde condensation. Therefore, in each temperature "zone" of the gradient, local heating caused preferential condensation of less soluble ester components thereby enriching the fluid phase with respect to more soluble components present at a given point in the fractionation. Ester-laden fluid emerging from the top of the column was expanded through a heated expansion valve to isolate the extract.

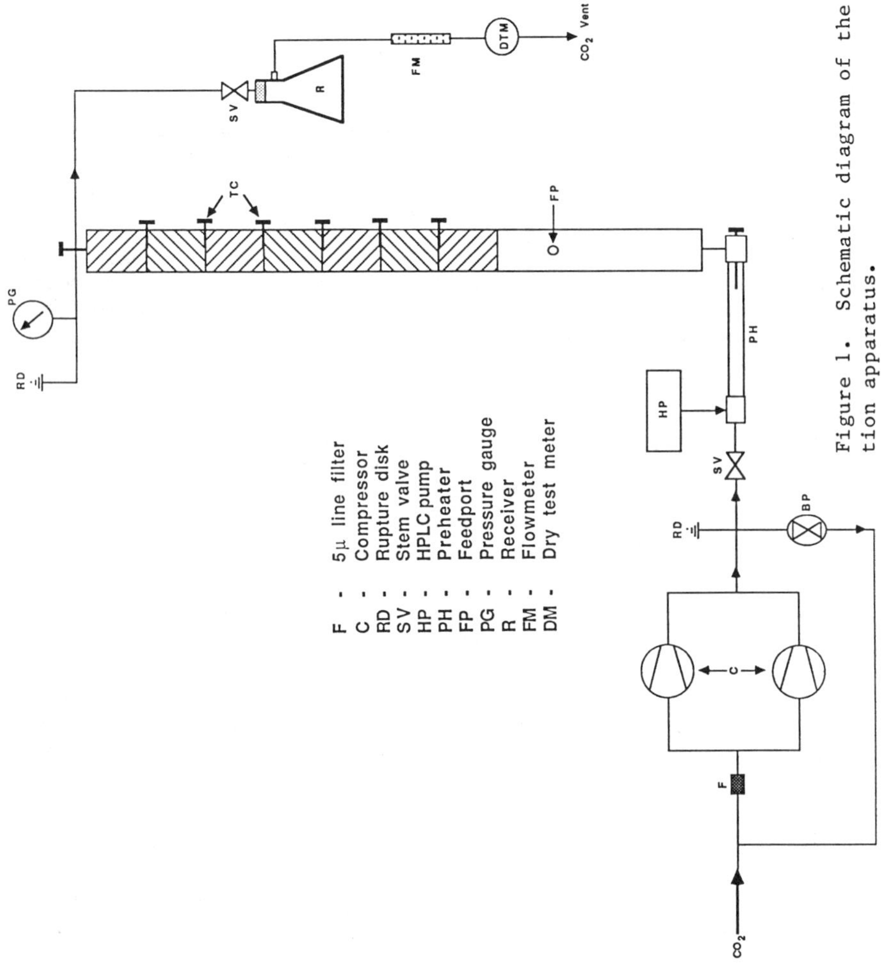

Figure 1. Schematic diagram of the supercritical fluid extraction apparatus.

Triglyceride Synthesis. The synthetic procedure was a modification of that discussed previously (18). Triglycerides were synthesized from glycerine and ethyl esters via the glyceroxide formed in situ from catalytic amounts (<5%) of sodium dispersion. Mole ratios up to 14 moles of ester per mole of glycerol were used to drive the reaction equilibrium toward the desired triglyceride. Reduced pressure also served to drive the reversible reaction by removal of ethanol as it was formed. For trieicosapentaenoylglycerol (tri-EPA), the reaction is written as:

$$\begin{matrix} H_2COH \\ | \\ HCOH \\ | \\ H_2COH \end{matrix} + 3RCOOEt \underset{in\ vacuo}{\overset{Na,\ \Delta}{\rightleftharpoons}} \begin{matrix} H_2COOCR \\ | \\ HCOOCR \\ | \\ H_2COOCR \end{matrix} + 3EtOH \qquad (1)$$

$$R = CH_3(CH_2CHCH)_5(CH_2)_2CH_2-$$

Anhydrous conditions are essential to avoid hydrolysis to the fatty acids which consume the catalyst through formation of sodium salts of the acids. The stirred reagents and catalyst were evacuated gradually to 3.3×10^{-4} bar before heating in a silicone bath at temperatures of 84 to 105 °C, mainly at 95-100 °C, for 21-24 hours.

Caution: To avoid possible fire or explosion, toward the end of the heating the neck of the vessel was examined for excess sodium. If unreacted sodium was present, the vessel was cooled to room temperature before admitting nitrogen and pushing the sodium into the reaction mixture.

At the end of the heating period, the mixture was cooled and ice cold dilute HCl (to neutralize the NaOH formed) and hexane were added to the vessel. After shaking to extract the organic compounds, the hexane solution was separated, washed with sodium sulfate-saturated water, dried, and evaporated to give the crude product.

Experiments were performed initially with mixtures of unsaturated esters and, when conditions were worked out, with 90% EPA and 90% DHA esters. Typical fatty acid profiles of starting esters are shown in Table I.

Table I. Major fatty acid esters present in three ester starting materials used in this work. Values are given in units of GC peak area %

Ester	90% EPA	PUFA	90% DHA
16:3ω4	---	5.0	---
16:4ω1	---	5.6	---
18:4ω3	2.0	7.5	---
20:4ω6	2.6	1.5	---
20:5ω3	90.1	47.7	2.7
21:5ω3	1.0	1.4	1.5
22:6ω3	3.3	22.5	89.5

SC-CO_2 Purification of Reaction Mixture. For purification of smaller batches (20 g) of the crude reaction mixture, a scaled down version of the large vessel, a 6-foot length of 1.0" OD x 11/16" ID SS pipe, was used. In addition, several purifications were performed using 4 ± 0.5 wt% ethanol (EtOH) as a cosolvent. The EtOH was injected at a constant volumetric flow rate into the preheater where indicated in Figure 1 using a Shimadzu LC-6A HPLC pump.

Analytical Procedures. Glyceride fractions from purification of reaction mixtures were analyzed by thin layer silica gel chromatography (TLC) and HPLC. In addition, several preparative TLC's were performed on the isolated triglycerides. The purified triglycerides were esterified directly to methyl esters by a modification of the method of Morrison and Smith (19), and the fatty acid profile determined by gas chromatography using conventional techniques described elsewhere (10).

Results and Discussion

Isolation of Ester Concentrates. A detailed discussion of the method used to obtain concentrates of EPA and DHA has been presented in two previous reports (10-11). Those data were collected from 20-g charges in the smaller 6' long, 11/16" ID column. Figure 2 shows fractionation curves for selected components of the feedstock mixture with a 100-g charge in the larger 6' long, 1.25" ID column. The maximum temperature of the gradient was 80 °C, the SC-CO_2 flow rate was ca. 50 standard liters/min, and the run was pressure programmed as indicated by the step-curve scaled on the right-hand ordinate. Of the EPA present in the feed, ca. 65% was recovered in the 90% concentrate; similarly ca. 75% of the DHA was recovered in the 90% DHA concentrate.

Isolation of tri-EPA Using SC-CO_2. EPA ester of 90% purity obtained in the process described above was used as a reactant to synthesize triglycerides. We will refer to the product as tri-EPA in subsequent discussion although it should be emphasized that triglycerides synthesized from 90% pure EPA are not expected to contain trieicosapentaenoylglycerol in 90% purity. In fact, if all components present in the original ester starting material have equal probability of incorporation into a triglyceride, random statistics predict that the final product would contain $(0.9)^3$ x 100 or 73% tri-EPA. The remainder would mainly be comprised of mixed triglycerides containing two EPA moieties.

In all syntheses of tri-EPA, the crude product mixture contained mono- and diglyceride intermediates and a significant quantity of unreacted esters. In addition, free fatty acids and other uncharacterized byproducts were present. This finding is essentially in line with that observed by Lehman and Gauglitz (15), who purified their crude reaction mixture by molecular distillation at temperatures above 250 °C. Fractionation of highly unsaturated glyceride mixtures using supercritical fluid CO_2 is an attractive alternative to molecular distillation because of the possibility of isolating the product at much lower temperatures. Brunner and Peter (20) used a bench scale countercurrent continuous apparatus with an unspecified

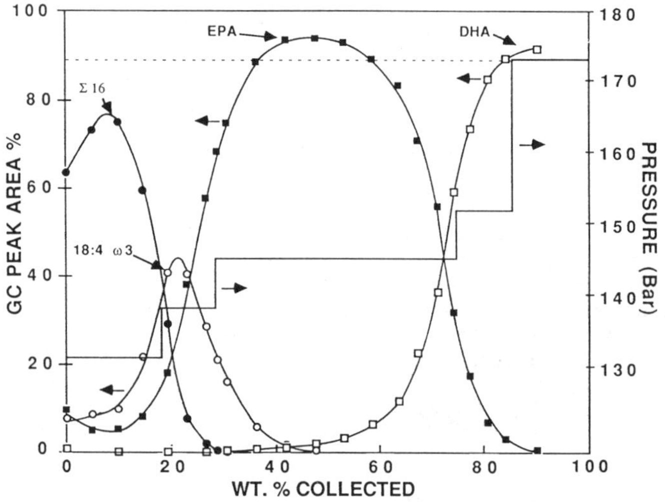

Figure 2. Fractionation curves generated in a pressure programmed fractionation of 100 g of urea-fractionated ethyl esters (see 'PUFA, Table 1). Column temperatures from top to bottom (see Figure 1) were 80, 71, 63, 56, 50, 45, 40 °C. Pressures used are represented by the step-curve which is scaled on the right-hand ordinate. The left-hand fractionation curve, designated Σ16, is the sum of 16:3ω4 and 16:4ω1.

supercritical fluid to obtain monoglycerides of ca. 99% purity from a mixture containing 40% diglycerides by weight. Panzer et al. (21) stated that little separation of glyceride mixtures could be achieved using supercritical fluid carbon dioxide, although no experimental evidence was given to support this claim.

While it may be true that SC-CO_2 can effect little separation of mono- and diglycerides, we have found it to be useful in isolating unreacted esters and triglycerides from our crude mixtures. Table II provides a summary of experimental conditions and results for the fractionation of a crude tri-EPA reaction mixture.

Table II. Data from the fractionation of the crude tri-EPA product mixture with SC-CO_2. All heated zones at 60 °C

Fraction number	Pressure bar	Fraction wt (g)
Feed	---	20.00
1	138	3.92
2	186	1.34
3	207	1.56
4	241	1.58
5	310	4.16
6	310	4.11
7	310	0.31

Figure 3 is a photograph of a TLC of the crude product mixture and each of the seven fractions (an ester and mixed glyceride standard of 18:3ω3 was spotted on right-hand side of the plate). As is apparent by visual inspection of the TLC of fraction 1, unreacted ethyl esters, accounting for about 20% of the mixture by weight, are cleanly extracted early in the fractionation. By gas chromatography fraction 1 was found to have a composition essentially identical to that of the original ester reactants (see '90% EPA' Table 1), thus providing good evidence that for this mixture the composition of the synthetic triglycerides is the same as that of the starting material. It is also useful to point out that the recovered unreacted esters, which in themselves have significant value, can be recycled in a subsequent synthesis. The TLC's of fractions 2-3 show them to contain the bulk of the mono- and diglyceride intermediates as well as the free fatty acids and other unidentified byproducts. Fractions 4-6 accounting for nearly 50% of the feed by weight are seen to contain most of the synthetic triglyceride product. HPLC analyses indicate that taken together, these three fractions contain triglycerides of at least 93% purity. Since the HPLC analyses were performed a few weeks after the fractionation and some autoxidation is known to have occurred, 93% should be considered a conservative figure.

Isolation of Tri-EPA Using SC-CO_2 and 4 wt% Ethanol as a Cosolvent. Although the purity of tri-EPA obtained in the above fractionation is satisfactory the solvent-to-feed ratio, defined as the weight of CO_2 necessary to fractionate a unit weight of feedstock, was 490. Reduction of the S/F would be desirable. One approach to increase the

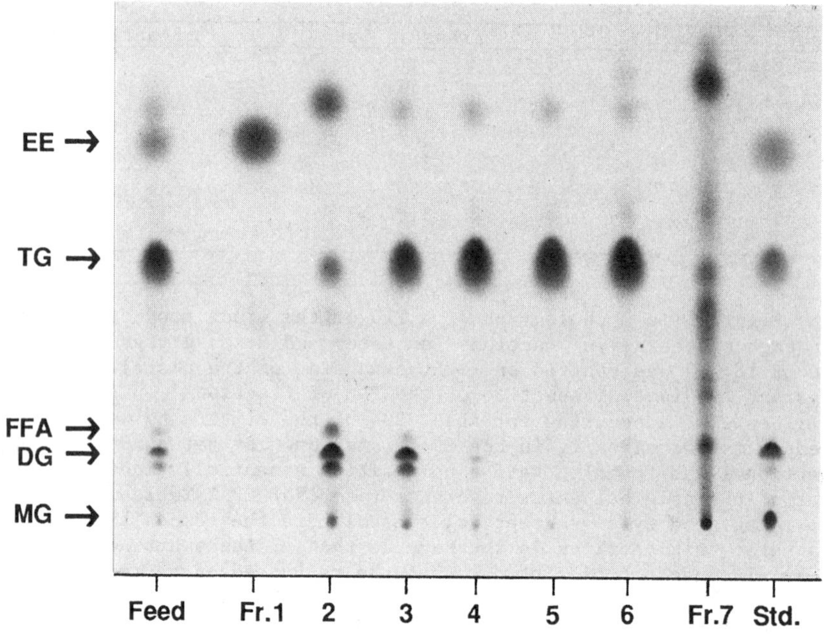

Figure 3. Thin layer silica gel chromatogram of the feed and fractions obtained in the fractionation of the tri-EPA reaction mixture using SC-CO_2. MG = monoglyceride, DG = diglyceride, FFA = free fatty acid, TG = triglyceride, EE = ethyl ester. Other spots are unidentified byproducts.

solubility of solutes in SC-CO_2 and thus reduce the S/F is to add a small quantity (ca. 10% or less by weight) of a miscible compound to CO_2 as a "cosolvent". In their work on the fractionation of glyceride mixtures with SC-CO_2, Brunner and Peter (20) investigated a number of cosolvents including acetone, benzene, methylene chloride, and ethanol (EtOH). Panzer (21) selected hexane as a cosolvent in similar work. All of these solvents are toxic to some degree and therefore must be removed from the product. Since it is difficult to guarantee the complete removal of solvent residues, EtOH is an attractive choice due to its relatively low toxicity.

Preliminary data on the solubility of soybean triglycerides in CO_2 at 60 °C without cosolvent and with 2 and 4 wt% EtOH between 207 and 310 bar are shown in Figure 4. Values obtained with pure CO_2 are in good agreement with those reported by Friedrich (22). The data in Figure 4 suggested that small quantities of added EtOH could be used to increase the solubility of trigycerides (and presumably mono- and diglycerides) in CO_2 at pressures similar to those used in the fractionation with pure CO_2, thereby reducing the S/F. However, in order that a comparably successful fractionation be realized, the increased loading of the solute in the fluid phase must not be accompanied by a large decrease in the selectivity of the fluid for one component over another. Brunner and Peter (20) present evidence that the selectivity of SC-CO_2 for free fatty acids in palm oil is actually enhanced by use of certain cosolvents including EtOH.

To confirm this result, a crude tri-EPA reaction mixture containing 90% tri-EPA was fractionated using SC-CO_2 with a 4 $\pm$ 0.5 wt% EtOH cosolvent. The rate at which EtOH was introduced by the HPLC pump was predetermined based upon a CO_2 flow rate of ca. 10 standard liters per minute. Table III provides a summary of the results of the test.

Table III. Data from the fractionation of the crude tri-EPA product mixture with SC-CO_2 containing 4 wt% ethanol. All heated zones at 60 °C

Fraction number	Pressure (bar)	Fraction wt (g)
Feed	---	23.53
1	138	4.68
2	186	3.58
3	207	1.84
4	310	10.17
5	431	2.38

The pressures were essentially the same as those in the fractionation without cosolvent (see Table II). As anticipated from the data of Figure 4, the observed S/F of 200 for the test with 4 wt% EtOH is significantly less than that found using CO_2 alone. Visual inspection of the TLC's shown in Figure 5 indicates that fraction 4, accounting for about 45% (w/w) of the feed is triglyceride of high purity. HPLC of fraction 4 indicated that it contains triglycerides of at least 92% purity. Again, because of delays in performing the HPLC analyses, this is considered a conservative estimate.

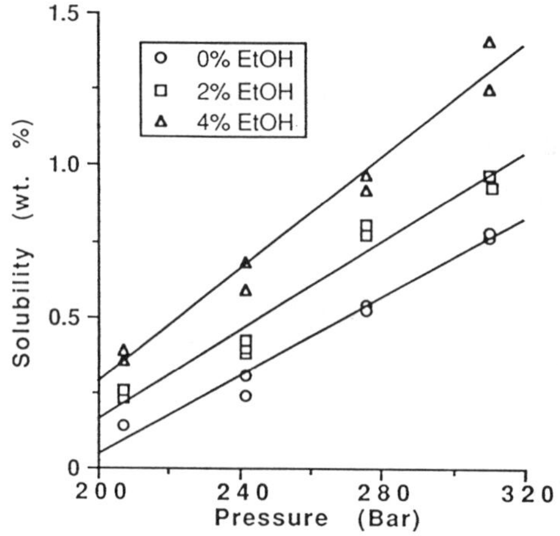

Figure 4. Solubility of soybean triglycerides in pure CO_2 as well as with a 2% and 4% (w/w) ethanol cosolvent at 60 °C.

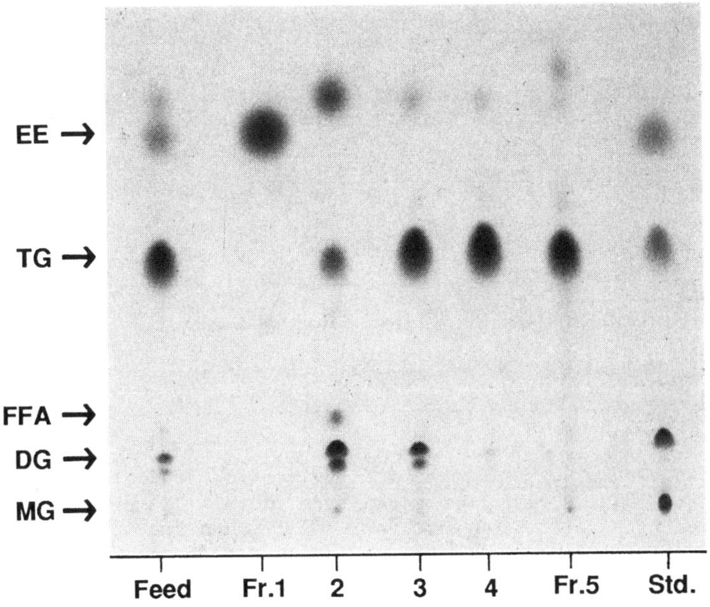

Figure 5. Thin layer silica gel chromatogram of the feed and fractions obtained in the fractionation of the tri-EPA reaction mixture using SC-CO_2 and a 4 wt% EtOH cosolvent. MG = monoglyceride, DG = diglyceride, FFA = free fatty acid, TG = triglyceride, EE = ethyl ester. Other spots are unidentified byproducts.

Incidentally, the mono- and diglycerides present in the reaction mixture were miscible with the EtOH recovered in the receiver while the highly unsaturated triglyceride was not. This observation provided a useful visual indicator of where to cut fractions.

Another advantage of using an EtOH cosolvent suggested by Brunner and Peter (20) was that increased solute solubility allows reduction of processing pressures. Table IV summarizes the results of the final fractionation of the crude tri-EPA mixture. The pressures used in this test are generally ca. 35-70 bar lower than those in the previous two fractionations. Fractions 6-7 comprise ca. 60 wt% of the feed. HPLC analyses estimated the purity of both to be better than 98%. As expected, the S/F was considerably higher (ca. 450) than in the previous test at elevated pressures. On the other hand, the yield and purity of the product were apparently improved.

Table IV. Data from the fractionation of the crude tri-EPA product mixture with SC-CO_2 containing 4 wt% ethanol at reduced pressures. All heated zones at 60 °C

Fraction number	Pressure (bar)	Fraction wt (g)
Feed	---	20.00
1	138	2.76
2	159	1.26
3	172	1.14
4	172	0.95
5	172	0.48
6	241	5.95
7	241	6.18
8	241	0.62

Fractionation of Other Synthetic Triglyceride Product Mixtures. To obtain tri-DHA, a mixture derived from the 90% DHA esters was fractionated with 4 wt% EtOH cosolvent. A tri-DHA fraction accounting for 44% of the feed by weight was isolated. The purity of this fraction with respect to triglycerides was estimated by HPLC to be 93% or better.

In all of the fractionations discussed to this point, the composition of the unreacted esters recovered early in the test was found to be very similar to that of the ester reactants. From this evidence, it can be concluded that all components in the 90% EPA and DHA reactants are incorporated into the triglyceride product with equal probability. This is probably not surprising, as the components in both concentrates differ little in chain length or degree of unsaturation. To provide a more rigorous test, the product mixture from a synthesis starting with mixed chain-length polyunsaturated fatty acid esters ('PUFA' in Table I) was fractionated using CO_2 with 4 wt% EtOH. As observed in previous tests, the first fraction contained essentially all of the unreacted ethyl esters. Gas chromatographic analysis of this fraction gave the following composition in GC peak area %: 16:3ω4, 3.9%; 16:4ω1, 4.2%; 18:4ω3, 7.3%; 20:4ω6, 1.5%; 20:5ω3, 49.3%; 21:5ω3, 1.4%; 22:6ω3, 21.3%. Although some differences in composition vis-à-vis the starting material do exist, they are certainly not large and may well fall within experimental error. Lehman and Gauglitz (15) reported similar observations for the procedure they developed using free fatty acids. Fractionation of this mixture proved to be somewhat more difficult than fractionation of the tri-EPA. The wider distribution of chain lengths in the glycerides synthesized from the PUFA mixture required the more difficult separation of, for example, C_{22}-containing diglycerides from C16-containing triglycerides.

Conclusions

Isolation of triglycerides from mixed glycerides can be accomplished using pure supercritical fluid CO_2 at moderate temperatures by pressure programming. A few wt% of added ethanol can significantly reduce the solvent-to-feed ratio if comparable pressures are used or alternatively permit reduction of pressures during the fractionation. Regardless of the composition of esters used as reactants, there appears to be little or no selective incorporation of individual components into the triglyceride product.

Acknowledgement

We wish to express appreciation to Jeanne Joseph, Bob Ernst, and coworkers of the National Marine Fisheries Service, Charleston, SC, for supplying urea-fractionated menhaden oil esters.

Literature Cited

1. Lands, W. E. M. Fish and Human Health; Academic Press, Inc.: Orlando, FL, 1986.
2. Kinsella, J. E. Seafoods and Fish Oils in Human Health and Disease; Marcel Dekker, Inc.: New York, 1987.
3. Leaf, A. New Eng. J. Med. 1988, 318, 549.
4. Dehmer, G. J.; Popma, J. J.; van den Berg, E. K.; Eichhorn, E. J.; Prewitt, J. B.; Campbell, K. D.; Jennings, L. J.; Willerson J. T.; Schmitz, J. M. New Eng. J. Med. 1988, 319, 733.
5. Miller, D.; Leong, K. C.; Knobl, G. M.; Gruger, E., Jr. Proc. Soc. Exp. Biol. Med. 1964, 116, 1147.
6. Boggio, S. M.; Hardy, R. W.; Babbitt, J. K.; Brannon, E. L. Aquaculture. 1985, 51, 13.
7. Stout, V. F. J. Am. Oil Chem. Soc. 1963, 40, 40.
8. McHugh, M. A.; Krukonis, V. J. Supercritical Fluid Extraction Principles and Practice; Butterworks: Boston, 1986, p. 178.
9. Eisenbach, W. Ber. Bunsenges. Phys. Chem. 1984, 88, 882.
10. Nilsson, W. B.; Gauglitz, E. J., Jr.; Hudson, J. K.; Stout, V. F.; Spinelli, J. J. Am. Oil Chem. Soc. 1988, 65, 109.
11. Nilsson, W. B.; Gauglitz, E. J., Jr.; Hudson, J. K.; Teeny, F. M. Paper presented at the 194th ACS National Meeting, New Orleans, LA, Paper AGFD 0051.
12. Krukonis, V. J. In Supercritical Fluid Extraction and Chromatography; Charpentier, B. A., and Sevenants, M. R., Eds.; ACS Symposium Series 366; American Chemical Society: Washington, DC, 1988, pp. 34-36.

13. El-Boustani, S.; Colette, C.; Monnier, L.; Descomps, B.; Crastes de Paulet, A.; Mendey, F. Lipids. 1987, 22, 711.
14. Takahashi, K.; Hirona, T.; Saito, M. Nippon Suisan Gakkaishi, 1988, 54, 523.
15. Lehman, L. W.; Gauglitz, Jr., E. J. Jr. J. Am. Oil. Chem. Soc. 1964, 41, 533.
16. Miyashita, K.; Takagi, T. J. Am. Oil Chem. Soc., 1986, 63, 1380.
17. Sumerwell, W. N. J. Am. Chem. Soc., 1957, 79, 3411.
18. Stout, V. F. J. Am. Oil Chem. Soc., 1988, 65, 499.
19. Morrison, W. R.; Smith, L. M. J. Lipid Res., 1964, 5, 600.
20. Brunner, G.; Peter, S. Sep. Sci. and Tech., 1982, 17, 199.
21. Panzer, F.; Ellis, S. R. M.; Bott, T. R. Inst. Chem. Eng. Symp. Ser., 1978, 54, 165.
22. Friedrich, J. P. U.S. Patent 4 466 923, 1984.

RECEIVED June 9, 1989

Chapter 28

Supercritical Carbon Dioxide Extraction of Lipids from Algae

J. T. Polak, M. Balaban, A. Peplow, and A. J. Phlips

Food Science and Human Nutrition Department, University of Florida, Gainesville, FL 32611

Supercritical CO_2 (SC CO_2) extraction of lipids from two freeze dried microalgae species (Skeletonema costatum, a marine diatom and Ochromonas danica, a freshwater phytoflagellate) between 17 and 31 MPa and 40°C was studied. Extracted and remaining lipids were analyzed with gas chromatography. For Skeletonema having relatively low levels of oil, there was no difference in yields between extractions at 24 and 31 MPa. Treatment of samples with phospholipase C slightly increased SC CO_2 extracted oil yield. Three-week-old cultures exhibited higher amounts of lipids. Two-week-old cultures contained a larger proportion of polyunsaturated fatty acids. SC extracts at higher pressures had about 25 % eicosapentaenoic acid (EPA). Ochromonas had a higher oil level than Skeletonema, but the relative fraction of polyunsaturated fatty acids such as EPA was lower. Oil yield from Ochromonas increased with pressure at all levels. Chlorophylls in algae were not extracted by SC CO_2.

Possible health benefits of consuming fish oils and omega-3 fatty acids have been reviewed by various authors (1-5). These oils can also be obtained from phytoplankton (6-8). Oils obtained from phytoplankton do not contain cholesterol. Phytoplankton require very simple nutrients in easy-to-maintain growth conditions. They can be harvested in short culture cycles. They can be manipulated by genetic engineering to increase their yield and their resistance to environmental stresses.

An algal cell contains approximately 90% water and 10% solids by weight (9). The solid composition consists of organic compounds and minerals. Lipids are stored within the cell and in cell membranes. Storage lipids are made up of mainly acylglycerols and free fatty acids (9).

Skeletonema costatum, a marine diatom is reported to have relatively high proportions of polyunsaturated fatty acids (6,7,10). It is also relatively easy to culture, requiring light, CO_2, simple minerals and organic nutrients (11).

0097–6156/89/0406–0449$06.00/0

Ochromonas danica is reported to contain higher levels of lipids compared to Skeletonema (12). It is a freshwater species, and its nutritional requirements are richer than those of Skeletonema.

It is possible to obtain lipids from microalgae by conventional solvent extraction. However, extraction with supercritical (SC) CO_2 is emerging as a potential alternative to obtain lipids from natural sources (13-19).

The objective of this study was to determine the lipid extraction yields from two species of microalgae by SC CO_2, and to investigate the factors that influence these yields.

Materials and Methods

Growth and Harvesting of Microalgae. Skeletonema costatum starter cultures were obtained from Bigelow Laboratory for Ocean Sciences, West Boothbay Harbor, Maine. The culture conditions are described by Polak (11). The cells were harvested when they were 2 or 3 weeks old by continuous centrifugation. The biomass obtained was immediately frozen at -28°C, freeze-dried, flushed with N_2 and stored at -28°C in darkness. A total of four batches were grown and used in experiments.

Ochromonas danica culture (code L1298) was obtained from the culture collection of the University of Texas at Austin. The cells were cultured in a modified Hoaglands medium (11) for 10 days. For both species, the culture conditions were : T = 22°C; pH = 8.2; light intensity : 100 μEinstein/m^2-sec; light/dark cycle of cool-white fluorescence light : 16/8hr; CO_2 source : bubbling with filter sterilized air.

Ochromonas cells were harvested and samples prepared as described above for Skeletonema.

Composition of cells. Moisture content of the cells was determined before and after freeze-drying by the vacuum oven method (20). Ash content was determined by ashing at 600°C for 2 hours (20).

Total lipids were solvent extracted from freeze dried algae (21,22) for the initial lipid profile, and for the analysis of remaining lipids after SC extraction.

Lipid Analysis. Fatty acids in lipids were first converted to methyl esters by reacting them with METH-PREP II (ALLTECH-Applied Science, Deerfield, IL) and then analyzed by GC. Methyl esters were analyzed by a Tracor GC equipped with a flame ionization detector (Tracor, Austin, TX) and a RSL 500, 30 meter FSOT column (ALLTECH, Deerfield, IL). Details of GC analysis conditions can be found elsewhere (11).

Supercritical Carbon Dioxide Extraction. The flow diagram of the equipment used for SC extraction is shown in Figure 1. The maximum operating pressure was 34.5 MPa (5000 psi). Liquid CO_2 (Coleman instrument grade, 99.99% purity) was passed through a 7μm particulate filter and entered two pumps of 0-460 mL liquid/hr capacity, with cooling heads. The extraction vessel was a 150 ml capacity, 316 stainless steel high pressure cylinder with an electrical heating jacket. Temperature for all experiments was set at 40° C. A freeze-dried sample of about 6 grams was tightly packed into a pre-weighed pure cellulose extraction thimble (Whatman,22 mm x 80

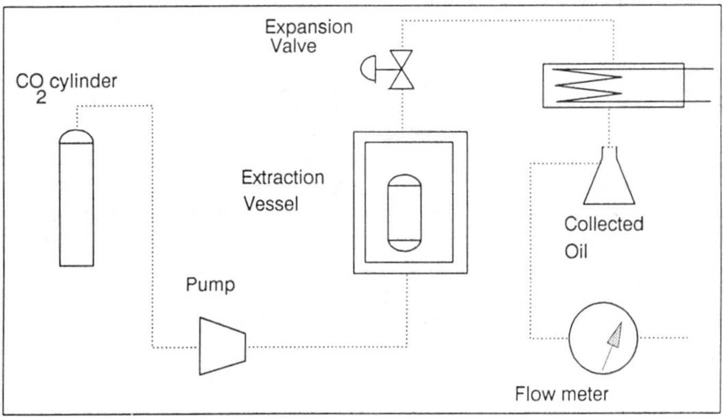

Figure 1. Flow diagram of Supercritical Extraction Apparatus.

mm) and was sealed by sliding half of a thimble over the first. The net sample weight was determined by weight difference. The enclosed thimble was placed in the pre-heated extractor vessel. The system was allowed to reach the experimental pressure : 14.7 MPa, 24 MPa or 31 MPa (2500 psi, 3500 psi and 4500 psi). The CO_2 rich phase in the extractor vessel was expanded across a backpressure valve to 1 atm, and passed through a cold trap at -15°C to collect extracted matter. The quantity of CO_2 used was determined with a wet test flow meter (Singer American Meter Division, Philadelphia, PA). Initial experiments confirmed that 155 L CO_2 were used during a 4 hr experiment. It was arbitrarily decided to use 155 L CO_2 for all subsequent experiments. After the experiment both the weight of the sample remaining in the thimble and the weight of the collected SC extracted oil were determined gravimetrically.

Since the thimble and the freeze-dried mass could hinder mass transfer, another set of experiments were conducted at 24 and 31 MPa in the cups of an aluminum sample holder with six cups (0.5 cm x 0.5 cm). It was reasoned that if the thimble was eliminated from the extraction, and if only the lipids were exposed to the SC CO_2, there would be negligible mass transfer resistance to extraction. Solvent extracted lipids were placed in the cups, and extracted with SC CO_2.

Effect of Phospholipase C on lipid yield. Phospholipase C from *Clostridium perfringens* (Sigma Chemical Company, St. Louis, MO) was used to cleave the phosphate group of Skeletonema lipids, and to change the polarity of phospholipids to increase the SC CO_2 extraction yield. Six grams of freeze dried Skeletonema were extracted with solvents (21). The lipids obtained were split into three fractions. The first fraction was used to determine the fatty acid profile of the lipid to serve as a basis. The second fraction was deposited onto the sample holder. The lipids were extracted with SC CO_2 as control at 31.0 MPa. The third fraction was treated with phospholipase C in Trizma buffer (Sigma Chemical Company, Saint Louis, MO) and incubated at pH 3.7 at room temperature for 3 hours (22). The treated lipids were deposited onto the sample holder and extracted with SC CO_2 at 31 MPa. The extracted lipids were compared with the profile lipids and with the lipids from the control experiment. In both SC extraction cases, the SC CO_2 extracted lipids and the lipids remaining on the sample holder were compared by GC analysis.

Results and Discussion

Moisture, and Ash Content. For Skeletonema, the wet basis moisture content in batches one through four were 88.9%, 87.5%, 90.3%, and 87.6%, respectively. The average wet basis moisture content of the freeze dried material was 5%. The average ash content of all batches was 30% dry weight basis. The high ash level is due to the silica skeleton of *Skeletonema costatum*.

For Ochromonas, the moisture content of the cells was 80.2 % wet basis. Upon freeze drying, the moisture content was reduced to 1.5 % dry basis. The ash content was 4 % dry basis.

Total Lipid Content and Fatty Acid Composition. Dry basis lipid content of the freeze dried Skeletonema samples were 7.55%, 11.5%, 5.63% and 10.3% by weight for batches one through four, respectively.

This is in agreement with 8.6 % total lipids reported for Skeletonema costatum (7). The lipid content of freeze dried Ochromonas was 28 % dry weight basis. The fatty acid profile of Skeletonema obtained through solvent extraction for each batch is given in Table I. Identification of most of the peaks was based on comparison with fatty acid standards. Due to the lack of availability of some standards, some C16 peaks could not be positively identified. Peaks labeled as Peak 1, 2, 3, and 4 in Table I most likely represent fatty acids C16:2w7, C16:2w4, C16:3 and C16:4.

The experimental average fatty acid composition of the lipids in Ochromonas is shown in Table II and compared with data given in the literature (12). In Figure 2, the comparison of Ochromonas fatty acids with those of Skeletonema is shown. Obviously, Ochromonas has more lipids. Although the percentage of EPA is higher in Skeletonema, its absolute level is greater in Ochromonas. Conventional solvent extraction of lipids from both freeze dried algae resulted in a dark green extract due to the extraction of chlorophyll together with lipids. In contrast, the SC CO_2 extracted oils appeared golden yellow in color. Carotenoids are soluble in SC CO_2. β-carotene has a solubility of 0.01-0.05% (wt) in liquid CO_2 which is about one tenth of that of lipids (23). The presence of carotenoids in the SC CO_2 extracted oils from krill has also been reported (24).

Supercritical CO_2 Extraction Results. The extent of lipid extraction from freeze dried algae was influenced by the total CO_2 volume passed through the system, by the operating pressure, and by the ratio of non-polar to polar lipids present in the batches.

Effect of Pressure. The maximum solubility for stearic acid (18:0) under increasing pressure was found around 35.5 MPa in the literature (25). Schmitt and Reid reported that from 14.2 to 37.0 MPa there was no significant increase in the solubility of C18:0. The solubility was higher at 14.2 MPa than at 37 MPa (25). They could not explain this behavior. Another study also found that the maximum solubility of C18:0 in SC CO_2 was around 36.5 MPa (26). This suggests that for some fatty acids, an increase in pressure may not always result in higher solubility in SC CO_2. This conclusion is valid for the Skeletonema results of the present study, as shown in Figure 3, where extracted levels of identified fatty acids are shown at two pressures. The amount of lipids extracted by CO_2 at 24 MPa is not significantly different than the amount extracted at 31 MPa. The difference in yields between extractions at 17 MPa and 31 MPa is shown in Figure 4. About one fifth of the identifiable fatty acids of Skeletonema were of the C18 class. This may partially explain the leveling of fatty acid yield at 24 MPa. Therefore, extraction of Skeletonema at 31.0 MPa is unnecessary under the conditions of the experiments. Extraction yields of sample holder experiments show that there was no difference between extractions at 31 and 24 MPa (Figure 5). In the case of Ochromonas, the yield of SC extracted oils increased with increasing pressure, as shown in Figure 6.

Total Lipid Mass Balances. Tables III and IV represent overall total lipid mass balances for SC CO_2 extractions of Skeletonema carried out in thimbles and in sample holder at different pressures. The amount

Table I. Fatty Acid Composition of <u>Skeletonema costatum</u>

	Batch 1 (mg/g)*	%	Batch 2 (mg/g)*	%	Batch 3 (mg/g)*	%	Batch 4 (mg/g)*	%
C14:0	4.53	12.9	10.57	21.7	3.72	15.4	7.27	16.8
C14:1	0.10	0.3	0.32	0.7	0.25	1.0	0.16	0.4
C16:0	2.67	7.6	4.10	8.4	1.24	5.1	2.28	5.3
C16:1	4.97	14.1	9.78	20.1	3.48	14.4	7.24	16.8
PEAK 1	0.70	2.0	0.96	2.0	0.51	2.1	0.68	1.6
PEAK 2	2.16	6.1	3.47	7.1	1.25	5.1	2.31	5.6
PEAK 3	3.87	11.0	3.81	7.8	3.38	14.0	5.90	13.7
PEAK 4	2.64	7.5	4.06	8.3	3.50	14.5	5.68	13.1
C18:0	0.34	1.0	0.27	0.6	0.14	0.1	0.04	0.1
C18:1w9	1.62	4.6	0.38	0.8	0.17	0.7	0.16	0.4
C18:1w7	0.78	2.2	1.28	2.6	0.45	1.9	1.14	2.6
C18:2	1.31	3.7	1.06	2.2	0.71	2.9	0.72	1.7
C18:3w6	1.50	4.3	1.90	3.9	1.15	4.7	0.52	1.2
C18:3w4	0.67	1.9	0.87	1.8	0.57	2.4	0.96	2.2
C18:4	0.24	0.7	0.31	0.6	0.19	0.8	1.04	2.4
C20:4	0.03	0.1	0.35	0.7	0.05	0.2	0.05	0.1
C20:5w3	5.50	15.6	3.55	7.3	2.44	10.1	5.49	12.7
C22:6w3	1.53	4.4	1.62	3.3	1.00	4.1	1.54	3.6
	35.16	100.0	48.66	100.0	24.20	100.0	43.18	100.0
original lipid (mg/g)	75.50		115.00		56.30		103.00	

* = freeze dried Skeletonema costatum

C18:1w9 indicates a fatty acid with 18 carbon atoms, 1 double bond, and the double bond is at w9 position.

Table II. Fatty acid profile of Ochromonas danica. Percentage of identifiable fatty acids

Fatty acid	Literature	Experimental
C14:0	15	18.06
C14:1		0.31
C16:0	10	10.99
C16:1		1.50
C18:0	3	3.23
C18:1w9	8	6.42
C18:2	16	17.55
C18:3w6	2	
C18:3w4	10	14.77
C20:0	1	1.15
C20:2		0.25
C20:3	5	4.83
C20:4	11	11.50
C20:5w3		3.83
C22:0		0.95
C22:1		4.64
C22:5	2-4	
	-------	--------
	85	99.98

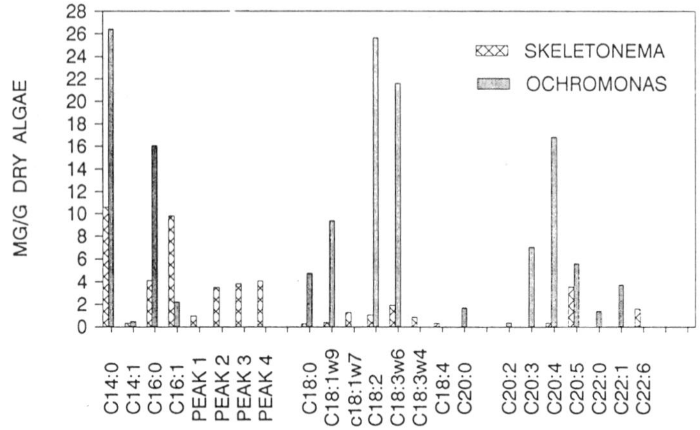

Figure 2. Comparison of lipid profiles of Ochromonas danica and Skeletonema costatum.

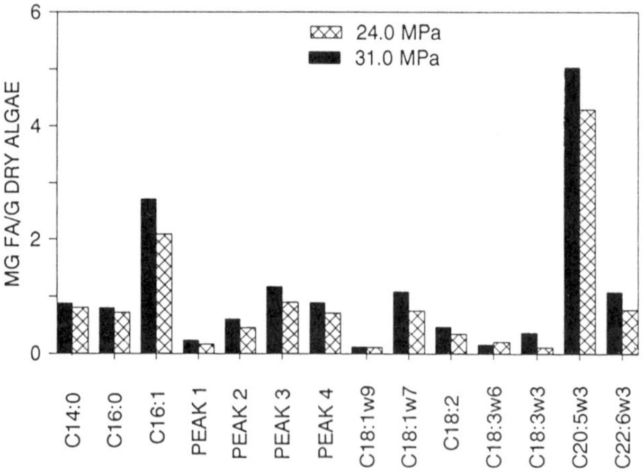

Figure 3. Comparison of SC CO_2 Extracted Fatty Acids From Skeletonema Costatum for Extractions at 24.0 and 31.0 MPa.

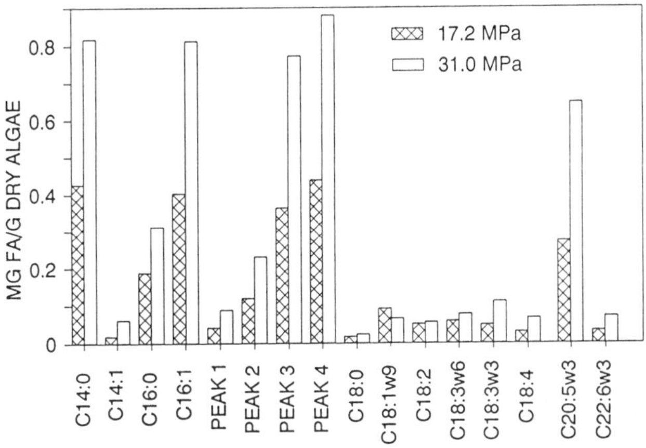

Figure 4. Comparison of SC CO_2 Extracted Fatty Acids From <u>Skeletonema costatum</u> for Extractions at 17.2 and 31.0 MPa.

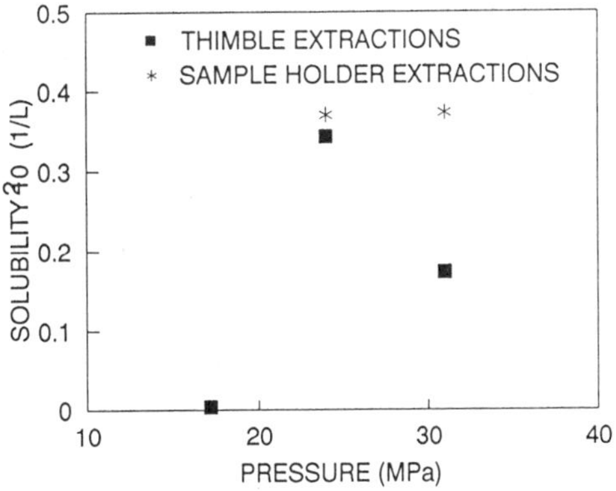

Figure 5. SC CO_2 Extraction Yield From <u>Skeletonema costatum</u> as a Function of Pressure. Solubility is expressed as g of lipid extracted per gram of original oil, per liter of CO_2 used.

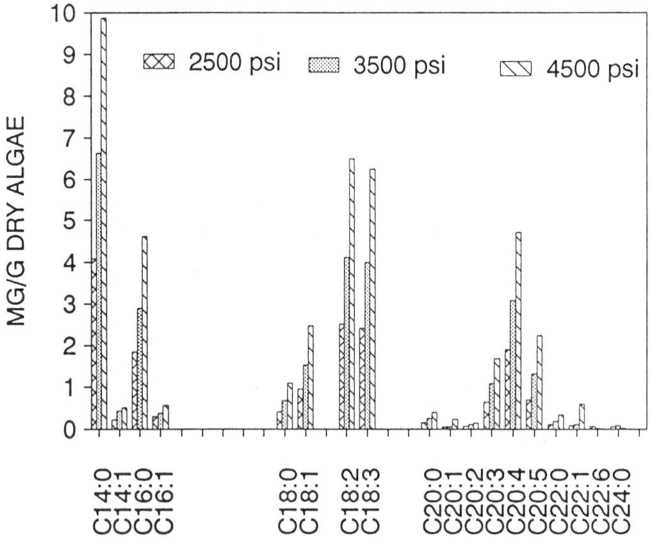

Figure 6. Comparison of SC CO_2 Extraction Yields From Ochromonas danica at Different Pressures.

Table III. Overall Total Lipid Mass Balance: Extraction of freeze dried <u>Skeletonema costatum</u>. Thimble experiments

	Initial algae wt(g)	Initial oil wt (mg/g)	Final algae wt(g)	Final oil wt (mg/g)	Oil Removed Exp. (mg)	Oil Removed Calc. (mg/g)	Oil Removed Exp. (mg/g)	CO_2 used (L)
				Experiments at 17.2 MPa				
1a.	6.17	56.3±1.6	6.13	50.8±1.6	38.4±1.	5.5±0.3	6.2±1.	155
1b.	4.94	75.5±3.2	4.83	63.0±1.1	81.7±1.	12.5±0.6	16.5±1.	475*
				Experiments at 24.0 MPa				
2a.	6.45	115.0±1.1	6.03	64.6±2.4	164.0±1.	50.4±1.9	25.4±1.	155
2b.	6.37	115.0±1.1	6.01	65.1±2.4	177.8±1.	49.8±1.9	27.9±1.	155
				Experiments at 31.0 MPa				
3a.	6.09	75.5±3.2	5.98	56.1±1.7	107.4±1.	19.4±1.0	17.1±1.	165
3b.	6.37	75.5±3.2	6.20	58.3±1.2	108.8±1.	17.2±0.8	17.1±1.	155

Exp. : Experimental values
Calc.: Calculated difference between initial and final weights
mg/g : based on 1 g freeze dried algae
* : not considered for solubility calculations

Table IV. Total Lipid Mass Balance: experiments in sample holder for freeze dried <u>Skeletonema costatum</u>

Pressure (MPa)	Initial wt (mg)	Final wt (mg)	Oil extracted calc. (mg/g)	Oil extracted exp. (mg/g)	CO_2 used (L)
24.0	216.2	95.3	540.0±1.0	451.0±1.0	155
31.0	269.5	114.5	581.2±1.0	428.9±1.0	155

Exp : experimental
Calc : calculated

of total saponifiables was around 40% to 50% in both the SC CO_2 extracted and chloroform-methanol extracted oils.

The amounts of oil collected in the cold trap and the amount of oil absorbed by the thimble did not always add up to the total amount of oil extracted by the CO_2. During extraction and depressurization the oils in the SC phase deposited onto the extractor vessel walls, and onto the inner surfaces of the apparatus. This was confirmed by flushing the system with CO_2 between experiments. Tiny droplets of oil most likely from previous extractions were collected in the cold trap. Therefore, the experimental amount of SC extracted oil was not totally reliable for mass balance purposes. Instead, total amounts of lipid extracted by SC CO_2 were calculated by subtracting the amount of residual lipids (lipids that were not extracted by CO_2) from the amount of lipids present prior to SC CO_2 extraction. Both the initial lipid content and the amount present after SC CO_2 were obtained through similar solvent extraction treatments having small experimental errors.

In the sample holder extraction experiments, the overall total lipid mass balances for the extraction of 1 gram of lipid (Table IV) indicated only a small difference between the actual and the theoretical amounts of oil collected.

At 24 MPa, the calculated amount extracted was 54.4% of the available lipid versus 45% for the experimental amount. At 31 MPa, the values were 58% and 42%. Lee et al. (7) indicated that 57% of the total lipid composition in <u>Skeletonema costatum</u> consisted of acylglycerols. The calculated yields of 58% and 54% in the present study indicated that almost all of the nonpolar lipids were extracted by CO_2. Friedrich et al. (14) claimed that no significant difference was observed in the quantity of total saponifiables between SC CO_2 and hexane extracted soybean oils. This is in agreement with the results for SC CO_2 extracted and solvent extracted oils of Skeletonema.

Table V shows the total mass balance for Ochromonas extractions carried out at different pressures. The lipid levels obtained experimentally are lower than those obtained by the difference between the initial and final sample weights.

Since the trend of amounts of lipids extracted at different pressures in both the sample holder and thimble experiments were very similar (Figure 6), it can be concluded that the thimble did not create a resistance to mass transfer in Skeletonema. The general trend for greater extraction yield at 31.0 MPa versus 17.2 MPa is shown in Figure 4. Friedrich (14) indicated an increase in extraction efficiency of soybean oils with an increase in pressure from 35 to 55 MPa. Yamaguchi et al. (24) claimed that the extraction yield of krill meal oils remained almost constant with increase an in pressure from 24.5 to 39.0 MPa at constant temperature. Figures 7 to 9 show individual fatty acids in the original Skeletonema, those left after SC extraction, and those in the SC extracted oil, at different pressures.

For Ochromonas, Figures 10 to 12 show the effect of pressure on the levels of SC extracted oils, and those of lipids remaining in the algae. As pressure increased, the relative fraction of long chain fatty acids in the SC extract increased. It can be seen that, in general the sum of the residue and the SC extract was close to the level of original fatty acid. Any discrepancy can be explained by the fact that there is some lipid left in the inner surfaces of the equipment.

Table V. Overall Total Lipid Mass Balance: Extraction of freeze dried <u>Ochromonas danica</u>

	Initial algae wt(g)	Initial oil wt (mg/g)	Final algae wt(g)	Final oil wt (mg/g)	Oil Removed Exp. (mg)	Oil Removed Calc. (mg/g)	Oil Removed Exp. (mg/g)	CO_2 used (L)
			Experiments at 17.2 MPa					
1a.	4.73	275	4.48	206.2	157.0	68.8	33.16	155
1b.	5.05	275	4.8	234.2	137.0	40.8	27.14	156
			Experiments at 24.0 MPa					
2a.	6.95	275.0	6.37	207.4	178.0	67.6	25.6	155
2b.	5.82	275.0	5.51	219.8	215.0	55.2	36.96	155
			Experiments at 31.0 MPa					
3a.	5.59	275.0	4.995	174.6	335.0	100.4	59.94	155
3b.	7.01	275	6.58	197.1	282.5	77.9	50.5	155

Exp. : Experimental values
Calc.: Calculated difference between initial and final weights
mg/g : based on 1 g freeze dried algae

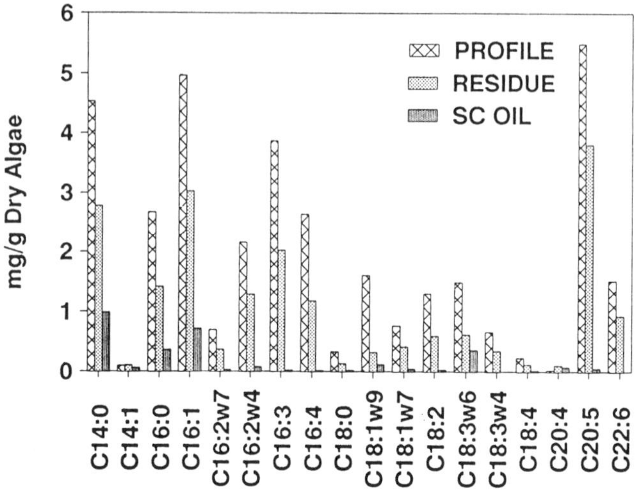

Figure 7. Fatty Acids in the original, and residual <u>Skeletonema costatum</u>, and those in the SC Extracted Oil. 17.2 MPa.

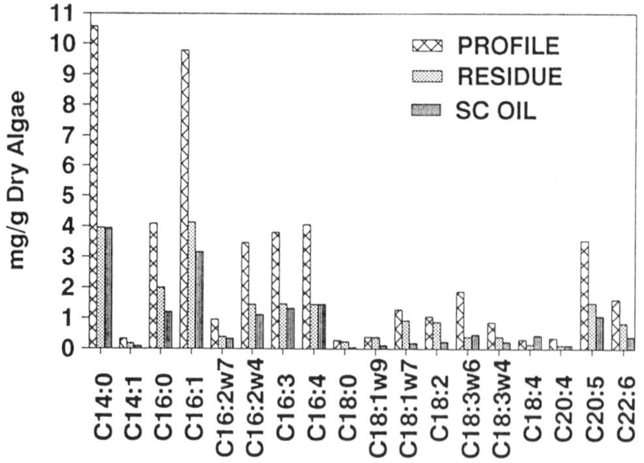

Figure 8. Fatty Acids in the Original, and Residual <u>Skeletonema costatum</u>, and those in the SC Extracted Oil. 24.0 MPa.

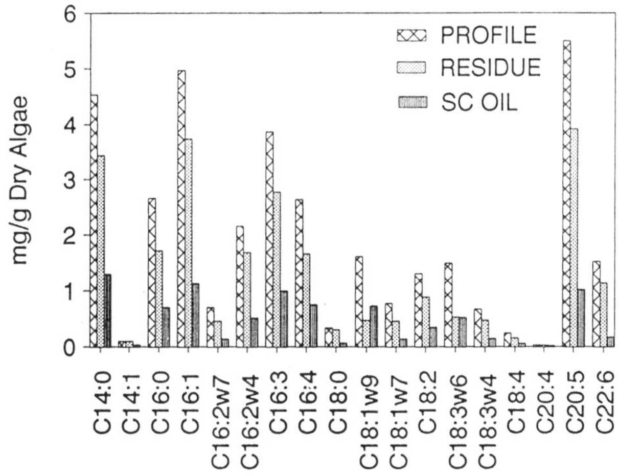

Figure 9. Fatty Acids in the Original, and Residual Skeletonema costatum, and those in the SC Extracted Oil. 31.0 MPa.

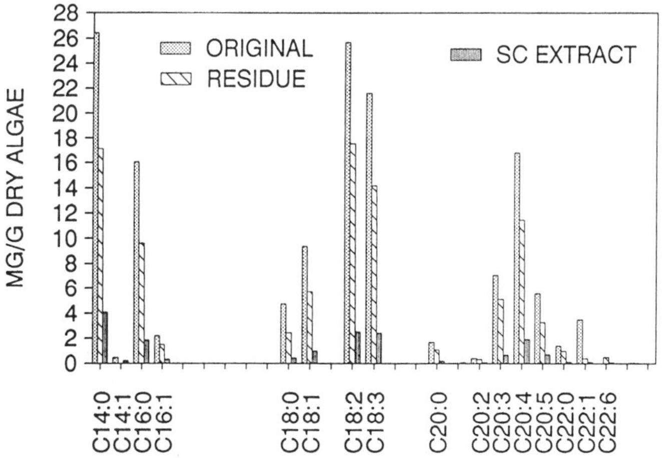

Figure 10. Fatty Acids in the Original, and Residual Ochromonas danica, and those in the SC Extracted Oil. 17.2 MPa.

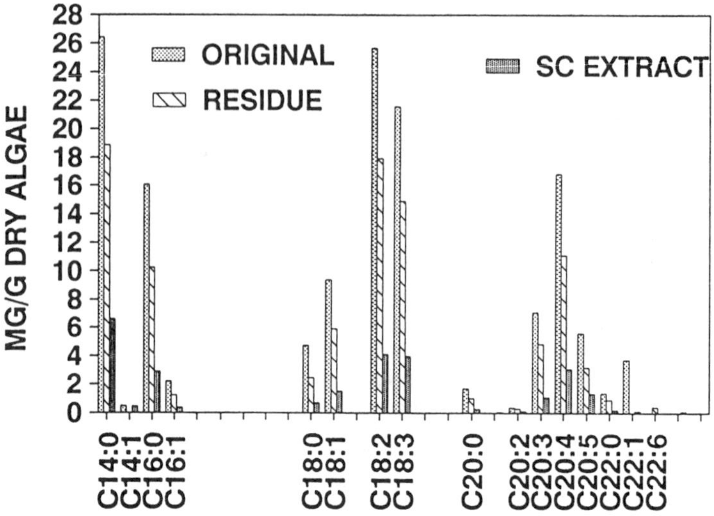

Figure 11. Fatty Acids in the Original, and Residual _Ochromonas danica_, and Those in the SC Extracted Oils. 24 MPa.

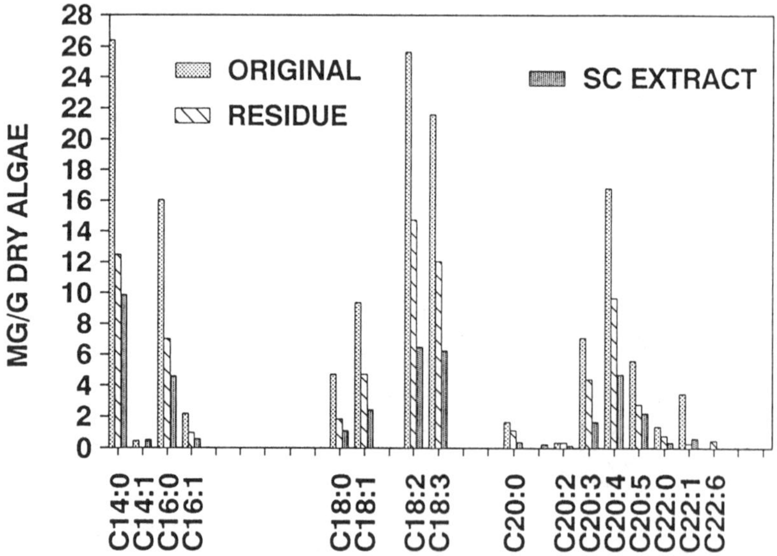

Figure 12. Fatty Acids in the Original, and Residual _Ochromonas danica_, and Those in the SC Extracted Oils. 31 MPa.

Effect of Phospholipase C Treatment. Since it was expected to extract more lipids by SC CO_2 if the level of phospholipids were lower, Skeletonema oil was treated with phospholipase C. The results indicated a small increase in the yield of all major fatty acids for the treated oil (13%) in the sample holder experiment. Figure 13 shows the amounts of the original, and the SC extracted fatty acids in significant amounts with and without enzyme treatment. The total amount of lipids not extracted decreased by about 50% as a result of enzymatic treatment.

Effect of Culture Age. Sample age of Skeletonema affected the fatty acid composition of batch 2 compared to batches 1, 3, and 4 (Table I). At harvest time batches 1, 3, and 4 were two weeks old whereas batch 2 was three weeks old. The total fatty acid content was 35.2, 48.7, 24.2 and 60.1 mg fatty acid/g of dried algae for batches one through four, respectively. Figure 14 indicates the difference in the fatty acid profiles between two and three week old cultures. A sharp increase was observed in the production of C14:0 (260%) and C16:1 (230%) for the three week old culture compared to the two week old culture. The relative proportion of C20:5w3 for the three week old culture was smaller compared to the two week culture (0.16 versus 0.07). Ackman et al.(6) reported a decrease in the degree of unsaturation in the lipids of Skeletonema costatum with maturity. Figure 14 indicates that the overall level of lipids were 75% higher in the three week old cultures compared to the two week old cultures. It is also evident from the same figure that the proportion of polyunsaturated lipids decreased with time, and that of the saturated lipids increased. Age comparison on the lipid levels were not studied in the case of Ochromonas.

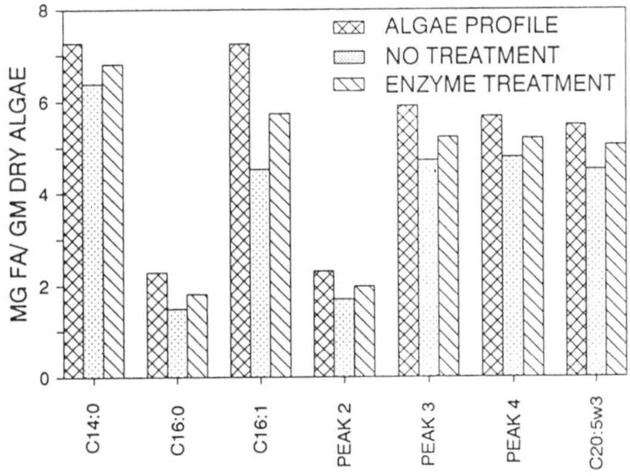

Figure 13. Amounts of Fatty Acids extracted with SC CO_2 from Skeletonema costatum With and Without Enzymatic Treatment.

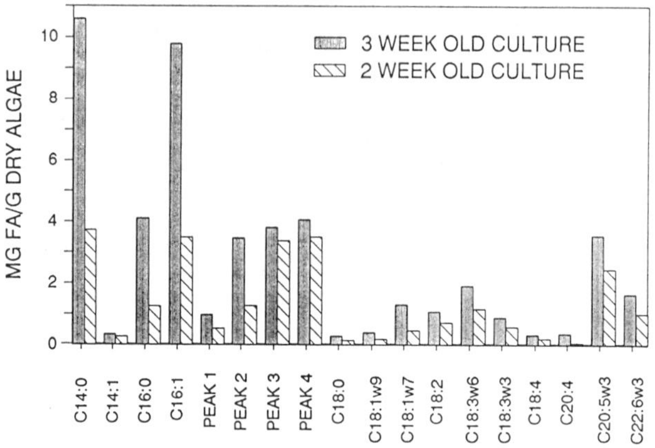

Figure 14. Comparison of Fatty Acid Profiles of Two and Three Week Old Skeletonema costatum Cultures.

Conclusions

The results of this study indicate that it is technically feasible to extract oils from the microalgae Skeletonema costatum and Ochromonas danica by SC CO_2 .

Under the conditions of this study there was no need to extract Skeletonema at 31.0 MPa since there was negligible increase in the yield of oils that can be obtained from SC CO_2 extraction at 24.0 MPa. However, an increase in pressure improved the yield of oils extracted from Ochromonas. Chlorophylls were not extracted by SC CO_2; the bleaching procedure in conventional solvent extraction can be eliminated. Treatment with phospholipase C improved the yield slightly.

Culture age affected the production of specific fatty acids. In a culture period of two weeks adequate amounts of EPA can be synthesized by Skeletonema costatum.

Acknowledgments

Partial financial assistance was provided by the Gulf and South Atlantic Fisheries Development Foundation, by the Organization of American States, and by the Wenthworth Scholarship Foundation. Their contributions are greatly appreciated.

Literature cited

1. Bang, H. O.; Dyerberg, J. Acta Medic. Scand. 1972, 192, 85-94.
2. Kromhout, D.; Bosschieter, E. E.; Coulander, C. D. L. N. Engl. J. Med. 1985, 312, 1205-1209.
3. Yamori, Y.; Nara, Y.; Iritani, N.; Workman, R. J.; Inagami, T. J. Nutr. Sc. Vitaminol. 1985, 31, 417-422.
4. Goodnight, S. H. Proc. Seafood and Health'85: issues, questions and answers. 1985 p 76-78.
5. Kinsella, J. E. Food Technol. 1986. 40, 89.
6. Ackman, R. G.; Jangaard, P. M.; Hoyle, R. J.; Brockerhoff, H. J. Fish. Res. Bd. Canada. 1964, 21, 747-756.
7. Lee, R. F.; Nevenzel, J. C.; Paffenhofer, G. A. Marine Biology. 1971, 9, 99-108.
8. Ben-Amotz, A., Tornabene, T. G.; Thomas, W. H. J. Phycol. 1984, 21, 72-81.
9. Solar Energy Research Institute. Fuels From Microalgae Technology Status, Potential, and Research Requirements. Golden, CO. 1986
10. Chuecas, L.; J. P. Riley. J. Mar. Biol. Ass. 1969, 49, 97-116.
11. Polak, J. T. M.Sc. Thesis. University of Florida. FL. 1988
12. Haines, T. H.; Aaronson,S.; Gellerman, J. L.; Schlenk, H. Nature 1962, 194, 1282-1283.
13. Williams, D. F. Chem. Eng. Sci. 1981, 36, 1769.
14. Friedrich, J. P.; List, R. G. J. Agric. Food. Chem. 1982, 30, 192.
15. Paulaitis, M. E.; Krukonis, V. J.; Kurnik, R. C.; Reid, R. C. Rev. Chem. Eng. 1983, 1, 179.
16. Stahl, E.; Quirin, K. W.; Blagrove, R. J. J. Agric. Food Chem. 1984, 32, 938.
17. McHugh, M. A., and Krukonis, V. J. Supercritical Fluid Extractions: Principles and Practice. Butterworths Publ., Stoneham, MA. 1986
18. Rizvi, S. S. H.; Benado, A. L.; Zollweg, J. A.; Daniels, J. A. Food Technol. 1986a, 40, 55.
19. Rizvi, S. S. H., Benado, A. L.; Zollweg, J. A.; Daniels, J. A. Food Technol. 1986b 40, 57.
20. A.O.A.C. Official Methods of Analysis. 13th ed. Association of Official Analytical Chemists, Washington, DC. 1980
21. Bligh, E. G.; Dyer, W. J. Can. J. Biochem. Physiol. 1959, 86, 3016-3021.
22. Kates, M. In Laboratory Techniques in Biochemistry and Molecular Biology. T. S. Work and E. Work (Ed), American Elsevier Publishing Co., Inc. New York. 1972. p. 351.
23. Hyatt, J. A. J. Organic Chem. 1984. 49, 5097.
24. Yamaguchi, K.; Murakami, M.; Nakano, H.; Konosu, S.; Kokura, T.; Yamamoto, H.; Kosaka, M.; Hata, K. J. Agric. Food Chem. 1986, 34, 904.
25. Schmitt, W. J. and R. C. Reid. Presented at the Annual Meeting of the AIChE, November 1986, Miami, FL.
26. Czubryt,J.J.; Meyers, M.N.; Giddings, J.C. J. Phys. Chem. 1970, 74, 4260-4265.

RECEIVED May 1, 1989

Chapter 29

Supercritical Extraction of Pollutants from Water and Soil

Robert K. Roop, Richard K. Hess, and Aydin Akgerman[1]

Department of Chemical Engineering, Texas A&M University, College Station, TX 77843

Phenol, a common priority pollutant, was extracted from two environmental matrices, soil and water, using near critical and supercritical carbon dioxide. The primary objective of this study was to determine the distribution of the contaminant between the soil or water and the supercritical phase, and the effect of soil moisture and co-solvents on the distribution coefficients. Static equilibrium extractions were performed on dry and wetted soil contaminated with 1 wt.% phenol and on water containing 6.8 wt.% phenol. Supercritical carbon dioxide (with and without entrainers) was chosen as the solvent for the study. An appropriate entrainer for dry soil extractions (methanol) differed from that found for aqueous extractions (benzene). However, soil moisture was found to have a significant impact on the effectiveness of entrainers for soil extractions of phenol. Entrainers appropriate for extracting wetted soil were found to be the same as those advantageous for aqueous extractions. Benzene was also extracted from dry and wetted soil to investigate the extractability of a hydrophobic compound.

There are presently over 26,000 uncontrolled waste sites in the United States as reported by the Environmental Protection Agency (EPA) (1). Of these sites, 951 are on the National Priority List (NPL) which makes these sites eligible for funds from the Superfund. One common characteristic of the NPL sites is that the site ground water is contaminated by the leaching of pollutants from contaminated soil. Therefore, the task of cleaning up such a site involves the detoxification of both solid (soil) and liquid (water) matrices contaminated with a common pollutant.

Current methods available for cleaning up contaminated soil include burial and incineration of the soil. Burial is inexpensive and simple in technology, but due to legislation and litigation, burial is no longer considered a permanent solution. Incineration is a proven technology but is energy intensive for soil cleanup and still requires the burial of the incinerated solid product (ashes). Also, gaseous products of incomplete combustion present an immediate environmental risk. Methods available for cleaning up water contaminated with over one percent organics include distillation, incineration, and liquid extraction. Distillation is energy intensive for dilute aqueous solutions and incineration

[1]Address correspondence to this author.

0097–6156/89/0406–0468$06.00/0

is both energy intensive and suffers from the previously mentioned problems. Liquid extraction is a viable technique but has had limited use due to concern over residual solvent present in the processed water. One alternative technique applicable to both contaminated water and contaminated soil is supercritical extraction (2). The ability of supercritical fluids (with and without entrainers) to solublize heavy molecular organics is well documented. However, limited data are available in the literature dealing with the extraction of organic contaminants from soil or water. Capriel, et al. (3) used supercritical methanol to extract bound pesticide residues from soil and plant samples. Brady, et al. (4) extracted PCBs, DDT, and toxaphene from soil with carbon dioxide. Dooley, et al. (5) extracted DDT from soil with carbon dioxide and mixtures of methanol/carbon dioxide and toluene/carbon dioxide. Kuk and Montagna (6) extracted ethanol from water using supercritical carbon dioxide. Panagiotopoulos and Reid (7,8) also used carbon dioxide to extract acetone from water. Schultz and Randal (9) extracted aroma constituents from fruit using liquid carbon dioxide.

The fundamental thermodynamic parameter of interest for the extraction of contaminants from soil or water is the distribution coefficient, defined as the ratio of the weight fraction of contaminant in the supercritical phase to the weight fraction of contaminant in the soil or water phase. As a model system for the extraction of toxic contaminants from soil and water, we have investigated the supercritical extraction of phenol (a priority pollutant) using carbon dioxide as the primary solvent. Supercritical carbon dioxide is the preferred solvent since it is environmentally acceptable, inexpensive, and readily available. We have successfully extracted phenol from water using supercritical carbon dioxide at various pressures and temperatures (Roop, et al., Journal of Supercritical Fluids, in review). We have also developed a thermodynamic model which allows us to predict *a priori* the qualitative effect of adding a given entrainer to the aqueous system (Roop and Akgerman, Ind. Eng. Chem. Res., in review). The purpose of this work was to extend the theory developed to the extraction of organics from soil containing some degree of moisture. It is anticipated that the entrainer selection for the extraction of dry soil versus wetted soil would be significantly different, and our studies on aqueous phase extractions may bring some explanation to the phenomena. Methanol and benzene were added to the supercritical carbon dioxide as entrainers for the extraction of both the contaminated dry soil and the contaminated soil containing 10 wt.% water and the results compared to those obtained in the aqueous study.

In addition, benzene was used as a model hydrophobic contaminant and extracted from soil (dry and wetted) using pure carbon dioxide and a carbon dioxide /methanol mixture. It was expected that the presence of water as soil moisture could influence the extraction extent and entrainer effect.

EXPERIMENTAL

Standard liquified phenol (10% water) was purchased from Fisher Scientific (lot # 745855). Radioactively labeled ^{14}C phenol was purchased from ICN Radiochemicals (43 μCi/mmole) and added to standard phenol to create a labeled phenol solution. Carbon dioxide was purchased from Conroe Welding Supplies with a purity of at least 99.8%. Methanol was purchased from American Scientific Company (lot # 3174KAHA) and standard benzene from Fisher Scientific Company (lot # 724430). Radioactively labeled ^{14}C benzene was purchased from Sigma Chemical Company (0.02 mCi/mmole). All materials were used as purchased. Soil used in the study was composed of 48.2 % sand, 15.2 % silt, and 36.6 % clay, with 1.4 wt.% organic carbon.

A schematic of the experimental apparatus used for the batch equilibrium soil extractions is shown in Figure 1. The high pressure contacting vessel (300 mL Autoclave) is initially charged with 100 grams dry soil containing 1 wt. % contaminant. A small amount of entrainer and/or water is added, if desired. The high pressure compression cylinder (1 L Welker), the syringe pump (Isco), and the contacting chamber are filled with carbon dioxide from supply, and then isolated through shut off valve V1. The carbon dioxide is compressed in the compression cylinder with nitrogen to fill the pump with liquid carbon dioxide at 10 MPa. The syringe pump is then isolated through shut off valve V2 and used to pump the carbon dioxide into the extraction vessel to the desired pressure. Once charged with supercritical carbon dioxide, the vessel contents are stirred for one hour with a specially designed impeller followed by a one hour period for the solids to settle. The fluid phase of the solid - fluid system is then sampled by opening valves V3 and allowing 2.6 gram sample of the carbon dioxide phase to be bled through an organic solvent trap. A dry test meter is used to determine the amount of carbon dioxide sampled. The contaminant from the sample remains in the traps, which are analyzed in a liquid scintillation counter (Beckman 3801). The syringe pump maintains a constant pressure in the system during sampling. Once the concentration of the contaminant in the supercritical phase is measured, the amount remaining on the soil is calculated by material balance and distribution coefficients determined.

A detailed description of the experimental apparatus and procedure used for the aqueous study are given elsewhere (Roop and Akgerman, Ind. Eng. Chem. Res., in review) Static equilibrium extractions were carried out in a high pressure equilibrium cell (300 mL Autoclave). After the vessel is initially charged with 150 mL of water containing 6.8 wt.% phenol and supercritical carbon dioxide (and a small amount of entrainer, if desired), the contents were mixed for one hour followed by a two hour period for phase separation. Samples from both the aqueous phase and the supercritical phase were taken for analysis and the distribution coefficient for phenol calculated.

RESULTS

The distribution coefficients of phenol obtained for the aqueous system as a function of pressure and temperature using pure supercritical carbon dioxide are shown in Figure 2. The values increase proportionately with pressure for each isotherm, but decrease overall at higher temperatures. At 298 K, reproduction of the distribution coefficients yielded an average standard deviation of 1.5 %. At 323 K the data are somewhat more scattered due to fluctuations in the temperature caused by control. Figure 3 shows the effect of various concentrations of benzene and methanol used as entrainers in supercritical carbon dioxide at 17.3 and 27.6 MPa at 298 K. Methanol, a commonly used entrainer in studies concerning solid organics, was found to have little effect on the distribution of phenol in the aqueous system. The presence of a small amount of benzene, however, did increase the distribution coefficient up to 50 % over those obtained with pure carbon dioxide.

Figure 4 shows the results of the extraction of water (figure 4a) and dry and wetted soil (figure 4b and 4c, respectively) using pure carbon dioxide, carbon dioxide with 2 mol % benzene, and carbon dioxide with 2 mol % methanol at 15 MPa and 298 K. For the extraction of dry contaminated soil, the distribution coefficient of phenol between the soil and pure carbon dioxide was 0.35, as shown in Figure 4b. The presence of benzene increased the distribution coefficient almost 100 %, while the presence of methanol resulted in the removal of essentially all of the phenol from the soil (within experimental accuracy, K-values > 7 correspond to almost complete removal). For the extraction of wetted soil (10 wt.% water) with pure carbon dioxide the distribution coefficient of phenol was again 0.35. The effect of the entrainers on this system, however,

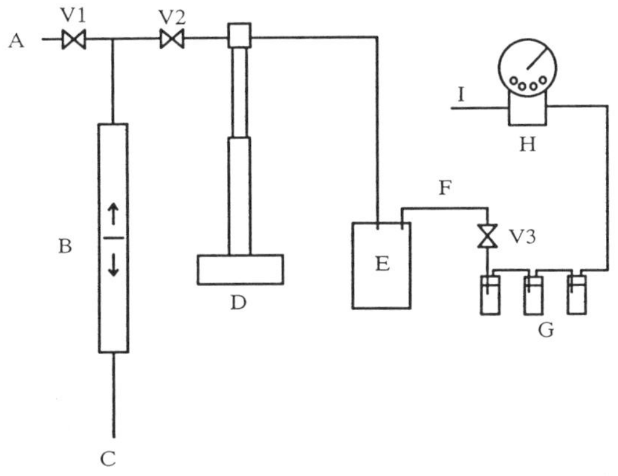

A. Carbon Dioxide Supply
B. High Pressure Compression Cylinder
C. Nitrogen Inlet
D. High Pressure Syringe Pump
E. Contacting Chamber
F. Narrow Bore Tubing
G. Organic Solvent Trap
H. Dry Test Meter
I. Exit Vent
V1–V3 Valves

Figure 1. Experimental apparatus for supercritical soil extractions.

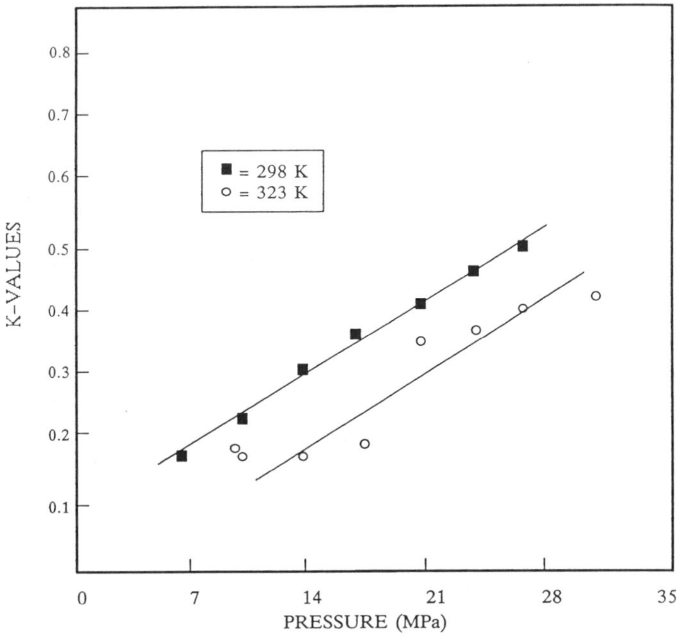

Figure 2. Distribution coefficients of phenol from aqueous extractions at 298 and 323 K up to 31 MPa.

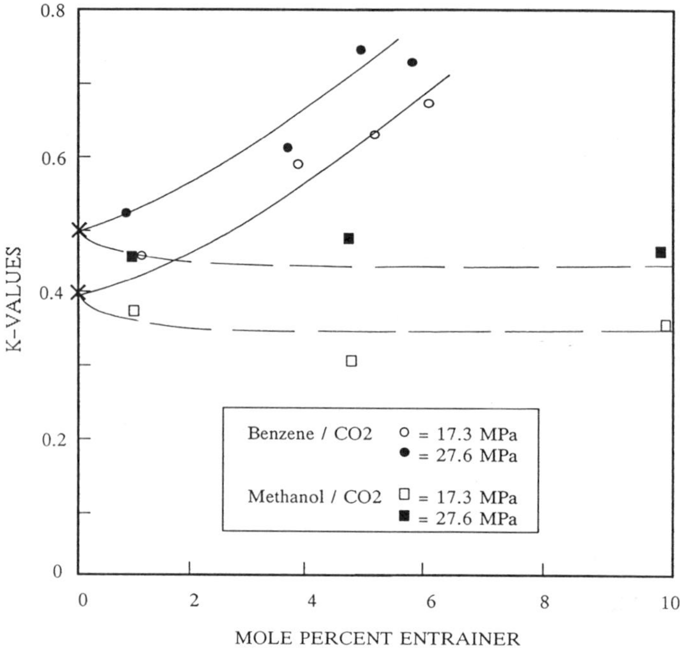

Figure 3. Entrainer effect for aqueous extractions at 298 K.

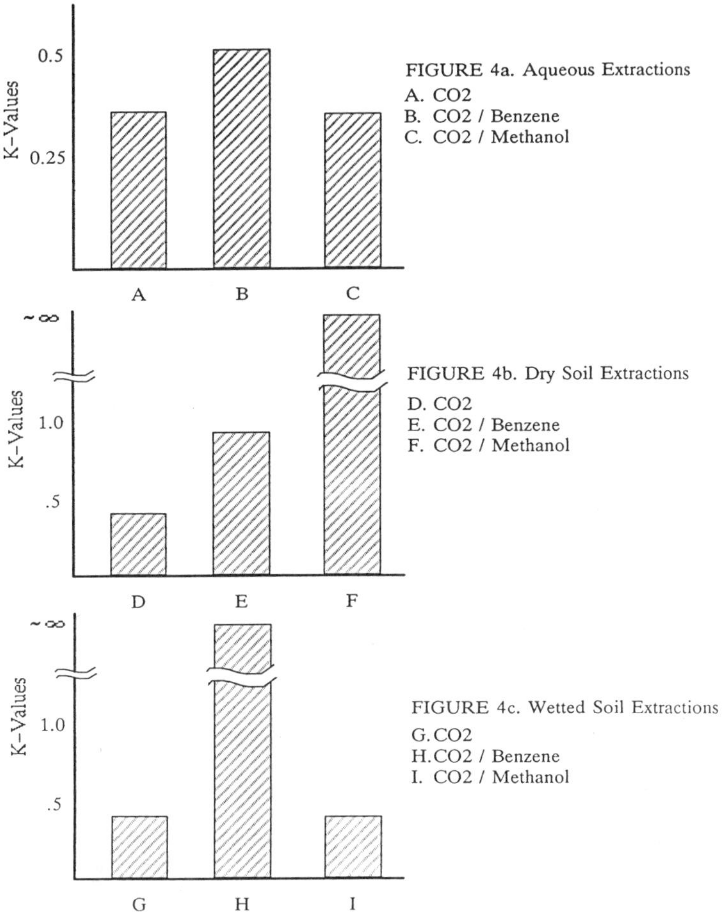

Figure 4. Extraction capability of supercritical carbon dioxide and carbon dioxide containing 2 mole percent benzene or methanol for the removal of phenol from water and soil (dry and wetted) at 15 MPa and 298 K.

was quite different. The presence of 2 mol % benzene resulted in the removal of almost all of the phenol from the soil, whereas the presence of methanol in this case had no effect.

Figure 5 shows the distribution coefficients for benzene extraction from soil at 15 MPa and 298 K. Pure carbon dioxide was able to remove almost all of the benzene from the dry soil. The presence of water in the soil is expected to increase the distribution of the benzene into the supercritical phase since benzene is hydrophobic, however, we were unable to distinguish this effect from carbon dioxide extraction of dry soil within our experimental accuracy. When methanol was added to the wetted system as an entrainer, it did not enhance the distribution of the benzene. In fact, the presence of a small amount of methanol served to decrease the amount of benzene in the supercritical phase.

CONCLUSIONS

Phenol was successfully extracted from water using pure supercritical carbon dioxide at pressures up to 31 MPa for two isotherms; 298 and 323 K. The distribution coefficient increased with increasing pressure, but decreased with increasing temperature. This is expected since increasing the temperature severely drops the carbon dioxide density and hence the solubility of the phenol in it. Increased volatility at the higher temperature is not sufficient to off-set the density effect, since phenol has a low vapor pressure. Benzene was found to be a suitable entrainer since its solubility in water is very small and it enhances the distribution of phenol into the supercritical phase. The presence of methanol was found to have no effect. Since methanol is polar and completely soluble in water, it favors the aqueous phase and therefore does not change the characteristics of the supercritical phase. Others have found that the distribution of short chain alcohols between water and supercritical carbon dioxide highly favors the aqueous phase (10).

Two soil systems were considered; contaminated dry and wetted soil. Pure supercritical carbon dioxide was able to remove phenol from both systems equally effectively. For the contaminated dry soil, both entrainers increased the distribution coefficient of phenol. However, methanol was by far the most effective. The presence of 2 mol % methanol (based on carbon dioxide) provided almost complete removal of the phenol from the soil. This is probably due to an increase in the polarity of the supercritical phase with the addition of methanol (11). The benzene/carbon dioxide mixture doubled the distribution coefficient, which can be attributed to structural similarities between the benzene and the phenol.

The presence of moisture in the soil was found to have no effect on the extraction of phenol using pure carbon dioxide. The effect of the presence of water may have been masked, since phenol distributes between water and supercritical carbon dioxide the same as it does between soil and supercritical carbon dioxide. However, the presence of water did have a dramatic impact on the effectiveness of the entrainers. The benzene/carbon dioxide mixture was able to remove essentially all of the phenol from the wetted soil. Since benzene is virtually insoluble in water, it highly favors the supercritical phase over the wetted soil phase. Hence the supercritical phase is able to attract almost all of the phenol, possibly due to chemical similarities between benzene and phenol. The methanol/carbon dioxide mixture, however, offered no enhancement over that of the pure carbon dioxide. As seen in the aqueous extractions, the methanol highly favors water and therefore is probably staying with the wetted soil. Therefore, the supercritical phase polarity is not increased as in the dry soil extractions and no solubility enhancement occurs.

In order to investigate a hydrophobic species, benzene was extracted from soil by supercritical carbon dioxide. Pure carbon dioxide was able to remove

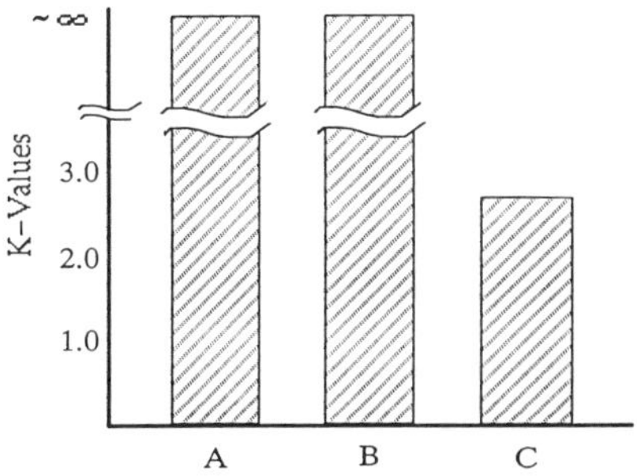

Figure 5. Extraction of benzene from soil (dry and wetted) with pure carbon dioxide and carbon dioxide / methanol mixture at 15 MPa and 298 K: A, dry soil with CO_2; B, wetted soil with CO_2; C, wetted soil with CO_2 / methanol mixture.

almost all of the benzene from both the dry and wetted soil. The presence of methanol as an entrainer decreased the distribution of the benzene into the supercritical phase for the extraction of wetted soil.

In summary, phenol (a model toxic pollutant) was successfully extracted from both water and soil (dry and wetted) using supercritical carbon dioxide. It was found that entrainers greatly enhance the distribution coefficient of phenol for each system. However, the choice of a good entrainer is not independent of the contaminated matrix or, in the case of soil, its moisture content. Benzene was successfully extracted from both dry and wetted soil using pure carbon dioxide. Entrainers such as methanol could serve to decrease the distribution of such hydrophobic systems, indicating that further investigation in this area is warranted.

LITERATURE CITED

1. Environmental Protection Agency, "National Priorities List, Proposed Update No. 6," HW-10.1, 1987.
2. Groves, F. R.; Brady, B.; Knopf, F. C. *CRC Crit. Rev. Environ. Control*, 1985, 15, 237.
3. Capriel, P.; Haisch, A.; Khan, S. U. *J. Agric. Food Chem.*, 1986, 34, 70.
4. Brady, B. O.; Kao, C.; Dooley, F. C.; Knopf, F. C. *Ind. Eng. Chem. Res.*, 1987, 26, 261.
5. Dooley, K. M.; Kao, C.; Gambrell, R. P.; Knopf, F. C. *Ind. Eng. Chem. Res.*, 1987, 26, 2058.
6. Kuk, M. S.; Montagna, J. C. in *Chemical Engineering at Supercritical Fluid Conditions*; Paulaitis, M. E., et al., Eds., Ann Arbor Science, Ann Arbor, Mich., 1983, p 101.
7. Panagiotopoulos, A. Z.; Reid, R. C. *ACS Division of Fuel Chem. Preprints*, vol 30, No 3, 1985, p 46.
8. Panagiotopoulos, A. Z.; Reid, R. C. *ACS Symposium Series*, No. 329, 1987, p 115.
9. Schultz, W. G.; Randall, J. M. *Food Tech.*, 1970, 24, 94.
10. McHugh, M. A.; Krukonis, V. J. *Supercritical Fluid Processing: Principles and Practice*, Butterworths, Stoneham, MA, 1986, p 139.
11. Dobbs, J. M.; Wong, J. M.; Lahiere, R. J.; Johnston, K. P. *Ind. Eng. Chem. Res.*, 1987, 26, 56.

RECEIVED May 2, 1989

DESIGN OF COMMERCIAL PLANTS

Chapter 30

Current State of Extraction of Natural Materials with Supercritical Fluids and Developmental Trends

Rudolf Eggers[1] and Uwe Sievers[2]

[1]Technische Universität Hamburg-Harburg, Arbeitsbereich Verfahrenstechnik II, Postfach 90 14 03, D–2100 Hamburg 90, Federal Republic of Germany
[2]Krupp Maschinentechnik GmbH, Werk Harburg, Postfach 90 08 80, D–2100 Hamburg 90, Federal Republic of Germany

Supercritical fluid extraction in large-scale industrial plants has concentrated upon natural materials, e.g. coffee, tea, hops and tobacco, and is being extended towards specialities which would require smaller production plants. Process engineering aims at improved scale-up from experimental to production scale and at improved economic viability, e.g. by saving energy and by continous transport of solids through pressure vessels. Equipment design investigations are concerned with closure systems and the relevance of pressure release and solvent recovery for the design of pressure vessels.

The present state and the developmental trends concerning the extraction of natural materials with supercritical fluids may be described under the following aspects:

The large-scale plants which have so far been realized for the decaffeination of tea and coffee and for the extraction of hops and spices are now established. These are in general one-stage plants in which the extract is separated by adsorption on activated carbon or by pressure reduction. The tendency now is to replace adsorption by absorption in washing towers. As an example, the annual production of 20,000 t/a coffee with activated carbon caffeine separation involves the loss of more than 200 t of caffeine, since this potential by-product is destroyed during the thermal regeneration of the adsorbent. The questions of energy-saving and of the economic recovery of CO_2 are also of interest for large-scale plants which already exist or which are still to be designed.

Investigations on laboratory and pilot-plant scale have already been performed with many natural products.

0097–6156/89/0406–0478$06.25/0

Here a distinction must be made between those substances which are already extracted commercially in profitable extraction processes (Table I (Quirin, K.-W.; Gerard, D., FLAVEX Naturextrakt GmbH, Rehlingen, personal communication, 1988)) and those substances which are at present still obtained by conventional methods, e.g. commodities such as seed oils.

Table I. Natural Products of Commercial SCF Extraction

Active components in pharmaceuticals and cosmetics	
Ginger	Calmus
Camomile	Carots
Marigold	Rosemary
Thyme	Salvia
Spices and aromas for food	
Basil	Cardamon
Coriander	Ginger
Lovage root	Marjoram
Vanilla	Myristica
Paprika	Pepper
Odoriferous substances for perfumes	
Angelica root	Ginger
Peach and orange leafs	Parsley seed
Vanilla	Vetiver
Oil of spices	
Aromas for drink	
Angelica root	Ginger
Calamus	Juniper
Further applications	
Separation of pesticides	
Refinement of raw extract material	
Separation of liquids	
Extraction of cholesterol	

In this area, research work is in progress on the development of continuous supercritical fluid (SCF) extraction processes for solid feeds. Finally, for numerous natural substances the construction of smaller production plants is expected in the near future. The de-fatting and cholesterol reduction of egg yolk powder is shown as an example in Figure 1 (Quirin, K.-W.; Gerard, D.; FLAVEX Naturextrakte GmbH, Rehlingen, personal communication, 1988).

Few investigations are available on liquid natural materials. One example is the extraction of caffeine from aqueous phases. Phase transitions as well as the hydraulic behaviour of packed and unpacked columns have been investigated (1-4).

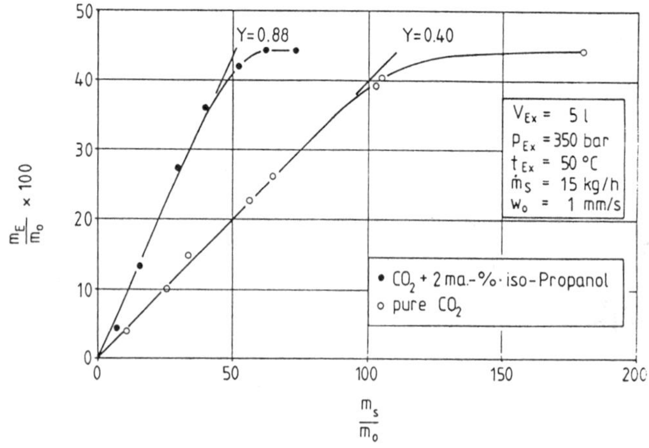

Figure 1. Extraction yield $E = m_E/m_O$ as function of specific solvent mass (m_S/m_O) for SCF extraction of spray-dried egg yolk.

When designing plants for the extraction of natural materials with supercritical fluids, the scale-up factor from pilot plants to large-scale plants is often very large. One trend is therefore towards the development of reliable scale-up rules.

Of increasing interest are structural changes in or coatings of natural materials. Thus the volume increase of tobacco resulting from controlled pressure release is already used industrially. The discontinuous extraction of natural materials requires the plant to be depressurised after completion of the process, whereby the non-stationary behaviour of pressure and temperature in the extraction vessel is of great importance for the design of the pressure vessel.

Finally, in addition to the proven quick-release clamp closure, the finger-pin closure and the segment-ring closure have proved themselves with an opening and closure cycle of less than 5 minutes in large-scale plants.

This contribution will report upon:
- Scale-up possibilities
- Optimization of the energy requirement
- CO_2 pressure release and CO_2 recovery
- Investigations on the continuous extraction of solids.

Scale-up possibilities

Particularly, the extraction of natural materials with compressed gases involves the problem of applying results obtained at laboratory or pilot scale to production-scale conditions. In accordance with the required production rate, increases in scale between 1:10 and 1:1000 may result.

During the extraction phase the whole mass transfer may be regarded as a quasi-stationary process. Scale-up rules therefore take account of external mass transfer from the solid surface to the supercritical fluid only. If pilot and production plants are required to display the same mass transfer properties, then

$$Sh = f\,(Re,\ Sc) \qquad (1)$$

demands equal flow velocities w_o in the extractor under the following conditions:
- identical material to be extracted, and the same pretreatment
- identical extraction parameters in terms of pressure, temperature and type of solvent.

Under these conditions, the ratio of the specific solvent mass flow ($\dot{m}_s/m_o$) between the production plant (index 2) and pilot plant (index 1) is fixed.

$$\left(\frac{\dot{m}_s}{m_o}\right)_2 = \left(\frac{\dot{m}_s}{m_o}\right)_1 \cdot \frac{H_1}{H_2} \qquad (2)$$

with the mass flow of solvent

$$\dot{m}_s = A \cdot w_o \cdot \rho_s \qquad (2a)$$

and the feed mass of solid material

$$m_o = A \cdot H \cdot \rho_o \qquad (2b)$$

If equal extraction yields

$$E = \frac{m_E}{m_o} \qquad (3)$$

in pilot and production plant are defined to be the scale-up target then different extraction times

$$\tau_{Ex} = E \cdot \left(\frac{m_s}{m_E}\right) \cdot \left(\frac{m_o}{\dot{m}_s}\right) \qquad (4)$$

result from the fulfilment of Equation 2 in pilot and production plants, since the extract loading (m_E/m_S) cannot increase linearly with length H of the solid material (= flow length),

$$\left(\frac{m_E}{m_S}\right)_2 \neq \left(\frac{m_E}{m_S}\right)_1 \cdot \frac{H_2}{H_1} \qquad (5)$$

because the loading capacity of the solvent with extract is limited by the phase equilibrium.

As an example, measurements are tabulated in Table II which were obtained from a pilot plant (extraction volume 200 l) and a production plant (extraction volume 2000 l) during the extraction of oenothera (evening primrose) seed oil (Gehrig, M., Hopfenextraktion HVG Barth, Raiser & Co., Wolnzach, personal communication, 1988) at identical flow velocities w_O. An extract yield of 20 % was achieved after 1.6 h in the pilot plant, and after 4.6 h in the production plant, cf. Figure 2. When the flow length was quadrupled, the extract loading of the solvent increased from 0.63 % to 0.83 % (ratio of ordinate m_E/m_O and abscissa m_S/m_O in Figure 3). The theoretical solubility (phase equilibrium) is c_O = 0.9 % for $H \rightarrow \infty$ (4). For the purpose of experimental scale-up, the relationship

$$\left(\frac{m_E}{m_S}\right) = c_o \cdot \left(1 - \exp\left(-k \cdot H\right)\right) \qquad (5a)$$

with k > 0 must be determined in a pilot plant for different charge heights H.

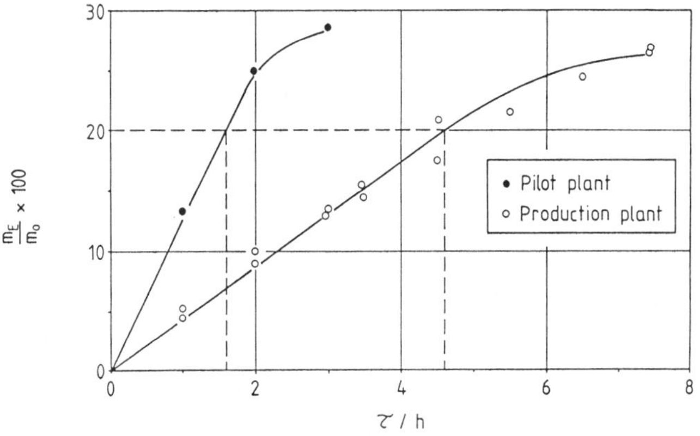

Figure 2. Extraction yield $E = m_E/m_o$ as function of time τ for SCF extraction of oenothera seed with CO_2.

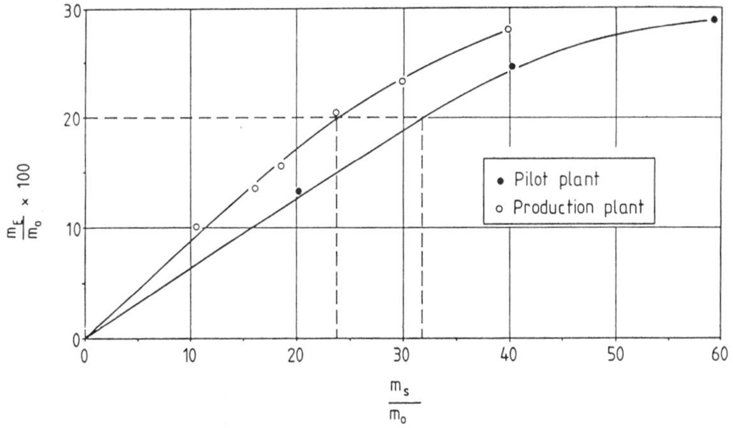

Figure 3. Extraction yield $E = m_E/m_o$ as function of specific solvent mass (m_s/m_o) for SCF extraction of oenothera seed with CO_2.

Table II. Scaling up the SCF Extraction of Oenothera Seed Oil

		Pilot	Production
Extractor volume	V	200 l	2000 l
Inner diameter	d	400 mm	800 mm
Inner height	$\bar{H}$	1.59 m	4.0 m
Solvent flow velocity	w_o	2.5 mm/s	2.5 mm/s
Mass flow CO_2	m_s	1000 kg/h	4000 kg/h
Raw material mass	m_o	50 kg	800 kg
Bulk density	o	450 kg/m^3	450 kg/m^3
Filling height	H	0.9 m	3.6 m
Pressure	p	300 bar	300 bar
Temperature	T	40 °C	40 °C

$$\left(\frac{\dot{m}_s}{m_o}\right)_1 \cdot \frac{H_1}{H_2} = \left(\frac{1000\,kg/h}{50\,kg}\right) \cdot \frac{0.9\,m}{3.6\,m} = \left(\frac{\dot{m}_s}{m_o}\right)_2 = \left(\frac{4000\,kg/h}{800\,kg}\right)$$

$$(W_o)_1 \quad = \quad (W_o)_2$$

In scale-up calculations, Equation 4 requires knowledge of the product of mean extract loading (m_E/m_S) and specific solvent mass flow ($\dot{m}_S/m_O$) and their mutual dependence. When the extraction efficiency E/τ_{Ex} is plotted qualitatively against the specific solvent mass flow ($\dot{m}_S/m_O$), the decrease in extract loading with increasing $\dot{m}_S/m_O$ must yield an optimum specific solvent mass flow, cf. Figure 4 above (5). If a scale-up were to be made on the basis of such an optimised extraction curve, Equation 2 with equal solvent mass flow

$$\left(\frac{\dot{m}_s}{m_o}\right)_2 = \left(\frac{\dot{m}_s}{m_o}\right)_1 \qquad (6)$$

would require omission of the condition $(w_O)_2$ = $(w_O)_1$ and gives

$$(w_o)_2 = (w_o)_1 \cdot \frac{H_2}{H_1} \qquad (7)$$

for equal residence times $\tau_{o2} = \tau_{o1}$ of the solvent in the solid material. Equal residence time at higher velocity w_O must in view of Equation 1, however, lead to overall higher mass transfer for scale-up, resulting in the dashed curves in Figure 4 for the production plant.

The following conclusions can be drawn for the solution of scale-up problems which arise in the extraction of natural materials with supercritical fluids:

- Scaling at constant specific solvent mass flow is not useful, since the solvent flow velocities increase linearly with the height, whereby the pressure loss across the filling increases so greatly that there is a risk of compacting the starting material.

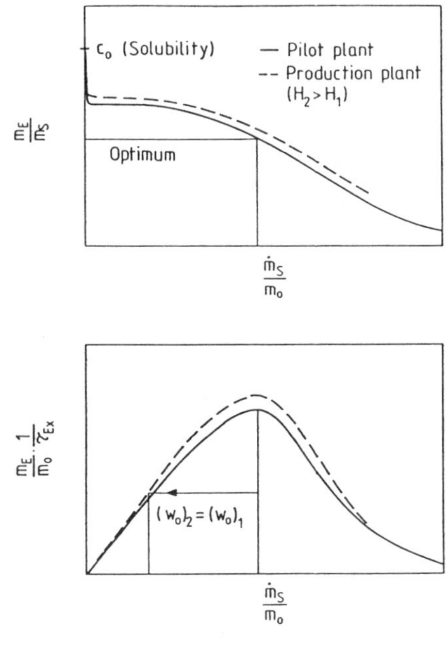

Figure 4. Optimization and scale-up for SCF extraction of solid material. Extract loading (m_E/m_S) (above) and extraction efficiency $E/\tau_{Ex} = (m_E/m_o)/\tau_{Ex}$ (below) as function of specific solvent mass flow $(\dot{m}_S/m_o)$.

- The relationship between extract loading and flow length should be determined during the experimental phase with the flow velocity as a parameter.
- In accordance with Figure 4, the extraction efficiency in a production plant is poorer than in the pilot plant if scale-up is performed with the flow velocity w_o corresponding to the optimum specific solvent mass flow rate for the pilot plant.
- If scale-up is made in accordance with the relationship between extraction efficiency and specific solvent mass flow rate gained from pilot operation, the large-scale plant will give better results than expected (dashed curves in Figure 4).

Optimization of the energy requirement

With CO_2 as a solvent, an investigation by Eggers (6) indicates that high-pressure extraction with supercritical separation is energetically superior over subcritical separation in a wide range of extraction pressures and temperatures. In the CO_2 solvent extraction cycle, compressor pressurization is followed by heat-exchanger cooling to extraction temperature and then pressure reduction in a throttle valve between extractor and separator. Energetic calculations are based upon thermodynamic data from (7). The specific mechanical work w_{t12} to be performed upon the compressor is plotted in Figure 5 in terms of the separator pressure p_{Se} for extraction pressures of 250 bar and 300 bar and an extraction temperature of 80 °C. The larger the difference between separation pressure p_{Se} and extraction pressure p_{Ex} to be overcome by the compressor, the greater is the energy expenditure required. In a tea-decaffeination plant with a CO_2 mass flow of 200,000 kg/h operating at an extraction pressure of 300 bar, the power consumption of the compressor is 2,500 kW at a separation pressure of 80 bar, and 183 kW at a separation pressure of 280 bar. These numbers and Figure 5 make clear the reason for aiming at separation with a small pressure drop for processes involving large CO_2 mass flow. Tea decaffeination therefore employs caffeine separation by adsorption on activated carbon or absorption in water at pressures little lower than the extraction pressure.

In Germany, production plants for extracts, e.g. hop extract, are operated with subcritical separation (Figure 6). In these plants, heat recovery measures may significantly reduce the energy consumption. Energy calculations were performed for the example of a SCF extraction plant for processing 15 t hops/24 h with extraction at 300 bar and 60 °C, separation at 60 bar and 40 °C, a specific solvent mass flow of 10 kg CO_2 / (h kg hops) and an extraction time of 4 h. A comparison of different process variants for

SCF extraction is given in Figure 7. It shows theenergy requirements for mechanical work w_t, for heating q_h and for cooling q_c, for 1 kg hops in 3 cases. Case A gives the values for SCF extraction without heat recovery, case B the values for SCF extraction with heat recovery via a heat-pump between condenser WT3 and evaporator WT2, and case C applies to SCF extraction with heat recovery by means of a heat-pump between condenser WT3, heat exchanger WT1 and evaporator WT2. Heat recovery reduces the quantity of heating and cooling at the cost of an increase in mechanical work. Case C involves less heating and cooling than does conventional solid/liquid extraction, but the amount of mechanical work is significantly greater.

Figure 8 shows the energy costs of the different variants of SCF extraction. Energy cost were estimated on the basis of: 0.10 DM/kWh for mechanical work, 27 DM/MJ for heating and 50 DM/MJ for cooling. The inclusion of a heat-pump between condenser WT3 and evaporator WT2 (Case B) approximately halves the energy costs which however remain somewhat higher than for conventional solid/ liquid extraction. When heat-exchanger WT1 is also included in the heat-recovery process, the energy costs are less than in industrial processes for conventional solid/liquid extraction.

CO_2 pressure release and CO_2 recovery

Figure 9 shows a load-change cycle which is typical for discontinuous SCF extraction. When the pressure release phase following completion of the extraction is considered, the question arises as to the true temperature course. It is wellknown that the pressure-dependent equilibrium temperature of CO_2 falls to -79 °C under atmospheric conditions. This relationship leads to short-term thermal stresses within the inner surfaces of the pressure vessel, particularly in the lower part where dry ice may form. There is an additional risk to the process that the charge may freeze within the pressure vessel. When designing equipment for the extraction of natural substances, definition of the non-stationary courses of pressure and temperature during pressure release is therefore of especial importance in the choice of materials and for the geometry of the pressure vessel.

Within the pressure release times common in practice, the actual temperature conditions inside the extractor and within the walls of the pressure vessel do not follow the adiabatic course, but depend upon the filling and upon the size of the pressure vessel as well as upon the pressure release time itself. Knowledge of these temperatures allows better design of the pressure vessel (8), simplifies the commissio-

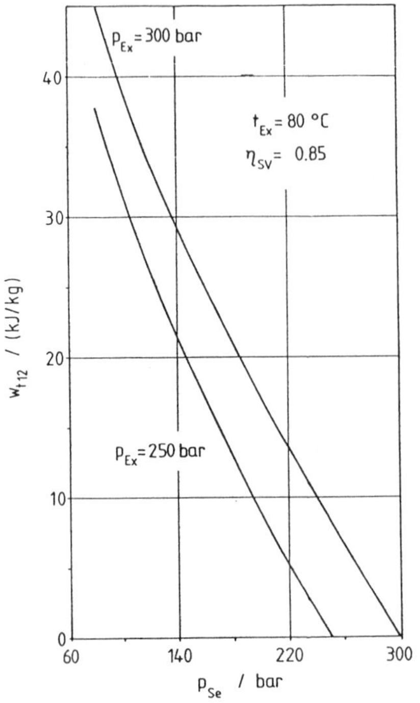

Figure 5. Specific mechanical work w_t for SC-CO_2 extraction with supercritical separation.

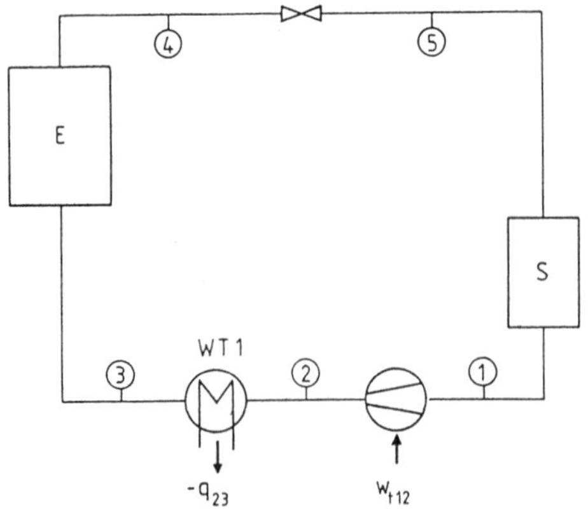

Figure 6. SCF extraction process with subcritical separation.

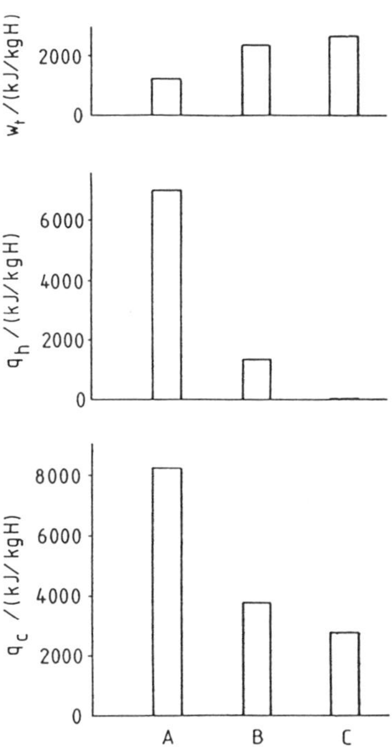

Figure 7. Energy consumption for various SCF extraction processes for hops.

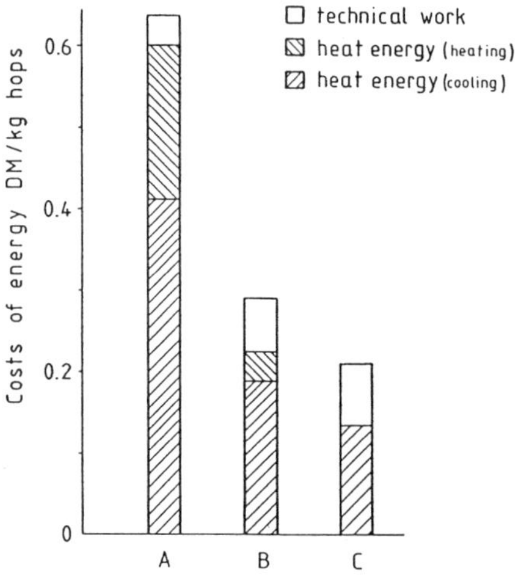

Figure 8. Cost of energy for various SCF extraction processes for hops.

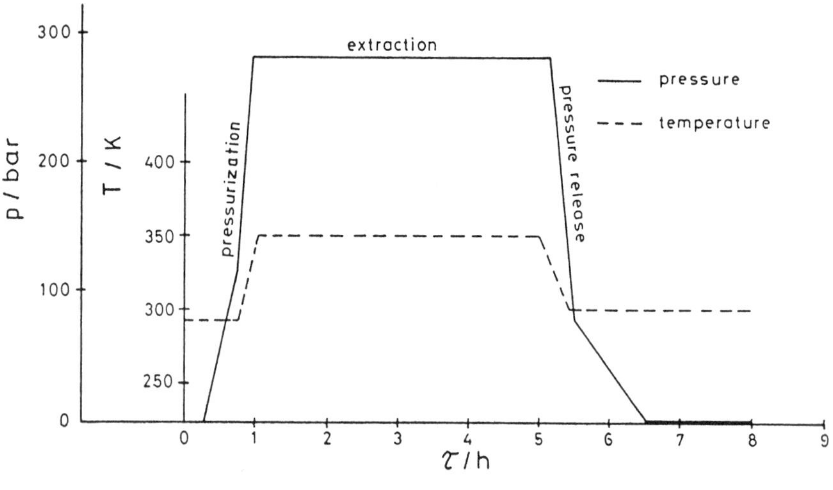

Figure 9. Cycle of pressure p and temperature T in an extractor vessel.

ning tests and allows the operator to choose the shortest possible safe time interval between the extraction phase and the emptying of the pressure vessel.

A theoretical estimate of the temperature course at the inner wall surfaces of a 200 l extraction vessel (solution via Laplace transformation) is shown in Figure 10. The recognisable deviations from the experimentally-measured course are attributed both to the decrease in heat transfer with time and to the lack of consideration of the enthalpy of solidification which is released during the formation of dry ice. It has also proved necessary to record the local temperatures which differ greatly during pressure release.

Figures 11 and 12 likewise show results of pressure release experiments under near-operating conditions on the above-mentioned 200 l pressure vessel with and without a solid charge. The mitigating effect of the internal energy of the solid charge on the position of the minimum temperature of the inner walls is clearly recognizable.

The following implication for natural substance extraction may be stated: if the pressure release times are sufficiently long, and if the temperatures of pressure vessel and charge are adjusted beforehand, the lower part of the pressure vessel need not be constructed of low-temperature steel.

In the case of large batch-operated production plants, extensive CO_2 recovery from the extractor during pressure reduction is necessary for economic reasons. Recovery of CO_2 from the extractor after completion of the extraction phase is made in three steps:

- Reduction of pressure in the extractor by blow-off via a reducing valve and a condenser into a CO_2 working tank, beginning at the extraction pressure and ending at the equilibrium pressure of about 60 bar.
- Reduction of the pressure in the extractor by evacuation of gaseous CO_2 via a recovery compressor with decrease of suction pressure and compression to the pressure in the working tank, the gaseous CO_2 being condensed after compression.
- After achievement of the lower recovery pressure between 2 and 8 bar, reduction of pressure in the extractor to the equilibrium pressure, by blowing off CO_2 into the surroundings.

Apart from the state of the CO_2 at the end of the extraction phase and the change of state of the CO_2 which exchanges heat with the solid in the extractor and with the extractor walls, the design of the recovery compressor determines the time duration and energy demand of the second recovery step. Calculations were performed for an extractor in which a volume of 10 m^3 is available to the CO_2, with an

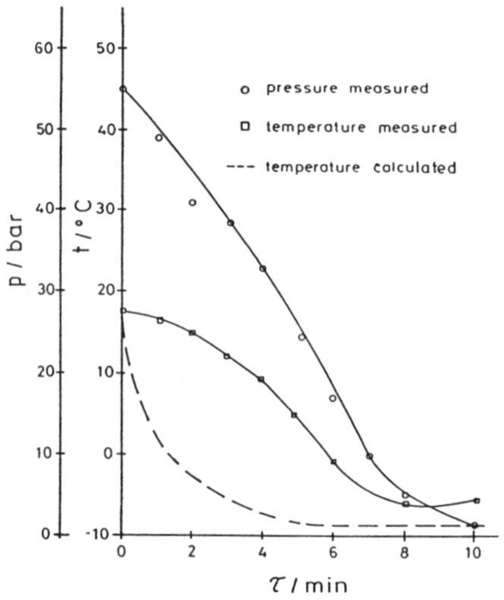

Figure 10. Pressure p and inner wall temperature t of a 200 l pressure vessel during CO_2 pressure release.

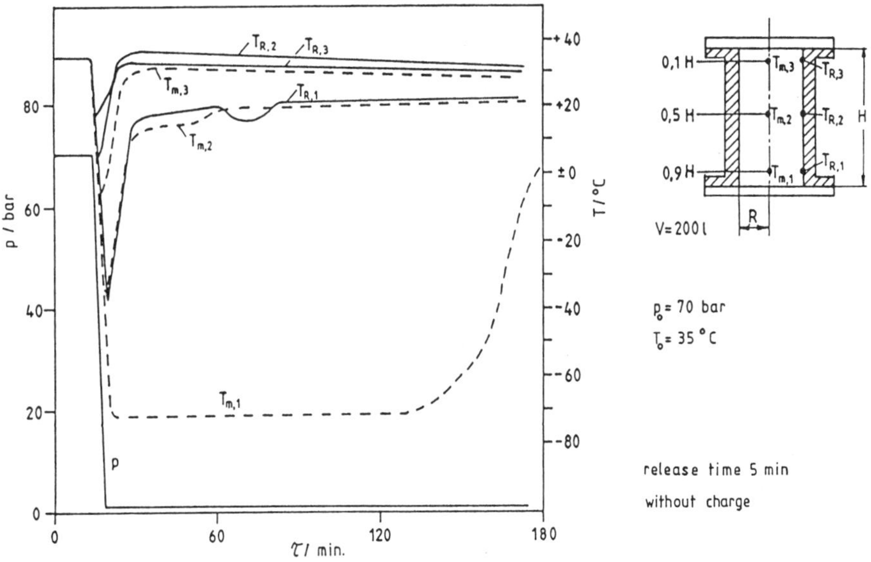

Figure 11. Pressure p and temperature T in a 200 l pressure vessel without charge during CO_2 pressure release.

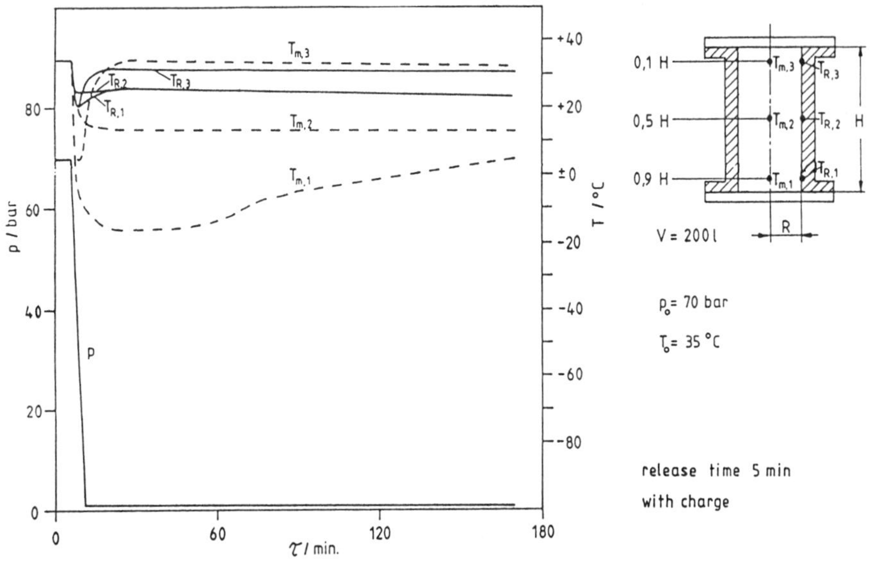

Figure 12. Pressure p and temperature T in a 200 l pressure vessel with charge during CO_2 pressure release.

initial pressure in the extractor of 60 bar for two limiting cases:

A. Change of state of CO_2 along the dew point line
B. Change of state of CO_2 in the superheated gas region starting at 60 bar, 26 °C, and ending at 2 bar, 0 °C.

Case A sets a lower limit, case B an upper limit to the pressure reduction time and to the energy expenditure required during the second recovery step. The CO_2 temperature will in reality lie above the saturation temperature at the given pressure, cf. Figure 12.

The compressor employed is a dry-running piston compressor with intermediate cooling. The delivered volume flow is to decrease linearly from 250 m^3/h at 60 bar suction pressure down to 200 m^3/h at 2 bar suction pressure, an isentropic compressor efficiency of η_{sv} = 0.8 is assumed, and the opposing pressure in the working tank is taken to be 65 bar.

The CO_2 mass flow m_{CO2} evacuated from the extractor by the recovery compressor likewise decreases sharply with decreasing suction pressure. Figure 13 shows the dependence of m_{CO2} upon time. The area below the curve indicates the mass of CO_2 recovered from the extractor up to a particular time. In case A, of the CO_2 mass (assumed to exist entirely as saturated vapor) present at the start of the second recovery step, 91.8 % has been recovered after 6.7 min when the pressure has fallen to 6.5 bar, and a further 6.5 % has been recovered after 10 min when the pressure has fallen to 2 bar. The lower the suction pressure p, the greater is the specific mechanical work w_t expended in the recovery compressor and the greater is the specific heat q expended for intermediate cooling in the recovery compressor and for complete condensation of gasous CO_2 if CO_2 is passed as a boiling liquid into the working tank, cf. Figure 14. The specific mechanical work and the specific heat are higher for case B than for case A.

Investigations on the continuous extraction of solids

In order to extend the range of applications of SCF extraction of natural substances, there are developmental trends towards the construction of systems for the continuous transport of solid materials through pressure vessels. According to the natural substance concerned, a decision must be made between:

- Materials whose geometry must not be destroyed during the process (e.g. coffee, tea, tobacco).
- Materials whose geometry may be destroyed, or which are to be pre-treated (e.g. oilseed, hops).
- Materials which may be suspended in a transport fluid.

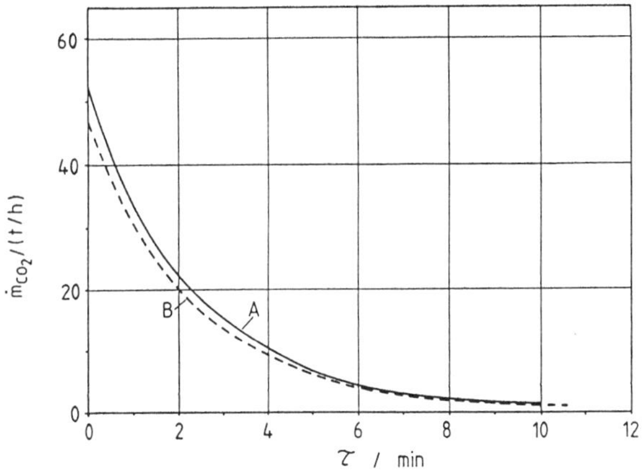

Figure 13. CO_2 mass flow $\dot{m}_{CO2}$ from the extractor as function of time τ during CO_2 recovery by compressor.

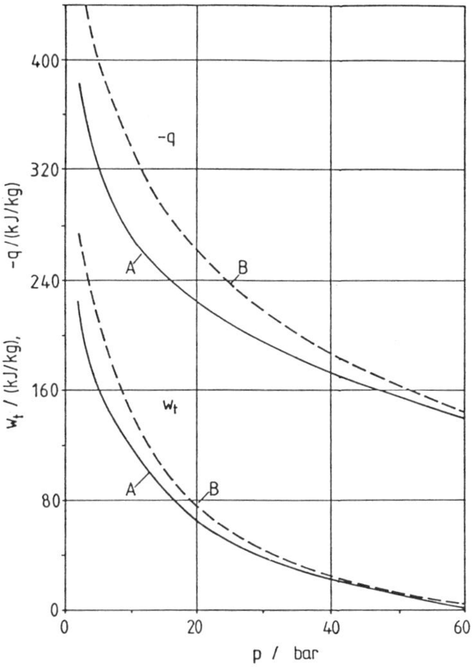

Figure 14. Specific mechanical work w_t and specific heat q for CO_2 recovery by compressor.

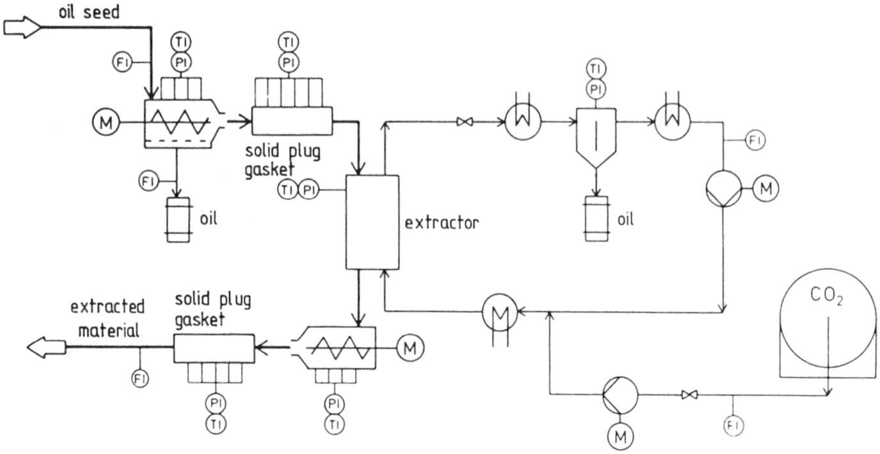

Figure 15. Flow sheet of a pilot plant for continuous SCF extraction.

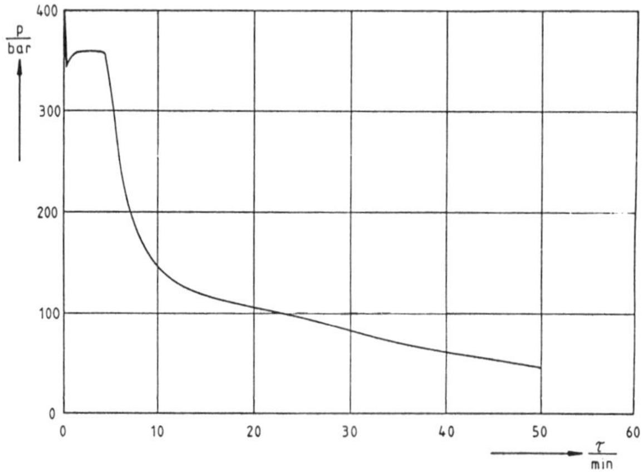

Figure 16. Pressure difference p at the gas seal section as function of time τ .

A continuous system with screw presses for the extraction of oilseed is currently being built, cf. Figure 15. Here, economic pre-deoiling is possible within the feed screw press itself (a so-called cagescrew press) and at the same time the pressure can be built up so as to form a gastight plug. Cage screw press, gas seal section and extraction vessel are arranged serially. The plant is designed for a turnover of 2 to 25 kg/h. The plug of material for extraction which is formed within the press is heated to 60 - 90 °C by the work of compression and by wall friction. Under these conditions the plug is fully plastically deformable whilst being completely gas-tight. The permeation of CO_2 through the seal formed by such a plug is investigated in a pressure-tube in terms of various parameters (9). Adequate gastightness has so far been demonstrated for pressure differences up to 400 bar. For example, Figure 16 shows a sealing time of 5 min. This time is much longer than the residence time whilst moving the material through the gas seal section.

Legend of Symbols

A	:	flow cross-sectional area
c_o	:	mass fraction of extract in solvent at phase equilibrium
E	:	extraction yield
H	:	filling height for solid material, flow length
m_E	:	mass of extract
m_s	:	mass of solvent
$\dot{m}_s$	:	mass flow of solvent
m_o	:	feed mass of solid material
q	:	specific heat
p	:	pressure
p_{Ex}	:	pressure in extractor
p_{Se}	:	pressure in separator
Re	:	Reynolds number
Sc	:	Schmidt number
Sh	:	Sherwood number
T,t	:	temperature
w_t	:	specific mechanical work
w_o	:	solvent flow velocity in empty cross-sectional area
η_{sv}	:	isentropic compressor efficiency
ρ_s	:	densitiy of solvent
ρ_o	:	bulk density of solid material
τ	:	time
τ_{Ex}	:	extraction time

Indices

1	pilot plant	2	production plant

Literature Cited

1. Tiegs, C. Dissertation, Universität Erlangen, 1985.
2. Rathkamp, P.L.; Bravo, J.L.; Fair, J.R. Solvent Extraction and Ion Exchange 1987, 5, 367.
3. Seibert, A.F.; Moosberg, D.G.; Bravo, J.L.; Johnston, K.P. Proc. Int. Symp. Supercritical Fluids, Nice, 1988, p 561.
4. Quirin, K.-W. Dissertation, Universität des Saarlandes, Saarbrücken, 1984.
5. Eggers, R. Gasextraktion, Essen, 1985.
6. Eggers, R. Chem.-Ing.-Techn. 1981, 53, 551-554.
7. Sievers, U. Die thermodynamischen Eigenschaften von Kohlendioxid; Fortschr.-Ber. VDI-Z. Reihe 6, Nr. 155. VDI-Verlag: Düsseldorf, 1984.
8. AD-Merkblätter, Verein TÜV; Beuth Verlag: Berlin, 1986.
9. Eggers, R. Proc. Int. Symp. Supercritical Fluids, Nice, 1988, p 595.

RECEIVED May 1, 1989

Chapter 31

Design, Construction, and Operation of a Multipurpose Plant for Commercial Supercritical Gas Extraction

F. Böhm[1], R. Heinisch[2], S. Peter[3], and E. Weidner[3]

[1]Norac Technologies, Inc., 4222–97 Street, Greystone Pavillion, Edmonton, Alberta T6E 5Z9, Canada

[2]Kasyco GmbH, Dammannstraße 61, D–4300 Essen 1, Federal Republic of Germany

[3]Lehrstuhl für Technische Chemie II Egerlandstraße 3, D–8520 Erlangen, Federal Republic of Germany

The design of a multi-purpose plant for the continuous extraction of liquids with supercritical fluids is presented. To provide flexibility in order to treat different feedstocks, a modular concept was developed based on experience gained in the operation of bench-scale and pilot plants. Four test systems were chosen in order to determine the proper dimensions for the equipment. Based on experimental data, e.g. measurements of flooding points and maximum flows for various column internals, the design pressure and temperature and heat exchange requirements were determined. The plant was built by a German manufacturer and was operated successfully by a Canadian company in Edmonton, Alberta.

As reported in a lot of reviews, extraction with supercritical solvents has very promising commercial potential. Until now the commercialization has been restricted mainly to batchwise extraction of solids with carbon dioxide (e.g. decaffeination of coffee and tea, extraction of hops). Although laboratory and pilot-plant experiments have indicated very good economics for continuous extraction of liquids with carbon dioxide and other gases, so far this technique has been applied industrially only for the production of 2-butanol by Idemitsu Petrochemical Corp. in Japan.

A Canadian group of entrepreneurs and scientists decided in 1985 to install an industrial extraction centre in Edmonton, Alberta, for research, product development and plant design. The centre is equipped with 5 units from laboratory size to commercial plants. Not only are batch extractors installed, but also columns for treating liquid materials are available. The maximum capacity for continuous extraction is about 50 tons/yr.

In order to achieve a maximum degree of flexibility, a modular concept was developed for the column, based on experience gained at the University Erlangen in Germany.

0097–6156/89/0406–0499$06.00/0

Balances and Height Calculations

In the last ten years about 12 separations of technical interest were studied extensively both by doing phase equilibrium measurements and continuous operations. Four typical separation problems were chosen as basis for designing the plant:

a) Deacidification of palm oil with ethane as extractant and acetone as entrainer.
b) Separation of monoglycerides (C18:1) from a mixture with Di- and Triglycerides with carbon dioxide as extractant and propane as entrainer.
c) Separation under point b) where the column is operated with a temperature gradient.
d) Deoiling of lecithin with a mixture of CO_2 and propane.

The mass flows and purities obtained by experiments in a pilot plant with a height of 4 m were used as a basis. By applying a modified Redlich-Kwong equation of state (1,2), the heat and mass balances were calculated for the new plant, which contains an extractor with an inner diameter of 7 cm and a height of 6.5 m.

In the course of process design and operation, it was found that phase equilibrium measurements and pilot plant data were required in order to develop a reliable computer simulation. The theoretical prediction or even a description of the mass transfer was often impossible, because of a lack of knowledge of the thermodynamic properties of the co-existing phases, especially if large molecules were involved. A practical method for process design and scaleup is to determine HETP values by stagewise computer simulation of pilot plant experiments, based on phase equilibrium calculations. If the range of the HETP values for a certain set of experiments is known, an optimization of the separation can be carried out by varying pressure, temperature, solvent to feed ratio, reflux and so on. The results of the optimization must be verified by experiment.

This optimization strategy was applied to the separation problems listed above. Based on these results the extraction column was designed.

In Figure 1 a typical flow sheet including the mass and heat flows is given for the separation of Monoglycerides from Di- and Triglycerides using a mixture of carbon dioxide (45 wt.%) and propane as extractant (55 wt.%). As shown in the figure, a product with 99% purity of Monoglycerides can be obtained.

Based on the heat and mass balances the required dimensions of pumps, heat exchangers, piping were defined. Additionally the energy and solvent consumption were calculated.

Layout

In general, a multi-purpose plant should be able to treat many kinds of different feedstocks. In order to achieve this requirement, a flexible design of the mechanical and electrical parts is required. Nevertheless it is not possible to run every feedstock with a fixed arrangement of apparatus and control units. Therefore a modular

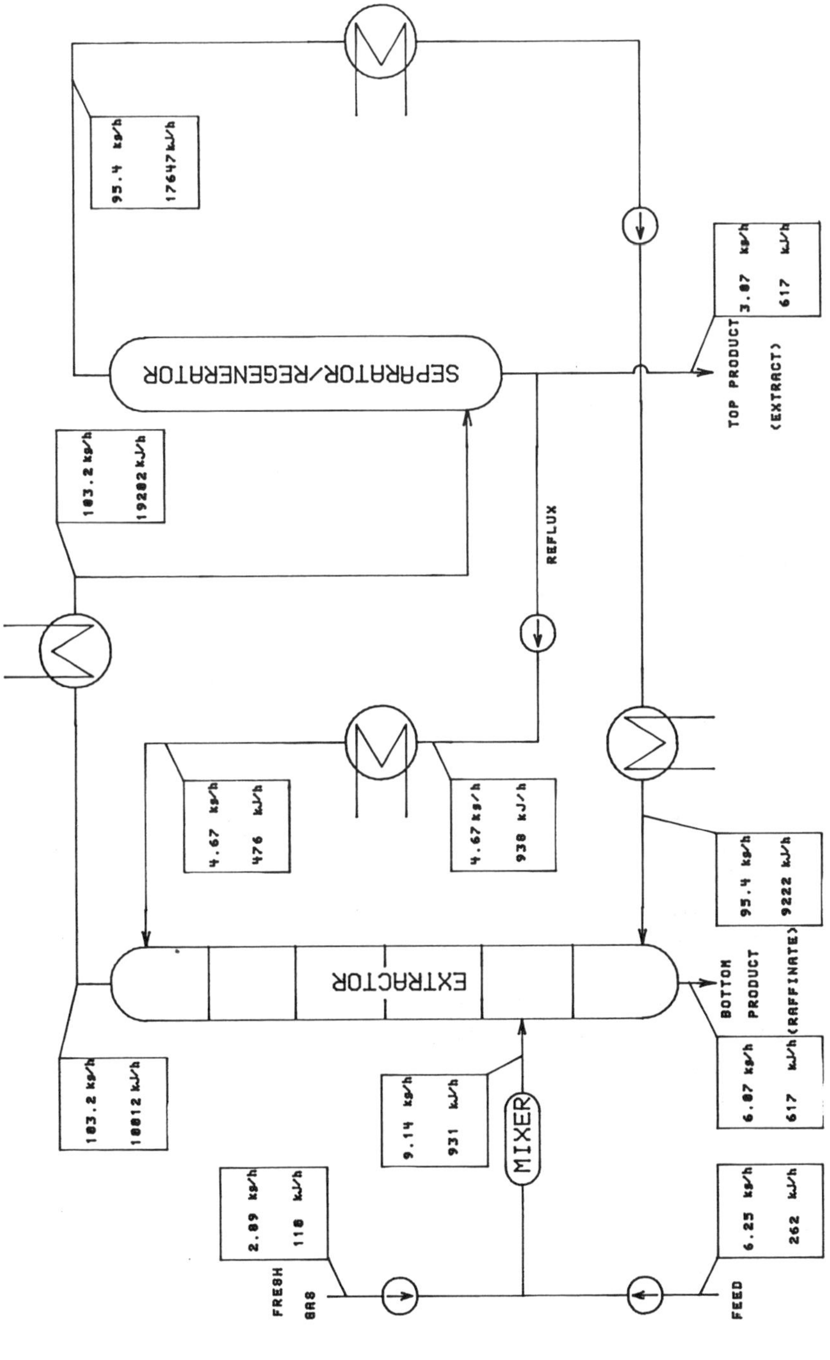

Figure 1. Mass and heat flows in the free fatty acids-palm oil-ethane-acetone system.

concept was developed, so that the plant could be adapted quickly to a given feedstock.

Column Design The extraction column consists of six modules with a length of 1 m each (Fig. 2). The packing is fixed in each module by supports. The modules are linked together with connecting elements with a length of 0.25 m each. Each connecting element has 6 ports in two levels. These ports can be used for the feed, for temperature and pressure measurement, for the reflux, and for taking samples from the extraction column (Figure 3).

The modular concept with this type of connecting elements allows for relatively easy changes in the column height and in the feed position along the length of the column. This is important for the flexibility because different separation problems can require different lengths of the stripping and enriching sections. Additionally it can be necessary, e.g. in the separation of Glycerides, to apply a reflux in order to improve the efficiency of the column. In this case, a part of the bottom product collected in the first regenerator is pumped to the connecting element situated at the top of the column.

Packings and Flooding As pointed out above, optimized mass and heat balances have been derived from a combination of experimental results with a computer simulation of the process. The optimized balances can be used for the layout of a production plant. A multi-purpose plant should be able not only to produce samples, but also to determine scaleup parameters. The scaleup parameters depend on the type of packing and its specific flooding point. The ability to measure flooding points or to test different packings is restricted mainly by the range of flow rates.

For sizing the gas circulating pump, feed pump and reflux pump the results of the optimizations were used. The determination of the maximum flows was based on a flooding point diagram. As an example, the optimization of the Monoglyceride process required a gas flow of the regenerated CO_2/C_3H_8 mixture of 97 kg/h. The flooding point occurs at about 130 kg/h. From comparison with other separation processes, it was determined that a gas circulation pump with a capacity of about 140 kg/h should meet most of the requirements of a multi-purpose plant. Naturally it must be taken into account, that normal metering pumps work volumetrically. It the density of the media changes, the delivered mass flow changes too. A pump that is designed for 130 kg/h of a certain solvent delivers a different mass flow if the density of the extraction solvent is changed.

The flooding point and the column efficiency are functions of the type of packing. In order to choose a useful packing for different feedstocks, a comparison was carried out in two pilot plants with inner diameters of 2.5 cm and 6.5 cm.

As indicated in Figure 4, the HETP values for different packings decrease with increasing liquid flow. The endpoint of the plotted curves are the flooding points where liquid droplets are withdrawn with the gas. The best separation

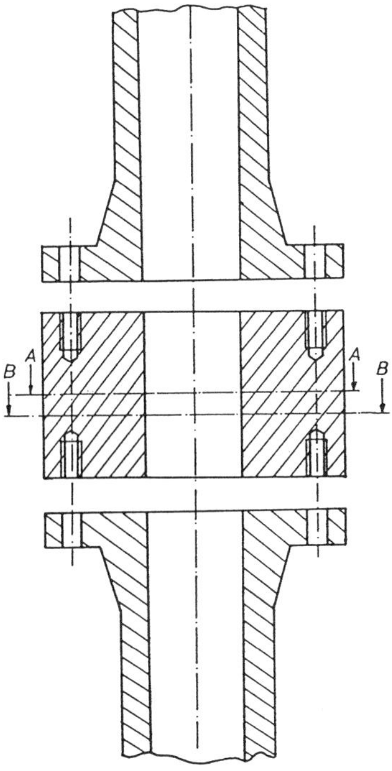

Figure 2. Column Module and Connecting Element

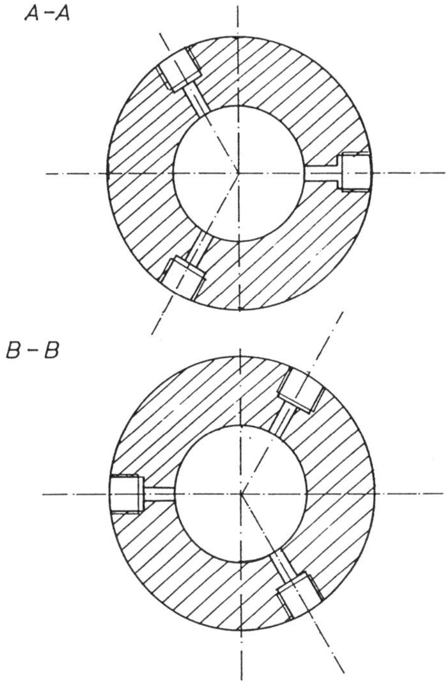

Figure 3. Detail of Connecting Element (cut A-A, B-B)

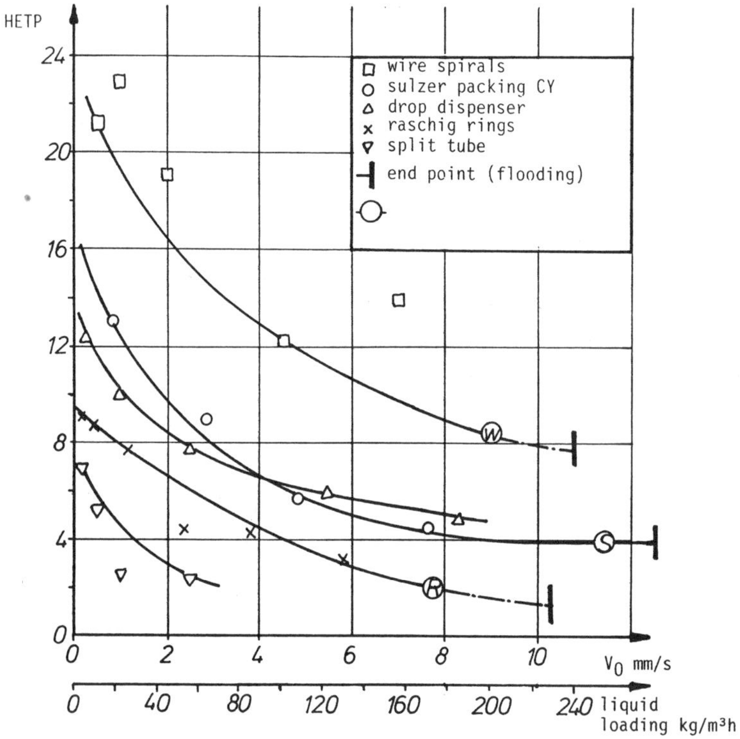

Figure 4. Separation Efficiency of Several Packings

efficiency was found for wire spirals, followed by a wire mesh packing (Sulzer CY) and drop dispensers (Figure 5). From this diagram it can be deduced that a wire mesh packing should provide a good combination of capacity (very high flooding point) and separation efficiency (1).

The test system for the packing was a glyceride separation with acetone and CO_2. The results for the Sulzer packings were confirmed by three further systems. All the investigated separations have a feature in common, that is the packing material is not wetted by the liquid phase under the applied conditions. There are some indications by other authors (2) and our own experiments that the HETP values might change if the packing is wetted.

Regeneration For fractionating the extract three regenerators were designed for the new plant.

In pilot plant experiments it was found that the regeneration efficiency has a very strong influence on the quality of the bottom product of the extractor. The reason is that in supercritical fluid extraction relative high solvent to feed ratios (5 to 25) must frequently be applied. In such cases already small concentrations of extracted product in the regenerated solvent cause a non-negligible backmixing. At the bottom of the extraction column the regenerated gas comes in contact with the raffinate. If the regeneration quality is not sufficient, contamination of the raffinate occurs. A sufficient regeneration quality is essential to obtain a pure raffinate. The main parameters influencing the regeneration are pressure and temperature (phase equilibrium) and the complete separation of liquid droplets.

In Figure 6 the residual concentration of Triglycerides in a mixture of propane and CO_2 is plotted versus mean residence time of the gas in the regenerator. The regenerator was equipped with a wire mesh for improving the droplet separation. The right hand points are the inlet concentration of the Triglycerides in the gas. The points on the left side indicate the remaining outlet concentration. It can be seen from the plotted curve that the regeneration efficiency is improved by increasing the mean residence time. The influence gets smaller with higher residence times, which correspond to lower linear gas velocities. The inlet concentration has no significant influence on the outlet concentration.

Reducing the pressure to 54 bar at 100°C gave about half the residual concentration in the regenerated gas. Removing the wire mesh from the regenerator doubled the residual outlet concentration at constant pressure and temperature. The regenerator and separators in the new plant were sized for a residence time of about 400 sec at 45 bar, 80°C. The inner diameter was increased in order to reduce the linear gas velocity at a given mean residence time.

Pressure, Temperature, Extraction Solvents The plant was designed for extracting a variety of natural products. For most of these products a temperature of 120°C

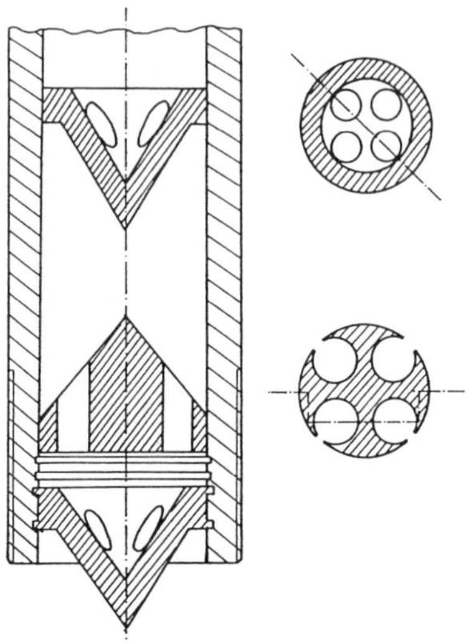

Figure 5. Drop Dispenser Packing

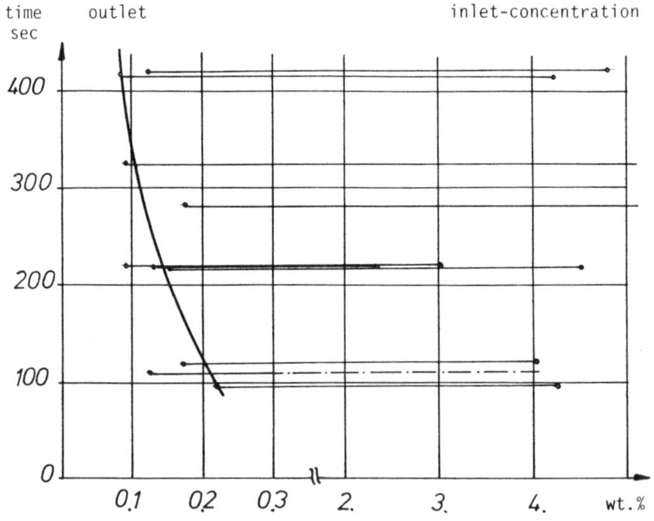

Figure 6. Regeneration Efficiency vs. Mean Residence Time of Gas

should meet all requirements. At higher temperatures these products can easily degrade.

The maximum pressure is not restricted by issues such as thermal lability but by the investment costs. The maximum operation pressure was fixed at 350 bar. This value seems rather high. For economical extractions the pressure should be reduced to an optimum value. This can be done by adding entrainers. As an example the pressure for extracting lecithin from soya oil can be reduced from 350 bar with CO_2 to 80 bar by using propane as an entrainer (3). Supercritical extraction should not be restricted to CO_2 as solvent. Some separation problems, e.g. the glyceride separation cannot be solved economically with CO_2 as a pure extractant. Therefore the plant was designed also for other solvents and for the use of entrainers. Design characteristics are:

- explosion-proof installation
- sandwich membrane pumps to avoid leakage
- installation of a flare
- sufficient air circulation
- installation of gas sensors

Heat exchangers A relatively difficult task is the calculation of heat transfer coefficients for near-critical fluids, especially when entrainers are used. Normally the heat exchangers are designed for heating, cooling or condensing carbon dioxide. The difficulties in calculating the heat transfer results mainly from the large changes in thermo-physical properties in the critical region. With the generally acknowledged design methods, this effect is not treated in a satisfactory manner. Therefore relatively large safety factors of up to 100% are required.

The question is, whether the calculated heat transfer area is sufficient, when entrainers are used. Fig. 7 shows measured heat transfer coefficients in an existing shell and tube heat exchanger for a mixture of propane and CO_2. It can be seen, that the behavior, especially at pressures around the critical pressure of CO_2, is relatively complex and needs further experimental study. A conclusion from Figure 7 is that a heat exchanger designed for pure CO_2 cannot meet all requirements when an entrainer is used.

Some additional considerations have been made concerning the heat exchanger between the extraction column and the regenerator. Some products which are dissolved in a supercritical gas can be precipitated by increasing the temperature and reducing the pressure, which is equal to reducing the density and thus the solvent power. Often at higher pressure, the situation is reversed. Here temperature reduction is required to precipitate the solid. Therefore the heat exchanger in front of a regenerator should be designed for heating and cooling. This demand causes additional investment costs. In the case of the plant under consideration the heat exchanger is only a heater. This limits the flexibility only a small amount, since cooling may be obtained by expansion. Normally, the pressure reduction of a

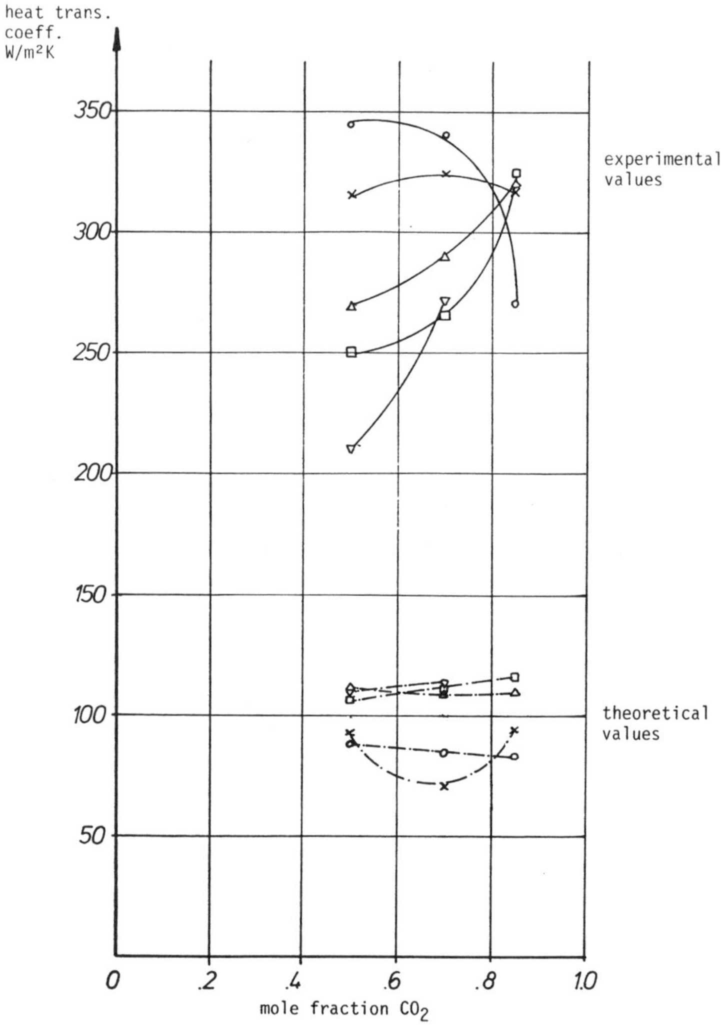

Figure 7. Heat Transfer Coefficients for CO_2/C_3H_8-Mixtures

supercritical fluid in an expansion valve causes a significant temperature reduction due to the Joule-Thompson effect.

Automation Total automation is not favorable for a multi-purpose plant, because the operation conditions for various feedstocks can be quite different. Each control circuit has only a certain dynamic range which cannot cover all the requirements. Therefore complete automation is recommended only for a production plant, when the properties of the treated product change only over a relative small range.

Different feedstocks and/or solvents require different solvent to feed ratios. Each packing has a specific flooding point. These two effects can lead to a great difference in mass flow. Because of this, pressures and temperatures should be controlled automatically. The flow must be measured exactly, preferably with a mass flow meter, but automatic flow control is not suitable for a multi-purpose plant. The valves which are used for pressure and temperature control should be installed so that they can be replaced easily or adapted to new conditions. A very important requirement for the continuous operation of the plant is liquid level indication and control in the extractor and regenerator. Therefore the plant is equipped with capacitive level sensors which are part of a control circuit. The suitability of these sensors for measuring the level of oily products, vitamins and some type of hydrocarbons in supercritical systems have been tested in the lab previously.

Measurement of the loading of the gas at the top of the column and at the top of the regenerator is required for establishing the mass balance. The ability to monitor the concentration-profile along the separation column is desirable but not necessary for technological process development.

Construction and Operation

The plant was built by UHDE in Germany in late 1986. Before shipping, test runs were carried out by circulating pure CO_2 to ensure that the heaters, coolers and pumps were sized properly. The skid-mounted plant was partially disassembled and shipped to Edmonton, Alberta. It was built at the NORAC extraction centre. It was in operation by mid-1987 and tested with two systems.

At first it was attempted to deacidify and deodorize natural, non-esterified fish oil. The main components of the feedstock were Triglycerides, which are quite similar to the products studied during the plant design stage.

The first objective was to prove that all the components work sufficiently well at the maximum operation conditions. To do this, fish oil Triglycerides are a good test system due to their relatively high flooding point. The plant was operated at 300 bar with a pure CO_2 flow of 110 kg/h, corresponding to 85% of the maximum capacity of the gas pump. The components of the plant fulfilled the requirements. As an example of some minor problems, the removal of the bottom product is mentioned. This product is enriched in Triglyceride and the free fatty acid concentration is reduced.

If the level sensor indicated an upper limit of the liquid level the product removal valve was opened automatically until the lower switch point of the sensor is reached by the liquid. Then the valve was closed. Due to the gas that dissolved in the Triglycerides (about 35 wt.%) a relative large Joule-Thompson effect occurred. The temperatures in the expansion valve became so low, that the crystallization point of the Triglycerides was reached (5 - 10 °C). The solid Triglycerides in the expansion valve caused damage to the stem during closing of the valve. This problem could be solved by reducing the distance between the upper and lower switch points of the level sensor and by heating the product removal pipe and the valve body.

In a second set of experiments fish oil esters were used to test the plant behavior. In this case the flooding point is significantly lower. The plant was operated at CO_2 flows between 40 and 50 kg/h. Evaluation of phase equilibria measurements indicated that an extraction temperature between 60°C and 70°C is required. The reduction in flow and the increase in temperature resulted in problems with the preheating of the gas before the extraction column.

At a temperature of 90°C for the heating fluid, the lower end of the dynamic range of the temperature regulating valve was reached. The gas temperature after the preheater was no longer constant but fluctuated by about ± 4°C. Replacing the stem of the regulation valve at the heating fluid inlet with a larger stem solved the problem. These examples described the kind of problems that may occur in the operation of multi-purpose pilot plants.

Literature Cited

1. Tiegs, C., Ph.D. Thesis, University Erlangen 1984.
2. Seibert, A. F., Bravo, J. L., Johnston, K. J., Int. Symp. on Supercritical Fluids, Nice 1988, Vol. 2.
3. Weidner, E., Ph.D. Thesis, University Erlangen 1985.
4. S. Peter, C. Tiegs, Preprints of the International Symposium High Pressure Chemical Engineering October 8.-10. 1984 Erlangen West Germany.
5. S. Peter, M. Schneider, E. Weidner, R. Ziegelitz; Chem. Eng. Technol. 10 (1987) 37-42.

RECEIVED May 1, 1989

Chapter 32

Supercritical Fluid Extraction of Flavoring Material

Design and Economics

Richard A. Novak and Raymond J. Robey

Supercritical Processing, Inc., 966 Postal Road, Allentown, PA 18103

A battery limits capital cost of $2.8 million and an operating cost of $1.10 per kilogram of feed are estimated for a supercritical extraction plant with a capacity of 0.8 million kg/year of feed spices. Production cost declines significantly as plant size increases, reaching $0.50 per kilogram for a 3.1 million kg/year plant. Process costs for 12 spices and 8 herbs are estimated, ranging from $0.60 to $4.30 per kilogram feed. Costs are based on a factored estimate with probable accuracy of +/- 50%. A preliminary process design for a multiproduct plant is described, based on proprietary pilot plant data. The advantages of supercritical extraction over conventional solvent extraction are discussed, including: natural, nontoxic solvent; control of flavor profile; mild process temperatures; high quality products.

Plant materials, such as spices and herbs, can be supercritically extracted for flavor, fragrance, and pharmaceutical applications. Using nontoxic carbon dioxide as a solvent, supercritical extraction (SCE) leaves no harmful residues. Food materials produced with SCE are viewed as natural and have been shown to be of high quality, often with superior properties not obtainable with other separation techniques. The purpose of this paper is to discuss the advantages of SCE for flavor applications, to describe a preliminary design for a commercial plant, and to present economics for this application.

Background

SCE is a powerful separation tool for many applications in the chemical, food, and pharmaceutical fields (1-7). The process has proven economical in large scale commercial application. For

0097-6156/89/0406-0511$06.00/0

example, General Foods is now operating a 50 million pound per year (23 million kg) coffee decaffeination plant using SCE (8).

Some of the advantages of SCE with carbon dioxide are:

- No Harmful Residue - Carbon dioxide is odorless, tasteless, inert and nontoxic. It is easily and completely removed from the extracts and processed materials.
- Low Temperature Process - SCE temperatures are low enough to prevent degradation of sensitive materials, often important in flavor applications.
- High Quality - Extracts prepared by SCE retain more top and back notes, with no off flavors. Flavors and fragrances are closer to the natural material, and are often judged superior to conventional extracts (5, 9, 11).
- Flexible Process - The selectivity of SCE can be adjusted by proper choice of operating conditions and procedures. The process can be fine tuned to achieve a desired flavor or fragrance profile.

Extracts of a wide variety of spices and herbs have been prepared with liquid and supercritical carbon dioxide (5, 10). The techniques of fractional extraction and fractional separation (5, 11) allow the separation of flavor from aroma components during SCE. Controlled blending can then be used for standardization of extracted product. Supercritical Processing, Inc.'s technology includes both fractional extraction and fractional separation. The choice of approach depends on the feed material and product requirements.

In fractional extraction, the feed material is extracted in two or more stages. The selectivity for essential oils, fatty oils, and resins is controlled in each stage through selection of extraction pressure, temperature, or cosolvent addition. With the first extraction stage at subcritical temperatures, sensitive essential oils are removed at mild conditions and at short processing times. A second stage extraction at higher temperatures can then isolate flavor components. For example, in the extraction of black pepper (11), a first stage extraction at 300 bar, 30°C (31 MPa, 303 K) produces an essential oil yield of 2.1% and a piperine (hot flavor component) yield of 0.6%. The second stage extraction at 312 bar, 58°C (32 MPa, 331 K) produces a lower essential oil yield (0.7%) but a much higher piperine yield (5.2%).

Conversely, conditions in the extractor may be held constant and the extracted components separated in stages. Optimized extraction and separation conditions for a variety of spices and herbs have also been developed using fractional separation technology (Henkel KGaA, High-Pressure Extraction of Spices By Means of Carbon Dioxide, Düsseldorf, 1982, proprietary technology package.). The consistent extractor conditions of fractional separation simplifies plant operation where multiple batch extractors are used. Also, the supercritical extraction fluid is always fully loaded with extracted material, making a more efficient process.

Design Principles

The design of commercial SCE plants has been discussed by several authors (1-2, 5, 7). The following unit operations and design specifications are important for SCE of flavor materials:

- Raw material preparation - particle size, moisture content, cell disruption
- Extraction conditions - pressure, temperature, time, solvent to feed ratio, solvent flow
- Extractor operation - batch or continuous, constant conditions or staged (fractional extraction)
- Separation conditions - pressure, temperature, disengagement design, volatiles recovery, water removal
- Separator operation - batch or continuous, single stage or multiple conditions (fractional separation)
- Supercritical solvent recycle and treatment, if any
- Extract recovery and treatment, including degassing, filtration, testing, dehydration, homogenization, and blending

Appropriate values of these parameters must be determined for each raw material to achieve optimum yield and quality of extract. Lab and pilot plant optimization studies are required.

The following factors must also be considered in the design of SCE plants for natural materials:

- Equipment must be designed for pressures between 80-400 bar (8-40 MPa), as well as for the stress of repeated cycles.
- Food plant design procedures must be used. Surfaces in contact with food must be corrosion resistant, smooth, and easy to clean and decontaminate. Equipment and piping must be designed with no "dead ends" where material may collect and stagnate, possibly causing contamination.
- The difficulty and expense of continuous feed of solids into high pressure vessels means that batch operation is probable (except, perhaps, for very large, single product plants). Low bulk density, often difficult to handle solid feeds may require use of pre-loaded baskets. Full bore openings are used on vessels with quick, high pressure closures.
- Heat tracing must be used where extracted material may deposit in lines and valves.
- Appropriate alarms, interlocks, and emergency systems must be used to allow safe operation at high pressure.

Preliminary Plant Design

Design Basis. A preliminary design for a multiproduct spice extraction plant was prepared, based on SCP's proprietary process (Henkel KGaA, op. cit.) and plant design data (Stearns Catalytic Corporation, Tolling/Demo SCE Plant, Philadelphia, PA, 1984, proprietary data.). Key parameters in the "base case" design are:

Feedstock:	Spices and herbs (solid)
Extractor volume:	Total 974 L, Basket 695 L, Solids level 90%

Number of extractors:	2
Extraction:	Pressure 60-300 bar (6-30 MPa)
	Temperature 20-80°C (293-353 K)
Separation:	Pressure 45-150 bar (4-15 MPa)
	Temperature 15-40°C (288-313 K)
Extraction solvent:	Supercritical carbon dioxide
Solvent flow rate:	10,000 lb/hr (4,550 kg/hr)
Annual capacity:	For an average spice, 1.7 MM lb (0.8 million kg).
Operation:	24 hr/day, 7 days/week, at 85% onstream

The assumed 85% availability is probably a maximum. For a multiproduct plant, additional cleaning time may be needed, depending on feed mix.

Process Description. A simplified flowsheet for the process is shown in Figure 1. Feed will be placed in extraction baskets and loaded into the extraction vessels (R-1). The two extractors will operate batchwise and in a staggered cycle, so that solvent circulation and extract removal is carried out continuously. Except the case of fractional extraction, solvent will flow through the extractors in series, so that the extractor containing the most depleted charge always receives the freshest solvent. The extraction conditions and cycle times will vary, depending on the feed. When fractional extraction is desired, the extractors will operate batchwise, in parallel, at different conditions. Each would feed a different separator, in parallel. When an extractor has reached the desired extraction time, it will be taken off line, depressurized, the baskets unloaded, and the processed material sent on to disposal or by-product sale. A basket containing fresh feed will be loaded, and the extractor repressurized and returned to operation.

The supercritical solvent leaving the extractors will flow through a filter (F-1) to capture entrained solids. One or two separation stages will be used, in series (fractional separation) or in parallel (fractional extraction). For each stage, the pressure of the supercritical fluid will be reduced by flow through a throttling valve (PCV-1, PCV-2) and the temperature adjusted in a heat exchanger (E-1, E-2) to the conditions desired for the stage of separation. Any extracted material which precipitates in each stage will be collected in separator vessels (V-1, V-2).

Gaseous solvent from the separators will flow through a filter (F-2), an air cooler (E-3), and a molecular sieve drier (X-1) to remove moisture. The dried solvent will be condensed to a liquid in a heat exchanger (E-4), and sent to a holding tank (V-5). Liquid carbon dioxide will be maintained in the hold tank at about 900 PSIG (6.3 MPa) and ambient temperature. Solvent will be added to the holding tank to make up for process losses. A high pressure reciprocating pump (P-3) will compress the liquid solvent to extraction pressure, and a heat exchanger (E-5) will heat it to extraction temperature, before it flows back to the extractor vessels.

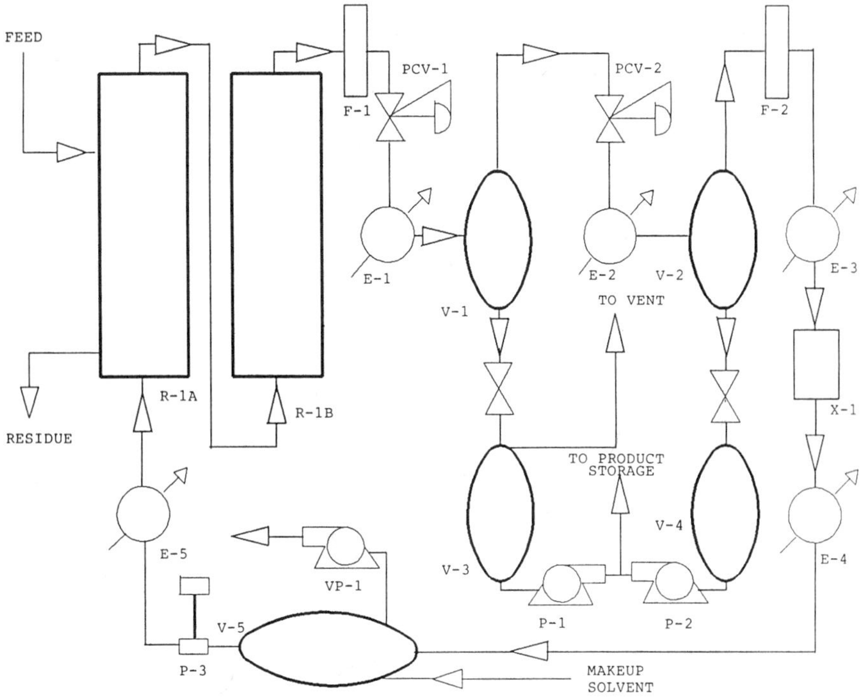

Figure 1. Process Flowsheet: Supercritical Extraction of Spices.

Extracted material will be continuously withdrawn from the pressurized separators (V-1, V-2) to atmospheric holding tanks (V-3, V-4). Gaseous solvent that flashes out of the extracted material in the hold tanks will be sent to the vent system for disposal. Pumps (P-1, P-2) will transfer the degassed extracts to product finishing and storage.

Economics

Scope. Based on the preliminary process design, battery limits capital and operating costs were estimated for mid-1988. The battery limits plant comprises equipment and systems directly associated with the supercritical extraction operation, as shown in the process flowsheet (Figure 1). Offsites, such as material shipping and handling, buildings, land, etc., were not included. For a supercritical plant installed at an existing facility, these services may already be available. For a new or "grassroots" plant, additional capital investment, perhaps 25% to 75% of battery limits capital, will be needed for offsites.

Capital Cost. The estimated capital cost for the base case plant is $2.8 million, as shown in Table I. All equipment in contact with process fluids is 300 series stainless steel, or stainless steel lined where appropriate. Heat exchangers are shell and tube, with stainless steel tubes for process fluids and carbon steel shells for heat transfer fluids.

The capital cost includes cost of equipment, installation materials and labor, and engineering and construction costs. The capital cost excludes costs of land, buildings, utilities (e.g., steam or electrical supply equipment and distribution), feed and product storage, feed preparation and handling, receiving and shipping, other offsites, contingencies, escalation, working capital, and royalties.

Operating Cost. The estimated operating cost for the base case plant is about $ 115 per hour of extraction time, as shown in Table II.

Included in the operating cost are the costs of utilities supplied to the battery limits, such as power, fuel, and water (excluding any capital charges for the utilities investment), makeup supercritical solvent, labor (operating and production supervision), maintenance material and labor, depreciation on battery limits capital, and capital related taxes and insurance. Operating costs for plant equipment outside the battery limits (such as feed preparation) are excluded. Also excluded from the operating cost are the costs of raw materials, plant overhead and support services, financing, corporate fees (sales, general and administrative, research and development), contingencies, profit, and royalties.

Production Costs. Data on extraction times and yields were used to estimate production costs for extracts of twelve spices and eight herbs, as shown in Tables III and IV (Supercritical Processing,

Table I. Battery Limits Capital Cost, Base Case Design

Tag #	Description	Design Pressure PSIG	Size	Units	Estimated Cost $1000
R-1A	Extractor	4800	973	L	166.9
R-1B	Extractor	4800	973	L	166.9
X-2	Ext. Baskets (10)		695	L	10.0
E-1	1st Sep. Preheater	3500	43	FT2	11.0
V-1	1st Separator	3500	208	L	74.0
V-3	1st Prod. Rec.	15	757	L	7.5
P-1	1st Prod. Trans.		10	GPH	1.3
E-2	2nd Sep. Preheater	3500	117	FT2	20.0
V-2	2nd Separator	3500	208	L	74.0
V-4	2nd Prod. Rec.	15	757	L	7.5
P-2	2nd Prod. Trans.		10	GPH	1.3
F-1	Ovhd Filter	4800	20	CFM	3.8
F-2	Recyc. Filter	1000	30	CFM	3.4
E-3	Solv. Cooler	1000	1100	FT2	28.0
X-1	Solv. Drier	1000	25	PPH	60.0
E-4	Solv. Condenser	1000	600	FT2	28.0
V-5	Solv. Hold Tank	1000	600	GAL	60.0
VP-1	Vacuum Pump		30	PPH	6.1
P-3	Solv. Recyc.	4800	20	GPM	38.1
E-5	Solv. Preheater	4800	137	FT2	24.6
Total, Equipment Cost ($1000)					792.4
Installation Factor		3.5			
TOTAL, Battery Limits Capital ($Million)					2.8

Table II. Battery Limits Operating Cost, Base Case Design

Item	Comments	$K/yr	$/hr operation	% total
Labor				
Operating labor	2 per shift, $15/hr	287	39	33
Supervision	25% operating labor	72	10	8
Subtotal		359	48	42
Utilities				
Electricity	134 kW, 4.3 cents/kWh	43	6	5
Steam	0.9 MMBTU/hr, $4.9/MMBTU	32	4	4
Cooling Water	6.9 MGAL/hr, 7.4 cents/MGAL	4	1	< 1
Carbon Dioxide	110 kg/hr, 10 cents/kg	83	11	10
Subtotal		162	22	19
Capital				
Depreciation	15 years, $2.8 MM	185	25	22
Taxes & Ins.	1.5% B.L. Capital	42	6	5
Maintenance M&L	4% B.L. Capital	111	15	13
Subtotal		337	45	39
TOTAL	Battery Limits Operating Cost	858	115	100

Table III. Battery Limits Production Cost: Spices
Base Case Design

Raw Material	Bulk Density kg/L	Extract. Charge kg	Batch Time hours	Process Rate kg/hr	Extract Yield Percent		Process Cost: Feed $/kg	Process Cost: Extract $/kg	
					Low	High		Low	High
Black Pepper	0.36	225	6	75	10	12	1.5	12.8	15.3
Clove	0.36	225	3	150	19	22	0.8	3.5	4.0
Clove stems	0.36	225	2.5	180	6	8	0.6	8.0	10.7
Nutmeg	0.34	213	6	71	38	41	1.6	4.0	4.3
Mace	0.37	231	6	77	34	38	1.5	3.9	4.4
Nutmeg Oil	0.37	231	4	116	18	22	1.0	4.5	5.5
Ginger	0.40	250	6	83	4	6	1.4	23.0	34.5
Allspice	0.40	250	4	125	7	9	0.9	10.2	13.2
Cassia	0.40	250	4	125	2	3	0.9	30.7	46.0
Caraway	0.35	219	4	109	21	23	1.1	4.6	5.0
Coriander	0.32	200	4	100	20	23	1.2	5.0	5.8
Cinnamon	0.38	238	5	95	2	3	1.2	40.4	60.6
Average Spice	0.37	231	4.5	103	15	18	1.1	6.2	7.5

Table IV. Battery Limits Production Cost: Herbs
Base Case Design

Raw Material	Bulk Density kg/L	Extract. Charge kg	Batch Time hours	Process Rate kg/hr	Extract Yield Percent		Process Cost: Feed $/kg	Process Cost: Extract $/kg	
					Low	High		Low	High
Thyme	0.19	119	4	59		2.1	1.9	92.3	
Camomile	0.185	116	5	46		4.3	2.5	57.9	
Fennel	0.18	113	2	113		16	1.0	6.4	
Peppermint	0.22	138	6	46		3.9	2.5	64.4	
Eucalyptus	0.23	144	4	72		3.7	1.6	43.3	
Valerian	0.25	156	4.5	70		4.7	1.7	35.3	
Arnica	0.15	94	7	27		3.8	4.3	113.1	
Vanilla	0.37	231	4.5	103		6	1.1	18.7	
Average Herb	0.22	139	4.625	60		5.6	1.9	34.5	

Inc., proprietary data). Costs are also reported for an average spice and herb (having average values of density, extraction time, and yields). Yield are net, after removal of extracted moisture.

For spices, costs based on feed ranged from $0.6/kg ($0.29/lb) for clove stems to $1.6/kg ($0.74/lb) for nutmeg. The cost for the average spice was $1.1/kg ($0.51/lb). The feed bulk density, which sets the mass charge per extractor, and the required extraction time determine the production cost on a feed mass basis. With the addition of yield data, production cost per unit of extract was calculated. This ranged from $3.5-4.0/kg ($1.60-1.80/lb) for clove extract to $40-60/kg ($18-27/lb) for cinnamon extract. For the average spice, the values were $6.2-7.5/kg ($2.8-3.4/lb) of extract.

For herbs, costs based on feed ranged from $1.0/kg ($0.47/lb) for fennel to $4.3/kg ($1.95/lb) for arnica. The cost for the average herb was $1.9/kg ($0.87/lb). The herb densities are lower than those of the spices, reducing extractor loading and increasing costs on a feed mass basis. Herbs also have lower average yields relative to spices, increasing costs per unit of extract. These ranged from $6.4/kg ($2.90/lb) for fennel extract to $113/kg ($51/lb) for arnica extract. For the average herb, the value was $35/kg ($16/lb) of extract.

Sensitivity To Capacity. The base case plant, with two extractors of 973 L, can process 770 Mg (1.7 MM lb) per year of a spice with average density. Costs were also estimated for plants with extractor volumes of 2, 3, and 4 times the base case, as shown in Table V. Note that in Table V, 1 Mg = 1000 kg = 1 metric ton.

Capital cost per unit of capacity declines by over 40% as plant size is increased to 4X base case. The reduction is about 25% of the base value for the 2X case, but only 8% of the base value between the 3X and 4X cases. This shows diminishing returns in the increase of capital efficiency as plant size increases.

Operating cost per hour increases by 87% of the base value as capacity increases to 4X. For the range of plant sizes studied, staffing requirements were held constant at two operators per shift. This caused the distribution of operating costs to change as plant size increased. For example, in the base case, operating costs per hour are $48 labor (42%), $22 utilities (19%), and $45 capital related (39%). For the 4X case, operating costs are $48 labor (22%), $66 utilities (31%), and $101 capital related (47%). Operating costs for the small base case plant are dominated by labor, while costs for the largest plant design are capital intensive.

Production cost declines by 54% as capacity is increased to 4X base case. Operating cost per unit of feed declines from $1.12/kg ($0.51/lb) to $0.52/kg ($0.24/lb). For materials with average properties, cost per unit of extract declines to $2.9-3.5/kg ($1.3-1.6/lb) for spices, and $16/kg ($7.3/lb) for herb extracts, at the 4X capacity level. This indicates a significant advantage to building a larger plant. Larger plant sizes can be achieved in a multiproduct facility processing several spices and herbs, possibly for different customers. Or, smaller spice volumes can be

Table V. Battery Limits Process Costs, Sensitivity To Plant Size

Extractor Volume L	Annual Capacity Mg	Annual Capacity MMlb	Capital Cost $MM	Operating Cost $/hr	Capital Cost $/Mg/y	Operating Cost based on feed $/kg	$/lb
970	770	1.7	2.8	115	3620	1.10	0.50
1950	1530	3.4	4.1	153	2700	0.75	0.35
2880	2270	5.0	5.2	184	2280	0.60	0.30
3890	3060	6.8	6.2	215	2010	0.50	0.25

"piggybacked" on a large volume SCE application, such as coffee or tea decaffeination.

Potential for Cost Reduction. The process and plant design discussed here has not yet been optimized for cost efficiency. Selection of process pressure, solvent/feed ratio, extractor configuration and cycle, removal of extracted components from solvent, and many other factors should be done using a combination of technical and economic impact.

Methods and Accuracy. The estimates presented here are "order of magnitude", with a possible accuracy of +/- 50%. No contingency has been included in the capital and operating cost estimates.

Cost data were obtained from SCP's Tolling/Demonstration Plant design package. Costs in the reference package were based on vendor quotes for equipment. Factors were used to estimate the costs of installation materials and labor, based on the experience of the engineering construction firm. Plant size in the present base case is the same as that in the reference design, so that the same installation material and labor factor for the battery limits plant was used (3.5X equipment cost) as was used in the reference. Installation factors in the range of 3X to 5X are possible for a plant of this size.

Capital costs for larger plants were estimated by scaling the equipment costs, using exponents appropriate for each type of equipment. The overall exponent was 0.57, which is close to the 0.6 rule-of-thumb often used for cost scaling with plant capacity. This method is not highly accurate, but is useful in giving a rough idea of cost sensitivity to capacity. Greater accuracy would require plant redesign and recosting for each capacity case.

Major changes in process conditions, especially changes to the assumed pressure or solvent/feed ratio, could have a significant impact on costs. Any handling problems with the feed or products, or significant changes to the assumed bulk density (and hence the extractor loading) could also be important. Also, the many project uncertainties, such as plant location, availability of existing facilities, etc., have at least as important effect on costs as the process design.

Comparison With Conventional Extraction. A comparison of SCE costs with costs of conventional solvent extraction is under study, but results are not yet available. SCE extracts may have a distinctly different character than conventional products, so that a direct cost comparison may not tell the entire story. However, some comments can be made. While SCE is expected to be more capital intensive, because of high pressure equipment, it may enjoy cost advantages due to higher yield, more rapid processing, and low solvent cost. An important cost advantage may be due to lower environmental control and waste disposal costs, since SCE with carbon dioxide does not cause toxic emissions or residues.

Conclusions

1. Supercritical extraction (SCE) has significant advantages in the production of flavor extracts from spices and herbs. High quality extracts are produced in high yield, by a process perceived as "natural," leaving no solvent residues.
2. Flavor and aroma components can be separated during the extraction, allowing controlled blending for product standardization. Process flexibility allows control over flavor or fragrance profile.
3. Process steps include feed preparation, extraction, separation, and extract finishing. Process parameters must be determined for each raw material to achieve the optimum yield and quality of extract. The plant design must take into account high pressure operation, food processing concerns, and repeated batch operation on low bulk density solids.
4. A preliminary process design was prepared for a multiproduct plant. The base case design, with two 973 L extractors, could process 770 Mg (1.7 MM lb) of an "average" spice feed per year. The high capacity design would have an annual feed rate of 3060 Mg (6.8 MM lb).
5. A battery limits capital investment of $2.8 million was estimated for the base case, and $6.2 million for the high capacity design. A battery limits production cost for an average spice of $1.1/kg ($0.50/lb) was estimated, based on raw material, for the base case plant. Production cost declines significantly as plant size increases, reaching $0.5/kg ($0.24/lb) for the high capacity case.

Literature Cited

1. Deutscher, V., Ed. Preprints: International Symposium on High Pressure Chemical Engineering, VDI-Gesellschaft Verfahrenstechnik und Chemieingenieurwesen, Düsseldorf, 1984.
2. Schneider, G. M.; Stahl, E.; Wilke, G. Extraction with Supercritical Gases; Verlag Chemie: Deerfield Beach, Florida, 1980.
3. McHugh, M. A.; Krukonis, V. J. Supercritical Fluid Extraction: Principles and Applications; Butterworths: Boston, 1986.
4. Squires, T. G.; Paulaitis, M. E., Eds. Supercritical Fluids - Chemical Engineering Principles and Applications; ACS Symposium Series No. 329; American Chemical Society: Washington, DC, 1987.
5. Stahl, E.; Quirin, K. W.; Gerard, D. Dense Gases For Extraction and Refining; Springer-Verlag: New York, 1988.
6. Charpentier, B. A.; Sevenants, M. R., Eds. Supercritical Fluid Extraction and Chromatography, Techniques and Applications; ACS Symposium Series No. 366; American Chemical Society: Washington, DC, 1988.

7. Perrut, M., Ed. Proceedings: International Symposium on Supercritical Fluids, Societe Francaise de Chimie: Nice, France, October 17-19, 1988, Vols. 1&2.
8. "Chementator," Chem. Eng. Sept. 26, 1988, 95 (13), 21.
9. Caragay, A. B. Perfumer & Flavorist 1981, 6 (4), 43.
10. Moyler, D. A. Perfumer & Flavorist 1984, 9 (2), 109.
11. Behr, N., et. al. U.S. Patent 4 490 398, 1984.

RECEIVED May 2, 1989

Chapter 33

Selection of Components for Commercial Supercritical Fluid Food Processing Plants

Rodger T. Marentis and Samuel W. Vance

Pitt-Des Moines, Inc., Neville Island, Pittsburgh, PA 15225

Criteria for selection of equipment and components for commercial supercritical fluid processing plants for the food processing industry are listed and discussed. Unique features and designs for SCF food processing are specified. Requirements for vessels, heat exchangers, instrumentation, piping, fluid transport devices and typical ancillary equipment are reviewed.

Supercritical fluid (SCF) food processing plants have become one of the more robust technologies for new applications within the food industry in recent years. The announcement of the construction and start up of a coffee decaffeination plant in Houston, Texas (1) has markedly heightened interest, resulting in increased awareness of the unique factors that apply to the design of the SCF processing plant and, more importantly, the considerations necessary to select equipment and components for installation in a SCF processing plant.

Figure 1 shows a flow schematic for such a plant and process. The major equipment items are extraction vessels, separation vessels, heat exchangers and pumps or compressors which are interconnected with piping, flanges and couplings. Valves specially designed for the operating conditions permit sequencing of flows and control of process variables. A data acquisition and logging system and process control loops make the process far more workable and efficient.

Figure 2 reveals the increase in piping complexity and valve quantity required for multiple extractor vessels, with an increase in valve quantities to include the necessary diverting valves for efficient system operation. Only two valves are required for a one extractor vessel system, but 22 valves are required for a four extractor vessel system.

In addition, process plants need utilities and auxiliaries, such as cooling water, a refrigeration system, steam, hot water and compressed air. These items may be available at an existing plant site, may be added for a new plant site, or may be expanded if existing units do not provide the needed quantities and qualities of utilities for the supercritical fluid process.

0097–6156/89/0406–0525$06.00/0

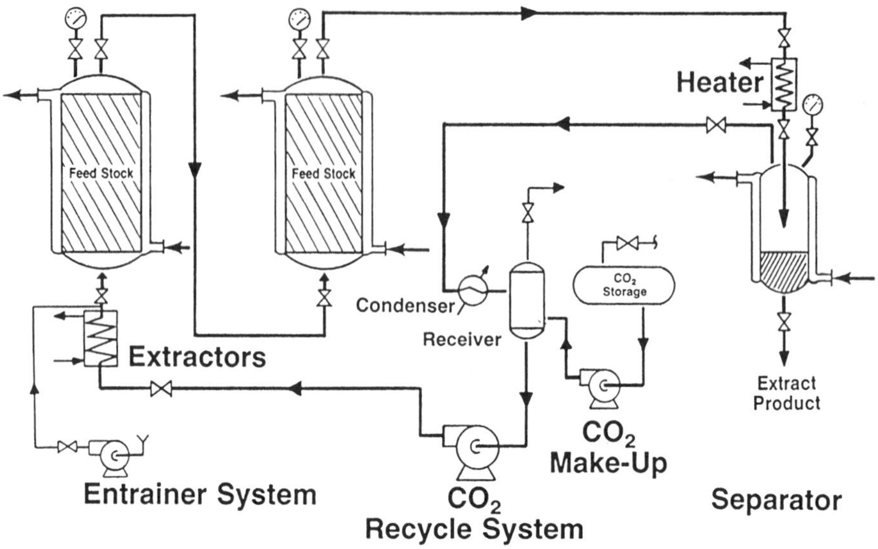

Figure 1. PDM pilot plant unit schematic diagram. (Reprinted from ref. 9. Copyright 1988 American Chemical Society.)

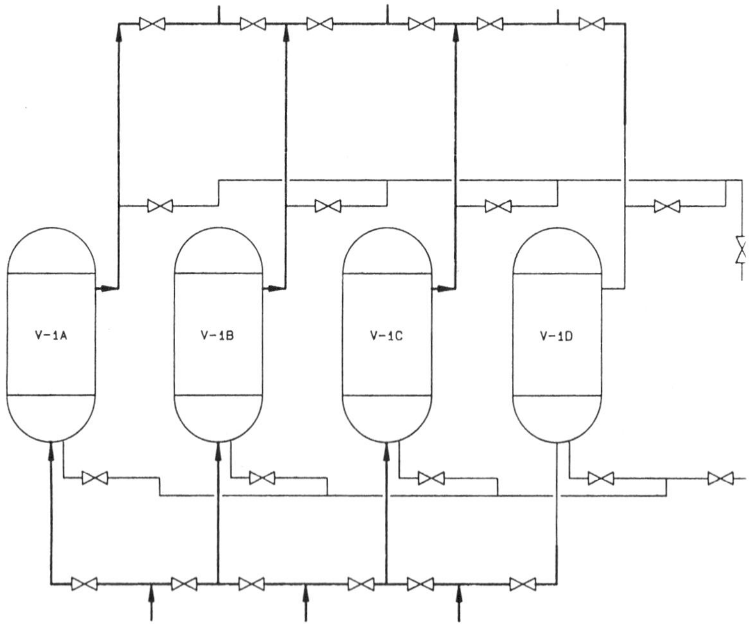

Figure 2. PDM schematic drawing of four extraction vessels with interconnecting piping and valves.

For small plants, with extractor volumes of less than 200 liters, equipment can be mounted on a single skid. For medium size plants (200 liter to 2,500 liter extractor volume) the plant arrangement consists of an interconnected group of modular functional skids. Very large plants (over 2,500 liter extractor volume) involve erection in the field, rather than prior assembly of modules and skids.

The following sections describe the criteria and selection factors for the various components. Desirable features and alternatives are addressed.

GENERAL CRITERIA

Criteria for consideration throughout the SCF Plant are sanitation requirements, capital and operating costs, maintenance requirements, operating reliability and plant safety. These criteria are heavily weighted in the equipment selection process for the entire plant. Several standards, codes and manufacturing practices (2), (3) are considered in the selection of equipment for a food processing facility to ensure a safe, sanitary, clean and efficient plant; they specify design, cleaning techniques, sterilizing and cleaning compounds and apparatus, and product changeover requirements to avoid contamination.

The American National Standards Institute (ANSI)/American Society of Mechanical Engineers (ASME) codes for pressure piping (4) and boilers and pressure vessels, (5) establish design criteria, standards and methods, as well as fabrication and testing methods to be followed during construction and start-up, to ensure safe designs and selection of proper materials for vessels and interconnecting piping and fittings. In addition, national, state and local governmental regulations must be satisfied before a plant can be started up. The final criterion concerns industrial insurance company requirements for design, construction and testing of pressure vessels and piping which must be met to obtain liability insurance for an operating plant facility.

SPECIFIC CRITERIA

There are a number of specific criteria which must be satisfied for each component.

VESSELS. Factors which are important in vessel design and fabrication are materials of construction, method of construction, pressure and thermal cycling, charging and unloading mechanisms and ease of cleaning.

All vessels must be designed for the specific most demanding temperature and pressure combinations which may be encountered. Cycling of pressure and temperature must be considered, and a fatigue analysis performed based on the anticipated operating life of the vessel. This operating environment governs the vessel design. Both thermal and pressure stress analyses are required and a fault tree analysis of the operating system should be performed.

Four manufacturing methods are used for commercial vessel construction: forging; casting; rolling and welding heavy plate; and multilayering of lighter plate. The most economic method depends upon vessel diameter and height, design operating pressure and temperature, and choice of materials for contact with the process and for the remainder of the vessel body. In general, forgings and castings should be considered for smaller diameter vessels (under 0.5 meter), welded heavy plate for mid-sized vessels (up to 2 meters diameter), and multilayered construction for large vessels (1.5 meter diameter vessels and larger).

The vessel may require one or more full diameter quick opening closures to permit rapid filling and emptying of feed stock or cleaning of the vessel if commodity changeover is needed. All closures must meet the ASME Boiler and Pressure Vessel Code design and testing requirements. Four types of quick opening closures (Figure 3) are presently in use: breech lock for small vessels; clamp type or segment ring type for midsized diameter vessels; and pin type for large and very large diameters. Examples of clamp type and segment ring type closures on commercial SCFE vessels are shown in Figures 4 and 5.

All metal surfaces in contact with process fluids and solids must be suitable for foodstuffs, usually by polishing the stainless steel surface to a No. 4 finish. A compatible high strength alloy will be used for the balance of the vessel wall. Alloys used may include 304 SS, 316 SS, 304L SS, 316L SS, quenched and tempered high strength alloys and nickel-alloy steels. The least expensive vessel may depend upon the relative availability and cost of several possible alloy combinations at the time of vessel fabrication. For example, within the past year, the cost of nickel varied by a factor of ten, and thus all alloys containing nickel reflected marked cost fluctuations.

Penetrations and nozzles must be carefully located and designed since they affect the vessel body design. The number of penetrations and nozzles should be minimized without sacrificing operational efficiency and safety.

Seals and gaskets suitable for contact with human food products and compatible with the process fluids must be specified and evaluated carefully. Many elastomers and gasket materials which are suitable at conventional operating temperatures and pressures may be totally unsatisfactory at SCF processing conditions, requiring additional evaluation and testing. Examples of elastomers and gasket materials used in SCF CO2 extraction are: selected and specially formulated fluorocarbon polymers, perfluoroelastomers, synthetic rubbers and polymers, and metallic gasketing and seals.

The vessel design must also facilitate loading and unloading of contents and movement of the closures and vessel heads. Vessel foundations must withstand operating vibration.

HEAT EXCHANGERS. Specific considerations for heat exchangers are selection of materials of construction, susceptibility to fouling and ease of cleaning. The minimization of fouling during operation and the ability to clean the heat exchange surfaces when excessive fouling occurs or if contamination is to be prevented from the previous product, are of prime importance. These requirements limit the designs to relatively large tube diameters, single pass, with the (high pressure) extraction fluid on the tube side. SCF pressures are well above the "standard" exchanger design. Materials of construction (both metallurgy for tubes and tube sheets and seal and elastomer choices) must be carefully selected and evaluated. Inclined and vertical heat exchanger orientations to eliminate clogging must also be considered. Two types of exchangers are most often selected for these services: double-pipe heat exchangers for smaller plants, and single pass shell-and-tube heat exchangers for larger facilities.

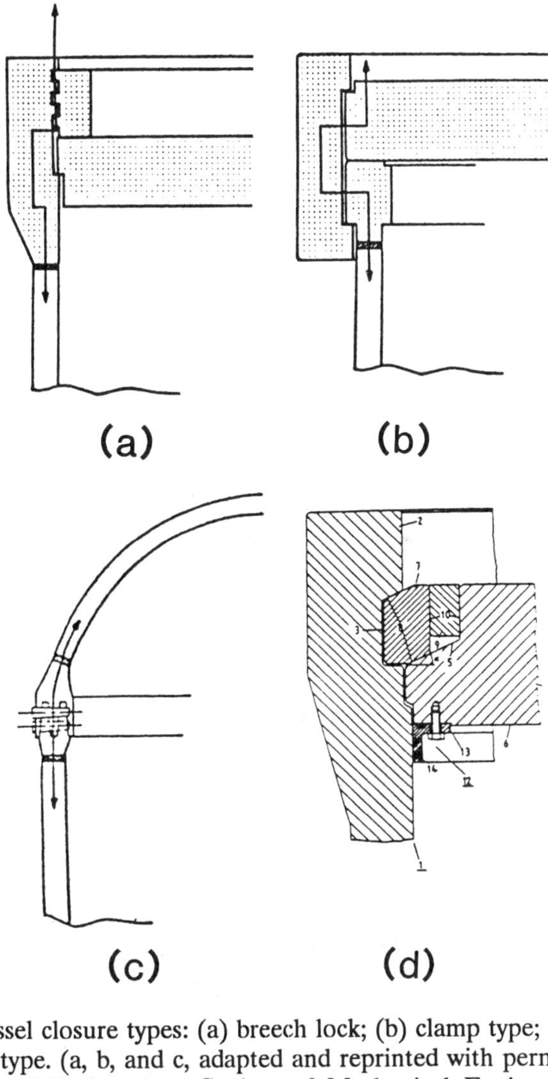

(a) (b)

(c) (d)

Figure 3. Vessel closure types: (a) breech lock; (b) clamp type; (c) pin type; (d) segment ring type. (a, b, and c, adapted and reprinted with permission from ref. 10. Copyright 1978 American Society of Mechanical Engineers. d, reprinted with permission from ref. 11. Copyright 1988 American Society of Mechanical Engineers.)

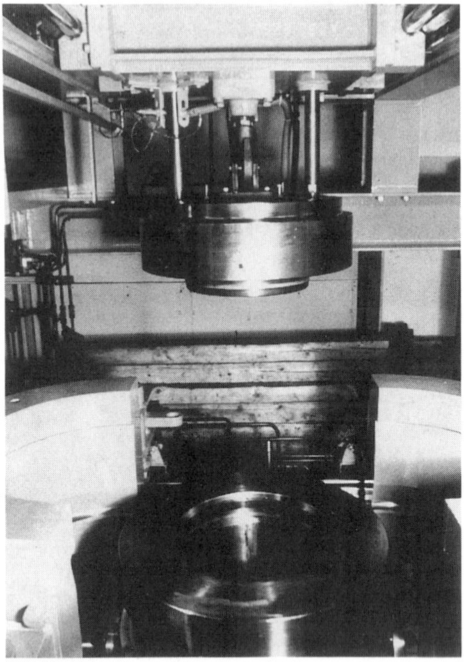

Figure 4. Automatic clamp type closure. (Reprinted with permission from ref. 11. Copyright 1988 American Society of Mechanical Engineers.)

Figure 5. Automatic segment ring closure. (Reprinted with permission from ref. 11. Copyright 1988 American Society of Mechanical Engineers.)

The double pipe heat exchanger has the advantage of simplicity and large diameter tubes for cleaning. Inspection after cleaning can be easily done. Differential thermal expansion between the inner and outer pipes can be accommodated by means of a relatively simple expansion joint. No high pressure tube sheet or large diameter end cap or cover is required. Cleaning of the inner tube surface can be done by chemicals or mechanical scraping.

The single pass shell-and-tube heat exchanger can provide more surface area in a smaller length and volume. Tube diameter is usually smaller than for a double pipe exchanger, but the tube surfaces can still be chemically or mechanically cleaned. Tube sheets inherently make the design more complicated than the double pipe exchanger. A thermal expansion joint may be installed on the exchanger shell. As plant size increases, the shell-and-tube heat exchanger becomes more cost effective and will be the exchanger of choice. At intermediate sizes, both types must be evaluated for process and economic reasons.

COMPRESSORS AND PUMPS. The fluid transport device may be either a compressor (operating in the gas phase) or a pump (operating in the liquid phase). Significant factors in selecting the equipment are cost and cleaning.

Two types of compressors are available for commercial plants: reciprocating piston and centrifugal. Each may be operated as single-stage or multi-stage units. Compressors are generally supplied complete with necessary intercoolers and aftercoolers, either air or water-cooled, and with necessary lubricating systems, interlocks and operating and safety controls. Capital costs, operating costs, maintenance costs and process limitations (temperature limits, etc.) must be determined. As capacity increases, centrifugal equipment becomes more economic.

Two types of pumps may be considered: reciprocating packed plunger and multistage centrifugal. Selection is primarily based on cost and influenced strongly by capacity. Multistage centrifugal pumps generally are less costly for very large capacities. The selection of materials in contact with the fluid, including seals, gaskets and lubricants, must be carefully evaluated. For example, a ceramic-coated plunger or highly polished hardened metal plunger surface may increase pump reliability and operating life substantially.

PIPING AND FITTINGS. The specific considerations for piping and fittings are the ability to clean components and the selection of materials of construction. Pipe or tubing for the required diameter, pressure and temperature may exceed the wall thickness usually available. Metallurgy must be suitable for the contacting fluid, yet with sufficient strength to withstand the pressure. This, as for the vessels, may require some trade-offs to determine the most economical selections. Surface finish must meet good manufacturing practices for cleaning. The ANSI, API and 3-A Standards must be considered. Flange ratings required may exceed the ANSI flange standards (2500 lb Class maximum). Special proprietary coupling systems may be needed (for example, Grayloc (Figure 6), UHDE (Figure 7), or Dur O Lok (Figure 8)). These special couplings have a weight and cost advantage over ANSI flanged connections. Comparisons of weight for similar service are shown in Table I.

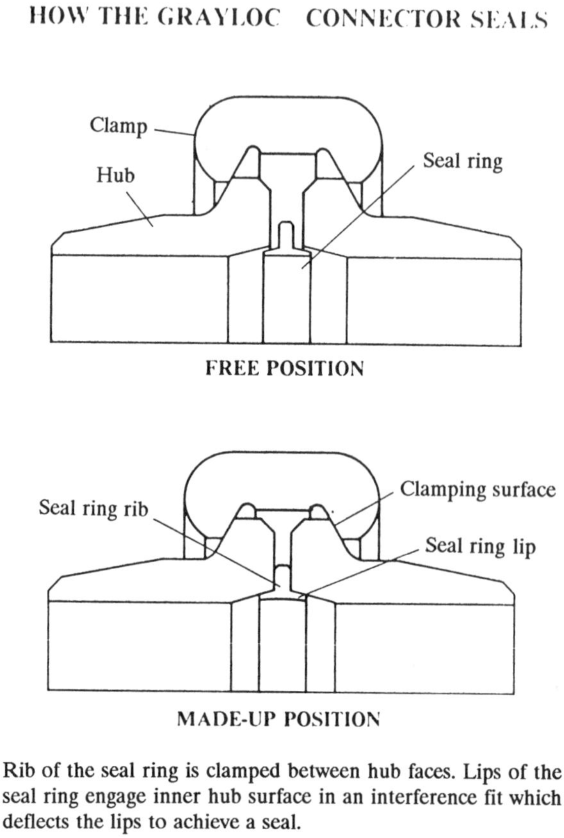
HOW THE GRAYLOC CONNECTOR SEALS
Clamp
Hub
Seal ring
FREE POSITION
Seal ring rib
Clamping surface
Seal ring lip
MADE-UP POSITION
Rib of the seal ring is clamped between hub faces. Lips of the seal ring engage inner hub surface in an interference fit which deflects the lips to achieve a seal.

Figure 6. Grayloc high-pressure pipe coupling. (Reprinted with permission from ref. 6. Copyright 1985 Gray Tool Company.)

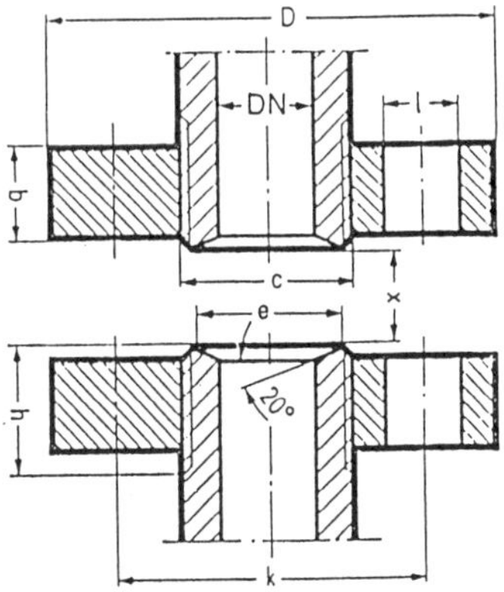

Figure 7. Lens high-pressure pipe coupling. (Reprinted from ref. 7. Courtesy UHDE GMBH.)

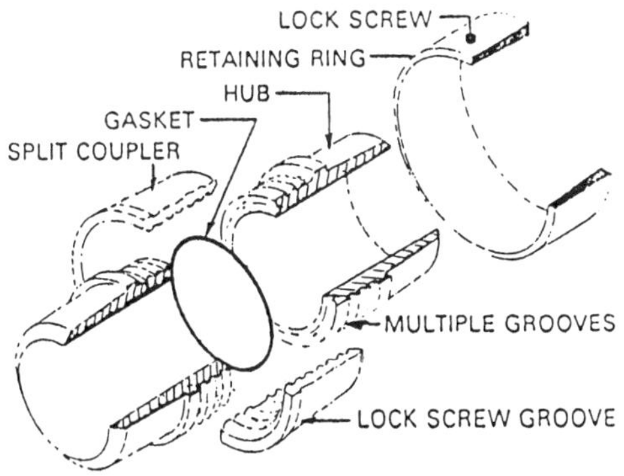

Figure 8. Dur O Lok high-pressure pipe coupling. (Reprinted with permission from ref. 8. Copyright 1982 American Society of Mechanical Engineers.)

Table I. Comparison of Allowable Working Pressures and Weights for 3-Inch Steel High Pressure Pipe at 100°F

Connector Description	Pressure psig	Weight, lbs.
ANSI Class 1500 Flange (6)	3600	121
API Class 5000 Flange (6)	5000	92
UHDE PN 325 Lens Joint (7)	4713	72
Grayloc Sch. 80 (6)	7797	37
Dur O Lok Sch. 80 (8)	3385	12.4

Welded fittings must be examined in the piping design to ensure that surface finishes are not compromised and that piping and equipment are accessible for maintenance, repair and cleaning.

VALVES. Valving must be carefully selected to provide process control, minimum leakage, and remote operation. Even on-off service may require special designs to meet the pressure differentials when closed. Special valves designed to meet the rigorous operating environment are well worth the additional investment to ensure reliable operations and to minimize part replacement, down time and product damage. High pressure ball valves are frequently chosen, which feature low leakage rate, resistance to corrosive chemicals and to abrasion and erosion by solids, and low pressure drops. Repetitive cycling has been demonstrated in severe temperature, pressure and particulate environments.

Throttling valves must be selected for the high pressure drops encountered with maximum unrestricted open area to avoid clogging. Valve noise level must be investigated, with provisions made to reduce the noise level if necessary. As with other equipment choices, the location, design and type of seals, packing and gaskets must be carefully assessed.

INSTRUMENTATION. Instrumentation criteria include: determining accuracy limits, assuring the design will meet the pressure requirements and providing for system flexibility. Primary sensing elements for measurement of temperature, pressure, level, and flow must be carefully evaluated for mechanical integrity at the operating conditions and for suitability of seals, elastomers, metals, and gaskets. Some types of sensors are not available for the pressures and temperatures encountered in SCF processing. Differential pressure cells and flow meters are two examples where only a limited number of types and vendors are available for the required operating conditions. Test programs may be required to ensure the element selected will meet the process specifications.

The cyclical and semicontinuous operating modes for these plants make a data acquisition and data logging system most desirable. Many of the controls are sequential and on-off; thus, the control philosophy is particularly amenable to PLC (Programmable Logic Controller), PC (Personal Computer), or Distributive Microprocessor Controls. Proportioning controls can be easily incorporated into such a system. A Graphic Panel Display with digital indicating and recording modes will add clarity for the operators of the plant.

ANCILLARIES. The equipment and components mentioned thus far are directly connected with the supercritical fluid food processing plant, but other ancillary and auxiliary equipment must also be specified, procured and installed. Heating, cooling, and other utilities must be integrated with the SCF process components and operate efficiently and reliably.

Insulation for process piping, vessels and equipment will be specified based on process energy conservation or personnel protection. Standard piping and fittings for steam, hot water, refrigerants, and instrument air will be utilized in most cases.

CONCLUSIONS

The selection of the components for a SCF food processing plant requires careful evaluation of the operating requirements from the perspectives of cleanliness, sanitation, maintenance, accessibility and safety for the pressures and temperatures to be encountered. Specific areas must be carefully considered to ensure the satisfactory commissioning of the SCF processing plant. This listing is certainly not all-inclusive, but provides an overview of key technical areas which must be considered for succesful design, construction, commissioning and operation of these plants. Attention to these details will result in a cost effective, efficient and safe facility.

ACKNOWLEDGEMENTS

The authors wish to acknowledge the several equipment suppliers who have provided information for this paper: Brown Fintube Co., Dur O Lok, Gray Tool Co., Hahn & Clay, Inc., Ingersoll-Rand, Inc., Kammer Valves, Inc., Mogas Industries, Inc., UHDE GMBH, and Valvtron Industries. These companies are intended to be representative of component suppliers and types. No warranty or endorsement of these products is made or implied.

The authors also appreciate the efforts of Mrs. Sandra Martin in the typing of this manuscript and Mrs. Christine James for editorial assistance.

LITERATURE CITED

1. Anon. Chem.Eng.(N.Y.) 1988, 95(13), p.21
2. Code of Federal Regulations,Title 21, Part 110, Office of the Federal Register, National Archives and Records Administration, 1987
3. Index of Published 3-A and E-3-A Sanitary Standards and Accepted Practices, Journal of Food Protection, 1984.
4. ASME Code for Pressure Piping, B31, An American National Standard, Chemical Plant and Petroleum Refinery Piping, The American Society of Mechanical Engineers: New York, 1986
5. ASME Boiler and Pressure Vessel Code, Section VIII, Pressure Vessels, American Society of Mechanical Engineers: New York, 1986
6. Anon., Grayloc Products Catalog, Gray Tool Co.: Houston, TX, 1985.

7. Anon., High-Pressure Fittings: Pressure Rating 325, UHDE GMBH: Hagen, FRG, 1980; p. 10.
8. Van Tassel, D. H., and Hitz, G. L. "Design of Compact High Pressure Couplings and Closures for Pressure Vessels and Piping", In High Pressure Engineering and Technology for Pressure Vessels and Piping Systems: Pohto, H. A., Ed.; Vol. 61, The American Society of Mechanical Engineers: New York, 1982, p. 2.
9. Marentis, R. T. "Steps to Developing a Commercial Supercritical Carbon Dioxide Processing Plant". In Supercritical Fluid Extraction and Chromatography, Charpentier, B. A.; Sevenants, M. R., Eds.; ACS Symposium Series No. 366; American Chemical Society: Washington, DC, 1988; p. 138.
10. Pechacek, R. "High Pressure, Quick Acting Closure for Large Diameter, Full Opening Nuclear and Petro-Chem Pressure Vessels". The American Society of Mechanical Engineers: New York. ASME 78-PVP-74; 1978, p. 2.
11. Koerner, J. P. "A New High Pressure Automated Closure System", In High Pressure Technology: Material, Design, Stress Analysis and Applications, Khare, A. F., Ed.; Vol. 148, The American Society of Mechanical Engineers: New York, 1988; pp. 89-93.

RECEIVED May 2, 1989

Author Index

Affiliation Index

Subject Index

C

Production: Raymond L. Everngam, Jr.
Indexing: Janet S. Dodd
Acquisition: Cheryl Shanks

Elements typeset by Hot Type Ltd., Washington, DC
Printed and bound by Maple Press, York, PA